Advances in Intelligent and Soft Computing

149

Editor-in-Chief

Prof. Janusz Kacprzyk
Systems Research Institute
Polish Academy of Sciences
ul. Newelska 6
01-447 Warsaw
Poland
E-mail: kacprzyk@ibspan.waw.pl

For further volumes:
http://www.springer.com/series/4240

David Jin and Sally Lin (Eds.)

Advances in Electronic Commerce, Web Application and Communication

Volume 2

 Springer

Editors
David Jin
Wuhan Section of ISER Association
Wuhan
China

Sally Lin
Wuhan Section of ISER Association
Wuhan
China

ISSN 1867-5662   e-ISSN 1867-5670
ISBN 978-3-642-28657-5   e-ISBN 978-3-642-28658-2
DOI 10.1007/978-3-642-28658-2
Springer Heidelberg New York Dordrecht London

Library of Congress Control Number: 2012932758

Printed on acid-free paper

Springer is part of Springer Science+Business Media (www.springer.com)

Preface

In the proceeding of ECWAC2012, you can learn much more knowledge about Electronic Commerce, Web Application and Communication all around the world. The main role of the proceeding is to be used as an exchange pillar for researchers who are working in the mentioned field. In order to meet high standard of Springer, the organization committee has made their efforts to do the following things. Firstly, poor quality paper has been refused after reviewing course by anonymous referee experts. Secondly, periodically review meetings have been held around the reviewers about five times for exchanging reviewing suggestions. Finally, the conference organization had several preliminary sessions before the conference. Through efforts of different people and departments, the conference will be successful and fruitful.

During the organization course, we have got help from different people, different departments, different institutions. Here, we would like to show our first sincere thanks to publishers of Springer, AISC series for their kind and enthusiastic help and best support for our conference.

In a word, it is the different team efforts that they make our conference be successful on March 17–18, 2012, Wuhan, China. We hope that all of participants can give us good suggestions to improve our working efficiency and service in the future. And we also hope to get your supporting all the way. Next year, In 2013, we look forward to seeing all of you at ECWAC2013.

January 2012

ECWAC2012 Committee

Committee

Honor Chairs

Prof. Chen Bin	Beijing Normal University, China
Prof. Hu Chen	Peking University, China
Chunhua Tan	Beijing Normal University, China
Helen Zhang	University of Munich, China

Program Committee Chairs

Xiong Huang	International Science & Education Researcher Association, China
LiDing	International Science & Education Researcher Association, China
Zhihua Xu	International Science & Education Researcher Association, China

Organizing Chair

ZongMing Tu	Beijing Gireida Education Co. Ltd, China
Jijun Wang	Beijing Spon Technology Research Institution, China
Quanxiang	Beijing Prophet Science and Education Research Center, China

Publication Chair

Song Lin	International Science & Education Researcher Association, China
Xionghuang	International Science & Education Researcher Association, China

International Committees

Sally Wang	Beijing Normal University, China
LiLi	Dongguan University of Technology, China
BingXiao	Anhui University, China

Z.L. Wang	Wuhan University, China
Moon Seho	Hoseo University, Korea
Kongel Arearak	Suranaree University of Technology, Thailand
Zhihua Xu	International Science & Education Researcher Association, China

Co-sponsored by

International Science & Education Researcher Association, China
VIP Information Conference Center, China
Beijing Gireda Research Center, China

Reviewers of ECWAC2012

Contents

Design of the Control System for High-Altitude Descent Control Device

Guangyou Yang[1,2] and Xiaofei Liu[1,2]

[1] School of Mechanical Engineering, Hubei University of Technology,
Wuhan, China
[2] Hubei Key lab of Modern Manufacture Quality Engineering,
Wuhan, China
pckka@126.com, liuxf85@yahoo.cn

Abstract. Traditional descent control device is generally just mechanical structure and lack of real-time control of running status for descent control device, having a lot of inconvenience to high-altitude rescue. The paper presents a control system of descent control device based on SCM (Single Chip Microcomputer), realizing the effective control of descent control device. The system use PWM to control the speed of the DC motor, use electromagnetic brake to realize stop in midway, and use digital remote-control system to realize real-time control of running status for descent control device. Meanwhile, this paper gives a function module for itself charge-discharge, increasing energy efficiency. The structure of the system is simple, reliable and sensitive.

Keywords: Descent control device, Single Chip Microcomputer, Control system.

1 Introduction

At present, the world faces a fire control problem that the capability of fire emergency rescue and the development of the high-rise building are not harmonious. The working range of aerial ladder and high pressure gun is limited, which increases the difficulty to rescue in case of fire. Particularly in big city, high-rise building exist hidden trouble. As a kind of life preserver, descent control device has many advantages, such as simple structure, convenient maintaining, settings flexible, rescuing quickly and so on. The descent control device has been generally installed in all kinds of high-rise building and elevating platform of the fire trucks [1]. However, the current descent control device exist many shortcomings, for instance, relying too much on mechanical structures, too simple in functions, lack of effective control of mechanical system [2] . Most of the descent control devices can't realize brake in midway, are not flexible in falling and rising, can't realize repeat rescue, and is always the low reliability [3]. Based on the above problems, this paper adopts the method of electromechanical, and puts forward a kind of control system of descent control device with the application of microcomputer control technology.

D. Jin and S. Lin (Eds.): Advances in ECWAC, Vol. 2, AISC 149, pp. 1–7.
springerlink.com

2 Structure of Control System

The descent control device consists of mechanical systems and control system. The mechanical system is shown in Fig. 1. This system can realize slowing down in decline, parking at any position, vertical motion and safety protection function. In descent stage, the weights through wire rope drive the winder to rotate, use damper to provide the function of resistance to realize rise and fall in stable. When the electromagnetic brakes work, it can stop at any position. When the system is out of control, according to the speed, emergency brake can realize automatic closing to brake. The control system of descent control device takes 89S51SCM as a controlling core to control the DC-motor through motor driver module to provide power for rising, and regulate the speed of DC-motor with PWM. In ascent stage, DC-motor charge the battery through charging circuits to supply the power for lifting vacant load. This system inputs control commands in two ways, wireless remote control and keyboard. The structure of control system is shown in Fig. 2.

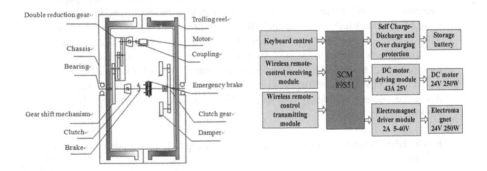

Fig. 1. The structure of mechanical system   Fig. 2. The structure of control system

3 Hardware Design of Control System

3.1 DC Motor Driving Circuit

The driving circuit of DC motor utilizes BTS7960B chips [4]. Two BTS7960B combine to be a H-Bridge to drive the DC motor, and the drive current can reach 43A, and the drive voltage range 5.5 ~ 25V. DC motor driving circuit is shown in Fig. 3[5].

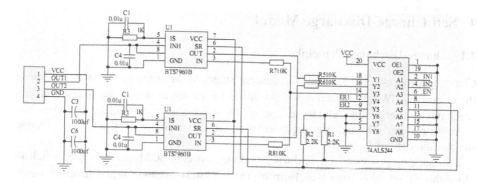

Fig. 3. DC motor driving circuit

3.2 Digital Remote-Control Module

This system uses the PT2262/PT2272 to transmit data. PT2262/PT2272 is a type of low power integrated circuit to encode and to decode produced by CMOS technology [6]. It has many characteristics, just as long data-transmitting distance, fast speed, reliable transmitting, high anti-jamming capability and low cost etc.

3.3 Power Module

This system will adopt two models of power supply, the first is through switching power supply; the second use the storage battery. Two models of power supply can be more applicably at the usage of different region environment. When the external power sources are usable, we can choose switching power supply directly. In order to guarantee portability and persistence of the system, we use the 24volt storage battery to be the second power sources. Especially in the particular surroundings, when being short of the outside power sources, storage battery can make the system under normal operating condition. The power-supply structure is shown in Fig. 4.

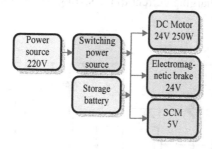

Fig. 4. The structure of power module

4 Self Charge-Discharge Module

4.1 Charge-Discharge Principle

Charging process: Fig. 5,during the descent stage, SCM controls the switch to close the circuit ①。 When the heavy weight drives the motor to rotate, the motor converts mechanical energy into electrical energy by electromagnetic induction. Then the current produced, through the regulator rectifier circuit and detection circuit, charge the storage battery.

    Discharging process: Fig. 5, during braking process, SCM controls the switch to close the circuit ②。 Then the electromagnetic clutches realize brake function based on the supply of power. During the ascent stage, SCM controls the switch to close the circuit ③. And the motor lifts heavy weights with the power supplied by the storage battery.

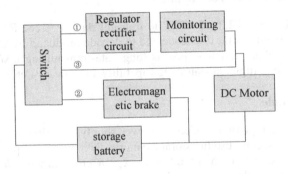

Fig. 5. The structure of Self Charge-Discharge circuit

4.2 Over Charging Protection Circuit

When storage battery charging voltage exceeds a preset value, as is shown in Fig. 6, this circuit will reduce charging current to the setting value, in case of overcharge for the storage battery.

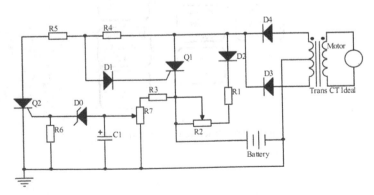

Fig. 6. Over charging protection circuit

5 Software Design of Control System

The Systems software is mainly constituted by following several parts: the main program, DC motor speed adjustment with PWM subroutine module, electromagnetic brake drive subroutine module, digital remote-control subroutine module, Self Charge-Discharge subroutine module.

5.1 System Main Program

The system main program realizes timer initialization, interrupting initialization and calls the subroutines of all functional modules. DC motor speed adjustment with PWM subroutine module is used in controlling positive and negative rotation and speed regulation of the DC motor; electromagnetic brake drive subroutine module realizes safety brake for the system; digital remote-control subroutine module realize wireless control with the low power integrated circuit, T2262/2272, to encode and to decode. The work flow chart of the system is shown in Fig. 7.

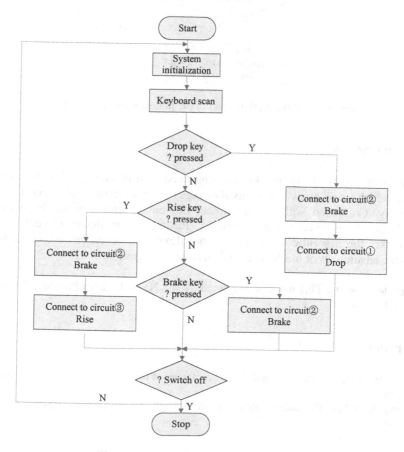

Fig. 7. The work flow chart of the system

5.2 DC Motor Speed Regulation with PWM Subroutine Module

During the motor speed adjustment, by changing the proportion of on-off time that means to change the value of duty cycle to change the average of the input voltage, the system realizes DC motor speed regulation [7]. The program flow chart of the interrupt service routine is shown in Fig. 8.

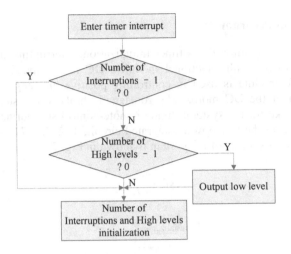

Fig. 8. The program flow chart of the interrupt service routine

6 Conclusions

In this paper, we put forward a kind of improved control system of descent control device, which overcomes some disadvantages of the traditional control system effectively. Compared with the traditional descent control device, the improved control system has some unique superiority. In this system, design of controlled by SCM makes the descent control device more flexible, quicker, and effective, which shows the advantages of mechanical and electrical integration.

Acknowledgments. This work is supported by the NSF of Hubei Province Grant No. 2010CDB02504.

References

1. Kong, W., Liu, Y.: A method and the device design of fire rescue for high-rise construction (2009)
2. Zhang, R., Chen, H., Liu, S.: Research on A High—Rise Buildings Rescue Apparatus (2010)

3. Zhou, R., Hu, Y.: Study and Application of Automatic and Continuous Escape Device (2010)
4. Infineon. BTS7960B_datasheet (2004)
5. Information (2003), http://www.znczz.com
6. He, X.: New practical electronic circuit 400 cases. House of Electronics Industry publishing, Beijing (2000)
7. Zhang, J., Lin, F., Cui, H.: Design of PWM Speed Regulation for DC-motor Based on MCS-51 SCM (2005)

Graph Based Representations SAT Solving for Non-clausal Formulas

Ruiyun Xie[1] and Benzhai Hai[2,3]

[1] Dep. of Computer Science and Technology,
Henan Mechanical and Electrical Engineering College, Xinxiang, China
[2] School of Information Engineering Wuhan University of Technology, Wuhan, China
[3] School of Computer Science &Technology Henan Normal University, Xinxiang, China
haibenzhai@gmail.com

Abstract. Boolean satisfiability (SAT) solvers are used heavily in hardware and software verification tools for checking satisfiability of Boolean formulas. Most state-of-the-art SAT solvers are based on the DPLL algorithm and require the input formula to be in conjunctive normal form (CNF). We represent the vhpform of a given NNF formula in the form of two graphs called vpgraph and hpgraph. The input formula is translated into a 2-dimensional format called vertical-horizontal path form (vhpform). In this form disjuncts (operands of ∨) are arranged orizontally and conjuncts (operands of ∧) are arranged vertically.The formula is satisfiable if and only if there exists a vertical path through this arrangement that does not contain two opposite literals (l and ¬l). The input formula is not required to be in CNF.

Keywords: SAT, NNF, DPLL.

1 Introduction

The Boolean satisfiability (SAT) problem decides whether a given Boolean formula is satisfiable or unsatisfiable. The SAT problem is of central importance in various areas of computer science, including theoretical computer science, hardware and software verification, and artificial intelligence. The SAT problem is NP-complete [1] and no provably efficient algorithms are known for it.Most state-of-the-art SAT procedures are variations of the Davis-Putnam-Logemann-Loveland (DPLL) algorithm and require theinput formula to be in conjunctive normal form (CNF). The design and implementation of SAT solvers becomes much easier if the input formulas are restricted to CNF. Given a truth assignment s to a subset of variables occurring in a formula, a Boolean constraint propagation (BCP) algorithm determines if s falsifies the given formula, else it provides the set of implied assignments (unit literals). Modern SAT solvers spend about 80%-90% of the total time during the BCP steps. For formulas in CNF, BCP can be carried out very efficiently using the two-watched literal scheme [2].

Typical formulas generated by the industrial applications are not necessarily in CNF.We refer to these formulas as non-clausal formulas. In order to check the satisfiability of a non-clausal formula f using a CNF based SAT solver, f needs to be

D. Jin and S. Lin (Eds.): Advances in ECWAC, Vol. 2, AISC 149, pp. 9–14.
springerlink.com

converted to CNF. This is done by introducing new variables [4, 3]. The result is a CNF formula f0 which is equi-satisfiable to f and is polynomial in the size of f. This is the most common way of converting f to a CNF formula. Conversion of a non-clausal formula to a CNF formula destroys the initial structure of the formula, which can be crucial for efficient satisfiability checking. The advantage of introducing new variables to convert f to f ' is that it can allow for an exponentially shorter proof than is possible by completely avoiding the introduction of new variables [5]. However, the translation from f to f ' also introduces a large number of new variables and clauses, which can potentially increase the overhead during the BCP steps and make the decision heuristics less effective. In order to reduce this overhead modern CNF SAT solvers use pre-processing techniques that try to eliminate certain variables and clauses [6]. The disadvantage with pre-processing is that it does not always lead to improvement in the SAT solver performance. It can also fail on large examples due to significant memory overhead.

We propose a new SAT solving framework based on a representation known as vertical-horizontal path form (vhpform) due to Peter Andrews [7, 8]. The vhpform is a two-dimensional representation of formulas in NNF. We represent the vhpform of a given NNF formula in the form of two graphs called vpgraph and hpgraph. The vpgraph encodes the disjunctive normal form and the hpgraph encodes the conjunctive normal form of a given NNF formula. The size of these graphs is linear in the size of the given formula.

2   Preliminaries

A Boolean formula is in negation normal form (NNF) iff it contains only the Boolean connectives "$\wedge$" (.AND.), "$\vee$" (.OR.) and "$\neg$" (.NOT.), and the scope of each occurrence of "$\neg$" is a Boolean variable. We also require that there is no structure sharing in a NNF formula, that is, output from a gate acts as input to at most one gate. A NNF formula is tree-like, while a circuit can be DAG-like.

Conversion of Boolean Circuits to NNF

In our work Boolean circuits are converted to NNF formulas in two stages. The first stage re-writes other operators (such as xor, iff, implies, if-then-else) in terms of $\wedge$, $\vee$, $\neg$ operators. In order to avoid a blowup in the size of the resulting formula we allow sharing of sub-formulas. Thus, the first stage produces a formula containing $\wedge$, $\vee$, $\neg$ gates, possibly with structure sharing. The second stage gets rid of the structure sharing in order to obtain a NNF formula.

Vertical-Horizontal Path

It is known that every propositional formula is equivalent to a formula in NNF. Furthermore, a negation normal form of a formula can be much shorter than any DNF or CNF of that formula. More specifically, we use a two-dimensional format of a nnf formula, called a vertical-horizontal path form (vhpform) as described in [9]. In this form disjunctions are written horizontally and conjunctions are written vertically. For example Fig. 1(a) shows the formula f= $(((p \lor \neg r) \land \neg q \land r) \lor (q \land (p \lor \neg s) \land \neg p))$ in vhpform.

Vertical path: A vertical path through a vhpform is a sequence of literals in the vhpform that results by choosing either the left or the right scope for each occurrence of $\lor$. For the vhpform in Fig. 1(a) the set of vertical paths is {<p, ¬q, r>,<¬r, ¬q, r >,< q, p, ¬p >,< q, ¬s , ¬p >}.

Horizontal path: A horizontal path through a vhpform is a sequence of literals in the vhpform that results by choosing either the left or the right scope for each occurrence of $\land$. For the vhpform in Fig. 1(a) the set of horizontal paths is {<p, ¬r,q>, <p, ¬r, p,¬s>, <p, ¬r, ¬p>, <¬q,q>, ,<¬q, p,¬s>, <¬q, ¬p >, <r,q>, <q, p,¬s>,<r, ¬p>} The following are two important results regarding satisfiability of negation normal formulas from [2]. Let F be a formula in negation normal form and let σ be an assignment (σ can be a partial truth assignment).

Theorem 1. s satisfies F iff there exists a vertical path P in the vhpform of F such that s satisfies every literal in P.

Theorem 2. s falsifies F iff there exists a horizontal path P in the vhpform of F such that s falsifies every literal in P.

Example 1. The vhpform in Fig. 2.1 has a vertical path <p,¬q,r> whose every literal can be satisfied by an assignment s that sets q to true and p, r to false. It follows from Theorem 1 that s satisfies f. Thus, f is satisfiable. All literals in the vertical path <¬r, ¬q , r >cannot be satisfied simultaneously by any assignment (due to opposite literals r and ¬r).

$$\left[\left(\begin{array}{c} p \lor \neg r \\ \\ \neg q \\ \\ r \end{array} \right) \lor \left(\begin{array}{c} q \\ \\ p \lor \neg s \\ \\ \neg p \end{array} \right) \right]$$

Fig. 2.1. The vhpform for the formula f=$(((p \lor \neg r) \land \neg q \land r) \lor (q \land (p \lor \neg s) \land \neg p))$

3 Graph Based Representations

3.1 Graphical Encoding of Vertical Paths (Vpgraph)

A graph containing all vertical paths present in the vhpform of a NNF formula is called a vpgraph. Given a NNF formula f, we define the vpgraph Gv(f) as a tuple (V,R,L,E,Lit), where V is the set of nodes corresponding to all occurrences of literals in f, RV is a set of root nodes, LV is a set of leaf nodes, E⊆V ×V is the set of edges, and Lit(n) denotes the literal associated with node nV. A node nR has no incoming edges and a node nL has no outgoing edges.

The vpgraph containing all vertical paths in the vhpform of Fig. 3.1(a) is shown in Fig. 3.1(b). For the vpgraph in Fig. 3.1(b), we haveV ={1,2,3,4,5,6,7,8},R = {1,2,5}, L = {4,8}, E = {(1,3), (2,3), (3,4), (5,6), (5,7), (6,8), (7,8)} and for each nV, Lit(n) is shown inside the node labeled n in Fig. 3.1(b). Each path in the vpgraph Gv(f), starting from a root node and ending at a leaf node, corresponds to a vertical path in the vhpform of f. For example, path <1,3,4> in Fig.3.1(b) corresponds to the vertical path <p, ¬q, r>in Fig. 3.1(a) (obtained by replacing node n on path by Lit(n)).

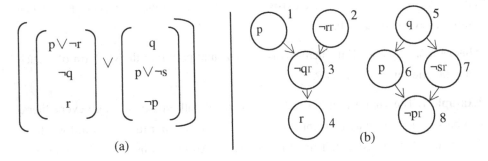

(a)   (b)

Fig. 3.1. (a) The vhpform for the formula f = (((p ∨¬r) ∧¬q ∧r) ∨ (q ∧ (p ∨ ¬s) ∧¬p))
(b) the corresponding vpgraph

3.2 Graphical Encoding of Horizontal Paths (Hpgraph)

A graph containing all horizontal paths present in the vhpform of a NNF formula is called a hpgraph. We use Gh(f) to denote the hpgraph of a formula f.The procedure for constructing a hpgraph is similar to the above procedure for constructing the vpgraph. The difference is that the hpgraph for f = f1f2 is obtained by taking the union of the hpgraphs for f1 and f2 and the hpgraph for f = f1f2 is obtained by concatenating the hpgraphs of f1 and f2.

The hpgraph containing all horizontal paths in the vhpform in Fig.3.2(a) is shown in Fig. 3.2(b). For the hpgraph in Fig. 3.2(b), we have V ={1,2,3,4,5,6,7,8}, R={1,3,4}, L={5,7,8}, E ={(1,2), (2,5), (2,6), (2,8), (3,5), (3,6), (3,8), (4,5), (4,6), (4,8), (6,7)} and for each n∈V, Lit(n) is shown inside the node labeled n.

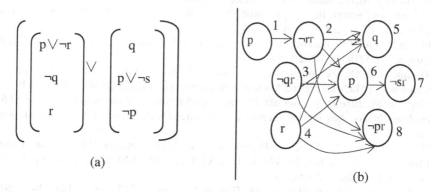

(a)   (b)

Fig. 3.2. (a) The vhpform for the formula f=(((p ∨¬r) ∧¬q ∧r) ∨ (q ∧ (p ∨ ¬s) ∧¬p))
(b) the corresponding hpgraph

It can be shown by induction that the vpgraph and hpgraph of a NNF formula are directed acyclic graphs (DAGs). One can also represent vpgraph and hpgraph as directed series-parallel graphs. Series-parallel graphs have been widely studied and many problems that are NP-complete for general graphs can be solved in linear time for series-parallel graphs [10].

When constructing an hpgraph/vpgraph from a NNF formula f each literal in f gets represented as a new node in the hpgraph and vpgraph of f. We assume that the node number corresponding to each literal l in f is the same in the hpgraph and the vpgraph of f. Thus, the set of nodes in the hpgraph and vpgraph are identical.

4 Conclusions

We presented a two-dimensional representation of NNF formulas called vertical horizontal path form (vhpform). The vhpform of an NNF formula contains vertical and horizontal paths. A vertical path is like a cube (term) in the DNF representation of a given formula, while a horizontal path is like a clause in the CNF representation of a given formula. The vpgraph encodes all vertical paths and the hpgraph encodes all horizontal paths. Both vpgraph and hpgraph can be obtained in linear time in the size of the given NNF formula. The NNF of formula is usually more succinct than the (pre-processed) CNF of the circuit in terms of number of variables.

14   R. Xie and B. Hai

References

1. Cook, S.A.: The complexity of theorem-proving procedures. In: STOC, pp. 151–158 (1971), 1.1
2. Moskewicz, M.W., Madigan, C.F., Zhao, Y., Zhang, L., Malik, S.: Chaff: Engineering an efficient SAT solver. In: Design Automation Conference (DAC 2001), pp. 530–535 (June 2001), 1.1, 1.2.1,1.2.4, 3.7, 3.8, 4.2, 4.4.1
3. Plaisted, D.A., Greenbaum, S.: A structure-preserving clauseform translation. J. Symb. Comput. 2(3) (1986), 1.1, 2.1, 3.8, 4.10.2
4. Tseitin, G.S.: On the complexity of derivation in propositional calculus. In: Studies in Constructive Mathematics and Mathematical Logic, pp. 115–125 (1968), 1.1, 2.1, 4.10.2
5. Järvisalo, M., Junttila, T., Niemelä, I.: Unrestricted vs restricted cut in a tableau method for boolean circuits. Annals of Mathematics and Artificial Intelligence 44(4), 373–399 (2005), 1.1
6. Eén, N., Biere, A.: Effective Preprocessing in SAT Through Variable and Clause Elimination. In: Bacchus, F., Walsh, T. (eds.) SAT 2005. LNCS, vol. 3569, pp. 61–75. Springer, Heidelberg (2005), 1.1, 2.1
7. Andrews, P.B.: Theorem Proving via General Matings. J. ACM 28(2), 193–214 (1981), 1.1.1, 2, 3
8. Andrews, P.B.: An Introduction to Mathematical Logic and Type Theory: to Truth through Proof, 2nd edn. Kluwer Academic Publishers, Dordrecht (2002), 1.1.1, 2, 2.2, 1, 2.2, 2.2
9. Davis, M., Putnam, H.: A computing procedure for quantification theory. J. ACM 7(3), 201–215 (1960), 1.1, 2, 4
10. Takamizawa, K., Nishizeki, T., Saito, N.: Linear-time computability of combinatorial problems on series-parallel graphs. J. ACM 29(3), 623–641 (1982), 2.3.2

Quantum-Behaved Particle Swarm Optimization Based on Comprehensive Learning

HaiXia Long* and XiuHong Zhang

School of Information Science Technology, Hainan Normal University, HaiKou HaiNan China
haixia_long@163.com

Abstract. This paper presents a variant of quantum-behaved particle swarm optimization (QPSO) that we call the comprehensive learning quantum-behaved particle swarm optimization (CLPSO), which uses a novel learning strategy whereby all other particles' historical best information is used to update a particle's local best position. This strategy enables the diversity of the swarm to be preserved to discourage premature convergence. The proposed QPSO variants also maintain the mean best position of the swarm as in the previous QPSO to make the swarm more efficient in global search. The experiment results on benchmark functions show that CLQPSO has stronger global search ability than QPSO and standard PSO.

Keywords: Quantum-behaved particle swarm optimization, Comprehensive learning, Local best position, Mean best position.

1 Introduction

Particle swarm optimization (PSO) is a population-based random optimization algorithms originally proposed by Kennedy and Eberhart [1]. It is motivated by simulation of social behavior of bird flock and fish schooling. During the past decade, PSO has been widely used in many real world applications and shown comparable performance with Genetic Algorithms (GA). However, as demonstrated by Bergh [2], PSO is not a global convergence guaranteed algorithm because the particle is restricted to a finite sampling space for each of the iterations.

In this paper, we present the comprehensive learning Quantum-behaved particle swarm optimizer (CLPSO) utilizing a new learning strategy.

2 CLQPSO Algorithm

2.1 Comprehensive Learning Strategy

In this new learning strategy, we use the following local attractor equation:

$$p_{ij}(t) = \varphi \cdot P_{f_i j}(t) + (1 - \varphi) \cdot P_{gj}(t) \qquad (1)$$

---

* Corresponding author.

where $f_i = [f_i(1), f_2(i), \cdots, f_i(D)]$ defines which particle' local best position ($pbest$)the particle i should follow. $P_{f_{ij}}$ can be the corresponding dimension of any particles' $pbest$, and the decision depends on probability P_{C_i} ,referred to as the learning probability, which can take different values for different particles. For each dimension of particle i , we generate a random number. If this random number is larger than P_{C_i} , the corresponding dimension will learning from its own $pbest$; otherwise it will learning from another particle's $pbest$. We employ the tournament selection procedure when the particle's dimension learns from another particle's $pbest$ as follows.

(1) We first randomly choose two particles out of the population which excludes the particle whose velocity is updated.
(2) We compare the fitness values of these two particles' $pbest$ and select the better one. In CLQPSO, we define the fitness value the smaller the better.
(3) We use the winner's $pbest$ as the exemplar to learn from for that dimension. If all exemplars of a particle are its own $pbest$, we will randomly choose one dimension to learn form another particle's $pbest$ corresponding dimension.

2.2 Learning Probability

In the procedure, different learning probability P_{C_i} yield different results on the same problem if the same probability P_{C_i} value was used for all the particles in the population. In order to address this problem in a general manner, we propose to set P_{C_i} such that each particle has a different P_{C_i} value. Therefore, particles have different levels of exploration and exploitation ability in the population and are able to solve diverse problems. We use the following expression to set a P_{C_i} value for each particle:

$$P_{C_i} = 0.05 + 0.45 * \frac{\left(\exp\left(\frac{10(i-1)}{ps-1} \right) - 1 \right)}{(\exp(10) - 1)} \qquad (2)$$

where ps is population size. Each particle has a P_{C_i} value ranging from 0.05 to 0.5.

3 Experiment Results and Discussion

To test the performance of the QPSO with comprehensive learning strategy, four widely known benchmark functions listed in Table 1 are tested for performance comparison with Standard PSO (SPSO), QPSO [3,4,5] and RQPSO[6]. These functions are all minimization problems with minimum objective function values zeros. The initial range of the population listed in Table 1 is asymmetry as used in [7] and [8].

Each function with three different dimensions (dimension=10, 20 and 30) was tested. The maximum number of iterations is set as 1000, 1500, and 2000 corresponding to the dimensions 10, 20, and 30 for four functions, respectively. In order to investigate the scalability of CLQPSO, 20, 40 and 80 particles were used for each function with dimension 10, 20 and 30. For SPSO, the acceleration coefficients were set to be $c_1=c_2=2$ and the inertia weight was decreasing linearly from 0.9 to 0.4 on the course of search as in [5] and [7]. In experiments for QPSO, the value of CE Coefficient varies from 1.0 to 0.5 linearly over the running of the algorithm as in [3] and [4], while in CLQPSO the CE decreases from 0.8 to 0.5 linearly. We had 50 trial runs for every instance and recorded mean best fitness and standard deviation

The mean values and standard deviations of best fitness values for 50 runs of each function are recorded in Table 2 to Table 5. The numerical results show that QPSO, RQPSO and CLQPSO are superior to SPSO.

On Shpere Function, RQPSO worse than QPSO; when the warm size is 40 and the dimension is 10 and when the warm size is 80 and the dimension is 20, the QPSO performs better than CLQPSO. Expect for the above two instances, the best result is CLQPSO. The Rosenbrock function is a mono-modal function, but its optimal solution lies in a narrow area that the particles are always apt to escape. The experiment results on Rosenbrock function show that the CLQPSO outperforms the QPSO and RQPSO except when the swarm size is 80 and dimension is 10 and RQPSO works better than QPSO algorithm in most cases. Rastrigrin function and Griewank function are both multi-modal and usually tested for comparing the global search ability of the algorithm. On Rastrigrin function, it is also shown that the CLQPSO generated best results than QPSO and RQPSO expect when the swarm size is 40 and dimension is 20. On Griewank function, CLQPSO has better performance than QPSO and QPSO-TS algorithm.

Figure 1 show the convergence process of the four algorithms on the four benchmark functions Sphere, Rosenbrock, Rastrigin and Griewank with dimension 30 and swarm size 40 averaged on 50 trail runs. It is shown that, although CLQPSO converge more slowly than the QPSO and RQPSO during the early stage of search, it may catch up with QPSO and RQPSO at later stage and could generate better solutions at the end of search.

Table 1. Expression of the FOUR tested benchmark functions

	Function Expression	**Search Domain**	**Initial Range**	**Vmax**
Sphere	$f_1(X) = \sum\limits_{i=1}^{n} x_i^2$	[-100, 100]	(50, 100)	100
Rosenbrock	$f_2(X) = \sum\limits_{i=1}^{n-1} (100 \cdot (x_{i+1} - x_i^2)^2 + (x_i - 1)^2)$	[-100, 100]	(15, 30)	100
Rastrigin	$f_3(X) = \sum\limits_{i=1}^{n} (x_i^2 - 10 \cdot \cos(2\pi x_i) + 10)$	[-10,10]	(2.56, 5.12)	10
Griewank	$f_4(X) = \dfrac{1}{4000} \sum\limits_{i=1}^{n} x_i^2 - \prod\limits_{i=1}^{n} \cos\left(\dfrac{x_i}{\sqrt{i}}\right) + 1$	[-600, 600]	(300, 600)	600

Table 2. Numerical results on Sphere function

M	Dim	Gmax	SPSO	QPSO	RQPSO	CLQPSO
			Mean Best (St. Dev.)	Mean Best (St. Dev.)	Mean Best (St. Dev.)	Mean Best (St. Dev.)
20	10	1000	4.61e-021 (1.03e-020)	3.19e-043 (2.20e-042)	1.43e-037 (7.10e-037)	2.84e-044 (1.64e-043)
	20	1500	9.02e-012 (3.54e-011)	1.91e-024 (7.15e-024)	9.23e-021 (2.56e-020)	1.51e-024 (2.75e-024)
	30	2000	3.96e-008 (7.04e-008)	7.37e-015 (2.17e-014)	1.49e-013 (3.89e-013)	1.16e-015 (1.23e-014)
40	10	1000	1.31e-024 (3.77e-024)	2.76e-076 (1.89e-075)	6.09e-053 (3.71e-052)	7.920e-076 (2.93e-075)
	20	1500	2.30e-015 (6.68e-015)	3.91e-044 (1.84e-043)	8.66e-031 (5.07e-030)	2.18e-044 (7.49e-043)
	30	2000	1.12e-010 (2.29e-010)	2.64e-031 (6.38e-031)	1.22e-021 (8.33e-021)	2.15e-032 (3.16e-031)
80	10	1000	1.60e-028 (7.10e-028)	2.56e-103 (6.48e-103)	4.29e-072 (2.12e-071)	1.87e-103 (1.83e-103)
	20	1500	6.38e-018 (1.48e-017)	1.41e-068 (7.44e-068)	1.17e-043 (2.63e-043)	1.55e-068 (2.02e-068)
	30	2000	3.27e-013 (7.59e-013)	6.57e-050 (3.60e-049)	1.9146e-031 (8.0241e-031)	2.57e-050 (1.93e-049)

Table 3. Numerical results on Rosenbrock function

M	Dim	Gmax	SPSO	QPSO	RQPSO	CLQPSO
			Mean Best (St. Dev.)	Mean Best (St. Dev.)	Mean Best (St. Dev.)	Mean Best (St. Dev.)
20	10	1000	51.0633 (153.7913)	9.5657 (16.6365)	8.6798 (11.4475)	5.3854 (0.8201)
	20	1500	100.2386 (140.9822)	82.4294 (138.2429)	45.6717 (56.0403)	35.5048 (31.4958)
	30	2000	160.4400 (214.0316)	98.7948 (122.5744)	78.5601 (99.5783)	72.9150 (55.2322)
40	10	1000	24.9641 (49.5707)	8.9983 (17.8202)	6.0319 (5.0777)	5.1936 (6.0371)
	20	1500	59.8256 (95.9586)	40.7449 (41.1751)	30.9714 (27.2429)	19.4821 (11.8921)
	30	2000	124.1786 (269.7275)	43.5582 (38.0533)	42.0232 (29.2780)	36.7373 (25.2817)
80	10	1000	19.0259 (41.6069)	6.8312 (0.3355)	5.0673 (3.7057)	5.0990 (6.8260)
	20	1500	40.2289 (46.8491)	33.5287 (31.6415)	30.8094 (29.0129)	16.9201 (0.2300)
	30	2000	56.8773 (57.8794)	44.5946 (31.6739)	44.8134 (28.8928)	20.1683 (0.2021)

Table 4. Numerical results on Rastrigin function

M	Dim	Gmax	SPSO Mean Best (St. Dev.)	QPSO Mean Best (St. Dev.)	RQPSO Mean Best (St. Dev.)	CLQPSO Mean Best (St. Dev.)
20	10	1000	5.8310 (2.5023)	4.0032 (2.1409)	4.1686 (3.5820)	3.7365 (2.7851)
	20	1500	23.3922 (6.9939)	15.0648 (6.0725)	16.0026 (13.4736)	14.9631 (11.5512)
	30	2000	51.1831 (12.5231)	28.3027 (12.5612)	23.8569 (6.3593)	23.7622 (5.0604)
40	10	1000	3.7812 (1.4767)	2.6452 (1.5397)	3.3741 (3.4335)	2.0154 (1.6440)
	20	1500	18.5002 (5.5980)	11.3109 (3.5995)	12.6376 (10.7444)	11.9271 (6.1596)
	30	2000	39.5259 (10.3430)	18.9279 (4.8342)	17.4264 (5.6821)	15.4428 (7.4389)
80	10	1000	2.3890 (1.1020)	2.2617 (1.4811)	1.8417 (2.6220)	1.4692 (2.5488)
	20	1500	12.8594 (3.6767)	8.4121 (2.5798)	8.7476 (2.7907)	7.7476 (3.7907)
	30	2000	30.2140 (7.0279)	14.8574 (5.0408)	15.6705 (5.0357)	13.8720 (3.7283)

Table 5. Numerical results on Griewank function

M	Dim	Gmax	SPSO Mean Best (St. Dev.)	QPSO Mean Best (St. Dev.)	RQPSO Mean Best (St. Dev.)	CLQPSO Mean Best (St. Dev.)
20	10	1000	0.0920 (0.0469)	0.0739 (0.0559)	0.0884 (0.0768)	0.0542 (0.0482)
	20	1500	0.0288 (0.0285)	0.0190 (0.0208)	0.0242 (0.0511)	0.0271 (0.0233)
	30	2000	0.0150 (0.0145)	0.0075 (0.0114)	0.0073 (0.0097)	0.0089 (0.0236)
40	10	1000	0.0873 (0.0430)	0.0487 (0.0241)	0.0863 (0.0950)	0.0541 (0.0781)
	20	1500	0.0353 (0.0300)	0.0206 (0.0197)	0.0108 (0.0168)	0.0129 (0.0372)
	30	2000	0.0116 (0.0186)	0.0079 (0.0092)	0.0066 (0.0080)	0.0014 (0.0213)
80	10	1000	0.0658 (0.0266)	0.0416 (0.0323)	0.0634 (0.0800)	0.0329 (0.0561)
	20	1500	0.0304 (0.0248)	0.0137 (0.0135)	0.0123 (0.0225)	0.0026 (0.0218)
	30	2000	0.0161 (0.0174)	0.0071 (0.0109)	0.0038 (0.0066)	1.02e-006 (4.62e-006)

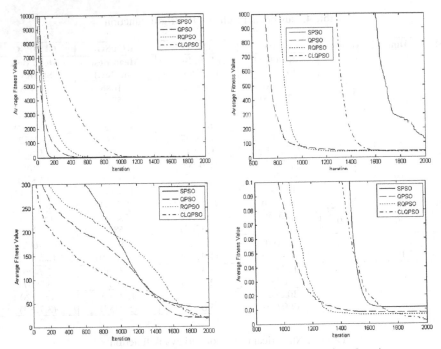

Fig. 1. Convergence process of the four algorithms on the four functions

Acknowledgment. This project was financially supported by National Science Fund of Hainan Province (No. 611131 and No. 610223).

References

1. Kennedy, J., Eberhart, R.: Particle swarm optimization. In: Proc. IEEE Int. Conf. Neural Networks, pp. 1942–1948 (1995)
2. Van den Bergh, F.: An Analysis of Particle Swarm Optimizers. University of Pretoria, South Africa (2001)
3. Sun, J., Feng, B., Xu, W.B.: Particle Swarm Optimization with Particles Having Quantum Behavior. In: Proc. 2004 Congress on Evolutionary Computation, Piscataway, NJ, pp. 325–331 (2004)
4. Sun, J., Xu, W.B., Feng, B.: A Global Search Strategy of Quantum-behaved Particle Swarm Optimization. In: Proc. 2004 IEEE Conference on Cybernetics and Intelligent Systems, Singapore, pp. 111–115 (2004)
5. Sun, J., Xu, W.B., Feng, B.: Adaptive Parameter Control for Quantum-behaved Particle Swarm Optimization on Individual Level. In: Proc. 2005 IEEE International Conference on Systems, Man and Cybernetics, pp. 3049–3054 (2005)
6. Sun, J., Lai, C.-H., Xu, W.-B., Ding, Y., Chai, Z.: A Modified Quantum-Behaved Particle Swarm Optimization. In: Shi, Y., van Albada, G.D., Dongarra, J., Sloot, P.M.A. (eds.) ICCS 2007. LNCS, vol. 4487, pp. 294–301. Springer, Heidelberg (2007)
7. Shi, Y., Eberhart, R.C.: A Modified Particle Swarm. In: Proc. 1998 IEEE International Conference on Evolutionary Computation, Piscataway, NJ, pp. 69–73 (1998)
8. Shi, Y., Eberhart, R.C.: Empirical Study of Particle Swarm Optimization. In: Proc. 1999 Congress on Evolutionary Computation, Piscataway, NJ, pp. 1945–1950 (1999)

Case Representation and Retrieval in the Intelligent RCM Analysis System

Zhonghua Cheng, Ling Chen, Xisheng Jia, and Huiyan Zeng

Department of Management Engineering, Mechanical Engineering College,
050003 Shijiazhuang, China
492870965@qq.com

Abstract. In order to improve the efficiency of drawing up RCM Analysis , we put forward to turning intelligence technique, especially CBR, which can be to applied to the process of RCM Analysis, we introduce the analysis process of the RCM Analysis and mainly study the means of RCM Analysis and organization mode, CBR of key on techniques such as the representation of case, the retrieval strategy of case, the study of case., finally make use of solid example analysis to prove the method is possibility, for further perfect develop the system of RCM Analysis base on case to lay a solid foundation.

Keywords: reliability centered maintenance (RCM) analysis, case retrieval mechanism, case-based reasoning (CBR).

1 Introduction

Reliability-Centered Maintenance (RCM) is a process to identify preventive maintenance (PM) requirements for complex systems, and has been recognized and accepted by many industries, such as aviation, military, energy, offshore oil production and management. The countries applying RCM include the United States, Britain, France, Canada, Japan, and Norway. The basic procedure of RCM is as follows, in order to ensure equipments to realize their design functions under the usage condition, based on failure data analysis and quantitative modeling for security, task and economy, functions and failure modes are systematically analyzed, and proper and efficient maintenance jobs are decided by failure data analysis.

RCM analysis is a complicated job and needs plenty of manpower and time. In order to improve the efficiency and applicability of RCM analysis, and to reduce the investment and time paid for developing RCM program, by collecting the existing RCM analysis cases, an intelligent RCM analysis system (IRCMAS) based on CBR was developed.

A RCM case includes all historical records in the RCM process, such as the list of functionally significant items (FSI), the Failure Mode and Effect Analysis (FMEA) information, and the RCM decision information. The idea for IRCMAS is based on the fact that the historical records of RCM analysis on similar items can be referenced and used for the current RCM analysis of a new item. Because many common or similar items may exist in the analyzed equipment, the repeated tasks of RCM

D. Jin and S. Lin (Eds.): Advances in ECWAC, Vol. 2, AISC 149, pp. 21–28.
springerlink.com

analysis can be considerably simplified or avoided by means of the improvement of similar RCM cases in the IRCMAS.

Case retrieval mechanism is the core part of IRCMAS, which has important effect on efficiency of RCM analysis. In this paper case retrieval mechanism is detailed discussed.

2 A Framework for IRCMAS

The framework is made up of function modules to realize the IRCMAS analysis.The basic input of the IRCMAS is the description of equipment requiring RCM analysis. The output is the preventive maintenance requirements of the equipment. Intelligent RCM analysis is driven by a reasoning process based on CBR.

A framework for IRCMAS is shown in Figure 3, which depicts the concept of the intelligent RCM Analysis.

From Fig. 1, we can find that according to user's description of RCM analysis requirement, IRCMA can quickly and accurately realize the RCM analysis on equipment by automatically retrieving the similar cases and adapting it. The adaptation of similar cases may be completed by identification of FSI, FMEA and RCM logic decision. If the adapted similar cases are evaluated and proved to be successful in practice, the new RCM analysis outcomes will be add to the case base as new case. Strictly speaking, the critical realization techniques of the IRCMAS include case acquisition and representation, formation of query case, case retrieval, adaptation, and learning.

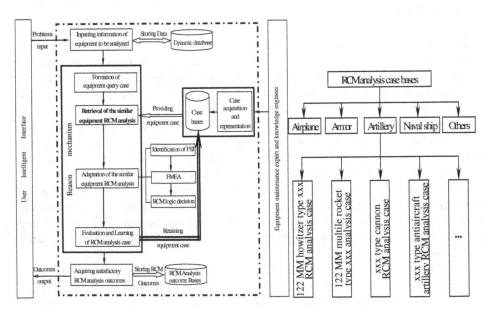

Fig. 1. A Framework for IRCMAS and Structures of RCM analysis case bases

3 Case Representation and Retrieval

It has been recognized that the retrieval mechanism plays an important role in IRCMAS. The key factors affecting the performance of the retrieval mechanism are case representation, indexing and similarity metric of parts. A good representation, indexing and similarity metric will enable the system to retrieve the most similar case rapidly and correctly.

3.1 Case Acquisition and Representation

The scale of case-base and the method of case representation have greater influence on the efficiency of reasoning. At present we have gathered and captured RCM analysis resources and records of about seventy kinds of equipment, including artillery, armor, etc. Those data are arranged, classified, integrated to form unique case base by knowledge engineers and equipment maintenance experts (Fig. 2).

A whole RCM analysis case = basic features information + RCM analysis information.

Some RCM analysis cases and their contents are as follows (see table 1).

There are many representation methods for case, such as First-Order Logic, Predicates, Semantic Nets, Frames, and Object Oriented Knowledge Representation etc[13]. Because frame representation can combine basic features information wiht RCM analysis of case, Frame representation is applied in this paper.

Table 1. Categories and contents of case

Case category	Basic features information	RCM analysis information
Equipment RCM analysis case	Name, application environment, category, and main structures of equipment	Functionally Significant Items (FSI) structure trees of equipment

Frame is made up of frame name and slots, which represent the features of case and include facets. Frame representation of equipment RCM analysis case is showed as table 2.

Table 2. Frame representation of equipment RCM case

Case Number:
Frame Name: the name of equipment RCM analysis case
Slot1: the description about equipment category   facet1: equipment category (artillery category, armor category, ..., other categories)
Slot2: the description of used environment of equipment facet1: application environment of equipment (island, desert ...)
Slot3: the description about main structures of equipment

Table 2. (*continued*)

facet1: main structure 1 (feature 1, feature value1, weight 1; feature 2, feature value2, weight 2...)
facet2: main structure 2 (feature 1, feature value1, weight 1; feature 2, feature value2, weight 2 ...)
...
facet n: main structure n (feature 1, feature value1, weight 1; feature 2, feature value2, weight 2 ...)
Slot4: collection of RCM analysis information
facet1: FSI structure trees of equipment

3.2 Formation of Query Case

According to the characteristics of equipment RCM analysis case, adoption template retrieve calculate way with according to add a power ash's connection an analytical case an retrieve the calculate way combine together of case retrieve strategy. Firstly Make use of a template retrieve strategy to work out the category or name and model number equipment RCM analysis case, then carry on the ash connection likeness degree analysis in same type or same case of name, find out the tallest likeness degree case, so this can shorten retrieve time, exaltation retrieve efficiency and accuracy, the retrieve of equipment RCM analysis case, can is divided into a classification, choice and confirm three stages:

(1) Categorizing stage is to search this type of with problem case related former or this name according to the characteristic of equipment problem case from the case database .Suppose the problem case is S, the case gathers for the C= (C1, C2, ..., Cp)in the equipment case database ,then categorizing stage can mean for find out same kind of case like S in the C, the case which gets initial sieving gathers S=(S1, S2, ..., Sm), and S ∈ C.

(2) Since Choose stage is to get closer to initial sieving gathers S , it can be divided into three steps as following:

Step 1: Adopt vector to return a turn formula to each main factor of the original data make to have no quantity key link to turn a processing and through vector return on turn processing, it is return on turn lately make reference to few rows with new of compare few row:

$$S_i'(k) = \frac{S_i(k)}{\sqrt{\sum_{i=0}^{m} S^2_i(k)}} ; 0 \le i \le m, 1 \le k \le n \tag{1}$$

Step 2: make use of The improved partial gray connection calculate way [13], work out the retrieval gray and alike matrix of alike RCM analysis case, process as follows: Firstly substitute reference progression and compare progression we get in the first step into the partial gray connection formula (2) which has been improved:

$$G_s(s_o(k), s_i(k)) = \frac{\min_{i \in m} \min_{k \in n}(w_k |s_o'(k) - s_i'(k)|) + \xi \max_{i \in m} \max_{k \in n}(w_k |s_o'(k) - s_i'(k)|)}{w_k |s_o'(k) - s_i'(k)| + \xi \max_{i \in m} \max_{k \in n}(w_k |s_o'(k) - s_i'(k)|)} \qquad (2)$$

(Among them, $\xi \in [0,1]$ is distinguish coefficient, generally take $\xi = 0.5$; The Wk is heavy for the power of different characteristic retrieve sign)

Secondly, from(2) can get the connection likeness of the attribute retrieve sign of a m × n degree, constitute of the case retrieval gray and alike matrix of the likeness allocation:

$$r = (G_s(s_o(k), s_i(k)))_{m \times n} \qquad (3)$$

Step 3: Combine Euclid to be apart from formula on the space of n Wei, build up RCMS case an retrieval gray connection theories likeness degree model, carry out model calculation to work out analytical case S0 and system case's Si compound gray likeness degree in the space of n dimension:

$$\begin{cases} G_d(s_o, s_i) = \sqrt{\sum_{k=1}^{n} G_d^2(s_o(k), s_i(k))} \\ G_s(s_o, s_i) = \dfrac{1}{1 + G_d(s_o, s_i)} \end{cases} \qquad (4)$$

(3) Confirm stage is the RCM analysis personnel confirms the most suitable case from the alternative program Q according to the likeness degree of equipment case, .Or can also the base on the former experience and realm knowledge, constitution system Yu value[14].When likeness degree is greater than the Yu value, regard this case is similar with analytical equipment case; If the search result has many individual case example of mutually greater than Yu value, then should consider recommendation likeness the degree is perhaps the biggest to use the case with the tallest frequency as a result case. When have no case whose likeness degree greater than the Yu value , then take this case as a new case.

Maintain the characteristic attribute of the analysis case of the resources need power heavy really settle count for much for the retrieve of the case. Equipment's maintaining the expert's knowledge and experience is the source which maintains the analysis knowledge of the resources need. Well make use of expert's knowledge, carry on beating a cent to the characteristic attribute of case according to the expert's experience, can make sure a case characteristic the power of the attribute is heavy. This literary grace uses the method that the expert beats a cent a certain case characteristic the power value of the attribute.

In the retrieval process of equipment RCM analysis case, because of analysis's familiar with the structure of equipment , the same kind of likeness equips are not that many, so the personnel almost can directly confirm likeness equipment case after the classification stage of case retrieve, and do not need the second stage again. Only just need to consider the adoption ash connection likeness degree under the situation that can't confirm a likeness equipment case by the model carries on the choice of the alike case of the second stage.

The concrete process of equipment RCM analysis case retrieve as follows , see figure 3.

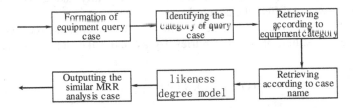

Fig. 2. Retrieval process of equipment RCM analysis case

4 Experiment

The XXX model cannon takes aim the problem case of the radar material antenna feedback line system to see table 2.

Table 2. XXX model cannon takes aim the problem case of the radar material antenna feedback line system

Treat analysis material	Follow speed K_1	The biggest power K_2	Install time K_3	Azimuth angle scope K_4
S_0	555	531	11	Have restriction

Retrieved equipment RCM analysis cases see Table 3.

Table 3. Retrieved equipment RCM analysis cases

Case name	Follow speed K_1	The biggest power K_2	Install time K_3	Azimuth angle scope K_4
S_1	480	454	9	Have restriction
S_2	565	538	10. 5	Have restriction
S_3	635	604	13. 5	no restriction
S_4	515	499	15	no restriction

Be beaten a cent by related realm expert, get 4 powers of attributes in form heavy respectively is:0.222, 0.222, 0.222, 0.334.Before carrying on connection analysis to calculate, the exploitation return a turn formula to the table 2, table 3 in each attribute value carry on having no quantity key link to return on turning a processing, and carry on gather to two forms. Return a turn a result such as the table 4 show:

Table 4. Return on turn stay analytical problem case and first step to sieve the attribute characteristic that the cases index sign

Case name	Follow speed K_1	The biggest power K_2	Install time K_3	Azimuth angle scope K_4
S_1	480	454	9	Have restriction
S_2	565	538	10. 5	Have restriction
S_3	635	604	13. 5	no restriction
S_4	515	499	15	no restriction

After a calculation:

$$\min_{i \in m} \min_{k \in n} \left(w_k \left| s_o'(k) - s_i'(k) \right| \right) = 0 \; ; \; \max_{i \in m} \max_{k \in n} \left(w_k \left| s_o'(k) - s_i'(k) \right| \right) = 0.053$$

Again from formula (2) with (3) can get to maintain the inspectional gray and alike matrix of the equipment RCM analysis cases:

$$r = \begin{bmatrix} 0.503 & 0.507 & 0.622 & 0.914 \\ 0.892 & 0.905 & 0.861 & 0.914 \\ 0.507 & 0.582 & 0.627 & 0.334 \\ 0.621 & 0.632 & 0.615 & 0.334 \end{bmatrix}$$

At last, use a model calculation to work out the analytical case S0 and system case Si compound gray likeness degree in the space of n-dimension: G(S0, Si)=(0.341, 0.623,0.294,0.313), we can get the biggest compound gray likeness degree from the calculation result: G(S0, S2)=0.623, then we know in the RCM analysis case database the No.2 case has the biggest likeness degree with analytical problem case. RCM analysis cases personnel can choose a big likeness degree of the No.2 case to be a reference, then carry on a modification at the foundation of the No.2 case , and shape a new case.

5 Conclusion

RCM analysis is a process to identify preventive maintenance (PM) requirements for complex systems, and has been recognized and accepted by many industrial fields. In the paper, the intelligent RCM analysis process is presented, the critical components of which, such as case representation, and case retrieval etc are discussed. Although the studies on the intelligent RCM analysis are still in the early stage, we can conclude that it is effective and feasible to apply the CBR in the RCM analysis process.

The study outcomes in this paper will further improve the accuracy and validity of RCM analysis, and have great significance for RCM popularization and application on military equipment and civilian facility. The next jobs of ours are to further perfect the intelligent RCM analysis system.

Acknowledgments. This paper is supported by National Natural Science Foundation (70971135).

References

1. Watson: Case-based reasoning is a methodology, not a technology. Knowledge-Based Systems (1999)
2. Shi, Y., et al.: Application of Case-Based Reasoning Technique on Gun Decision. Computer Mechanism (1999)
3. Cheng, Z.: Intelligent RCM analysis. PhD's degree thesis of Mechanical Engineering College, China (2006)
4. Cheng, Z., Jia, X., Wang, L., Bai, Y.: A framework for the case-based and model-based RCM analysis. In: 2010 The International Conference on Management Science and Artificial Intelligence (2010)
5. Gu, D.X., Li, X., Liang, C., Li, F.: Research on case retrieval with weight optimizing and its application. Journal of Systems Engineering (2009)
6. Wang, H., Ni, Z., Yan, J., Han, D.: Research on Application of Grey- Relational Theory in CBR. Computer Technology and Development (2010)

A Fuzzy Logic Based Approach for Crowd Simulation

Meng Li[1], ShiLei Li[2], and JiaHong Liang[1]

[1] College of Mechanical Engineering and Automation,
National University of Defence Technology, Changsha 410073, P.R. China
mengshuqin1984@163.com
[2] Department of Information Security, College of Electronic Engineering,
Naval University of Engineering, Wuhan 430000, P.R. China

Abstract. In this article, we present a new approach which incorporates fuzzy logic with data-driven crowd simulation method. Behavior rules are derived from the state-action samples obtained from crowd videos by MLFE algorithm. The derived rules complement the disadvantages of rule-based approach in the situation where the crowd behavior can't be reduced to rules, i.e., calibrate the behaviors produced by rule-based approach. During a simulation, the new derived rules can be combined with pre-defined ones. Then, the compositive rules are treated in the overview of fuzzy logic to simulate various crowd behaviors. It is noticeable that the derived rules can be used independently for data-driven crowd simulation or be incorporated with predefined rules. The advantages of our approach are obvious: interoperability, universality and behavior's diversity.

Keywords: crowd simulation, fuzzy logic, MLFE, rule extraction.

1 Introduction

In recent years, crowd simulation technologies have been gaining tremendous momentum. Various simulation models and architectures have been developed. Whether the behaviors of the crowd are natural and intelligent or not have many relations with the propriety of the crowd models. The most popular approaches are rule-based agent model and data-driven model. The agent-based approach is currently the dominant and most active approach to crowd modeling, in which the behaviors of each individual are represented by a set of predefined rules. A big advantage of agent-based approach is that the integration of individual rule exhibits collective group behavior of enormous complexity and subtlety and eventually some sustainable global behavior phenomena are induced [1]. Compared to data-driven model, the agent model provides the researchers a very convenient way to incorporate various behavior rules into crowd simulation. However, rule-based methods typically consider only the specified situations. When given a new situation, a new set of rules needs to be devised. Further more, the predefined rules are generally intuitive and specific to certain situations that can not be defined easily in some situations, thus defining the proper behavior rules is often an art rather than a science [2]. And it is difficult to produce anticipant styles of crowd behaviors by reason of a mass of relevant parameters in the rule-base needing to tune finely.

The data-driven model strut its stuff where the crowd behavior can't be reduced to rules in which tracking data from real crowds is used to create a database of examples

D. Jin and S. Lin (Eds.): Advances in ECWAC, Vol. 2, AISC 149, pp. 29–35.
springerlink.com

that is subsequently used to drive the simulated agents' behavior. By learning from real-world examples, the simulated agents display complex natural behaviors that are often missing in rule-based agent simulation approaches [3]. In data-driven approach, crowd models are built as representations of recurrent behaviors by analyzing video data of the crowd through vision methods. Crowd behaviors are achieved based on the extracted features which are denoted as state-action samples obtained from tracked videos and normally can be employed in crowd events inference. The state of each agent reflects its own position and internal characteristics, the influence of the stimulus it's affected including its orientation and type, the motion of nearby agents, and environment features. The action represents the adopted behavior strategies which includes the locomotion direction and corresponding velocity of the next time-step. During a simulation, the set of most suited state-action samples are obtained by matching the current state with the state-action-base which is used to predict the current action by fitting its elements according to regression-based learning method or neural network method. The observed state-action samples rely on two aspects: the specific crowd characteristics and crowd events which restrict their universality and reusability when facing a new circumstance, i.e., the capturing information from real crowds lacks flexibility. Therefore, we need to parameterize state-action samples from videos to build a parameterized behavior model that adapt to various circumstances. In some sense, our approach in which fuzzy behavior rules are derived from the state-action samples is similar to a parameterized behavior approach.

Fuzzy logic and fuzzy set are effective techniques for handling fuzzy uncertainties with well-developed mathematical properties. Fuzzy logic provides an excellent way to represent and process linguistic rules. In fact, in modelling and simulation of human behavior, we are not restricted to just absolute quantifier that represents a crisp value like one or two, but we are also concerned with relative quantifier that represents a fuzzy value, such as low, medium, high, most, or some [6]. In our approach, the first-line problem is to extract fuzzy rules from the state-action samples using "modified learning from examples" (MLFE) method. In this sense, we provide another method to define rule-base. The MLFE can be used in conjunction with the BLS to develop a more effective model. Later, the derived rules from state-action samples are combined with the predefined ones, no matter they are crisp or fuzzy rules in agent-based model. In a simulation, once the parameters of the membership functions of the rule-base have been specified, the compositive rules are used by BLS algorithms to predict the actions for the agents with given states in current circumstance. Together, they provide a contextual meaning to the underlying behaviors of the crowd. The framework of our approach is illustrated in figure 1.

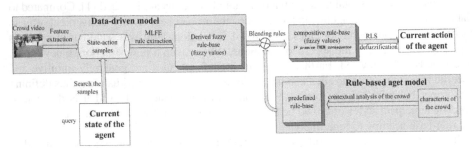

Fig. 1. The framework of our approach

2 Related Work

By now, substantial effort has been devoted to rule-based agent approaches. In his seminal work, Reynolds described a distributed behavioral model for simulating aggregate motion of a flock of birds, in which three local rules are predefined. The aggregate patterns are the result of the interaction of these relatively simple behaviors of the individual simulated birds. However, according to the same crowd, the rules defined by different users are not the same. In addition, definition of the rules is a subjective work to a certain extent, which needs some objective objects, i.e., the state-action samples to refine it.

Lerner et al. introduced a novel approach for rule-based simulation based on examples, in which tracking data from real crowds is used to create a database of examples that is subsequently used to drive the simulated agents' behavior [3][8]. Lee et al. proposed data-driven approaches which use learning algorithms based on locally weighted linear regression to simulate crowd behavior [1][8]. Musse presented a new model based on trajectories captured automatically from filmed video sequences. Lerner presented a data-driven approach for fitting behaviors to simulated pedestrian crowds in which the importance of each stimulus within a configuration is determined and compared with the examples using similarity function. However, these approaches were all based the fact that the feature extraction from real video is inerrant and accurate. And they were not applicable to various circumstances essentially. Our approach can deal with the situations when the quality of the feature extraction is not so good and satisfy diversified simulation requirements. The advantage of using MLFE for rule extraction is that it is not necessary to specify membership function parameters as MLFE can calculate both membership function parameter and rules. And in addition to synthesizing a rule-base, in MLFE we also modify the membership functions to try to more effectively tailor the rules to represent the data.

3 Extracting Rules from the State-Action Samples

We employ impulse membership function of zero width for the *consequence* (action of the samples, i.e., locomotion direction of the next time-step A_x, A_y and its corresponding velocity V_x, V_y) and Gaussian membership functions for the *premise* (state S of the samples including position and internal characteristics, the influence of the stimulus it's affected including its orientation and type, the position and velocity of nearby agents, and relative position of the environment obstacles). So, S-A_x, S-A_y, S-V_x and S-V_y provide basic data to rule extraction and action generation in run time.

Let b_i is the point in the output space at which the output membership function for the *ith* rule achieves a maximum, c_j^i and σ_j^i are respectively the center and spread of the membership function for the *jth* input and the *ith* rule. Let θ denote all the above parameters, where the dimensions of θ are determined by the number of inputs n and the number of rules R in the rule-base. The parameter vector θ to be chosen is

$$\theta = \left[b_1, \ldots, b_R, c_1^1, \ldots c_n^1, \ldots, c_1^R, \ldots c_n^R, \sigma_1^1, \ldots \sigma_n^1, \ldots, \sigma_1^R, \ldots \sigma_n^R \right]^T \tag{1}$$

Figure 2 presents how to construct the rule-base for the fuzzy estimator by choosing θ.

Rule extraction using MLFE algorithm

1.   Initialization, N=1, R=1, $b_1 = Z^N$, $c_j^1 = S_j^N$, $\sigma_j^1 = \sigma_0$ for $j = 1, 2, \ldots, n$ where the parameter $\sigma_0 > 0$, S_j^N denote the *N-th* state-tuple, the *j-th* input universe of discourse and Z^N is the *N-th* action-tuple of single value A_x, or A_y, V_x, V_y. $\varepsilon_f = \varepsilon_0$, where ε_0 characterize the accuracy with which we want the fuzzy system to approximate the output.

2.   N=N+1, for the *N-th* state-tuple S^N which consists of states S_j^N $j = 1, 2, \ldots, n$, computing the current estimated fuzzy output using equation (2)

$$f\left(S^N \mid \theta\right) = \frac{\sum_{i=1}^R b_i \prod_{j=1}^n \exp\left(-\frac{1}{2}\left(\frac{S_j^N - c_j^i}{\sigma_j^i} \right)^2 \right)}{\sum_{i=1}^R \prod_{j=1}^n \exp\left(-\frac{1}{2}\left(\frac{S_j^N - c_j^i}{\sigma_j^i} \right)^2 \right)} \tag{2}$$

If $\left| f\left(S^N \mid \theta\right) - Z^N \right| < \varepsilon_f$, return to step 2. Else, add a new rule to the rule-base to represent the current data-tuple (S^N, Z^N) by modifying the current parameters θ and letting R=R+1, $b_R = Z^N$, $c_j^R = S_j^N$ for all $j = 1, 2, \ldots, n$, then go to step 3.

3.   Modifying the widths σ_j^R to adjust the spacing between membership functions. Modification of the σ_j^R is done by determining the "nearest neighbour" n_j^* with the membership function center given by equation (3)

$$n_j^* = \arg\min\left\{ \left| c_j^i - c_j^R \right| : i = 1, 2, \ldots, R, \; i \neq R \right\} \tag{3}$$

n_j^* denotes the i index of the c_j^i that minimizes the expression. Then we update the σ_j^R by letting $\sigma_j^R = \sigma_j^{n_j^*}$, return to step 2.

Fig. 2. The framework of rule extraction algorithm

We construct four groups of fuzzy rules to represent the state-action samples. The *premise* and *consequence* of the derived rules are represented by fuzzy sets, which have the following forms:

if $x_1 \in A_1$ and $x_2 \in A_2$ and, $\ldots$, and $x_n \in A_n$

then A_x (or A_y, V_x, V_y) is b_i $i = 1, \ldots, R$

Where x_1, x_2 and x_n are variables of the state, $\underset{\sim}{A_1}$, $\underset{\sim}{A_2}$, $\underset{\sim}{A_n}$ are fuzzy sets whose parameters are determined by θ. A_x, A_y, V_x and V_y are the variables of the action, b_i is a crisp value owing to impulse membership function which also determined by θ. Later, the batch least squares algorithm (BLS) will be used in conjunction with the derived rules to predict the action in a run-time simulation.

If the number of the state-action samples is m, we can rewrite equation (2) as follows:

$$f\left(S^N \mid \theta\right) = \frac{\sum_{i=1}^{R} b_i \mu_i(S^N)}{\sum_{i=1}^{R} \mu_i(S^N)} \quad N = 1,...,m \quad j = 1,...,n \tag{4}$$

where $\mu_i(S^N) = \prod_{j=1}^{n} \exp\left(-\frac{1}{2}\left(\frac{S_j^N - c_j^i}{\sigma_j^i}\right)^2\right)$. We define the regression vector

$\xi_i(S^N) = \frac{\mu_i(S^N)}{\sum_{i=1}^{R} \mu_i(S^N)}$ and $\hat{\theta} = [b_1, b_2, ..., b_R]^T$. For each data-tuple we have R values

for $\xi_i(S^N)$, however it is noted that R can be modified by incorporating the predefined rules. Using equation (4) and the state-action samples, we determine $\hat{\theta}$,

$$\hat{\theta} = \left(\Phi^T \Phi\right)^{-1} \Phi^T Z \tag{5}$$

Where $\xi_i(S^N)$ is the N-th row, the i-th column elements of Φ, Z is $\left[A_x^1, ..., A_x^m\right]^T$, $\left[A_y^1, ..., A_y^m\right]^T$, $\left[V_x^1, ..., V_x^m\right]^T$, or $\left[V_y^1, ..., V_y^m\right]^T$. For a new state S^{N+1} in the run-time simulation, its according action is $\hat{\theta}^T \left[\xi_1(S^{N+1}), ..., \xi_R(S^{N+1})\right]^T$.

4 Simulation

Our approach requires the creation of an example database to extract fuzzy rules. The state-action samples are created from videos of real crowds which picture people's normal behaviors in a piazza. The durations of the videos are between three and six minutes with an average of 4-6 people per frame. The number of people in a group is about 2-4. In our simulation, 5213 samples are created from input video and 251 rules are extracted. The extracted rules are combined with several predefined rules to achieve more plentiful results.

The predefined rules which illustrated in figure3 depict people's abnormal behavior rules that happen when facing a danger or there is an urgent need required to satisfy. Also, we simulate group people's flocking behavior. Figure 4 illustrate our simulation results.

```
<Slow_down_danger>                                        <Urgent_requirement>
if      danger_in_front and μ(distance_to_danger) > 0.5    if      Have_ urgent_requirement
        and speed>0 and have no urgent requirement         then    change target to
then    Slow-down and become an onlooker                           satisfy new requriement
<Danger_x/y>
if      danger_front_x/y and μ(distance_to_danger) ≤ 0.5   <Flocking_behavior>
        and not(danger_front_y/x)                          if      when meeting intimate people
then    turn_y/x                                           then    form a new group
<Danger_in_front>                                                  and align
if      speed>0 and danger_in_front                                and avoid collision
        and μ(distance_to_danger) ≤ 0.5
then    turn around and flee
```

Fig. 3. Predefined rules in our simulation

(a)   (b)

Fig. 4. A typical image of the resulting simulation. (a) The simulation results in normal situation. The agents' behaviors are from derived rules extracted from examples. The two press-photographers in the triangle have normal requirements of reporting the scene of normal behaviors in the piazza. (b)The simulation results in the situation when a fire happens. The people's behaviors are from compositive rules. The two press-photographers in the triangle abandon normal requirement to report the scene of the fire. The people in the rectangle flee from the danger area. The people in the circle whose corresponding requirements are not urgent become stander-bys.

5   Conclusion and Future Work

In this paper, we have presented a fuzzy logic based approach for crowd simulation which extracts fuzzy rules from the realistic videos which can be seen as a parameterized behavior model. The derived rules can be used independently or be incorporated with predefined rules which extend and enrich our approach's applicability to new circumstances. The advantages of our approach are obvious: interoperability, universality and behavior's diversity based on the reason that we unite the virtues of data-driven model and rule-based agent model and fetch up each one's deficiency. And the weight of each model is determined by the users to adapt to various requirements.

In future work, we plan systematically to expand the behavioral and cognitive repertoires of our simulated agents. We also intend to develop a more plentiful set of reactive and deliberative behavior rules for our model.

References

1. Lee, K.H., Choi, M.G., Hong, Q., Lee, J.: Group Behavior from Video: A Data-Driven Approach to Crowd Simulation. In: Proc. ACM SIGGRAPH/ Eurographics Symposium on Computer Animation, San Diego, Eurographics Association Aire-la-Ville, Switzerland, pp. 109–118 (2007)
2. Zhou, S., Chen, D., Cai, W.: Crowd modeling and simulation technologies. ACM Transactions on Modeling and Computer Simulation 20(4), 1–39 (2009)
3. Lerner, A., Chrysanthou, Y., Lischinski, D.: Crowds by Example. Conputer Graphics Forum 26(3), 655–664 (2007)
4. Magnenat-Thalmann, N., Thalmann, D.: Virtual Humans: Thirty Years of Research, What Next? The Visual Computer 21(12), 997–1015 (2005)
5. Ross, T.J.: Fuzzy logic with engineering application, 3rd edn. John Wiley & Sons, Ltd. (2010)
6. Ghasem-Aghaee, N., Ören, T.I.: Towards Fuzzy Agents with Dynamic Personality for Human Behavior Simulation. In: Proceedings of the 2003 Summer Computer Simulation Conference, Montreal, PQ, Canada, pp. 3–10 (2003)
7. Thalmann, D., Raupp Musse, S.: Crowd Simulation. Springer, London (2007)
8. Pelechano, N., Allbeck, J., Badler, N.I.: Virtual Crowds: Methods, Simulation, and Control. Morgan&Claypool Publishers (2008)

Numerical Analysis for Stochastic Age-Dependent Population Equations with Diffusion

Jian He

Department of Basic Course, Beifang University of Nationalities, Yinchuan,
750021, People's Republic of China
hejian_0117@yahoo.com.cn

Abstract. A numerical method proposed to approximate the solution of a class of stochastic age-dependent populations with diffusion. We show that under certain conditions, weaker than Linear growth and Global Lipschits, the Euler scheme applied to Stochastic age-dependent population,converges to the analytic solution , and provide information on the order of approximation. An example is given to illustrate our theory.

Keywords: Stochastic age-dependent population, Numerical solution, Euler approximation.

1 Introduction

Recently, one of the most important and interesting problems in the analysis of stochastic age-structured population equations is their numerical solution. In this paper, we consider the convergence of stochastic age-structured population systems with diffusion

$$
\begin{cases}
\dfrac{\partial P}{\partial t} + \dfrac{\partial P}{\partial r} - k(r,x)\Delta P + \mu_1(r,t,x)P & \\
= f_1(r,t,x,P) + g_1(r,t,x,P)\dfrac{\partial W}{\partial t}, & in\ (0,A)\times\Gamma \quad (1.1) \\[2mm]
P(0,t,x) = \displaystyle\int_0^A \beta_1(r,t,x)\,P(r,t,x)dr, & in\ (0,T)\times\Gamma \quad (1.2) \\[2mm]
P(r,0,x) = P_0(r,x), & in\ Q_A = (0,A)\times\Gamma \quad (1.3) \\
P(r,t,x) & on\ \Sigma_A(0,A)\times(0,T)\times\partial\Gamma \quad (1.4) \\[2mm]
y(t,x) = \displaystyle\int_0^A P(r,t,x)dr & in\ Q \quad (1.5)
\end{cases}
$$

where $t\epsilon(0,T)$, $r\epsilon(0,A)$, $x\epsilon\Gamma$, $Q = (0,T)\times\Gamma$, $P(r,t,x)$ denotes the population density of age r at time r and in the location x, $\beta(r,t,x)$ denotes the fertility rate of females of age r at time t and in spatial position x, $\mu(r,t,x)$ denotes the mortality rate of age r at time t and in the location x, Δ denotes the Laplace operator with respect to the space variable, $k > 0$ (constant) is the diffusion coefficient. $f_1(r,t,x,P) + g_1(r,t,x,P)\dfrac{dW_t}{dt}$ denotes effects of external environment for population system, such

D. Jin and S. Lin (Eds.): Advances in ECWAC, Vol. 2, AISC 149, pp. 37–43.
springerlink.com

as emigration and earthquake and so on. The effects of external environment has the deterministic and random parts which depend on r, t, x and P.

There has been much recent interest in application of deterministic age-structures mathematical models with diffusion(when $g_1 = 0$)[1-3]. For example, Allen and Thrasher [1] considered vaccination strategies in age-dependent populations in the case of $g_1 = 0$. Pollard [2] studied the effects of adding stochastic terms to discrete-time age-dependent models that employ Leslie matrices.

In this paper, we will develop an numerical approximation method for stochastic age-dependent population equation of the type described by Eqs. $(1.1) - (1.5)$. e demonstrate that the Euler approximate solution will converge to the true solution if both remain within a bounded set. In particular, our results extend those of Zhang$[4-5]$.

2 Preliminaries and Approximation

Consider stochastic age-structured population system with diffusion $(1.1) - (1.5)$. A is the maximal age of the population species, so

$$P(r, t, x) = 0, \ \forall r \geq A.$$

By (1.5) integrating on $[0, A]$ to Eq. (1.1) and Eq. (1.3) with respect to r, then we obtain the following system

$$\begin{cases} \dfrac{\partial y}{\partial t} - k\Delta y + \mu(t, x)y - \beta(t, x)y \\ = f(t, x, P) + g(t, x, P)\dfrac{dW}{dt}, & in \ Q = (0, T) \times \Gamma, \\ y(0, x) = y_0(x) & in \ \Gamma, \\ y(t, x) = 0 & on \ \sum = (0, T) \times \partial\Gamma, \end{cases} \qquad (2.1)$$

where $\beta(t, x) \equiv \left(\int_0^A \beta_1(r, t, x)P(r, t, x)dr\right)\left(\int_0^A P(r, t, x)dr\right)^{-1}$, $\int_0^A P(r, t, x)dr$ is the total population, and the birth process is described by the nonlocal boundary conditions $\int_0^A \beta_1(r, t, x)P(r, t, x)dr$, clearly, $\beta(t, x)$ denotes the fertility rate of total population at time t and in the location x.

$\mu(t, x) \equiv \left(\int_0^A \mu_1(r, t, x)P(r, t, x)dr\right)\left(\int_0^A P(r, t, x)dr\right)^{-1}$, $\mu(r, t, x)$ denotes the mortality rate at time t and in the location x.

$$f(t, x, y) \equiv \int_0^A f_1(r, t, x, P)dr; \ g(t, x, y) \equiv \int_0^A g_1(r, t, x, P)dr,$$

let

$$V = H^1(\Gamma) \equiv \left\{ \varphi \middle| \varphi \in L^2(\Gamma), \dfrac{\partial\varphi}{\partial x_i} \in L^2(\Gamma), \ \text{where} \dfrac{\partial\varphi}{\partial x_i} \ \text{are generalized partial derivatives} \right\}$$

Then $\acute{V} = H^{-1}(\Gamma)$ the dual space of V. We denote by $|\cdot|$ and $\|\cdot\|$ the norms in V and $\acute{V}$ respectively; by $\langle \cdot \ \cdot \rangle$ the duality product between V, $\acute{V}$, and by $(\cdot \ \cdot)$ the scalar product in H. For an operator $B \in \mathcal{L}(MH)$ be the space of all bounded linear operators from M into H, we denote by $\|B\|_2$ the Hilbert-Schmidt norm, i.e.

$$\|B\|_2^2 = tr(BWB^T)$$

Let $(\Omega, \mathcal{F}, P)$ be a complete probability space with a filtrations $\{\mathcal{F}_t\}_{t \geq 0}$ satisfying the usual conditions (i.e., it is increasing and right continuous while $\mathcal{F}_0$ contains all P-null sets). Let $C = C([0, T]; V)$ be the space of all continuous function from $[0, T]$ into V with sup-norm $\|\psi\|_C = \sup_{0 \leq s \leq T} |\psi(s)|$, $L_V^P = L^P([0, T]; V)$ and $L_H^P = L^P([0, T]; H). T > 0, A > 0, f(t, x, \cdot): L_H^2 \to H$ be a family of nonlinear operators, $\mathcal{F}_t$-measurable almost surely in t. $g(t, x, \cdot): L_H^2 \to \mathcal{L}(M, H)$ is the family of nonlinear operator, $\mathcal{F}_t$-measurable almost surely in t.

Let $\Delta t = T/N$, for system (2.1) the discrete approximate solution on $t = 0$, $\Delta t, 2\Delta t, \cdots, N\Delta t$ is defined by the iterative scheme

$$y_{\Delta t}(t + \Delta t, x) - y_{\Delta t}(t, x) - k\Delta y_{\Delta t}(t, x)\Delta t + \mu(t, x)y_{\Delta t}(t, x)\Delta t - \beta(t, x)y_{\Delta t}(t, x)\Delta t$$
$$= f(t, x, y_{\Delta t}(t, x))\Delta t + g(t, x, y_{\Delta t}(t, x))\Delta W_t \tag{2.2}$$

with initial value $y_0 = y(0, x)$, $n \geq 1$, $y(t, x) = 0$, on $\Sigma = (0, T) \times \partial \Gamma$, where $\Delta y_{\Delta t}(t, x)$ denotes the Laplace of $y_{\Delta t}(t, x)$. for $t_n = n\Delta t$ the time increment is $\Delta t = \frac{T}{N} \ll 1$, and the Brownian motion increment is $\Delta W = W(t + \Delta t) - W(t)$.

For convenience, we shall extend the discrete numerical solution to continuous time. We first define the step function

$$\hat{y}_{\Delta t}(t, x) \equiv \Sigma_{k=0}^n y_{\Delta t}\big((k - 1)\Delta t, x\big) 1_{[((k-1)\Delta t, k\Delta t)]}(t, x) \tag{2.3}$$

where 1_G is the indicator function for the set G. Then we define

$$\Delta y_{\Delta t}(t, x) - y_0 - \int_0^t k\Delta y_{\Delta t}(t, x)ds - \int_0^t (\beta(s, x) - \mu(s, x)) \, \hat{y}_{\Delta t}(t, x)ds$$

$$= \int_0^t f\big(s, x, \hat{y}_{\Delta t}(t, x)\big) ds - \int_0^t g\big(s, x, \hat{y}_{\Delta t}(t, x)\big) dW_s \tag{2.4}$$

with $y_{\Delta t}(0, x) = y(0, x)$.

Let G be an open subset of V, and denote the unique solution of (2.1) for $t \in [0, T]$ given $y_0 \in G$ by $y(t, x) \in G$. Define $y_{\Delta t}(t, \alpha)$ as the Euler approximation (2.1) and let $\mathcal{D} \subseteq G$ be any bounded set. Assume the following conditions are satisfied:

$\mu(t, x)$, $\beta(t, x)$ are nonnegative measurable, and

$$\begin{cases} 0 \leq \mu_0 \leq \mu(t, x) < \infty & in \ Q \\ 0 \leq \beta(t, x) \leq \overline{\beta} < \infty & in \ Q \end{cases}$$

(i) (local Lipschitz condition)there exists a positive constant $K_1(\mathcal{D})$ such that $y_1, y_2 \in \mathcal{D}$

$$|f(t, x, y_2) - f(t, x, y_1)|^2 V\|g(t, x, y_2) - g(t, x, y_1)\|_2^2 \leq K_1(\mathcal{D})|y_2 - y_1|^2;$$

If condition (i) holds then there exists a positive constant $K_2(\mathcal{D})$ such that for $P \in \mathcal{D}$

$$|f(t,x,y)|^2 \vee \|g(t,x,y)\|_2^2 \le K_2(\mathcal{D}) \tag{2.5}$$

Since $y(t)$ is bounded, there exists a positive constant $K_3(\mathcal{D})$ such that

$$|y(t,x)|^2 \le K_3(\mathcal{D}) \tag{2.6}$$

In addition, Laplace Δy is a continuous and bounded linear operator, then there exists a positive constant $K_2'(\mathcal{D})$ such that for $y \in \mathcal{D}$

$$|\Delta y(t,x)|^2 \le K_2'(\mathcal{D}) \tag{2.7}$$

3 The Main Results

In this section, the objective is that under the conditions described above, we will prove the following convergence result. we set

$$y(t) := y(t,x), \ \Delta y_{\Delta t}(t) := \Delta y_{\Delta t}(t,x), \hat{y}_{\Delta t}(t) := \hat{y}_{\Delta t}(t,x).$$

Theorem 1. If τ is the first exist time of either the solution $y(t)$ or the Euler approximate solution $y_{\Delta t}(t)$ from a bounded region $\mathcal{D}$. $f(t,x,y(t))$ and $g(t,x,y(t))$ satisfy conditions (i) ,then for $\Delta tT < 1$

$$E\left[\sup_{0 \le t \le \tau \wedge T}|y_{\Delta t}(t) - y(t)|^2\right] \le C_1(\mathcal{D})e^{C_2(\mathcal{D})T}\Delta t = C(\mathcal{D})\Delta t,$$

where

$$C_1(\mathcal{D}) = 16\left[K_2'(\mathcal{D}) + |\mu_0 - \bar{\beta}|^2 K_3(\mathcal{D}) + K_2(\mathcal{D}) + CK_2(\mathcal{D})T\right]\left(|\mu_0 - \bar{\beta}| + 2K_1(\mathcal{D}) + 2KK_1(\mathcal{D})\right) C_2(\mathcal{D}) = 4[|\mu_0 - \bar{\beta}| + K_1(\mathcal{D}) + KK_1(\mathcal{D}) + 1].$$

Proof. Introduce the stopping time $\tau = \rho \wedge \theta$ where

$$\rho = \inf\{t \ge 0\colon y_{\Delta t}(t) \notin \mathcal{D}\} \text{ and } \theta = \inf\{t \ge 0\colon y(t) \notin \mathcal{D}\}$$

are the first time that $y_{\Delta t}(t)$ and $y(t)$, respectively, leave $\mathcal{D}$. We will define $\mathcal{D}$ more precisely later. Applying It $\hat{o}$'s formula to $|y_{\Delta t}(t) - y(t)|^2$, we obtain

$$\begin{aligned}
&|y_{\Delta t}(t) - y(t)|^2 \\
&\le -k\int_0^t \|y_{\Delta t}(s) - y(s)\|^2 \, ds \\
&+ \int_0^t |y_{\Delta t}(s) - y(s)|^2 ds + \int_0^t |f(s,x,\hat{y}_{\Delta t}(s)) - f(s,x,y(s))|^2 ds \\
&+ |\mu_0 - \bar{\beta}| \int_0^t |\hat{y}_{\Delta t}(s) - y(s)|^2 + |\mu_0 - \bar{\beta}| \int_0^t |y_{\Delta t}(s) - y(s)|^2 ds \\
&+ \int_0^t \|g(s,x,\hat{y}_{\Delta t}(s)) - g(s,x,y(s))\|_2^2 \, ds \\
&+ 2\int_0^t \left(y_{\Delta t}(s) - y(s), \left(g(s,x,\hat{y}_{\Delta t}(s)) - g(s,x,y(s))\right)\right) dw_s\big).
\end{aligned}$$

For $s \in [0, \tau \wedge T_1]$. let $T_1 \in [0, T]$ be an arbitrary time. We can derive that

$$E\left(\sup_{0 \le t \le \tau \wedge T_1} |y_{\Delta t}(t) - y(t)|^2\right) + k \int_0^{\tau \wedge T_1} \|y_{\Delta t}(s) - y(s)\|^2 ds$$

$$\le + \left|\mu_0 - \overline{\beta}\right| + 2K_1(\mathcal{D}) \int_0^{\tau \wedge T_1} E|\hat{y}_{\Delta t}(t) - y(t)|^2 dt$$

$$+ \left(\left|\mu_0 - \overline{\beta}\right| + 1\right) \int_0^{\tau \wedge T_1} E|y_{\Delta t}(t) - y(t)|^2 dt$$

$$+2E \sup_{0 \le t \le \tau \wedge T_1} \int_0^t \left(y_{\Delta t}(s) - y(s), \left(g(s, x, \hat{y}_{\Delta t}(s)) - g(s, x, y(s))\right)\right) dw_s. \quad (3.1)$$

Applying the Burkholder-Davis-Gundy's inequality to the last term of (3.1) leads to

$$E(\sup\_(0 \le t \le \tau \wedge T\_1) |y\_\Delta t(t) - y(t)|^2) + k \int_0^{\tau \wedge T_1} \|y_{\Delta t}(s) - y(s)\|^2 ds$$

$$\le 4\left[\left|\mu_0 - \overline{\beta}\right| + K_1(\mathcal{D}) + KK_1(\mathcal{D}) + 1\right] \times \int_0^{T_1} E\left[\sup_{0 \le r \le \tau \wedge s}|y_{\Delta t}(r) - y(r)|^2\right] ds$$

$$+4\left(\left|\mu_0 - \overline{\beta}\right| + 2K_1(\mathcal{D}) + KK_1(\mathcal{D})\right) \times E \int_0^{T_1} |\hat{y}_{\Delta t}(s) - y(s)|^2 ds \quad (3.2)$$

then applying the Gronwall's inequality leads to a bound on $E\left[\sup_{0 \le t \le \tau \wedge T}|y_{\Delta t}(t) - y(t)|^2\right]$. Inspection of (2.3) reveals that $\hat{y}_{\Delta t}(s) = y_{\Delta t}([s/\Delta t]\Delta t)$ where $[s/\Delta t]$ is the integer part of $s/\Delta t$. We can now use (2.4) to show that

$$|\hat{y}_{\Delta t}(s) - y_{\Delta t}(s)|^2 = |y_{\Delta t}([s/\Delta t]\Delta t) - y_{\Delta t}(s)|^2$$

$$\le 4\left|\int_{[s/\Delta t]\Delta t}^s k\Delta y_{\Delta t}([s/\Delta t]\Delta t) du\right|^2 + 4\left|\int_{[s/\Delta t]\Delta t}^s \left(\mu_0 - \overline{\beta}\right) y_{\Delta t}([s/\Delta t]\Delta t) du\right|^2$$

$$+4\left|\int_{[s/\Delta t]\Delta t}^s f(u, x, y_{\Delta t}([s/\Delta t]\Delta t)) du\right|^2 + 4\left|\int_{[s/\Delta t]\Delta t}^s g(u, x, \ y_{\Delta t}([s/\Delta t]\Delta t)) dw_u\right|^2,$$

$$(3.3)$$

whence applying the Burkholder-Davis-Gundy's inequality leads to

$$\left|\int_{[s/\Delta t]\Delta t}^s g(u, x, \ y_{\Delta t}([s/\Delta t]\Delta t)) dw_u\right|^2 \le CK_2(\mathcal{D})\Delta t,$$

where C is a constant. sing this result in (3.2) shows that

$$E\left(\sup_{0 \le t \le \tau \wedge T_1} |y_{\Delta t}(t) - y(t)|^2\right)$$
$$\le C_1(\mathcal{D})\Delta t + C_2(\mathcal{D}) \int_0^{T_1} E\left[\sup_{0 \le t \le \tau \wedge s}|y_{\Delta t}(r) - y(r)|^2\right] dr. \quad (3.4)$$

$$C_1(\mathcal{D}) = 16\left[K_2'(\mathcal{D}) + \left|\mu_0 - \overline{\beta}\right|^2 K_3(\mathcal{D}) + K_2(\mathcal{D}) + CK_2(\mathcal{D})T\right]\left(\left|\mu_0 - \overline{\beta}\right| + 2K_1(\mathcal{D}) + 2KK_1(\mathcal{D})\right)$$

and $C_2(\mathcal{D}) = 4\left[\left|\mu_0 - \overline{\beta}\right| + K_1(\mathcal{D}) + KK_1(\mathcal{D}) + 1\right]$

On applying the Gronwall's inequality we then have the following inequality

$$E\left(\sup_{0 \le t \le \tau \wedge T}|y_{\Delta t}(t) - y(t)|^2\right) \le C_1(\mathcal{D})e^{C_2(\mathcal{D})T}\Delta t = C(\mathcal{D})\Delta t.$$

To proceed further we define the bounded domain

$$\mathcal{D} = \mathcal{D}(r) \equiv \{y \in G \text{ such that } |y|^2 \le r\}.$$

Theorem 2. if θ is the first exist time of the solution $y(t)$ to equation (2.1) from the domain $\mathcal{D}(r)$, then the probability

$$P(\theta \geq T) \geq 1 - \varepsilon.$$

Proof. Applying It $\hat{o}'s$ formula to $|y|^2$ yields

$$d|y|^2 = 2\langle k\Delta y(t) \quad (\mu(\mathfrak{z},\mathfrak{n}) - \beta(\mathfrak{s},\mathfrak{v})y(\mathfrak{t})), y(\mathfrak{t})\rangle$$
$$+2\left(f(s,x,y(t)), y(t)\right) + \|g(s,x,y(t))\|_2^2 ds + 2\left(y(t), g(t,x,y(t))\right) dW_t.$$

Integrating from 0 to $t\wedge\theta$ and taking expectations gives

$$E|y(t\wedge\theta)|^2 + 2k\int_0^{t\wedge\theta} E\|y(s)\|^2 ds \leq y_0 + 2\left|\mu_0 - \overline{\beta}\right| E\int_0^{t\wedge\theta}|y(s)|^2 ds$$
$$+2E\int_0^{t\wedge\theta}\left(f(s,x,y(s)), y(s)\right) ds + E\int_0^{t\wedge\theta}\|g(s,x,y(s))\|_2^2 ds,$$

whence, applying condition(i) leads to

$$E|y(t)|^2 \leq |y_0|^2 + K_2(\mathcal{D})T + \exp\left(2\left|\mu_0 - \overline{\beta}\right| + K_2(\mathcal{D})\right).$$

Let $M = |y_0|^2 + K_2(\mathcal{D})T + \exp\left(2\left|\mu_0 - \overline{\beta}\right| + K_2(\mathcal{D})\right).$

On noting that $|y(\theta)|^2 = r$, since $y(\theta)$ is on the boundary of $\mathcal{D}(r)$, the probability $P(\theta < T)$ can now be bounded as follows.

$$M \geq E[|y(t\wedge\theta)|^2] \geq E[|y(t\wedge\theta)|^2 \, I_{\{\theta<T\}}(w)] \geq rE[\, I_{\{\theta<T\}}(w)] \geq rP(\theta < T), \quad (3.5)$$

whence rearranging (3.5) leads to

$$P(\theta < T) \leq M/r = \varepsilon. \qquad (3.6)$$

Here r can be made as large as required, to accommodate any $\varepsilon \in (0,1)$.

Theorem 3. let G be an open subset of V, and denote the unique solution of (2.1) for $t\in[0,T]$ given $y_0 \in G$ by $y(t) \in G$. Define $y_{\Delta t}(t)$ as the Euler approximation(2.4) and let $\mathcal{D} \subseteq G$ be any bounded set. Suppose condition (i) is satisfied. Then for any $\varepsilon, \delta > 0$ there exists $\Delta t^* > 0$ such that $P\left(\sup_{0\leq t\leq T}|y_{\Delta t}(t) - y(t)|^2 \geq \delta\right) \leq \varepsilon$, $\Delta t \leq \Delta t^*$ and the initial value $y_0 \in G$.
Proof of this result can be found in paper of Zhang[6].

References

1. Allen, L.J.S., Thrasher, D.B.: The effects of vaccination in an age-dependent model for varicella and herpes zoster. IEEE Transations on Automatic Control 43, 779–789 (1998)
2. Pollard, J.H.: On the use of the direct matrix product in analyzing certain stochastic population model. Biometrika 53, 397–415 (1966)

3. Renee Fister, K., Lenhart, S.: Optimal control of a competitive system with age-structure. Journal of Mathematical Analysis and Applications 291, 526–537 (2004)
4. Zhang, Q., Zhao, H.: Numerical analysis for stochastic age-dependent population equations. Applied Mathematics and Computation 176, 210–223 (2005)
5. Zhang, Q., Han, C.: Convergence of numerical solutions to stochastic age-structured population system with diffusion. Applied Mathematics and Computation 186, 1234–1242 (2007)
6. Zhang, Q., Zhang, W., Nie, Z.-K.: Convergence of the Euler scheme for stochastic functional partial differential equations. Applied Mathematics and Computation 155, 479–492 (2004)

An Improved Diagonal Loading Algorithm

Feng Wang, Jinkuan Wang, and Xin Song

School of Information Science & Engineering
Northeastern University, Shenyang, 110004, China
wangfeng_winner@163.com

Abstract. Traditional adaptive beamforming methods undergo serious performance degradation in the presence of mismatches between the assumed array response and the true array response. In this paper, we propose an algorithm which belongs to the class of diagonal loading approach, but the diagonal loading factor is computed fully automatically from the array observation vectors without the need of specifying any user parameters. Moreover, in order to decrease the computational cost, we adopt the gradient descent method to update weight vector. Some simulation results are presented to compare the performance of the proposed algorithm with sample matrix inversion (SMI) algorithm and loaded SMI (LSMI) algorithm.

Keywords: robust adaptive beamforming, steering vector mismatches, diagonal loading.

1 Introduction

Adaptive beamforming is used for enhancing a desired signal while suppressing noise and interference at the output of an array of sensors. In recent decades, adaptive beamforming has been widely used in wireless communications, microphone array speech processing, radar, sonar, medical imaging, radio astronomy, and other areas [1- 5].

In the practical applications, the performance of adaptive beamforming techniques may degrade severely because of underlying assumptions on the environments, sources, or sensor array can be violated and this may cause a mismatch between the assumed and actual signal steering vectors. Moreover, the covariance matrix may be inaccurately estimated due to limited data samples. Therefore, it is necessary to study the adaptive beamforming approaches.

In this paper, we propose an improved diagonal loading algorithm whose diagonal loading factor can be optimally chosen based on the array observation vectors. We show clearly how to efficiently compute our robust algorithm by using Lagrange multiplier method and gradient descent method. The excellent performance of the proposed algorithm is demonstrated as compared with other adaptive beamforming techniques via several examples.

D. Jin and S. Lin (Eds.): Advances in ECWAC, Vol. 2, AISC 149, pp. 45–52.
springerlink.com   © Springer-Verlag Berlin Heidelberg 2012

2 Background

2.1 Mathematical Formulation

Consider a uniform linear array(ULA) with M omnidirectional sensors spaced by the distance d and D narrowband incoherent plane waves, impinging from directions $\{\theta_0, \theta_1, \cdots, \theta_{D-1}\}$. The output of a narrowband beamformer is given by

$$y(k) = w^H x(k) \tag{1}$$

where $x(k) = \left[x_1(k), \cdots x_M(k)\right]^T$ is the complex vector of array observation, k is the time index, $w = \left[w_1, \cdots, w_M\right]^T$ is the complex vector of beamformer weights, and $(\cdot)^T$ and $(\cdot)^H$ stand for the transpose and Hermitian transpose, respectively. The observation vector is given by

$$\begin{aligned} x(k) &= s(k) + i(k) + n(k) \\ &= s_0(k)a + i(k) + n(k) \end{aligned} \tag{2}$$

where $s(k)$, $i(k)$ and $n(k)$ are the desired signal, interference, and noise components, respectively. Here, $s_0(k)$ is the signal waveform, and a is the signal steering vector. The weight vector can be found from the maximum of the signal to interference plus noise ratio (SINR)

$$\text{SINR} = \frac{\sigma_s^2 \left|w^H a\right|^2}{w^H R_{i+n} w} \tag{3}$$

where

$$R_{i+n} = E\left\{(i(k) + n(k))(i(k) + n(k))^H\right\} \tag{4}$$

is the $M \times M$ interference-plus-noise covariance matrix, and σ_s^2 is the signal power.

The problem of finding the maximum of (3) is equivalent to the following optimization problem

$$\min_{w} w^H R_{i+n} w \quad \text{subject to} \quad w^H a = 1 \tag{5}$$

From (5), the following solution can be found for the optimal weight vector

$$w_{opt} = \frac{R_{i+n}^{-1} a}{a^H R_{i+n}^{-1} a} \tag{6}$$

Inserting (6) into (3), we obtain the optimal SINR is given by

$$\text{SINR}_{opt} = \sigma_s^2 a^H R_{i+n}^{-1} a \tag{7}$$

Equation (7) gives an upper bound on the output SINR.

In practical applications, the exact interference-plus-noise covariance matrix R_{i+n} is unavailable. Usually, the sample covariance matrix

$$\hat{R} = \frac{1}{N} \sum_{n=1}^{N} x(n) x^{H}(n) \qquad (8)$$

is used in the optimization problem (5) instead of R_{i+n}, where N is the training sample size. In this case, (5) should be rewritten as

$$\min_{w} w^{H} \hat{R} w \quad \text{subject to} \quad w^{H} a = 1 \qquad (9)$$

The solution to this problem is commonly referred to as the SMI algorithm, i.e.,

$$w_{\text{SMI}} = \frac{\hat{R}^{-1} a}{a^{H} \hat{R}^{-1} a} \qquad (10)$$

However, using the sample covariance matrix $\hat{R}$ instead of the true interference-plus-noise covariance matrix R_{i+n} affects the performance of the algorithm dramatically. In this case, the SMI algorithm does not provide the robustness against the mismatch between the presumed and actual spatial signature vectors.

In order to improve the performance of SMI, several robust algorithms have been proposed. One of the most popular robust approaches is the loaded SMI (LSMI) algorithm [6,7], which the central idea is to replace the matrix $\hat{R}$ by the diagonally loaded covariance matrix

$$\hat{R}_{dl} = \gamma I + \hat{R} \qquad (11)$$

where γ is the diagonal loading factor. So, the weight vector of LSMI beamformer has the following form

$$w_{\text{LSMI}} = \left(\hat{R} + \gamma I \right)^{-1} a \qquad (12)$$

From equation (12), we obtain that the main drawbacks of the LSMI algorithm is follows: the high computational cost of the inversion of diagonally loaded sample covariance matrix and choice the accurate diagonal loading factor.

3 Improved Diagonal Loading Algorithm

In order to overcome the above problem, we propose an improved diagonal loading algorithm, which provides excellent robustness against signal steering vector mismatches and small training data size, improves the output performance dramatically.

3.1 The Choice of Diagonal Loading Factor

We consider a linear combination of $\hat{R}$ and R_{i+n}, which has the form [8]

$$\tilde{R} = \alpha R_{i+n} + \beta \hat{R} \qquad (13)$$

where $\alpha > 0$ and $\beta > 0$. In our paper, we assume the initial value of $\boldsymbol{R}_{i+n}$ is identity matrix $\boldsymbol{I}$ [9]. We have that $\tilde{\boldsymbol{R}} > 0$, and rewrite (13) as

$$\tilde{\boldsymbol{R}} = \alpha \boldsymbol{I} + \beta \hat{\boldsymbol{R}} \tag{14}$$

Then, we can obtain the enhanced covariance matrix as follow

$$\widehat{\boldsymbol{R}} = \hat{\boldsymbol{R}} + \frac{\alpha}{\beta} \boldsymbol{I} \tag{15}$$

Let $\widehat{\boldsymbol{R}}$ instead of $\hat{\boldsymbol{R}}$ in (9), the problem can be converted to

$$\min_{w} \boldsymbol{w}^{\mathrm{H}} \widehat{\boldsymbol{R}} \boldsymbol{w} \quad \text{subject to} \quad \boldsymbol{w}^{\mathrm{H}} \boldsymbol{a} = 1 \tag{16}$$

Using the Lagrange multiplier method to solve the problem, we obtain the Lagrange function $J(w, \lambda)$ as the following form:

$$J(w, \lambda) = \boldsymbol{w}^{\mathrm{H}} \widehat{\boldsymbol{R}} \boldsymbol{w} + \lambda \left(\boldsymbol{w}^{\mathrm{H}} \boldsymbol{a} - 1 \right) \tag{17}$$

The gradient vector of (17) is given by:

$$\boldsymbol{V} = 2 \widehat{\boldsymbol{R}} \boldsymbol{w} + \lambda \boldsymbol{a} \tag{18}$$

Let the gradient vector equal to zero, we can get the optimal weight vector

$$\boldsymbol{w}_{\mathrm{opt}} = -\frac{\lambda}{2} \left(\hat{\boldsymbol{R}} + \frac{\alpha}{\beta} \boldsymbol{I} \right)^{-1} \boldsymbol{a} \tag{19}$$

Contrast (12) and (19), we derive that the diagonal loading factor γ can be replaced by α / β. So the first step is to obtain α and β by minimizing the mean squared error of $\tilde{\boldsymbol{R}}$ [8]

$$\mathrm{MSE} = E \left\{ \left\| \tilde{\boldsymbol{R}} - \boldsymbol{R} \right\|^2 \right\} \tag{20}$$

where $\boldsymbol{R}$ denote the theoretical covariance matrix of the array output vector. The MSE minimization problem is derivationed as follows [10]

$$\begin{aligned}
& E \left\{ \left\| \tilde{\boldsymbol{R}} - \boldsymbol{R} \right\|^2 \right\} \\
&= E \left\{ \left\| \alpha \boldsymbol{I} - (1 - \beta) \boldsymbol{R} + \beta \left(\hat{\boldsymbol{R}} - \boldsymbol{R} \right) \right\|^2 \right\} \\
&= \left\| \alpha \boldsymbol{I} - (1 - \beta) \boldsymbol{R} \right\|^2 + \beta^2 E \left\{ \left\| \hat{\boldsymbol{R}} - \boldsymbol{R} \right\|^2 \right\} \\
&= \alpha^2 M - 2\alpha (1 - \beta) tr(\boldsymbol{R}) \\
&\quad + (1 - \beta)^2 \left\| \boldsymbol{R} \right\|^2 + \beta^2 E \left\{ \left\| \hat{\boldsymbol{R}} - \boldsymbol{R} \right\|^2 \right\}
\end{aligned} \tag{21}$$

The unconstrained minimization of (21) can be written as

$$\min_{\alpha} \quad \alpha^2 M - 2\alpha (1 - \beta) tr(\boldsymbol{R}) + \varphi(\beta) \tag{22}$$

where $\varphi(\beta) = (1 - \beta)^2 \left\| \boldsymbol{R} \right\|^2 + \beta^2 E \left\{ \left\| \hat{\boldsymbol{R}} - \boldsymbol{R} \right\|^2 \right\}$.

Computing the gradient of (22) and equating it to zero yields the optimal solution for α (β is fixed)

$$\alpha_0 = \frac{(1-\beta_0) tr(\boldsymbol{R})}{M} \tag{23}$$

Next, inserting (23) into (21), with β_0 replaced by β, we gain another unconstrained minimization

$$\min_{\beta} \quad \frac{(1-\beta)^2 \left[\|\boldsymbol{R}\|^2 M - tr^2(\boldsymbol{R})\right]}{M} + \beta^2 E\left\{\|\hat{\boldsymbol{R}} - \boldsymbol{R}\|^2\right\} \tag{24}$$

Computing the gradient of (24) and equating it to zero yields the optimal solution for β

$$\beta_0 = \frac{\eta}{\eta + \rho} \tag{25}$$

where $\rho = E\left\{\|\hat{\boldsymbol{R}} - \boldsymbol{R}\|^2\right\}$, $\eta = \frac{\|\boldsymbol{R}\|^2 M - tr^2(\boldsymbol{R})}{M}$.

To estimate α_0 and β_0 from the available data, we need an estimate of ρ and η. In practice, the exact covariance matrix $\boldsymbol{R}$ is unavailable. Therefore, $\boldsymbol{R}$ is replaced by $\hat{\boldsymbol{R}}$ to estimate η

$$\hat{\eta} = \frac{\|\hat{\boldsymbol{R}}\|^2 M - tr^2(\hat{\boldsymbol{R}})}{M} \tag{26}$$

Let $\hat{r}_m$ and r_m denote the mth columns of $\hat{\boldsymbol{R}}$ and $\boldsymbol{R}$ respectively. Consequently, we have

$$E\left\{\|\hat{\boldsymbol{R}} - \boldsymbol{R}\|^2\right\} = \sum_{m=1}^{M} E\left\{\|\hat{r}_m - r_m\|^2\right\} \tag{27}$$

According to [8], we can estimate ρ as

$$\begin{aligned}
\hat{\rho} &= \sum_{m=1}^{M}\left[\frac{1}{N^2}\sum_{n=1}^{N}\|x(n)x_m^*(n) - \hat{r}_m\|^2\right] \\
&= \frac{1}{N}\sum_{m=1}^{M}\left[\frac{1}{N}\sum_{n=1}^{N}\|x(n)\|^2 \cdot |x_m^*(n)|^2 - \|\hat{r}_m\|^2\right] \\
&= \frac{1}{N^2}\sum_{n=1}^{N}\|x(n)\|^4 - \frac{1}{N}\|\hat{\boldsymbol{R}}\|^2
\end{aligned} \tag{28}$$

where $x_m(n)$ denotes the mth element of $x(n)$. Applying (26) and (28), we can obtain the estimation of α_0 and β_0 as follows

$$\hat{\beta}_0 = \frac{\hat{\eta}}{\hat{\rho} + \hat{\eta}} \tag{29}$$

$$\hat{\alpha}_0 = \frac{(1 - \hat{\beta}_0) tr(\hat{\boldsymbol{R}})}{M} \tag{30}$$

Eventually, we obtain the estimation of $\hat{\boldsymbol{R}}$ as the following form

$$\hat{\boldsymbol{R}}_{RAB} = \hat{\boldsymbol{R}} + \frac{\hat{\alpha}_0}{\hat{\beta}_0} I \tag{31}$$

3.2 Recursive Implementation

In order to avoid computing the inversion of the covariance matrix in equation (19), we adopt the gradient descent method to update the weight vector

$$w(k+1) = w(k) - \mu_{\text{opt}}(k)\widehat{\boldsymbol{V}}(k) \tag{32}$$

where μ_{opt} is an optimal step size which determines the convergence speed of the algorithm. Let $\widehat{\boldsymbol{R}}_{\text{RAB}}$ instead of $\widehat{\boldsymbol{R}}$ in (18), we rewrite the gradient vector

$$\widehat{\boldsymbol{V}}(k) = 2\left(\widehat{\boldsymbol{R}} + \frac{\hat{\alpha}}{\hat{\beta}}\boldsymbol{I}\right)w(k) + \hat{\lambda}\boldsymbol{a} \tag{33}$$

The optimal step size is given by [11,12]

$$\mu_{\text{opt}} = \frac{\varepsilon \widehat{\boldsymbol{V}}^{\text{H}}(k)\widehat{\boldsymbol{V}}(k)}{\widehat{\boldsymbol{V}}^{\text{H}}(k)\widehat{\boldsymbol{R}}_{\text{RAB}}\widehat{\boldsymbol{V}}(k)} \tag{34}$$

The parameter ε is added to improve the numerical stability of the algorithm. For a practical system, it should be adjusted during the initial of the system and it should satisfy $0 < \varepsilon < 1$ [11,13].

In order to use (33) in the recursive implementation, we should known the value of $\hat{\lambda}$. So, inserting (19) into the linear constraint $w^{\text{H}}\boldsymbol{a} = 1$ and using $\widehat{\boldsymbol{R}}_{\text{RAB}}$ instead of $\widehat{\boldsymbol{R}}$ .We can get the expression of Lagrange multiplier

$$\hat{\lambda} = \frac{-2}{\boldsymbol{a}^{\text{H}}\widehat{\boldsymbol{R}}_{\text{RAB}}\boldsymbol{a}} \tag{35}$$

4 Simulation Results

In this section, we present some simulations to justify the performance of the improved diagonal loading algorithm. We assume a uniform linear array with $M = 10$ omnidirectional sensors spaced half a wavelength apart. For each scenario, 100 simulation runs are used to obtain each simulation point. In all examples, we assume two interfering sources with plane wavefronts and the directions of arrival impinging from $-50°$ and $50°$, respectively. The diagonal loading factor $\gamma = 10\sigma_n^2$ is taken in the LSMI algorithm, where σ_n^2 is the noise power.

Example 1: Exactly known signal steering vector
In this example, the plane wave signal is assumed to impinge on the array from $\theta = 0°$. Fig. 1 displays the performance of the three methods tested versus the number of snapshots for the fixed SNR=10dB. Fig. 2 shows the performance of these algorithms versus the SNR for the fixed training data size $N = 100$. In this scenario, we note that our improved diagonal loading algorithm outperforms the LSMI algorithm and makes the mean output SINR close to the optimal value.

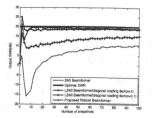 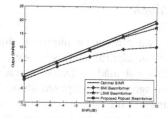

Fig. 1. Output SINR versus N **Fig. 2.** Output SINR versus SNR

Example 2: Signal look direction mismatch
We assume that both the presumed and actual signal spatial signatures are plane waves
impinging from the DOAs $0°$ and $3°$, respectively.

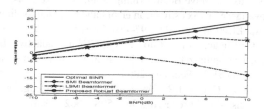

Fig. 3. Output SINR versus N

Fig. 3 displays the performance of the three methods tested versus the number of
snapshots for SNR=10dB. The performance of these algorithms versus the SNR for the
fixed training data size $N = 100$ is shown in Fig 4.

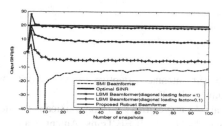

Fig. 4. Output SINR versus SNR

In this example, we note that the SMI algorithm is very sensitive even to slight
mismatches, and the LSMI algorithm can improve the performance of SMI algorithm
to some extent. Our improved diagonal loading algorithm consistently enjoys the best
performance among the methods tested, provides well robustness against signal
steering vector mismatches and the small training sample size.

5 Conclusions

In this paper, we present an improved diagonal loading algorithm. There are two advantages of our algorithm. Firstly, we adopt a computationally simple method to choose the diagonal loading factor instead of relaying on experimental experience. Another advantage of our proposed algorithm is that we adopt the gradient descent method to update the weight vector so that it decrease the computational cost. Simulation results demonstrate that our proposed algorithm provides well robustness against signal steering vector mismatches and small training data size, enjoys a significantly improved performance as compared with other algorithms.

Acknowledgment. The authors would like to thank the anonymous reviewers for their insightful comments that helped improve the quality of this paper. This work is supported by the National Nature Science Foundation of China under Grant no. 61004052.

References

1. Rapapport, T.S. (ed.): Smart Antennas: Adaptive Arrays, Algorithms, and Wireless Position Location. IEEE, Piscataway (1998)
2. Compton Jr., R.T., et al.: Adaptive arrays for communication system: An overview of the research at the Ohio State University. IEEE Trans. Antennas and Propagation 24, 599–607 (1976)
3. Gershman, A.B., Nemeth, E.: Experimental performance of adaptive beamforming in a sonar environment with a towed array and moving interfering sources. IEEE Trans. Signal Processing 48, 246–250 (2000)
4. Kameda, Y., Ohga, J.: Adaptive microphone-array system for noise reduction. IEEE Trans. Acoust., Speech, Signal Processing 34, 1391–1400 (1986)
5. Li, J., Stoica, P.: Robust Adaptive Beamforming. John Wiley & Sons, New York (2005)
6. Carlson, B.D.: Covariance matrix estimation errors and diagonal loading in adaptive arrays. IEEE Trans. Aerosp. Electron. Syst. 24, 397–401 (1988)
7. Vorobyov, S.A., Gershman, A.B., Luo, Z.-Q.: Robust adaptive beamforming using worst-case performance optimization: a solution to the signal mismatch problem. IEEE Trans. Signal Processing 51(2) (February 2003)
8. Stoica, P., Li, J., Zhu, X., Guerci, J.R.: On using a priori knowledge in space-time adaptive processing. IEEE Trans. Signal Processing 56, 2598–2602 (2008)
9. Ledoit, O., Wolf, M.: A well-conditioned estimator for large-dimensional covariance matrices. J. Multivar. Anal. 88, 365–411 (2004)
10. Li, J., Du, L., Stoica, P.: Fully automatic computation of diagonal loading levels for robust adaptive beamforming. In: IEEE International Conference. Acoustics, Speech and Signal Processing (2008)
11. Elnashar, A.: Efficient implementation of robust adaptive beamforming based on worst-case performance optimization. IET. Signal Processing 2, 381–393 (2008)
12. Elnashar, A., Elnoubi, S., Elmikati, H.: Further study on robust adaptive beamforming with optimum diagonal loading. IEEE Trans. Antennas Propag. 54, 3647–3658 (2006)
13. Attallah, S., Abed-Meraim, K.: Fast algorithms for subspace tracking. IEEE Trans. Signal Process. Lett. 8, 203–206 (2006)

Trajectory Planning for Industrial Robot Based on MATLAB

Xiancheng Fu, Guojun Wen*, and Han Chen

School of Mechanical and Electronic Information of China University of Geosciences (Wuhan), 430074
xcfuwh@yahoo.com.cn, wenguojun@126.com, nwx2578@163.com

Abstract. For advancing the design method, shortening the design time, and saving the costs for prototype, this paper builds the mathematic model for the simulation and gets the simulation results of the offline trajectory planning for six links which could be demonstrated intuitively and dynamically by using the MATLABfor the solid model of a welding Robot set up by using SolidWorks modeling software.

Keywords: Trajectory Planning, Industrial Robot, MATLAB.

1 Introduction

The operation precision of the industrial Robot depends on the matching between the location precision of itself and that of the workpiece although the former precision level is very high. In the actual application, the location precision of the workpiece is not satisfied with the expected with the limitation of the factory environment and the clamping apparatus precision, which must affect the final processing precision. At the same time, most of Robots conducts the reappearance running according to the processing trajectories determined through by the demonstration of an operator [1], so the processing precision is also influenced by the operator's proficiency degree to operate the Robot. For the best precision, therefore, it is essential to execute a fine trajectory planning to the industrial Robot, for example, welding Robot, by using the computer software. The main way is to figure out the expected trajectory in terms of the processing requirements in the relevant computer software, i.e., to work out the displacement, velocity, and acceleration, generate the motion trajectory and display the motion process of the trajectory just in time through by describing the assignment, motion route and trajectory of the Robot with the suitable computer language. Based on the theories of the joint space planning, the cartesian space planning, the kinematics and inverse kinematics [2], the simulation of trajectory planning is executed for the six-freedom welding Robot by using the MATLAB software on the computer, which supply the basic trajectory datum for the subsequent motion simulation analysis.

---

* Corresponding author. (wenguojun@126.com)

D. Jin and S. Lin (Eds.): Advances in ECWAC, Vol. 2, AISC 149, pp. 53–57.
springerlink.com   © Springer-Verlag Berlin Heidelberg 2012

2 Trajectory Planning Simulation

2.1 Kinematic Analysis of Robot

2.1.1 D-H Coordination System
Denavit and Hartenberg put forwards to a matrix method to build the attached coordinate system for each link in the joint chains of the Robot for describing the relationship of translation or rotation between the contiguous links in 1955 [3]. For the rotation joint, only the joint angle is the joint variable while the others are constant. For the translation one, only the offset is the variable while others are constant.

2.1.2 Adding the Attached Coordination System and List the Joint Parameters for the Robot
The simulation model is build through by using the SolidWorks software based on a kind of Panasonic welding Robot. The attached coordination systems are added on it respectively in terms of the D-H Coordination system above-mentioned illustrated in Figure 1. The parameters of all links and joints are illustrated in Table 1. Where, i is the number of each link; α_i represents the rotation angle about X_i axis from Z_i axis to Z_{i+1} axis; a_i represents the translation distance along X_i axis from Z_i axis to Z_{i+1} axis; d_i represents the translation distance along X_{i-1} axis from X_i axis to Z_{i+1} axis; and θ_i represents the rotation angle about Z_i axis from X_{i-1} axis to X_i axis.

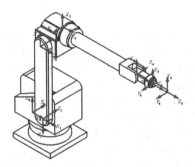

Fig. 1. Robot with attached coordination system

Table 1. Parameters of all links and joints

i	α_{i-1}	a_{i-1}	d_i	θ_i
1	-90°	155	0	θ_1
2	0	1000	0	θ_2
3	-90°	125	-25	θ_3
4	90°	0	1330	θ_4
5	-90°	0	0	θ_5
6	0	0	470	θ_6

2.2 Processing Trajectory Expected

Based on the PTP (i.e., point to point motion) planning [4] of the Robot, the starting point of the end of manipulator is supposed to be the point A (1955, -25, 1125), moves to the point B (1000, 200, 0) directly, then weld to an isosceles righttriangle route with the three vertexes of C (1200, 200, 0), D (1200, 0, 0), and B in turn. After the welding, the end of manipulator will go back the starting point A illustrated in Figure 2.

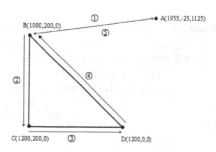

Fig. 2. Processing route of the manipulator

2.3 Simulation Modeling in MATLAB

The simulation model is built by the Link function in the Robotics Toolbox of MATLAB [5] illustrated as the following equation.

L = Link ([alpha A theta D]

Where alpha, A, theta, and D are α_i , a_i , θ_i , and d_i respectively. The whole mathematic model of the Robot with six links is in terms of the above-mentioned equation. This model could be examined by calling the function of drivebot(r) illustrated in Figure 3. The parameters of all links could be got by inputting r illustrated in Table 2 as same as Table 1.

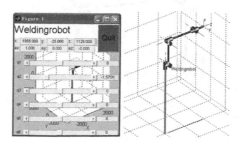

Fig. 3. Application of function drivebot(r) and model building

2.4 Motion Simulation of Trajectory Planning

The motion of the Robot model in MATLAB depends on the transformation matrixes according to the D-H Coordination system. There are some functions for the homogeneous transformations of translation and rotation such as transl(x, y, z), rotx(pi), roty(pi), and rotz(pi) in the Robotics Toolbox. The function ikine(ROBOT, T, Q, M) is used to get the variation of each joint variable, which is the inverse kinematic solutions process. In addition, the function JTRAJ(Q0, Q1,T) is used to realize the PTP trajectory planning in the joint space.

Table 2. Parameters of all links in MATLAB

r = Weldingrobot (6 axis, RRRRRR)					
grav = [0.00 0.00 9.81] standard D&H parameters					
alpha	A	theta	D	R/P	
-1.5708	155	0	0	R	(std)
0	1000	0	0	R	(std)
-1.5708	125	0	-25	R	(std)
1.570796	0	0	1330	R	(std)
-1.5708	0	0	0	R	(std)
0	0	0	470	R	(std)

2.5 Dynamic Demonstration of Trajectory Planning

For the intuition of the motion, it is convenient to show the dynamic motion process just in time by calling the function plot(r, q1). The Figure 4 and Figure 5 are the poses of the Robot while the end of its manipulator is at point A and B respectively.

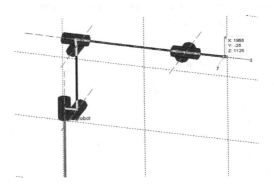

Fig. 4. Pose of the Robot at point A

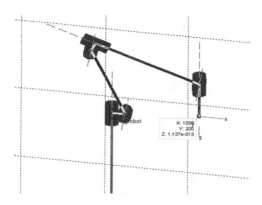

Fig. 5. Pose of the Robot at point B

3 Conclusions

This paper builds the mathematic model for the simulation and gets the expected simulation results of trajectory planning for six links which could be demonstrated intuitively and dynamically by calling the relevant function in MATLAB for a welding Robot set up by using SolidWorks modeling software.

Acknowledgments. This paper is supported by the Special Fund for Basic Scientific Research of Central Colleges, China University of Geosciences (Wuhan) and the project of Virtual Mechanical Machining Practical Educational Software of the China University of Geosciences (Wuhan).

References

1. Zhang, A., Zhang, Q.: Design and Study for the Robot Programming by Virtual Demonstration System. Manufacture Information Engineering of China, 79–81 (2003)
2. Craig, J.J., Yun, C.(trans.): Introduction to Robotics. China Machine Press (2006)
3. Chen, Y., Zhu, S., Luo, L., et al.: Kinematics analysis and simulation of QJ-6R welding robot based on Matlab. Mechanical and Electrical Engineering Magazine, 107–110 (2007)
4. Zhang, K., Liu, C., Fu, Z., et al.: Track Planning of 6R Robot and its Application to Welding. Journal of Machine Design, 20–23 (2002)
5. Luo, J., Hu, G.: Study on the Simulation of Robot Motion Based on MATLAB. Journal of Xiamen University (Natural Science), 640–644 (2005)
6. Chen, L.: Kinematics Analysis and ADAMS Application for Mechanical System. Tsinghua University Press (2005)

Feedforward Compensation Based the Study of PID Controller*

Shue Li and Feng Lv

Department of Electronics, Hebei Normal University
050031 Shijiazhuang, Hebei, China
lishue20106666@163.com

Abstract. This paper describes the conventional PID controller works, and feedforward control based on analysis of the PID controller works, and establish a feed-forward control of the PID controller the basic model, using MATLAB software and conventional PID control system based on the former the PID feedback control system for the simulation; and further analysis the feed-forward control based on the PID controller simulation waveforms, summed up the feed-forward control based on the superiority of the PID controller to achieve effective tracking of the input signal; this simulation system, the theoretical study of advanced PID, program design and failure analysis are of great help.

Keywords: PID, feedforward control, simulation, waveform analysis.

1 Introduction

PID (Proportion Integral Differential) controller is its simple structure, stable, reliable, easy to adjust to become one of the main techniques of industrial control. When the control object's structure and parameters can not fully grasp, or lack of a precise mathematical model and control theory is difficult to use other technologies, the system controller structure and parameters must rely on experience and on-site commissioning to determine when the application of PID control technology is most convenient. Use of the PID parameter adjustment, and feedforward compensation method to control the system, can achieve more satisfactory control effect.

2 PID Controller

Industrial PID controller in the control of the common law of the form:

$$u(t) = K_p [e(t) + \frac{1}{T_i} \int_0^t e(t)dt + T_d \frac{de(t)}{dt}] \tag{1}$$

* The paper supported by the National Natural Science Foundation (Project No.: 60974063) and School Youth Foundation (Project Number: L2009Q17).

Or written in the form of transfer function:

$$\frac{U(s)}{E(s)} = K_p[1 + \frac{1}{T_i s} + T_d s] \tag{2}$$

Where: K_p is the proportion coefficient; T_i is the integral time constant; T_d is the differential time constant; $u(t)$ is the controller output; $e(t)$ is for the deviation.

Written in differential equation (backward difference):

$$u(k) = K_p[e(k) + \frac{1}{T_i}\sum_{i=1}^{k} e(i)T + T_d \frac{e(k) - e(k-1)}{T}] \tag{3}$$

Where $u(k)$ is the output of the sampling time k; e (k), e (k-1) are deviation of the sampling time k, k-1 ; T is the sampling period.

In the computer system, using incremental PID control formula available to good control effect, it will type the formula (3) into the incremental form:

$$\Delta u(k) = u(k) - u(k-1)$$
$$= K_p\{e(k) - e(k-1) + \frac{T}{T_i} e(k) + \frac{T_d}{T}[e(k) - 2e(k-1) + e(k-2)]\} \tag{4}$$

PID controller, the percentage of deviation is mainly used for the "coarse", the control system to ensure "stability"; points mainly for the deviation "fine tune" the control system to ensure that "quasi"; differential is mainly used for deviation "fine-tune "to ensure that control systems" fast".

2.1 PID Parameter Tuning

PID controller parameter tuning controller design is the core content. It is based on the characteristics of the controlled process to determine the coefficient of PID controller proportional、 integral and derivative times the size of the time. PID controller parameter tuning of the many ways summed up in two categories: First, the entire theoretical titration. It is mainly based on the system mathematical model, the theoretical calculations to determine the controller parameters. This method of calculation of the data obtained may not be directly used, must also be adjusted by the actual construction and modifications. Second, the tuning method works, it mainly relies on engineering experience directly in the control experiments carried out, and the method is simple, easy to master, in engineering practice is widely used.

PID controller parameter tuning method works, there are the critical ratio method, the reaction curve and the decay curve. Three methods have their own characteristics, their common ground through trial and engineering experience in accordance with the formula to the controller parameter tuning. But no matter which way the resulting controller parameters are needed in the actual operation of the final adjustment and improvement. Now generally used is the critical ratio method and the decay curve.

3   Feedforward Compensation Based on the PID Control

The overall structure of the PID control based on feedforward compensation is shown in **Fig. 1**.

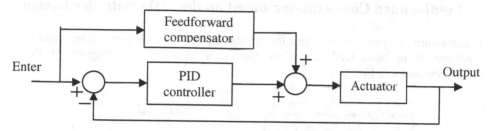

Fig. 1. The overall structure of the PID control based on feedforward compensation

In high-precision servo control, feedforward control can be used to improve system tracking performance. Classical control theory in the feedforward control design is based on the composite of thought control, closed-loop system for continuous system, the feedforward part of the closed-loop system transfer function of the product is 1, in order to achieve the output is completely re-enter the track. Feedforward control theory based on classical PID, based on the slip road parallel as part of feed-forward compensation. Design a feedforward compensation based on the overall structure of PID control as shown in **Fig. 3**.

3.1   Feedforward Compensation Based on the PID Controller Design Principles

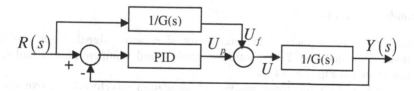

Fig. 2. Schematic of the PID with feedforward compensation

Using feedforward control theory, conventional PID control design for feedforward compensation based on the PID controller, the structure diagram shown in **Fig. 2**.
   Feedforward compensator is designed to:

$$u_f(\text{s}) = r(\text{s})\frac{1}{G(s)} \qquad (5)$$

Total control for the PID control output + output feedforward control output:

$$u(t) = u_p(t) + u_f(t) \qquad (6)$$

Written in discrete form:

$$u(k) = u_p(k) + u_f(k) \tag{7}$$

4 Feedforward Compensation Based on the PID Controller Design

Feedforward compensator will join the regular PID system, then based in simulink environment to build feed-forward compensation system block diagram of PID controller shown in **Fig. 3**.

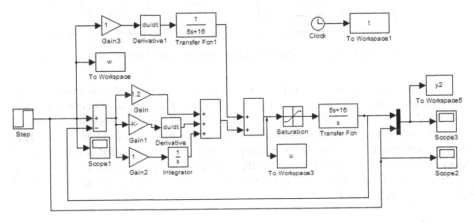

Fig. 3. Diagram of PID controller based on feed-forward compensation

4.1 Simulation Analysis

(1) Input sine wave signal: When the input sine wave signal based on the conventional PID and the PID controller feedforward compensation simulation waveforms shown in **Fig. 4** and **Fig. 5**.

Looking at **Fig. 4** for waveform analysis of simulation waveforms, the overshoot, settling time, steady-state error are:

$$\sigma = 6.38\%, \quad t_s = 1.85\text{ms}, \quad e(ss) = 0.0601$$

Looking at **Fig. 5** for waveform analysis of simulation waveforms, the overshoot, settling time, steady-state error are:

$$\sigma = 0.11\% \quad t_s = 1.25\text{ms} \quad e(ss) = 0.0001$$

(2) The unit step signal input: Unit step wave input signal with the conventional PID feedforward compensation based on the PID controller simulation waveforms shown in **Fig. 6** and **Fig. 7**.

Looking at **Fig. 6** to analyze the simulation waveform waveform overshoot, settling time, steady-state error are:

$$\sigma = 20.21\% \quad t_s = 3.25\text{ms} \quad e(ss) = 0.0001$$

Fig. 7 Waveform Analysis observe the simulation waveform of its overshoot, settling time, steady-state error are:

$$\sigma = 18.58\% \qquad t_s = 2.04\text{ms} \qquad e(ss) = 0.0001$$

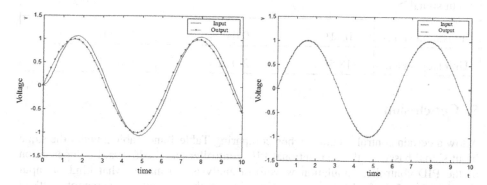

        Fig. 4. Conventional PID   **Fig. 5.** PID based on feedforward compensation.

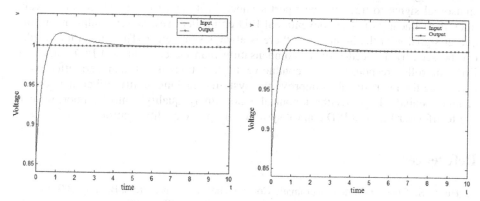

Fig. 6. Unit step input signal waveform simulation of conventional PID

Fig. 7. Unit step signal input simulation waveforms of the PID based on feedforward compensation.

Table 1. Conventional PID, under the unit step signal waveform analysis

Parameter / Input signal	Overshoot σ	Adjustment time t_s	Steady-state error $e(ss)$
Sinusoidal signal	6.38%	1.28ms	0.0601
Unit step signal	20.21%	3.25ms	0.0001

Table 2. PID sinusoidal feedforward compensation, unit step signal waveform analysis under

Parameter Input signal	Overshoot σ	Adjustment time t_s	Steady-state error $e(ss)$
Sinusoidal signal	0.11%	0.04ms	0.0001
Unit step signal	18.58%	2.04ms	0.0001

5 Conclusion

Allow a certain control accuracy when comparing Table 1 and Table 2 under the same input signal based on the conventional PID controller and feedforward compensation of the PID controller simulation waveform analysis, no matter what kind of input signal before PID feed-forward compensation in the amount of overshoot, settling time, steady-state error are superior to conventional PID controller. When the input sinusoidal signal to track the best performance in its unit step signal can effectively track their linear part. The conventional PID controller uses the error adjustment, the output signal with the input signal there is always some error. The simulation results can be seen by the feed-forward compensation than the conventional PID controller PID controller response speed conducive to fast tracking system and effectively reproduce the input signal to improve the system's tracking ability. When the control channel behind a large, feedback control is not timely, quality control is poor, we can use feedforward control PID control system to improve quality control.

References

1. Hu, S.-S.: Concise Guide to Automatic Control Theory. Science Press, Beijing (2008)
2. Li, S.: Automatic control theory. Xidian University Press, Xi'an (2007)
3. Xue, D.: Control System Computer Aided Design-MATLAB language and application. Tsinghua University Press, Beijing (2006)
4. Bai, J., Han, J.-W.: Based on MATLAB / Simulink environment, the PID parameter tuning. Harbin University of Commerce 12 (2007)
5. Liu, J.: Advanced PID control and MATLAB simulation. Electronic Industry Press, Beijing (2003)

The Application of Inventor Software in Design of Automatic Transplanting Robot

Xu Wen

Jilin Business and Technology College, Biological Engineering Division
Changchun 130062, China
xuwen2004@sina.com

Abstract. For the difficult problems existing in the subject " The design and motion simulation of seedling tray automatic transplanting robot" in this paper, Autodesk Inventor software as the physical design was introduced to this subject. Using rapid modeling characteristics of Inventor created mechanical model, analyzed the movement mechanism and calibrated mechanical properties, with the features of intuition, interactive operation, etc., and then calculate various properties of workpiece.

Keywords: Inventor physical design, associated design, parameter model.

1 Introduction

The size parameter design of robotic end effector is the key to seedling tray automatic transplanting robot. In accordance with the finger movement trajectory to complete the finger number, finger shape, fingers buried tilt and other key parameters in this end effector. Through the analysis of folder holding power and trajectory of robot fingers buried the design of the actuator structure in the end of robot operates Theoretical analysis and experimental study lead to the design of specific structural parameters in actuator finger. To this end, we used Autodesk Inventor, CAD software for three-dimensional design, and multimedia technology to assist the completion of the design task.

2 Application of Inventor in the Mechanical Design

Figure 1 is a seedling tray automatic transplanting robot machine model. Parts may be less commonandmany non-standard parts exist because the robot is new. Auxiliary parts for design features change with key parts and components. Using traditional way to design one by one is bound to need a lot of manpower, modify trouble, prone to error. This research used Autodesk Inventor 2008 to design and model using parametric modeling and adaptive aids, etc.

Through establishing contact with each other between parts and components other associated auxiliary part model size, components of the part model will be updated automatically to change with the key components in a model size.

D. Jin and S. Lin (Eds.): Advances in ECWAC, Vol. 2, AISC 149, pp. 65–70.
springerlink.com   © Springer-Verlag Berlin Heidelberg 2012

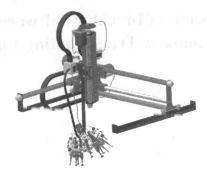

Fig. 1. Automatic transplanting robot machine model

2.1 Parts and Components Design

Taking automatic transplanting robot modeling as an example the structure size of machine finger is a critical to the robot, which affects many related parts model size. Such as (Figure 2), the link structure using parametric design is determined by spatial location and size of robot seat and fingers. So it is automatically updated when structure size of machine fingers change. Meanwhile, the cylinder used in the design of adaptive methods associates hinge side. It will be easy to calculate the cylinder stroke. This association modeling design reduced the repeated modified time and improved design efficiency and quality.

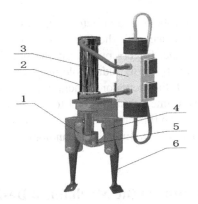

Fig. 2. Transplanting robot. 1 rod, 2 cylinder, 3 solenoid valve, 4 Manipulator Block, 5 hinged side, 6 Mechanical fingers.

2.2 Assembly and Inspection of Model Parts and Components

The Inventor interference checking function and less degrees of freedom constraints and other functions were used in the assembly process of component models to avoid ill-considered and other causes of errors when designed. The used function is a process of parts and components tests. A variety of design errors can be found during the model assembly. For example, the function includes part interference, lack of freedom constraints, structural analysis, kinematics analysis and analysis of quality characteristics and others.

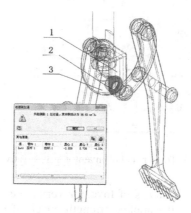

Fig. 3. Interferograms between two parts. 1 hinged side, 2 intervention part, 3 link.

Figure 3 shows interferograms between two parts. The negligence of size design in the beginning caused interference between two parts. The black part refers to intervention part and dialog box reports the specific interference volume.

Fig. 4. Diagram of parts and components freedom after assembly

Figure 4 shows the freedom diagram after transplanting robot assembly. A part due to constraints will cause too many freedom degrees and no action according to the design trajectory. So check the freedom degrees of part is an indispensable step.

3 Application of Inventor at Parts and Components Checking

In the new product design strength check often take up designers a lot of time. Although calculation can be followed, the process is cumbersome, complicated, heavy workload, error-prone and difficult to find. Inventor has a stress analysis functions

(Figure 5), including the equivalent stress, maximum principal stress, minimum principal stress, deformation and safety factor. Using the basic theories and methods of finite element analysis (FEA) can perform analysis of mechanical parts in the initial design phase, which help users with shorter time to design better products.

Fig. 5. Tools panel of Inventor stress analysis

Multiple using the stress analysis of Inventor verify the parts and components in this design. Figure 6 shows the applied strength check of robot. Because the robot uses light materials and wearing parts, so strength check is taken for it.

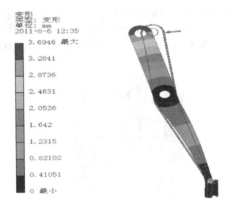

Fig. 6. Diagram of robot stress analysis

Parts can be checked and calculated efficiently for a variety ways by this method compared with manual checking, which is fast, efficient, intuitive, and comprehensive.

4 Applied Inventor for Agency Motion Simulation

To reduce the single-robot reciprocating time of abstract seedlings→ discard seedlings →abstract seedlings, combination of multi-robot body was designed in this system constructed by small stepper motor, lpomoea plate, a number of individual robots and pneumatic system. Small stepper motor and lpomoea plate are installed on the Z-axis arm and with a 60 ° angle. Small stepper motor control disc rotation direction and rotation angle of the lpomoea plate in order to let robot on the claw plate enter the work place and to non-working position away from the trays (to prevent injury seedlings). The robot in the working position with precise positioning coordinates removed non-healthy seedlings from the seedling tray out, then a small stepper motor driven Ipomoea rotating disk made next robot into the work place. Taking Inventor institutional motion simulation verified the feasibility of using a combination of multi-robot in the design.

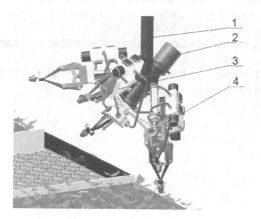

Fig. 7. Combination of multi-robot institution. 1 Z arm, 2 small stepper motor, 3 Ipomoea plate, 4 Single robot.

Inventor physical design can realize dynamic simulation. Computer simulation of movement can show mechanical movement after assembly on the screen presentation, which perform visually of the component functions. Users not only can use predefined animations to add animation quickly for parts and components according to design environment, but also can improve the existing animation by editing the properties, then broadcast by computer or media after edition and can be recorded as a variety of media file formats for archiving the record. The robot in the working position with precise positioning coordinates removed non-healthy seedlings from the seedling tray out, then a small stepper motor driven Ipomoea rotating disk made next robot into the work place.

5 Conclusion

Application of Inventor physical design software in this issue greatly simplifies the design process and improves efficiency, meanwhile saves a lot of time and extricates design staff from the tedious checking.

At present, Virtual simulation design relying on modern information technology has become a domestic hot and direction of scientific development. Mechanical model established by CAD technology can analyze product movement, checking mechanical properties and others. CAD software on the market today can be described as flourishing. CAD technology has been gradually from generation to mature today. Users can select or use combination according to their range of applications, preferences and habits.

References

1. Hu, R., Dong, Y., et al.: The mechanical design advanced applications of Internet 10 Chinese version, 2nd edn. Machinery Industry Press (February 2006)
2. Wu, X.: Students' comprehensive ability in the "Engineering Graphics" teaching course. Agricultural Research (4), 258–259 (2004)
3. Wu, P.: Design and motion simulation analysis of nonholonomic manipulator. Master's thesis in Wuhan University of Technology (May 2004)
4. Fang, J., Liu, S.: Dynamic simulation and mechanical engineering analysis. Chemical Industry Press (September 2004)

Modeling and Solving Optimum Intercept on Moving Objects for Robot

Dengxiang Yang

Sichuan Electric Power Corporation Xichang Electric Power Bureau, Xichang, Sichuan, China
tomgreen6@163.com

Abstract. Performance index of minimal time with constrains of successful intercepting on intercepting problem of moving objects for robot is built and it is changed into a programming problem with equality constrains to be solved. When acceleration of robot cart is considered, the model is changed in detailed and it show that the model is easy to expanded and corrected. Solving the models above using MATLAB and it is shown that the method of modeling is practical.

Keywords: moving objects, optimum intercept, robot, modeling, solving.

1 Introduction

Existing interception of moving objects for robot mainly adopted maximum velocity interception to intercepted object. References main discuss action and strategy of interception, rarely modeling to motion process of intercept action and solving optimum performance index to this model that widely used in military and air interception. In order to improve the accuracy and decreasing intercept time, this paper brings up a modeling and solution method to moving objects for robot intercept issue, and also optimizes performance index and relevant constraint condition. It has been found that the method of modeling is practical.

2 Description of Optimum Intercept

(1) The site and coordinate of moving objects for robot as shown in figure 1.

The length and width of site are respectively 359 and 312, and original point is located in top left of screen. Angle value is (0~+180)(-180~0). General size of robot is 20x20, diameter of soccer is 10, size of four corners is 20x20.

(2) Coordinate mark of intercept model (for the convenience of interface with original system, angle discussed blow is all crossing angle with the horizontal axis, shown in figure 2.)

D. Jin and S. Lin (Eds.): Advances in ECWAC, Vol. 2, AISC 149, pp. 71–76.

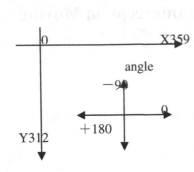

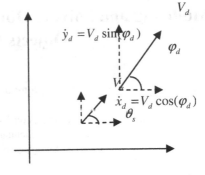

Fig. 1. Field and coordinate of moving objects for robot

Fig. 2. Coordinate of intercepting model

When intercepted:

(1) Coordinate position of intercepted object: (x_d, y_d)

Angle between direction of motion and horizontal axis is φ_d, uniform speed, speed is V_d;

(2) Coordinate position of interceptor: (x_s, y_s)

Angle between direction of motion and horizontal axis is θ_s, uniform speed, speed is V_s.

3   Modeling and Solution of Optimum Intercept

3.1   Objective Functionbject Function

$$\min J = \int_{t_0}^{t_f} dt = \min(t_f - t_0),$$

$$\text{when } t_0 = 0, \quad \min J = \min t_f \tag{1}$$

3.2   Intercept Model

When interception successful, interceptor and intercepted object converge, that is, x-coordinate is equal to y- coordinate. First the time of model car with maximum speed from 0 to θ_s is $\Delta t_r = \theta_d / w_{\max}$.

Where $w_{\max}$ is maximum rotational angular velocity; θ_d is crossing angle between θ_f (relative to positive x-axis) and θ_s, that is, $\Delta t_r = \angle(\theta_f - \theta_s) / w_{\max}$, $\angle$ means the minimum angle between θ_f and θ_s (ensure model car turns with smallest angle). So the constraint condition is

$$\begin{cases} x_d + V_d \cdot \cos \varphi_d \cdot (t_f + \Delta t_r) = x_s + V_s \cos \theta_s \cdot t_f \\ y_d + V_d \cdot \sin \varphi_d \cdot (t_f + \Delta t_r) = y_s + V_s \sin \theta_s \cdot t_f \end{cases} \tag{2}$$

Simplify, and then:

$$\begin{cases} x_d - x_s + V_d \cdot \cos \varphi_d \cdot (\angle(\theta_{f\_}\theta_s)/w_{max}) = (V_s \cdot \cos \theta_s - V_d \cdot \cos \varphi_d) \cdot t_f \\ y_d - y_s + V_d \cdot \sin \varphi_d \cdot (\angle(\theta_{f\_}\theta_s)/w_{max}) = (V_s \cdot \sin \theta_s - V_d \cdot \sin \varphi_d) \cdot t_f \end{cases} \tag{3}$$

Because the two sub-formulas in formula (3) have symmetry, without loss of generality, in view of four kinds of situations solve the formula.

(1) When $y_d - y_s \neq 0, V_s \sin \theta_s - V_d \sin \varphi_d \neq 0$,

Two sub-formulas divided, if $\Delta x = x_d - x_s$,

$\Delta y = y_d - y_s$, we have

$$\frac{\Delta x + V_d \cdot \cos \varphi_d \cdot (\angle(\theta_{f\_}\theta_s)/w_{max})}{\Delta y + V_d \cdot \sin \varphi_d \cdot (\angle(\theta_{f\_}\theta_s)/w_{max})}$$
$$= \frac{(V_s \cdot \cos \theta_s - V_d \cdot \cos \varphi_d)}{(V_s \cdot \sin \theta_s - V_d \cdot \sin \varphi_d)} \tag{4}$$

Simplify, and then:

$$\Delta y \cdot V_d \cdot \cos \varphi_d - \Delta x \cdot V_d \cdot \sin \varphi_d$$
$$= \Delta y \cdot V_s \cdot \cos \theta_s - \Delta x \cdot V_s \cdot \sin \theta_s$$
$$- V_d \cdot \cos \varphi_d / w_{max} \cdot V_s \cdot (\angle(\theta_{f\_}\theta_s)) \cdot \sin \theta_s$$
$$+ V_d \cdot \sin \varphi_d / w_{max} \cdot V_s \cdot (\angle(\theta_{f\_}\theta_s)) \cdot \cos \theta_s$$
$$+ V_d^2 \cdot \cos \varphi_d \cdot \sin \varphi_d / w_{max} \cdot (\angle(\theta_{f\_}\theta_s))$$
$$- V_d^2 \cdot \cos \varphi_d \cdot \sin \varphi_d / w_{max} \cdot (\angle(\theta_{f\_}\theta_s)) \tag{5}$$

If let

$$c_1 = \Delta y \cdot V_d \cdot \cos \varphi_d - \Delta x \cdot V_d \cdot \sin \varphi_d, c_2 = \Delta y \cdot V_s, c_3 = \Delta x \cdot V_s, c_4 = V_d \cdot \cos \varphi_d / w_{max} \cdot V_s,$$
$$c_5 = V_d \cdot \sin \varphi_d / w_{max} \cdot V_s, \text{ thus,}$$
$$c_1 = c_2 \cdot \cos \theta_s - c_3 \cdot \sin \theta_s$$
$$- c_4 \cdot (\angle(\theta_{f\_}\theta_s)) \cdot \sin \theta_s + c_5 \cdot (\angle(\theta_{f\_}\theta_s)) \cdot \cos \theta_s \tag{6}$$

(2) When $y_d - y_s \neq 0$, from formula (3),

$V_s \sin \theta_s - V_d \sin \varphi_d = 0$, we have

$$\begin{cases} x_d - x_s + V_d \cdot \cos \varphi_d \cdot (\angle(\theta_{f\_}\theta_s)/w_{max}) \\ = (V_s \cdot \cos \theta_s - V_d \cdot \cos \varphi_d) \cdot t_f \\ y_d - y_s + V_d \cdot \sin \varphi_d \cdot (\angle(\theta_{f\_}\theta_s)/w_{max}) = 0 \end{cases} \tag{7}$$

From $y_d - y_s + V_d \cdot \sin\varphi_d \cdot (\angle(\theta_{f\_}\theta_s)/w_{max}) = 0$, obtain:

$$\angle(\theta_{f\_}\theta_s) = \frac{-(y_d - y_s)}{V_d \cdot \sin\varphi_d / w_{max}} \tag{8}$$

Substitute first sub-formula in formula (7)
  Simplify, and then:

$$t_f = \frac{(x_d - x_s - ctg\varphi_d \cdot (y_d - y_s))}{(V_s \cdot \cos\theta_s - V_d \cdot \cos\varphi_d)} \tag{9}$$

From formula (8), obtain:

$$\theta_s = \angle\theta_{f\_} \left(\frac{-(y_d - y_s)}{V_d \cdot \sin\varphi_d / w_{max}} \right) \tag{10}$$

(3) When $y_d - y_s = 0$, from formula (3),
$V_s \sin\theta_s - V_d \sin\varphi_d \neq 0$, we have:

$$\begin{cases} x_d - x_s + V_d \cdot \cos\varphi_d \cdot (\angle(\theta_{f\_}\theta_s)/w_{max}) = (V_s \cdot \cos\theta_s - V_d \cdot \cos\varphi_d) \cdot t_f \\ V_d \cdot \sin\varphi_d \cdot (\angle(\theta_{f\_}\theta_s)/w_{max}) = (V_s \cdot \sin\theta_s - V_d \cdot \sin\varphi_d) \cdot t_f \end{cases} \tag{11}$$

From $V_d \cdot \sin\varphi_d \cdot (\angle(\theta_{f\_}\theta_s)/w_{max}) = (V_s \cdot \sin\theta_s - V_d \cdot \sin\varphi_d) \cdot t_f$, obtain :

$$\angle(\theta_{f\_}\theta_s) = \frac{(V_s \cdot \sin\theta_s - V_d \cdot \sin\varphi_d) \cdot t_f}{V_d \cdot \sin\varphi_d / w_{max}} \tag{12}$$

Substitute first sub-formula in formula (11), we have:

$$t_f = \frac{x_d - x_s}{(V_s \cdot \cos\theta_s - V_d \cdot \cos\varphi_d) - ctg\varphi_d \cdot (V_s \cdot \sin\theta_s - V_d \cdot \sin\varphi_d)} \tag{13}$$

From $\min J = \min t_f$, obtain:

$$\begin{aligned} &\min t_f \\ &= \min_{\theta_s} \left(\frac{x_d - x_s}{(V_s \cdot \cos\theta_s - V_d \cdot \cos\varphi_d) - ctg\varphi_d \cdot (V_s \cdot \sin\theta_s - V_d \cdot \sin\varphi_d)} \right) \end{aligned} \tag{14}$$

If let:

$a_1 = x_d - x_s; \ a_2 = V_s; \ a_3 = V_d \cdot \cos\varphi_d;$

$a_4 = V_d \cdot \sin\varphi_d;$ we have : $ctg\varphi_d = \dfrac{a_3}{a_4}$,

Simplify formula (14), and then:

$$\min t_f = \min_{\theta_s} \left(\frac{a_1}{(a_2 \cdot \cos\theta_s - a_3) - \left(\dfrac{a_3}{a_4}\right) \cdot (a_2 \cdot \sin\theta_s - a_4)} \right) \tag{15}$$

(4) When $y_d - y_s = 0$, from formula (3),

$V_s \sin \theta_s - V_d \sin \varphi_d = 0$, we have:

$$\begin{cases} x_d - x_s + V_d \cdot \cos \varphi_d \cdot (\angle(\theta_{f\_}\theta_s)/w_{max}) = (V_s \cdot \cos \theta_s - V_d \cdot \cos \varphi_d) \cdot t_f \\ V_d \cdot \sin \varphi_d \cdot (\angle(\theta_{f\_}\theta_s)/w_{max}) = 0 \end{cases} \tag{16}$$

From $V_d \cdot \sin \varphi_d \cdot (\angle(\theta_{f\_}\theta_s)/w_{max}) = 0$, obtain:

$\angle(\theta_{f\_}\theta_s) = 0$, thus, $\theta_s = \theta_f$,

Substitute first sub-formula in formula (16), we have:

$$t_f = \frac{x_d - x_s}{V_s \cdot \cos \theta_s - V_d \cdot \cos \varphi_d} \tag{17}$$

3.3 Problem Solving

Based on discussion of 3.2, formulas demanding to solve are formula (15), use example to explain.

Without loss of generality, assuming $\Delta y = 0$,

$\Delta x = -10, V_d = 10$, $\varphi_d = \dfrac{\pi}{3}, V_s = 20$,

$w_{max} = 10, \theta_f = \dfrac{\pi}{6}$, it satisfies $y_d - y_s = 0$, $V_s \sin \theta_s - V_d \sin \varphi_d \neq 0$, it comes to solve formula (15).

Therefore,

$a_1 = x_d - x_s = \Delta x = -10$; $a_2 = V_s = 20$; $a_3 = V_d \cdot \cos \varphi_d = 10 \cdot \dfrac{1}{2} = 5$;

$a_4 = V_d \cdot \sin \varphi_d = 10 \cdot \dfrac{\sqrt{3}}{2} = 5\sqrt{3}$;

So, $ctg\varphi_d = \dfrac{a_3}{a_4} = \dfrac{\sqrt{3}}{3}$, formula (15) is:

$\min t_f$

$$= \min_{\theta_s} \left(\frac{-10}{(20 \cdot \cos \theta_s - 5) - \left(\dfrac{\sqrt{3}}{3}\right) \cdot (20 \cdot \sin \theta_s - 5\sqrt{3})} \right) \tag{18}$$

Use function z= fminbnd(fun,x1,x2) of matlab to solve it.

z=fminbnd('(-10)/((20*cos(x)-5)-(sqrt(3)/3)*(20*sin(x)-5*sqrt(x)))',0,1000)
z =618.3676rad.

Obtain: z =618.3676rad, that is, $150°$. It can be easily proved that time is minimum when use the intercept angle $150°$.

3.4 Result Analysis

In order to make convenient for programming of embedded system and simplify base calculation, we deduce and solve above-mentioned formula. After modeling, above-mentioned formula is equal to the problem with equality constraint extreme point, and we can solve the extreme point by matlab/ simulink.

References

1. Peng, Z., Chen, S.: Solve optimize interception by singular perturbation method. Journal of Nanjing University of Aeronautics & Astronautics 25(4), 438–444 (1993)
2. Wang, H.-L., He, X.-S., Zhang, Y.-F.: The design of the intercept optimum orbit in space warfare. Journal of Changchun University of Science and Technology 30(2), 51–54 (2007)
3. Wang, X.-Z., He, C.-Y., Wu, S.-C.: An improved algorithm for maneuverable target interception. Journal of Central South University of Forestry & Technology 28(5), 136–139 (2008)
4. Bai, B.-Y., Shen, Y.: Research and application of DR algorithm in heading-off strategy. Computer Simulation 26(3), 169–171, 255 (2009)
5. Li, H.-Q.: Strategy of coordinated hunting/intercepting by multiple mobile robots based on potential points. Automation & Instrumentation 5, 1–4 (2007)
6. Xing, Y.-B., Shi, H.-S., Zhao, H.-G., Ji, W.: Study of lead angle guidance method based on football robot intercept action. Journal of System Simulation 19(2), 393–395 (2007)

Project Supported the Fund

Hu Nan province Department of Education project (09C889).

Particle Swarm Optimization Based on Uncertain Knowledge for Dynamic Data Reconciliation

Jing Zhang, Congli Mei, and Guohai Liu

School of Electrical and Information Engineering, Jiangsu University
Zhenjiang, China, 212013
zhangjing8793110@126.com, clmei@ujs.edu.cn, ghliu@ujs.edu.cn

Abstract. Dynamic Data reconciliation (DDR) is the adjustment of a set of process data based on models of the process so that the derived estimates conform to natural laws. In this paper, a novel particle swarm optimization based on uncertain knowledge (PSO-UK) for DDR and outliers detection is proposed. The uncertain knowledge is introduced in particle swarm optimization (PSO) algorithm, which limited the search center and range of particles, making the particles fly toward the direction close to the optimal solution. So the confidence of particles is improved and the premature problem can be overcome. Simulation of a continuous stirred tank reactor (CSTR), verifies the effectiveness of the proposed algorithm.

Keywords: Particle Swarm Optimization, Uncertain Knowledge, Dynamic Data Reconciliation, Outliers.

1 Introduction

In the chemical industry processes, accurate process data is the foundation of on-line fault diagnosis, monitoring, dynamic optimization and advanced control. DDR can effectively eliminate random errors. However, in fact the industrial measurements are contaminated by outliers[1]. Because of the presence of outliers, the reconciled measurements may mostly deviate from their true values. It is quite necessary to carry out outliers detection and eliminate outliers in the data reconciliation.

Due to the complexity of dynamic data, dynamic data rectification method hasn't got enough development. Singhal, A.&Seborg, D. E. [2] proposes a probabilistic formulation which combines the extended Kalman filter (EKF) and the expectation-maximization algorithm to rectify data, but the application of EKF is restricted. The prediction accuracy will be decreased deeply when the state and measurement equations are highly non-linear. With the purpose of solving the nonlinear problem, David M.H. [3] proposes the artificial neural networks(ANN) which use historical data to distinguish system quickly and can be used in the condition with high nonlinear. But the construction of ANN needs a lot of training networks model, the accuracy of models will be low when the posterior distribution of the states is non-Gaussian. Romagnoli et al.[4] proposed the constrained nonlinear programming, it is common to use a time varying moving window to reduce the optimization problem to

D. Jin and S. Lin (Eds.): Advances in ECWAC, Vol. 2, AISC 149, pp. 77–82.
springerlink.com

manageable dimensions.But the linearization approach are inequality constraints and variable bounds cannot be used during the optimization step and time-consuming.

PSO algorithm often trapped in local optimum and caused the typical premature problem. To overcome this problem, some researchers have made many improved algorithms, according to the characteristics, parameters of PSO and the application background to carry out PSO parameter selection strategies[5], PSO hybrid strategy with other methods[6] and controlling population diversity strategy[7]. But these improved strategies are still under PSO algorithm framework, it is difficult to essentially overcome premature problems which caused by trapping in local optimum.

In this paper, the PSO-UK algorithm is proposed, which introduced the uncertain knowledge to basic PSO, and limited the particles' search center and search range, to ensure that the particles convergence to the optimal value. The confidence of particles is improved and the premature problem can be overcome. The influence of outliers to the estimation can be controlled, hence leading to more accurate results. Futher more, moving window is used in DDR, since it can decrease complexity of computation. As moving window moves across the data, the same value is being estimated multiple times. This can be treated as temporal redundancy. The study of CSTR shows that the proposed PSO-UK algorithm can reduce the influences of outliers and has a better performance than the PSO algorithm.

2   PSOUK for Dynamic Data Reconciliation

2.1   Nonlinear Dynamic Data Reconciliation

The general nonlinear dynamic data reconciliation can be formulated as a dynamic optimization problem, where the objective is to minimize the deviation and subjected to the dynamic model and nonlinear algebraic model and/or inequality constraints as follows:

$$\min \Phi = \sum_{i=0}^{c} [\hat{x}(t_i) - \tilde{x}(t_i)]^T Q^{-1} [\hat{x}(t_i) - \tilde{x}(t_i)] \tag{1}$$

Subjected to:

$$f[\frac{d\hat{x}}{dt}, \hat{x}(t)] = 0 , \quad h[\hat{x}(t)] = 0 , \quad g[\hat{x}(t)] \geq 0 \tag{2}$$

where, $\tilde{x}(t_i)$ are the measurements, $\hat{x}(t_i)$ are the estimated measurements at time step t, t_0 is the initial time, t_c is current time and c is the subscript indexing the current time, Q is the covariance matrix, f is differential equation constraints, h is algebraic equality constraints and g are inequality constraints including simple upper and lower bounds.

Moving time window technique can be used effectively reduce the optimization dimension, shorten the computing time and improve efficiency. With moving window technique, the objective function can be expressed as:

$$\min \Phi = \sum_{i=c-H}^{c} [\hat{x}(t_i) - \tilde{x}(t_i)]^T Q^{-1} [\hat{x}(t_i) - \tilde{x}(t_i)] \tag{3}$$

where, H is the length of the sliding window.

2.2 Particle Swarm Optimization Algorithm Based on Uncertain Knowledge

In the PSO algorithm, the new velocities and the positions of the particles for the next fitness evaluation are calculated using the following two equations:

$$v_{id} = v_{id} + c_1 rand1(\cdot)(p_{id} - x_{id}) + c_2 rand2(\cdot)(p_{gd} - x_{id}) \qquad (4)$$

$$x_{id} = x_{id} + v_{id} \qquad (5)$$

where, c_1 and c_2 are constants known as acceleration coefficients, and $rand1(.)$ and $rand2(.)$ are two different random numbers uniformly distributed on [0,1]. Empirical results have shown that a constant inertia of $w=0.7298$ and acceleration coefficients with $c_1=c_2=1.49618$ provide good convergent behavior. Theoretical analysis provided sufficient conditions that particles converge to a stable point, when $0<w<1$, which can be stated as $0<c_1+c_2<2(w+1)$ [8].

However, in equation (4) if values of r_1 and r_2 are very small, the particles will gained less experience information from the individual behavior and group behavior, and the particles mainly depend on inertia "flight", but the flight with a certain "blindness" ,at this point particle cognitive has uncertainty characteristics. Without cognitive, particle will take any point in the search space as flight target. In this paper, uncertain knowledge was introduced into the PSO algorithm to improve the integrity of particles knowledge structure, which is called particle swarm optimization based on uncertain knowledge, the evolution equation can be expressed as:

$$v_i^{k+1} = w \cdot v_i^k + c_1 \cdot l_1 \cdot (x_i^* - x_i^k) + c_2 \cdot l_2 \cdot (x^* - x_i^k) + c_3 \cdot l_3 \cdot (C_D - x_i^k) \qquad (6)$$

$$x_i^{k+1} = x_i^k + v_i^{k+1} \qquad (7)$$

where, $l_1 = r_1/(r_1+r_2+r_3)$, $l_2 = r_2/(r_1+r_2+r_3)$, $l_3 = r_3/(r_1+r_2+r_3)$, $C_D = sgn(r_4-0.5)*limit+p_c$, "sgn" represents that particles uncertainty under flight direction of uncertain knowledge, limit is the distance from search space border to center, p_c is the center of searching space, C_D represents the border of the uncertain knowledge.

2.3 Process of PSO-UK Algorithm for DDR

The process is shown below:

Step 1: According to the CSTR model parameters generated the true values of the concentration variable and temperature variable;

Step 2: Plus white noise and outliers based on the true value as measurements;

Step 3: Initialize the parameters, set the number of particles and the maximum number of iterations;

Step 4: Set sample number $k=i$ (i is from H to 100), take the objective function with data correction optimization as a particle of the fitness function;

Step 5: Save the excellent location of best particle, $t=t+1$, if the number of iterations $t >= 100$, then end the iteration, go to the next step; otherwise, return to step4;

Step 6: If $k > 100$, then end the loop, output the optimal location of the particles, that is the sample correction values; Otherwise, the implementation of $k=k+1$, and return to step3 and continue to run.

3   Case Study

3.1   Continuous Flow Stirred Tank Reactor Description

The performance of the proposed method has been tested using a simulated CSTR with a first order exothermic reaction [9]. There are four measured variables in CSTR: input concentration C_0, input temperature T_0, output concentration C_A and output temperature T. The process dynamic model is given by the following equations:

$$\frac{dC_A}{dt} = \frac{F}{V}(C_{A0} - C_A) - \alpha_d K C_A \tag{8}$$

$$\frac{dT}{dt} = \frac{F}{V}(T_0 - T) + \alpha_d \frac{(-\Delta H_r)CrKC_A}{\rho C_p Tr} - \frac{UA}{\rho C_p V}(T - T_J) \tag{9}$$

where, $K=k_0exp(-E/(T*Tr))$ is an Arrhenius rate expression. The process parameters are shown in Table 1.

Table 1. Parameters for CSTR dynamic model

Symbol	Value	Symbol	Value
F	$10.0cm^3s^{-1}$	A	$10\ cm^2$
V	$1000.0\ cm^3$	ΔHr	$2.7e4calgmol^{-1}$
TJ	$340\ K$	k_0	$7.86e12\ s^{-1}$
ρ	$0.001\ gcm^{-3}$	E	$1.409e4\ K$
Cp	$1.0\ cal(gK)^{-1}$	R	$8.317kJ/k\ mol\ k$
U	$5.0e\text{-}4cal(cm^2sK)^{-1}$		

3.2   Results and Discussion

Measurements for both state variables and input variables are generated at time steps of 2.5s by adding outliers and Gaussian noise to the true values as the measurements, where the true values were obtained through numerical integration of the dynamic equations. The CSTR simulation is initialized at a steady state operating point of C0=6.5mol·m⁻³, T0=3.5K, CA=0.1531mol·m⁻³, and T=4.6091K. At time step 30, the feed concentration was stepped up from 6.5to 7.5. In the PSO-UK algorithm, it limits the particles' search center and search range to make the particles fly toward the direction close to the optimal solution. Take the average of measurements in sliding window as search center p_c of particle moving, p_c plus or minus 50% of the measurements as the range of particle searching.

The simulation results shown in Fig 1and2 as below:

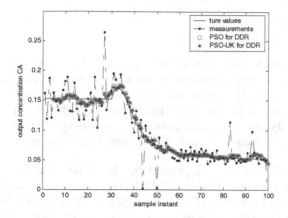

Fig. 1. Output concentration estimates response to time step change

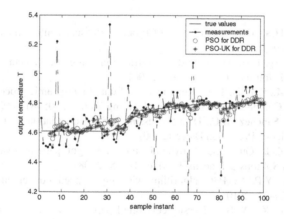

Fig. 2. Output temperature estimates response to time step change

Table 2. Performance comparison of different algorithms

Variables	MSE	
	PSO-UK	PSO
CA	0.000020662	0.000026487
T	0.000614595	0.001813698

From Figs it can be seen that PSOUK algorithm have better performance than the PSO algorithms. The reconciliation values of variables are closer to the true value, the impact of outliers are effectively reduced by the improved algorithm and obtain better correction results. As can be seen from Table 2, MSE of the proposed PSOUK algorithm are smaller than PSO algorithm, which further illustrates the improved algorithm's effectiveness.

4 Conclusion

In this paper, the particle swarm optimization based on Uncertain Knowledge method is proposed. The novel method introduced the search center of particles and range, to ensure that the particles fly toward the direction close to the optimal solution. It overcome the particle's premature problem and cut down the influence of outliers to estimations in the dynamic data rectification. A case study of CSTR process model typically used in the literature is used to demonstrate its effectiveness.

Acknowledgment. The work is supported by the Natural Science Foundation of Jiangsu University of China Grant 08KJD510011, the priority academic program development of Jiangsu higher education institutions (PAPD) and Natural Science Foundation for Qualified Personnel of Jiangsu University of China Grant 08JDG017.

References

1. Yuan, Y.G., Li, H.S.: Data Rectification of Process Measurements. China Petro Chemical Press, Beijing (1996)
2. Singhal, A., Seborg, D.E.: Dynamic data rectification using the expectation maximization algorithm. AICHE Journal 46, 1556–1565 (2000)
3. Himmelblau, D.M., Karjala, T.W.: Rectification of data in a dynamic process using artifical neural networks. Computers & Chemical Engineering 20(6-7), 805–811 (1996)
4. Romagnoli, J.A., Sanchez, M.C.: Data Processing and Reconciliation for Chemical Process Operations. Academic Press, San Diego (2000)
5. Jiao, B., Lian, Z.G., Gu, X.S.: A dynamic inertia weight particle swarm optimization algorithm. Chaos Solitons & Fractals 37(3), 698–705 (2008)
6. Zhu, H.M., Wu, Y.P.: A PSO algorithm with high speed convergence. Control and Decision 25(1), 20–24 (2010)
7. Fang, W., Sun, J., Xu, W.B.: Diversity-controlled particle swarm optimization algorithm. Control and Decision 23(8), 863–868 (2008)
8. Trelea, I.C.: The particle swarm optimization algorithm: Convergence analysis and parameter selection. Inf. Process. Lett. 85, 317–325 (2003)
9. Zhou, L.K., Su, H.Y., Chu, J.: A modified outlier detection method in dynamic data reconciliation. Chin. J. Chem. Eng. 13, 542–547 (2005)

An Approach Finding Multiple Roots of Nonlinear Equations or Polynomials

Junlong Peng[1] and Zhezhao Zeng[2]

[1] School of Traffic and Transportation Engineering,
Changsha University of Science and Technology, Changsha 410014, PR China
hncs6699@yahoo.com.cn
[2] College of Electric and Information Engineering,
Changsha University of Science and Technology, Changsha 410014, PR China
zengzhezhao@qq.com

Abstract. A new approach to find multiple roots of nonlinear equations or polynomials was developed, which were not well solved by the other methods. The method is especially suitable for finding the double roots of nonlinear equations or polynomials. Several examples were given to illustrate the efficiency of the new approach and to give the comparison with the recent cubic convergent methods. The results showed that the proposed approach can find the multiple roots of polynomials or polynomials at a very rapid convergence and very high accuracy with less computation.

Keywords: nonlinear equations, polynomials, multiple roots, algorithm.

1 Introduction

Finding rapidly and accurately the roots of nonlinear equations or polynomials is an important problem in various areas of control and communication systems engineering, signal processing and in many other areas of science and technology. The problem of finding the zeros of nonlinear equations has fascinated mathematicians for centuries, and the literature is full of ingenious methods, analysis of these methods, and discussions of their merits [1-3]. Over the last decades, there exist a large number of different methods for finding nonlinear equations roots either iteratively or simultaneously. Most of them yield accurate results only for small degree or can treat only special polynomials, e.g., polynomials with simple real or complex roots [4].

So far, some better modified methods of finding roots of nonlinear equations or polynomials cover mainly the Jenkins/Traub method [5], the Markus/Frenzel method [4], the Laguerre method [6], the Routh method [7], the Truong, Jeng and Reed method [8], the Fedorenko method [9], the Halley method[10], and some modified Newton's methods[11-13], etc. Although the Laguerre method is faster convergent than all other methods mentioned above, it has more computation. Among other methods, some have low accuracy, and some have more computation, especially to say is the modified Newton's methods must have a good initial value near solution. Furthermore, it is very difficult for the all methods mentioned above to find multiple real or complex roots of nonlinear equations or polynomials.

D. Jin and S. Lin (Eds.): Advances in ECWAC, Vol. 2, AISC 149, pp. 83–87.
springerlink.com   © Springer-Verlag Berlin Heidelberg 2012

In order to solve the problems above, Dong [14], Neta and Johnson[15], Neta [16], etc. developed some effective methods. In the paper, we proposed a new method finding multiple real or complex roots of nonlinear equations or polynomials. The approach can find multiple roots of nonlinear equations or polynomials with less computation, high accuracy and rapid convergence.

2　The Algorithm Finding Double Roots of Nonlinear Equations or Polynomials

Consider the nonlinear equation or polynomial of the type

$$f(x) = 0 \tag{1}$$

We assume that x is a simple or multiple root of the Eq. (1) and x_0 is an initial guess sufficiently close to x. Using the Taylor's series expansion of the function $f(x)$, we have

$$f(x_0) + (x - x_0)f'(x_0) + \tfrac{1}{2}(x - x_0)^2 f''(x_0) = 0 \tag{2}$$

This alternative equivalence formulation has been used to develop a class of iterative methods for solving nonlinear equation $f(x) = 0$. We can rewrite (2) in the following form:

$$f''(x_0)(x - x_0)^2 + 2f'(x_0)(x - x_0) + 2f(x_0) = 0 \tag{3}$$

This is a quadratic equation. From which we have

$$x - x_0 = \frac{-2f'(x_0) \pm \sqrt{4[f'(x_0)]^2 - 8f''(x_0)f(x_0)}}{2f''(x_0)} \tag{4}$$

or

$$x - x_0 = \frac{-f'(x_0) \pm \sqrt{[f'(x_0)]^2 - 2f''(x_0)f(x_0)}}{f''(x_0)} \tag{5}$$

This formula gives two possibilities for x, depending on the sign proceeding the radical term. Thus,

$$x = x_0 - \frac{f'(x_0) + \sqrt{[f'(x_0)]^2 - 2f(x_0)f''(x_0)}}{f''(x_0)} \tag{6}$$

$$x = x_0 - \frac{f'(x_0) - \sqrt{[f'(x_0)]^2 - 2f(x_0)f''(x_0)}}{f''(x_0)} \tag{7}$$

This fixed point formulation allows us to suggest the following method.

Algorithm 2.1. For a given x_0, compute approximate solution x_{n+1} by the iterative schemes
For the smaller root, using the following iterative formula

$$x_{n+1} = x_n - \frac{f'(x_n) + \sqrt{[f'(x_n)]^2 - 2f(x_n)f''(x_n)}}{f''(x_n)} \tag{8}$$

and for the larger root, using the following iterative formula

$$x_{n+1} = x_n - \frac{f'(x_n) - \sqrt{[f'(x_n)]^2 - 2f(x_n)f''(x_n)}}{f''(x_n)} \tag{9}$$

Since the denominator of both formula (8) and (9) does not include the term of $f'(x_n)$, they are specially fit to finding double roots of nonlinear equation or polynomial.

3 Numerical Examples

In order to confirm the validity of the algorithm proposed, we will give five examples to evaluate the polynomial or nonlinear equation at the initial values, and compared results with other methods.

Example 1. The first example is a polynominal having two double roots at $\xi = \pm 1$ [16]

$$f(x) = x^4 - 2x^2 + 1 \tag{10}$$

The iteration starts respectively with $x_0 = 0.8$ or $x_0 = 0.6$. The results are summarized in Table 1.

Example 2. The second example is a nonlinear equation with double root at $\xi = 0$ [16]

$$f(x) = x^2 e^x \tag{11}$$

Starting at $x_0 = 0.1$ or at $x_0 = 0.2$, the results are given in Table 2.

Table 1. Comparison of 7 methods for example 1

Method	$x_0 = 0.8$		$x_0 = 0.6$	
	n	$|f(x_n)|$	n	$|f(x_n)|$
Proposed method (8)	6	0	7	0
Proposed method (9)	9	0	15	0
Halley's method (2) in reference [16]	4	0	4	6.0e-15
Chebyshev's method (29) in reference [16]	4	0	4	5.8e-15
The modified Chebyshev's method (39) in reference [16]	3	0	3	2.0e-20
Neta's method (49) in reference [16]	5	0	5	0
The modified Osada method (51) in reference [16]	5	0	5	0

Table 2. Comparison of 7 methods for example 2

Method	$x_0 = 0.1$		$x_0 = 0.2$	
	n	$|f(x_n)|$	n	$|f(x_n)|$
Proposed method (8)	7	2.4e-35	8	9.4e-39
Proposed method (9)	7	1. 2e-36	8	6.2e-41
Halley's method (2) in reference [16]	2	4.0e-26	2	8.0e-21
Chebyshev's method (29) in reference [16]	2	2.0e-22	2	3.0e-17
The modified Chebyshev's method (39) in reference [16]	2	2.0e-22	2	3.0e-17
Neta's method (49) in reference [16]	4	3.0e-15	5	5.0e-26
The modified Osada method (51) in reference [16]	4	1.0e-14	5	1.0e-24

4　Concluding Remarks

We have developed two new methods to obtain multiple roots. We can know from the table 1 to table 2 that the numerical experiments showed the rapid convergence and high accuracy for finding the multiple roots of polynomials or nonlinear equations compared with other methods. Hence, the algorithm proposed will play a very important role in the many fields of science and engineering practice.

Acknowledgement. This work was supported by the project of National Natural Science Foundation of China (61040049) and the project of Natural Science Foundation of Hunan (11JJ6064).

References

[1] Burden, R.L., Douglas Faires, J.: Numerical ANALYSIS, 7th edn., pp. 47–103. Thomson Learning, Inc. (August 2001)

[2] Zeng, Z.-Z., Wen, H.: Numerical Computation, 1st edn., pp. 88–108. Qinghua University Press, Beijing (2005)

[3] Xu, C., Wang, M., Wang, N.: An accelerated iteration solution to nonlinear equation in large scope. J. Huazhong Univ. of Sci. & Tech.(Nature Science Edition) 34(4), 122–124 (2006)

[4] Lang, M., Frenzel, B.-C.: Polynomial root finding. IEEE Signal Processing Letters 1(10), 141–143 (1994)

[5] Jenkins, M.A., Traub, J.F.: A three-stage algorithm for real polynomials using quadratic iteration. SIAM Journal on Numerical Analysis 7(4), 545–566 (1970)

[6] Orchard, H.J.: The Laguerre method for finding the zeros of polynomials. IEEE Trans. on Circuits and Systems 36(11), 1377–1381 (1989)

[7] Lucas, T.N.: Finding roots of polynomials by using the Routh array. IEEE Electronics Letters 32(16), 1519–1521 (1996)

[8] Truong, T.K., Jeng, J.H., Reed, I.S.: Fast algorithm for computing the roots of error locator polynomials up to degree 11 in Reed-Solomon decoders. IEEE Trans. Commun. 49, 779–783 (2001)

[9] Fedorenko, S.V., Trifonov, P.V.: Finding roots of polynomials over finite fields. IEEE Trans. Commun. 50(11), 1709–1711 (2002)

[10] Cui, X.-Z., Yang, D.-D., Long, Y.: The fast Halley algorithm for finding all zeros of a polynomial. Chinese Journal of Engineering Mathematics 23(3), 511–517 (2006)

[11] Ehrlich, L.W.: A modified Newton method for polynomials. Comm. ACM 10, 107–108 (1967)

[12] Huang, Q.-L.: An improvement on a modified Newton method. Numerical mathematics: A Journal of Chinese Universities 11(4), 313–319 (2002)

[13] Huang, Q.-L., Wu, J.: On a modified Newton method for simultaneous finding polynomial zeros. Journal on Numerical Methods and Computer Applications(Beijing, China) 28(4), 292–298 (2006)

[14] Dong, C.: A family of multipoint iterative functions for finding multiple roots of equations. Int. J. Comput. Math. 21, 363–367 (1987)

[15] Neta, B., Johnson, A.N.: High order nonlinear solver for multiple roots. Comput. Math. Appl., doi: 10.1016/j.camwa. 2007.09.001

[16] Neta, B.: New third order nonlinear solver for multiple roots. Appl. Math. Comput., doi: 10.1016/j. amc.2008.01.031

A Path Planning Method to Robot Soccer Based on Dijkstra Algorithm

YunLong Yi and Ying Guan

Department of Information Engineering, Shenyang Institute of Engineering,
Shenyang 110136, China

Abstract. As a typical and very challenging multi-agent system, Robot Soccer System is considered by many researchers as an ideal research platform for multi-agent system, and its purpose is to study multi-robot (or agent) in a complex dynamic environment and multiple constraints, the completion of multi-task and multi-target for real-time reasoning and planning techniques which has far-reaching significance for the future society. Dijkstra algorithm is the most typical short-circuit algorithm used to calculate the shortest path of a node to all other nodes. Main features of a starting point as the center expand outward of layers, until the end of the extension. In this paper, the algorithm is applied to the robot soccer path planning and obstacle avoidance control and made a good effect. After Recognition the environment, Robot Soccer can determine the shortest path timely and the method has been applied to the actual robot control.

Keywords: path planning, Dijkstra algorithm, robot soccer.

Multi-agent soccer robot system as a test platform, is being studied extensively. Multi-agent system, each agent is able to sense the external environment (perceptible), can think (intelligent), and can interact with the outside world through acts (actable). Each agent is relatively simple, and each agent can complete a series of more complex tasks through the co-coordinated. Soccer robot system is the excellent test platform of multi-agent system, each robot is relatively simple, but can complement each other to finish the race through the decision-making system. Soccer robot system as a typical and very challenging multi-agent system, is used as an ideal research platform of multi-agent system by many researchers. The aim is to study the multi-robot (or multi-agent) of the far-reaching significance for the future society should have real-time reasoning and planning techniques in order to complete the multi-tasking and multi-purpose in a complex dynamic environment and multiple constraints.

Soccer robot path planning is to establish the basis for decision-making system of soccer robot. The path planning will directly affect the degree of intelligent decision system. All along, the soccer robot path planning problem is the the classic one in many disciplines including computer field,it can be described as a known starting point, destination point, and environmental information, to determine the line from a starting point to the target point[1]. In general, the line must meet the target of the minimum cost or the shortest distance Etc.

D. Jin and S. Lin (Eds.): Advances in ECWAC, Vol. 2, AISC 149, pp. 89–95.
springerlink.com   © Springer-Verlag Berlin Heidelberg 2012

1 Identify Obstacles

The premise of this study is that the robot soccer can identify the surrounding environment, that is able to determine obstacles and target coordinates. In order to obtain the shortest path, we must first determine the existence of obstacles. Figure 1 shows, the robot as a circle, the radius is marked by r, where the robot is called the starting point, and the robot moving target is called the goal point.Here,we do the robot's two parallel lines with the connection of the robot and the center of the target point, and the area between parallel lines is expressed as shade in Figure 1.There is an obstacle if the robot appears in the shaded area[2].

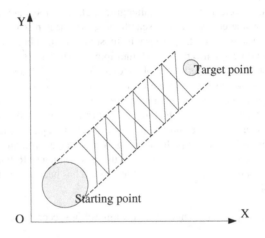

Fig. 1. Determine Obstacles's Regional

2 Find the Middle Point

Obstacles are determined, you should find the middle point to construct a directed graph. There is bound to be such a line between the robot and obstacles, this line is tangent to the robot and obstacles, as shown by dotted line in Figure 2.So it is called common tangent line in the paper.Assuming the robot avoids obstacles from the above, the robot first must go along the direction of the common tangent T1.The center track should be a straight line path L1. L1 should be parallel to the common tangent line(the same slope with the common tangent line), and the distance is r between two lines. When the robot reaches the middle point 2, it will move along the line L2. At this point, the line L2 must be parallel to the common tangent line T2,and the distance is r with T2. In this way, the shortest route that the robot can avoid obstacles in the direction is guaranteed.If the robot avoids obstacles from the bottom, it must find a middle point 1, and the principal is the same with above.

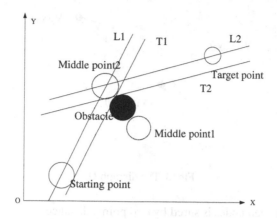

Fig. 2. Determine the intermediate point

The next,solving the middle point 2 (x2, y2) coordinates. Known conditions: starting point coordinate v0 (x0,y0), obstacles coordinate vs (xs, ys) and obstacles coordinate vs (xs, ys). First to solve the line L2. The slope of the line L1and the line T1 is the same, so we need solve the slope of the line T1 only. Because of the distances of the obstacle's circle coordinates and the robot' circle coordinate to the line T1 are both r, we can obtain the slope of the line T1(as k) through the distance formula of the point to the line. This way, we can get the line L1 equation.

$$L1: kx-y-kx_0+y_0=0 \tag{1}$$

The next,solving the slope of the line L2. The slope of the line L2 is the same with the line T2,so we can obtain the line L2's by solving the slope of the line T2. Because of the distances of the obstacle's circle coordinates and the target point' circle coordinate to the line T1 are both r, we can obtain the slope of the line T2(as k1) through the distance formula of the point to the line. This way, we can get the line L2 equation.

$$L2: k_1x-y-k_1x_3+y_3=0 \tag{2}$$

The intersection point coordinate of the line L1 and L2 is the coordinate of the intermediate point 2.

$$x_2=(kx_0-k_1x_3-y_0+y_3)/(k-k_1) \tag{3}$$

$$y_2=k(k_1x_0-k_1x_3-y_0+y_3)/(k-k_1)+y_0 \tag{4}$$

For the intermediate point coordinates (1 x1 y1), the process of the solution is the same as above. This will make the figure2 simplified to the digraph G with four vertices, and the vertex coordinates are v0(x0,y0),v1(x1,y1) ,v2(x2,y2) and v3(x3,y3),as shown in figure3.

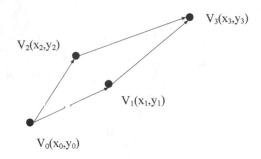

Fig. 3. The digraph G

The weight between nodes is sured by two points distance.

3 Apply Dijkstra Algorithm to Obtain the Shortest Path

After the attainment of directed graph G, Dijkstra algorithm can be applied to make a choice for the shortest path [3]. Dijkstra algorithm is the typical way of calculating the shortest path, mainly applied to calculate the shortest path from one node to all the rest nodes. It is the main feature of Dijkstra algorithm to expand from the starting point which is the centre to the outer layers until reaches the terminal point. Dijkstra algorithm can get the best answer for the calculation for the shortest path. However, it is inefficient for it traverses too many nodes. The premise is that an auxiliary value D needs being introduced. Every component value $D[i]$ refers to the length, found by now, of the shortest path from the starting point v to each terminal point v_i in directed graph G, among which $i=0,1,2.......n$.

According to the principle of algorithm, it can be described in steps as follows:

1) If weighted adjacency matrix arcs is employed to stand for weighted directed graph and, $arcs[i][j]$ refer to the value above $arc<v_i,v_j>$. Then if $<v_i,v_j>$ does not exist, $zrcs[i][j]$ is infinitely great. S refers to the collection of terminal points to which shortest paths have been found whose initial state is null set. Therefore, the initial value possibly reached from those who starts from v to each vertex v_i in graph is:

$$D[i]=arcs[LocateVex(G,v)][i]   v_i \in V$$

Among which V is the collection of all nodes in directed graph G.

2) Select v_i and make

$$D[j]=Min\{D[i]|v_i \in V\text{-}S\}$$

Then v_i is the terminal point of the shortest path started from v obtained for now. Make

$$S=SU\{j\}$$

3) Modify the lengths of the shortest paths starting from v to any vertex v_n in V-s collection. If

$$D[j]+arcs[j][n]<D[n]$$

Then modify D[n] as

$$D[n]=D[j]+arcs[j][n]$$

4) Repeat operations of 2) and 3) for n-1 times, thus obtaining a collection of the shortest paths starting from v to vertexes in the directed graph, and the collection shows an increasing sequence of lengths.

Next try to obtain the shortest path from v_0 to V_3 in directed graph G. Assume that crcs[0][1]=5, arcs[0][2]=3, arcs[2][3]=9, arcs[1][3]=6, then the weighted adjacency matrix is:

$$\begin{bmatrix} \infty & 5 & 3 & \infty \\ \infty & \infty & \infty & 6 \\ \infty & \infty & \infty & 9 \\ \infty & \infty & \infty & \infty \end{bmatrix}$$

If Dijkstra algorithm is conducted upon G, then the shortest path starting from v_0 to any other vertex obtained and the change of vector D in computing process are as shown in Table 1.

Table 1. Solutions of the shortest paths starting from v0 to any other vertex

Terminal point	i=1	i=2	i=3
v_1	5 (v_0,v_1)	5 (v_0,v_1)	
v_2	3 (v_0,v_2)		
v_3	∞	12(v_0,v_2,v_3)	11(v_0,v_1,v_3)
v_j	v_2	v_1	v_3
S	$\{v_0,v_2\}$	$\{v_0,v_1,v_2\}$	$\{v_0,v_1,v_2,v_3\}$

The shortest path starting from initial point v_0 to target point v_3 obtained from the above table is 11(v_0,v_1,v_3).

4   Results Obtained through Simulation

In practical path planning and obstacle avoidance control, at first the environment information around the mobile robot should be found out and tested in real time to make prerequisite for the robot to respond in time accordingly. Now make the real three-dimensional scene abstract to a multi-obstacle environment in two-dimensional space, i.e. spatial obstacle avoidance problem, and then conduct the path planning in two-dimensional space. While in the environment which is on the basis of Dijkstra algorithm, two-dimensional path planning has two obstacle avoidance problems which are weighted optimal path planning problem and non-weighted optimal path planning problem. However, this system discusses only into weighted optimal path planning. This path planning algorithm can make the requirements for optimal path planning and obstacle avoidance for real robot.

Results obtained through simulation are realized through programming[4]. The simulation for this path planning program can be realized by vc++ or by MATLAB. This paper adopts the simulated dynamic obstacle avoidance situations in MATLAB[5]. So the trace of action of the robot should be smooth curve as the results obtained through simulation shown in Table 4. In the table of results obtained through simulation, the initial point of the robot is origin of coordinates and the target point is the other one at the top right corner in Table 4. There are two obstacles in the whole process, staffed circles as shown in Table 4. The robot can avoid the obstacles skillfully and reach the target point with the help of Dijkstra algorithm. After the match begins, the speed of the robot is determined by the obstacle. When the robot approached the obstacles it will slow down, so the small round circles are densely set; when the robot is far away from the obstacles it will march forward at full speed, so the small round circles are scarcely set. Seen from Table 4, the results obtained through simulation, the results are very good.

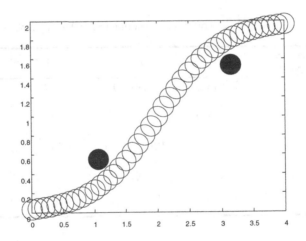

Fig. 4. Results Obtained through Simulation

5   Conclusion

This paper adopts Dijkstra algorithm, applying it to path planning and obstacle avoidance control of robot. There are nodes, i.e. control points which exist in robot's real path planning and obstacle-avoidance operations. And they can be shown by robot's real-time coordinates and length of the path. It is just these real-time coordinates and length of the path which the robot take as references to change its node at anytime that guarantee the shortest path. Good results are obtained through experiments as well as some conclusions as follows:

1) The robot can move to target points with the shortest path every time and its change trace of action would be a smooth curve according the change of target points.

2) Since the complexity of Dijkstra algorithm is O (n^2) the performance of the robot is reduced when obstacles are too densely set.

3) The method adopted in this paper can be expanded to be multi-obstacle avoidance. If, in real situations, there are too many obstacles, then the robot's operating speed will decline. Application of this method will prolong the robot's motion time, so it is not the optimal method. However, in real competition this kind of situation fulfilled with obstacles will not last long. In real situations, there are many factors, like frictim and Centrifugal force which will affect the speed of the robot. Therefore, further research should be conducted for in this paper there are no proposals deal with these factors.

Future Research Prospect

According to present trajectory planning methods, researches conducted in future may focus on the continuity of acceleration in which robot take the shortest path and smoothness of the acceleration curve. Thus the robot can touch the ball more gently and smoothly, and even complete more accurate actions.

References

1. Lv, C., Zhu, J.: The application of Dijkstra algorithm in mobile robot path planning and obstacle avoidance. Metallurgical Automation 1, 676–678 (2006)
2. Chen, B., Yang, Y.: The Study of Soccer robot obstacle avoidance's control. The Robot 26(2), 111–113 (2004)
3. Yan, W., Wu, W.: Data structure. Tsinghua university press, Beijing (2001)
4. Joues, J.L.: Robot programming technology. In: Yuan, K., Zhou, W. (trans.) Mechanical industry press, Beijing (2006)
5. Zheng, A.: MATLAB practical tutorial (Version 2). Electronic industry press, Beijing (2008)

A Novel Heuristic Search Algorithm Based on Hyperlink and Relevance Strategy for Web Search

Lili Yan[1], Wencai Du[2], Yingbin Wei[1], and Henian Chen[1]

[1] Department of Software Engineering Hainan College of Software Technology
Qionghai 571400, China
[2] College of Information Science & Technology Hainan University
Haikou 570228, China
softhainan@yahoo.com.cn

Abstract. With the rapid increase of information in Web, people have to waste much time to search the information they need. So the search engine has more and more become an absolutely necessary tool of Internet surfers. This paper brings forward a new Web search algorithm based on hyperlinks and content relevance strategy. The experimental results have showed that the proposed algorithm focuses on mining the potentially semantic relationship between hyperlinks and performs quite well in the topic-specific crawling.

Keywords: Web search, recall rate, precision rate.

1 Introduction

With the rapid development of Web technology, World Wide Web information leading to the explosive growth. World Wide Web cans all types of information without meaning seamless integration together to form a mass of information the world's largest library [1]. The face of the World Wide Web this scattered and disorderly, chaotic mass of information databases, Web users often find it difficult to find the information they need to meet and tap the potential of useful information, resulting in "information overload, lack of knowledge" problem. How to find valid users the information they need, the World Wide Web has become an urgent need to address the key issues, Web search engines to solve such problems is an effective way.

Most traditional Web search engines are based on keyword matching and return results containing the query terms the document. However, these search engine results are not satisfactory. In recent years, Web-based hyperlink topology of the Web search algorithms attracted extensive attention of many scholars, become more popular Web search technology [2]. Among the many findings, the more well-known algorithms are PageRank algorithm and HITS (Hyperlink Induced Topic Search) algorithm. Although the calculation of PageRank algorithm is simple and very efficient, to complete a hyperlink analysis can be used to sort all the results, but it completely ignores the content of the website. HITS algorithm is to use the page number and number of links referenced to determine the value of different pages. This method can get a higher recall rate, but ignoring the text, prone to drift theme. In this paper, based

D. Jin and S. Lin (Eds.): Advances in ECWAC, Vol. 2, AISC 149, pp. 97–102.
springerlink.com   © Springer-Verlag Berlin Heidelberg 2012

on the HITS algorithm, combined with hyperlinks and content relevance strategy, proposed a new CSHITS (Clonal Selected Hyperlink Induced Topic Search) Web search algorithm.

2 Web Search General Model

Each Web page is a link between the relationships can take advantage of important information. Based on this information technology is known as link analysis. Most link analysis algorithms have a common starting point: more page links by other pages are better quality pages, and more importantly, from the starting page with links to more weight [3].

HITS algorithm is the basic idea is: the query (topic) Q submitted to the traditional keyword-based search engine that matches the search engine returns many pages as the pages from which to take root set R. Then the entire page with the map position R to expand the root set, through to join the referenced web pages and references to expand into a larger collection called the base set T [4].

In this paper, the introduction of traditional text analysis methods to define pages P and topics Q related degree, and set the appropriate threshold to remove low page relevance. $Q = \{q_1, q_2, ..., q_i, ..., q_n\}$ Expressed as a theme, which q_i is the ith query keywords, $i = 1, 2, ..., n$. Topics Q related pages P and is defined as follows:

$$A_p = \sum_{i=1}^{m} H_i , \, i \in R$$

$$H_p = \sum_{j=1}^{n} A_j , \, j \in T$$

$$SIM(P,Q) = \gamma \times \sqrt{A_p^2 + H_p^2} + \frac{1}{\gamma} \times \sqrt{A_p^2 - H_p^2}$$

In which, $\gamma = \sum_{i=1}^{n} tf_i(P) \times ln(\frac{n}{i})$, $tf_i(P)$ is the frequency of the page P for all keywords in the number of occurrences of the number of n emerged.

3 The Structure of CSHITS

Genetic algorithm (GA) is a simulation of the natural environment of biological genetic and evolutionary process to optimize the probability of the formation of an adaptive search algorithm. It will solve the problem, said a "chromosome" of the survival of the fittest process, through the "chromosome" and evolving from generation to generation, including reproduction, crossover and mutation operations.

However, the genetic algorithm's ability to explore the new space is limited; it tends to converge to a local optimal solution.

GQA(Genetic Quantum Algorithm, GQA) will be the introduction of the quantum state vector to express the genetic code, the use of quantum rotation gate to achieve the evolution of chromosomes, the genetic algorithm to achieve a better than normal results. However, the algorithm is mainly used to solve the 0-1 knapsack problem. Coding scheme and the evolution of quantum rotation gate strategy does not have the versatility, especially since all individuals have evolved towards a goal, if there is no crossover, likely to fall into local optimum.

Biological immune system is a highly evolved biological systems, it is harmful to distinguish between external antigen and self organization, to clear the antigen and to maintain the stability of the organism. Clonal selection theory was first put forward by the Jerne, which roughly says: When the cells to achieve antigen recognition after, B cells are activated and copy the resulting B-cell clone proliferation, then cloned cells through mutation process, resulting in antigen specificity of antibodies[5]. Clonal selection theory describes the basic characteristics of acquired immunity, and declared that only a successful immune cell recognize antigen was able to proliferate. Through mutation of immune cells to differentiate into effecter cells (antibody) and memory cell two. The main features of clonal selection of antigen-stimulated immune cells produce clone proliferation, followed by the diversity of genetic variation to differentiate into effecter cells (antibody) cells and memory cells. Choose a clone corresponding to affinity maturation process, that is, a lower affinity for individual antigens in the role of clonal selection mechanism, through the proliferation of replication and mutation, its affinity to gradually increase the "mature" process.

According to the essential characteristics of information retrieval: that is bound to meet certain conditions (usually cost constraints) under the premise of system performance optimization. This is a typical combinatorial optimization problem. Genetic algorithm to solve combinatorial optimization problems an effective tool. Poor local search ability of genetic algorithm, but to grasp the overall search process ability, and clonal selection optimization algorithm itself has a strong local characteristics. , The organic integration of the three, can improve the algorithm efficiency and solution quality. In this paper, based on the HITS algorithm, combined with hyperlinks and content relevance strategies and integration into the parallel nature of quantum computing, proposes a new CSHITS search algorithm. The algorithm model is as follows:

$$CSHITS = (E, F, C, M, S)$$

In which, E is the quantum coding, F is antibody affinity functions, C is the cloning operation, M is the Gaussian mutation, S is the quantum selection operation. CSHITS described as follows:

Step 1: link analysis the collection T, so as to obtain the initial antibody population $A(k) = \{a_1(k), a_2(k), ..., a_i(k), ..., a_n(k)\}$ .According to the encoding E , and generate Chromosomal string. We elected to label the page, get set of Web pages $\{1, 2, ..., n\}$, And define a link matrix A($n \times n$), if there is a link to a page i

from a web page j link, then the matrix A in the first element (i, j) is 1, otherwise is 0.

Step 2: initialize the population size n (Web page number), the probability of mutation P_m , amplification factor α , correlation threshold β , the evolution of algebra $k = 0$, and the maximum evolution generation $MaxGen$;

Step 3: Calculated for each antibody population of antibodies affinity $f(a_i(k)), i = 1, 2..., n$;

Step 4: Groups of antibodies were cloned, crossover and mutation. Antibody population $A(k)$ size of each cloned antibody $a_i(k)$ by a new antibody population.

Step 5: $k = k + 1$; Delete pages P and topics Q of relevance is less than the threshold β of the page, and return to step (3).if $k > MaxGen$, the search is finished, and sort the search results displayed to the user.

4 Analysis of CSHITS Algorithm Convergence

In CSHITS algorithm, the initial population generation process is a random process, each new generation of population and its parent only the relevant populations, and populations in which the state is discrete state, CSHITS algorithms these features generally in line with Markov chain characteristics. So, here we will use the Markov chain of CSHITS convergence of the algorithm for analysis.

Theorem 1: CSHITS population series is finite homogeneous Markov chain $\{X_k, k \geq 0\}$

Proof. We used the gene m is expressed as the median of each antibody, each gene in binary digits is k , and the population size is N . As CSHITS binary encoding, antibody values are discrete 0 and 1, We represent $\Omega = \{\phi\}^{mk}$ as antibody population space, $\phi \in \{0,1\}$, $x_t^i \in \{\phi\}^{mk}$, State space where the population size is $N \times 2^{mk}$. Therefore, the population is limited.

$$P_{ij} = P\{X_{n+1} = j \mid X_n = j\}$$

We set $\{X_t, t \geq 0\}$ is a random variable of CSHITS sequence, then when the iterative algorithm to enough times, the following formula must be established,

$$\exists x_i \in X(t) : \lim_{k \to \infty} P\{X(t) \subset S^*\} = 1$$

Through the above derivation shows that the group contains the global optimal solution probabilities 1, CSHITS is the probability of a convergence.

5 Experiments

To test the performance of the proposed CSHITS algorithm, we use Matlab 7.0 platform CSHITS algorithm and the HITS algorithm with the classic PageRank algorithm simulation tests were carried out. Popular search engine evaluation criteria are recall and precision rate. Recall rate is defined as the relevant documents retrieved to meet the conditions of the document with the number of all proportion; precision rate is retrieved relevant documents and retrieval of all documents to the ratio.

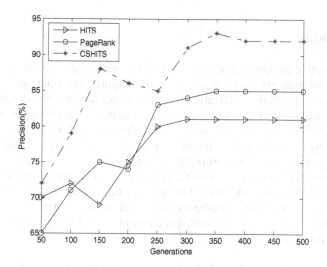

Fig. 1. Compare precision rate with three algorithms

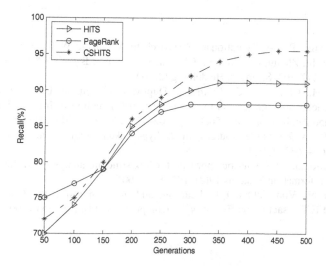

Fig. 2. Compare recall rate with three algorithms

Figures 1-2 shows the three search algorithms recall rate, precision of the experimental results of the comparison. It can be seen from Figures 1-2, CSHITS algorithm recall and precision rate than the HITS algorithm and the PageRank algorithm high. With the increase of the search Web page, the HITS algorithm based on a large number of database index has nothing to do with the theme of the page, seriously reducing the precision rate. CSHITS algorithm into the contents of the relevance evaluation mechanism, in time to determine the relevance of the theme page, and delete the page that has nothing to do with the topic, cloning, crossover and mutation operations can accelerate the algorithm convergence to the optimal solution.

6 Conclusions

With the amount of information on the Internet's explosive growth, it is increasingly dependent on information retrieval to obtain the required information. In this paper CSHITS search algorithm, HITS algorithm, based on the combined advantages of clonal selection algorithm, on the one hand by setting the appropriate threshold to remove low relevance pages; the other hand, the use of CSA algorithm for fast convergence, by increasing the iteration population diversity of individuals to improve the global search performance, and learning memory, etc., and tap the potential of hyperlinks between semantic relations can effectively guide the mining theme. How to further develop Web data mining, full use of Web resources and to further explore and optimize intelligent search algorithms, analysis of intelligent search algorithm's time complexity and space complexity of the problem, is the focus of future research work.

Acknowledgments. This research work was supported by the Natural Science Foundation of Hainan Province under Grant No. 610229.

References

1. Loia, V., Luongo, P.: An Evolutionary Approach to Automatic Web Page Categorization and Updating. In: Zhong, N., Yao, Y., Ohsuga, S., Liu, J. (eds.) WI 2001. LNCS (LNAI), vol. 2198, pp. 292–302. Springer, Heidelberg (2001)
2. Boughanem, M., Chrisment, C., Mothe, J., Dupuy, C.S., Tamine, L.: Connectionist and genetic approaches for information retrieval. In: Soft Computing in Information Retrieval: Techniques and Applications, vol. 50(1), pp. 102–121. Physica-Verlag, Heidelberg (2000)
3. Kleinberg, J.: Authoritative sources in a hyperlinked environment. Journal of the ACM 46(5), 604–632 (1998)
4. Lempel, R., Moran, S.: Stochastic approach for link structure analysis and the TKC effect. ACM Trans. Information Systems 19(2), 131–160 (2001)
5. De Castro, L.N., Von Zuben, F.J.: Learning and mization Using the Clonal Selection Principle. IEEE Transaction on Evolutionary Computation 6(3), 239–251 (2002)

Research of Image Encryption Algorithm Based on Chaotic Magic Square

YunPeng Zhang, Peng Xu, and LinZong Xiang

College of Software and Microelectronics Northwestern Polytechnical University,
710072, Xi'an, China
poweryp@163.com

Abstract. Due to the pseudo-randomness and sensitivity of the initial value of chaos theory, and the technology of magic square which can scramble image in an efficient way, this algorithm uses the magic square transformation for the original image preprocess firstly, this process is pretreatment on the basis of magic square algorithm improved ranks. Then use Arnold to a second image scrambling, finally using logistic, henon produce array and the image scrambling gray values, which makes the image's position and gray value transformed in the same time , get the final encrypted image. Through theoretical analysis and experiments show that the algorithm has great key space, good effect, and can resist common attack effectively, etc.

Keywords: encryption, chaos, magic square, scrambling, image.

1 Introduction

Along with the rapid development of Internet, an increasing number of network security cases contribute to the worldwide netizen great losses, information security has become the focus of attention from all walks of society [1]. Images play an important role in the process of people's access to information, the relevant industries for the security of the digital image is put forward higher request. As an old discipline, encryption has profound and common influence on today's information safety, but because of digital image file data with large capacity, encryption with real-time demand, the traditional encryption method hardly meet the requirements.

As a new discipline originated in 1970s, chaos theory constantly breakthroughs in many fields in the research of numerous scholars, chaos theory has been applied to mathematics, physics, astronomy, information technology and so on [2]. Chaos cryptography become the direction of the cryptography research [3][4]. In recent years, the chaos image encryption achieved all-around development, which provided numerous convinced examples for the image encryption that based on chaos theory[5]. At the same time, the use of magic square technology can scramble the digital image in an efficient way [6], this paper uses the diversity of magic square and the pseudo-randomness of chaotic sequences in encryption process, expecting this process can

D. Jin and S. Lin (Eds.): Advances in ECWAC, Vol. 2, AISC 149, pp. 103–109.

improve the image's safety, and can resist the plaintext attack and chosen-ciphertext attack more effectively.

2 Based on Chaotic Image Encryption Algorithm

2.1 The Preprocess of Image

This paper chooses the magic square encryption algorithm for the image preprocess, specific as follows:

A=[m1,m2,…,m128]→C=[m′1,m′2,…,m′128]

B×A=C

$$B=\begin{bmatrix} 0 & I_{128-n} \\ I_n & 0 \end{bmatrix}$$
(1)

Assume image with 128 * 128 pixels, let's transform the image into two-dimensional matrix and saved in matrix A, write matrix B which type is sparse matrix, I128-n and In, are respectively n advanced unit matrix.

1) Using sparse matrix B multiplies image matrix A, changing the rows of the matrix.

2) transposed matrix, repeat (1).

3) Repeat (1) and (2), iterative 10 times, get the final preprocess image, deposit C matrix

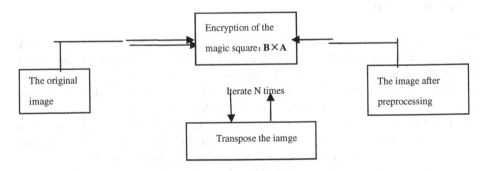

Fig. 1. The flow chart of image's preprocessing

2.2 Image Encryption Scheme

Using Mn×n represent the digital image Input_image,the colorful square image's length is N. Below the type used double key1 which values inputted by the users. Specific encryption process, as shown in figure 2:

Using the Input image for magic square transformation. Fig.2 shows the process.

Make the image that preprocessing Arnold transformation[7], input keys (key3, key4) into Arnold formula, that is assign the value of the image pixel array x to the X that appears in the formula , and assign the y appears in the pixel array to Y that in the formula, finally get the image that has position than formed.

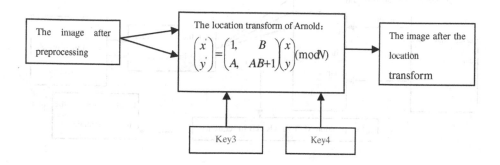

Fig. 2. The flow chart Arnold position transformation u (0-2)

Input keys (key1, key2) respectively, as the logistic mapping parameters u (0~2), and X0 (0-1). Every once a mapping, get the value of X, repeat until the quantity of the x is equivalent to the quantity of pixel, put these values into the array D. Get the values of array D as henon map. In order to get henon mapping, X, Y, and its positive integers, then the value X+Y to two-dimensions array Z.

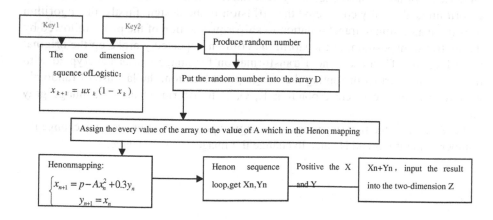

Fig. 3. The process of Logistic and henon sequence

Use of the image who's position has been transformed for the operation of XOR with the two-dimensions array Z to produce the new values of R,G,B, thus changing the value of image gray value, finally finished image encryption process.

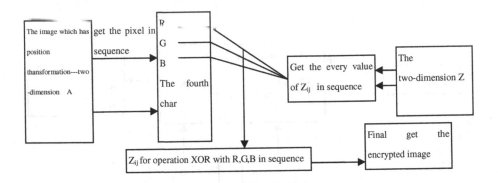

Fig. 4. The flow chart of the grey value transformation

3 The Analysis of the Algorithm Result

3.1 The Analysis of the Algorithm Diffusion

Diffusion algorithm is an important factor that ensures effective cover plaintext ciphertext. In addition to chaotic system itself provides some of the features, the algorithm is also fully considered the diffusion in the design. Firstly, the algorithm uses the magic square transformation to shift row position of the pixel, followed by the matrix transposition, to transform the column, repeatedly broke up the original pixel location. Then use Arnold transformation for image that has encrypted with magic square, further disrupting the image pixel location, the last use of randomly generated sequence to corresponding R, G, B transform, change the image gray value.

In addition, in the first step process of magic square encryption, you can change the number of iterations and N value to change the image of chaos degree.

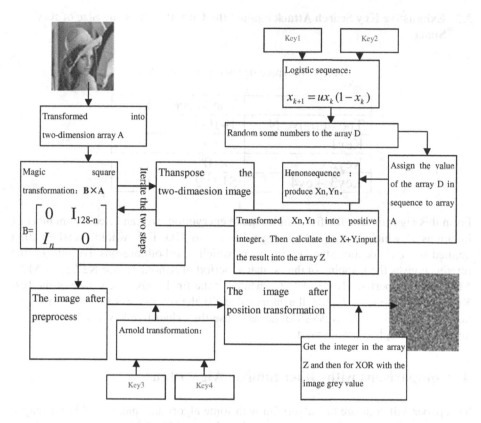

Fig. 5. The flow chart of the image encryption algorithm

3.2 The Analysis of Known Plaintext Resists

Known plaintext resist is a common tradition attack means. When the attacker obtained a certain number of ciphertext and corresponding plaintext, he can use known plaintext attack methods. Based on the stream cipher encryption algorithm, the cipher is produced by a series of the operation of plaintext and the key, thus the attacker can also through the corresponding inverse from plaintext encrypted sequence to launch the keys to repeat sequences, and then flow using the subkey sequence flow can be breached. So the key for re-use of the sub-sequence flow of all the ciphertext can be broken. The encryption algorithm is not carried out under the explicit key changes to the application is given by the user key or keys have been fixed, no random keys. Therefore, the randomness of the key should be improved, plaintext resist needs to be strengthened.

3.3 Exhaustive Key Search Attack against the Complexity of the Size of Key Space

Table 1. The space size of the algorithm keys

	The space of keys
The iterative N	1~100
Key1	2×10^8
Key2	1×10^8
Key3+key4	$512^2/2$

From this algorithm, in the first magic square encryption, we can select the number of iterations N which is a positive integer from 1 to 100. For Arnold transformation applied to pixel location, the key (a, b), in which a, b both are positive integer, the reference gives the mapping of the estimated period of Arnold $TN <= N2 / 2$, for 512 * 512 image, its period $TN <= 5122 / 2$, while for the final logistic sequence of the first Key u, the key space is 2×108, the initial value X0, the key space is 1×108. Exhaustive key space is 2.6×1023. It does can be seen; use the exhaustive key search attack, the key space enough to ensure safety.

4   Comparisons with Other Similar Algorithm

This paper will compare this algorithm with some algorithms mentioned in this paper. The comparison includes algorithm safety level control, and efficiency of the algorithm, the key space and the algorithm ability against the attack from four aspects. Comparison results shown in table 2. One of the test data is about efficiency obtained from 512 x 512 32-bit true color image Lena in PC P4 with 2.0 GHz CPU, and 256MB memory.

Table 2. The comparison of the similar algorithm

	Whethercan control the safety level	Efficiency	The capacity of the key space
The essay algorithm	Yes	0.703s	2.6×10^{23}
Algorithm [10]	No	0.75s	10^{23}
Algorithm [11]	No	6.516s	10^{18}

5 Conclusion

Image encryption algorithm based on chaotic magic square method makes full use of chaos random sequence. It can more effectively resist plaintext attack and chosen-ciphertext attack, and the encryption method for image size requirement is very low. The size can be M * N, are not necessarily N * N square. More important, from safety view, compared with traditional algorithm, the initial value of each encryption process is different in this algorithm, and each process using random sequence three times, and more random processes, more safety.

This algorithm can increase key space by modifying some fixed value to change keys. The efficiency of the algorithm is satisfactory. The complexity of the algorithm and the resist to the attack can be better. In this paper, the magic square algorithm used by author is two-dimensional encryption algorithm, the author intends to work in the next step will transform, two-dimension to three-dimension expects key space and the complexity of the algorithm are finally, further improved, and improve the difficulty of the decryption algorithm.

Acknowledgements. This work is supported by Aero-Science Fund of China (2009ZD53045), Science and Technology Development Project of Shaanxi Province Project(2010K06-22g), Basic research fund of Northwestern Polytechnical University (GAKY100101), and R Fund of College of Software and Microelectronics of Northwestern Polytechnical University (2010R001).

References

[1] Sheng, C., Zhang, H., Feng, D.: Information Security Review. Science in China E: Technological Sciences 37(2), 129–150 (2007)

[2] Stallings, W.: Cryptography and Network Security, 4th edn. Prentice Hall, Lodon (2005)

[3] Liu, J., Qiu, S.: Chaotic Systems in Cryptography Application Status and Prospects. Computer Engineering and Application 44(14), 5–12 (2008)

[4] Fridrich, J.: Image Encryption Based on Chaotic Maps. In: IEEE International Conference on System, Man, and Cybernetics, pp. 1105–1110 (1997)

[5] Zhao, L., Fang, Z., Gu, Z.: Novel Algorithm of Digital Image Scrambling and Encryption Based on Magic Cube Transformation. Journal of Optoelectronics Laser 19(1), 131–134 (2008)

[6] May, R.M.: Simple Mathematic Models With Very Complicated Dynamic. Nature 261, 459–467 (1976)

[7] Yi, K., Sun, X., Shi, J.: A series of images based on chaotic encryption algorithm (2005)

[8] Li, Q.: A Chaotic Encryption Method of Decoding of The Image. University of Defense Technology University 29(3), 45–47 (2007)

[9] Liu, J.: Image Encryption Based on Chaos Technology. Anhui University Press (April 2007)

[10] Wang, J.: The Application Status of the Chaos Technology In Digital Image Encryption. University of Science and Technology of China (2004)

[11] Bian, L.: Reasearch of Image Encryption Based on Chaos. Guang Dong University of Technology (2007)

Research on Lifetime Prediction Method of Tower Crane Based on Back Propagation Neural Network

Yang Yu[1], Yulin Tian[1], Naili Feng[2], and Ming Lei[2]

[1] School of Information Science & Engineering, Shenyang Ligong University,
Shenyang, China, 110159
[2] Shenyang Artillery Academy of PLA, Shenyang, China, 110162
yusongh@126.com, zhao.y.w@hotmail.com, kongkuchen@126.com

Abstract. The lifetime prediction of tower crane is very important in guiding the planning of use, making sure the lifetime of service about tower crane. By means of the abilities of neural network such as memory, induction & study, it can effectively build forecasting model for information which are difficult to be built into an exact math model like tower crane. This paper gains a relatively precision prediction result with BP neural network from the lifetime prediction of tower crane and describes the basic principle and the training work.

Keyword: BP neural network, service life of tower crane, prediction, Arithmetic.

1 Introduction

Tower crane is the key equipment which is widely used in building construction. Because of its high altitude and large coverage area, it is often a serious security accident in the event of an accident. In practice, due to the life of the tower crane influenced by its design life, overloading, corrosion, improper installation and removal, maintenance, accident damage and other factors, makes the tower crane large differences between individuals, which make it difficult to provide a uniform life. Therefore, this paper set the amount of tower crane's monitor as input, the life of the tower crane as output, we propose a method to predict the life of tower crane based on BP neural network.

2 Life Prediction Model of Tower Crane Based on BP Neural Network

BP neural network is also known as multilayer forward network, using back-propagation algorithm. In the forward calculation, enter the information from the input layer and then pass through hidden layers, processing layer by layer, transfer to the output layer finally. Each layer of neurons only influence the state of neurons in the next layer. If not get the desired output in the output layer, then turned to back-propagation, the error signal returns along the original connection channel; the error will be minimum by modifying the weights of each layer.

D. Jin and S. Lin (Eds.): Advances in ECWAC, Vol. 2, AISC 149, pp. 111–116.
springerlink.com   © Springer-Verlag Berlin Heidelberg 2012

Set input vector in the Input layer: $A = (a_1, a_2 ..., a_n)$, output vector in the hidden layers: $B = (b_1, b_2, ..., b_l)$,then

$$b_l = f\left(\sum_{h=1}^{n} V_{hi} a_h - \theta_i\right) \qquad (i = 1, 2, ..., p) \qquad (1)$$

Where V_{hi} is the weight from input layer to hidden layer, θ_i is the threshold of hidden layer.

Set output vector in the output layer: $C = (c_1, c_2, ...c_m))$, then

$$c_j = f\left(\sum_{j=1}^{p} W_{ij} b_i - \theta_j\right) \qquad (j = 1, 2, ..., m) \qquad (2)$$

Where W_{ij} is the weight from hidden layer to output layer, θ_j is the threshold of output layer. $f(*)$ is Sigmoid Activation function, that is $f(x) = 1/(1 + e^{-x})$.

3   BP Neural Network Algorithm

Set E_k to output cost function when the model couple provided to the network is (A_k, C_k), thus, the global cost function on the entire training set is

$$E = \sum_{k=1}^{m} E_k \qquad (3)$$

for the k-th model couple, the weighted input of output unit j is

$$netc_j = \sum_{i=1}^{p} W_{ij} b_i \qquad (4)$$

the actual output of this unit is

$$C_j = f(netc_j) \qquad (5)$$

The weighted input of hidden layer unit i is

$$netb_i = \sum_{h=1}^{n} v_{hi} a_h \qquad (6)$$

the actual output of this unit is

$$b_i = f(netb_i) \qquad (7)$$

for output units j , defined the generalized error as

$$d_j = -\frac{\partial E_k}{\partial netc_j} = -\frac{\partial E_k}{\partial c_j} f'(netc_j) \tag{8}$$

for output units i , defined the generalized error as

$$e_i = -\frac{\partial E_k}{\partial netb_i} = -\frac{\partial E_k}{\partial b_i} f'(netb_i)$$

$$= f'(netb_i)\left\{-\sum_{j=1}^{q} d_j \left[\frac{\partial\left(\sum_{i=1}^{p} w_{ij}b_i\right)}{\partial b_i}\right]\right\} = f'(netb_i)\sum_{j=1}^{q} d_j w_{ij} \tag{9}$$

The last formula can be regarded as the error which is back-propagated from front layer unit to this layer unit. Set the current connection weight to w_{ij} and v_{hi} , to reduce the cost function E_k , using gradient descent rule to change the value of weight, that is to say, the changes of connection weights proportional to the negative gradient. Thus, the formulas to calculate Δw_{ij} , Δv_{hi} are as follows:

$$\Delta w_{ij} = -\alpha\frac{\partial E}{\partial w_{ij}} = \sum_{k=1}^{m}\left[-\alpha\frac{\partial E_k}{\partial w_{ij}}\right] \tag{10}$$

$$\Delta v_{hi} = -\beta\frac{\partial E}{\partial v_{hi}} = \sum_{k=1}^{m}\left[-\beta\frac{\partial E_k}{\partial v_{hi}}\right] \tag{11}$$

It can be seen from the above two formula that changes in connection weights proportional to the sum of negative gradient which each model couple correspond on the mode set.

4 The Foundation and Training of BP Neural Network

The layer number of BP neural network is 3, the nodes number of Input layer is 8, corresponding to the 8 monitoring amount of the tower crane. Empirical formula used to determine the number of nodes in the middle layer, output layer nodes can be set to 1, training output is normalized values (greater than 0, tends to 1) of the tower crane life. The training life of tower crane will be got with a design life of the tower crane multiplied by the training value.

The monitoring data set QTZ160 model of the tower crane for the object of study, amplitude ranges are set as 45m, 25m, and 15m. For each amplitude range, the network take 7 time observation points, record eight monitoring value and record the observation time in the tower crane's rated life range. The last table take the situation that amplitude range is 45m for example to describe.

Table 1. The rest of life and the monitoring data of tower crane

Monitor-ingValue	lifting weight (M)	lifting torque (G)	lifting height (H)	lifting speed (V)	rotation angle (θ)	rotation speed (U)	amplitude range (L)	changing speed (K)	Remain-ing life (Y)
Units	t	tm	m	m/min	°	°/min	m	m/min	year
1	3.47	156.2	11.1	45	126	90	45	42	20
2	3.21	154.1	12.8	48	162	108	45	45	17.7
3	3.09	154.5	14.3	50	36	128	45	48	16.1
4	2.85	156.8	8.1	55	67	42	45	49	12.2
5	2.63	152.5	15.4	58	116	180	45	54	9.9
6	2.38	157.1	12.5	66	86	144	45	53	3.6
7	2.22	155.4	16	70	175	216	45	60	0.5

4.1 The Learning Steps of BP Neural Network

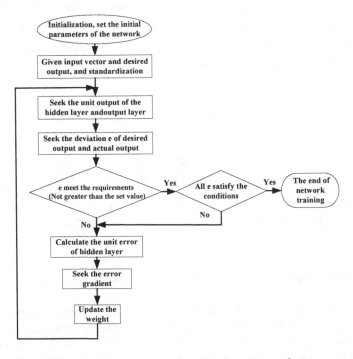

Fig. 1. The training process of BP neural network

Something to be noted is as following. We select a higher learning rate in the initial study while select a smaller learning rate late in the study.

4.2 The Process of Normalized

Input data must be normalized as training samples for the reasons of the transfer function when BP neural network get the mode recognized. The input sample normalized to [0, 1] in general case. The input volume normalized to [0.1, 0.9] in this network as the sigmoid function changes very slowly in [0, 0.1] and [0.9, 1].
    The Normalized formula is:

$$N(new) = 0.1 + \frac{0.8(N - N_{min})}{N_{max} - N_{min}} \tag{12}$$

N_{max}, N_{min} are the maximum and minimum input in training samples set.

4.3 Determination the Number of Nodes in Hidden Layer

According to reference [5], the following empirical formula can be used to determine the hidden nodes: $s = sqrt(0.43mn + 0.12nn + 2.54m + 0.77n + 0.35) + 0.51$. This paper uses the formula to determine the hidden layer nodes. Take $n = 8, m = 1$ into the above formula, the result is s=5.0011, as the result, the number of the hidden layer nodes is 5.

5 MATLAB Simulation

This paper use *newff* function of *Matlab* Toolbox to create BP network. The neuron transfer function of hidden layer and output layer are *tansig* and *logsig*. The training function is *traingdx*.

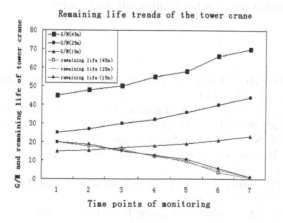

Fig. 2. Remaining life trends of the tower crane

    The above figure shows that the tower crane would provide greater torque to bring the same quality at certain amplitude as time goes by. Those three different downward curves show the remaining life change trends of the tower crane in different amplitude. It shows that the remaining life of tower crane is reduced along with time.

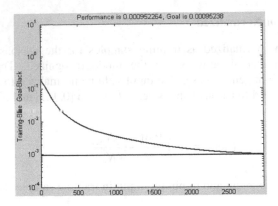

Fig. 3. The training error trends of BP neural network

As is shown in the above figure, the error rate of the network is 0.001 after 3000 times of training.

6 Conclusion

Life prediction of tower crane based on BP neural network has the characteristic of flexible information processing. It also shows a very strong ability of knowledge acquisition and solves the knowledge acquisition problem perfectly of traditional expert systems. It also has the advantage of high learning efficiency and powerful ability of fault-tolerant. The error rate of can be as small as 0.001. The downside is that training speed is a little slowly. This shortcoming can be overcame after take the improved BP algorithm and Structural optimization algorithm.

References

1. Wang, B.: Talking about Remaining Life Assessment of Tower Crane. Construction Mechanization 12(12), 32–33 (2008)
2. Zhang, Y., Tong, D.: Auto's Fault Diagnosis System Based on BP Neural Network. Automation Instrumentation 04(4), 11–12 (2009)
3. Zhu, D.: Artificial Neural Network Structure Learning Algorithm and Problem-Solving Research, vol. 5. Institute of Computing Technology Chinese Academy of Sciences, Beijing (1999)
4. Mi, J., Ji, G.: Application of Improved BP Neural Network in Fault Diagnosis of Fans. Noise and Vibration Control 4(2), 95 (2011)
5. Gao, D.: On Structures of Supervised Linear Basis Function Feedforward Three-Layered Neural Networks. Chinese Journal of Computers 1(1), 8 (1998)

An Ant Colony Algorithm Based on Interesting Level

Yu Jin-ping, Zhou Chun-hong, and Mei Hong-biao

Jiangxi University of Science and Technology

Abstract. An ant colony optimization algorithm with Interesting Level(IL) is presented. We couple a group of parameters with the basic ant colony approach This approach narrates the pheromone increasing style with IL, and the parameter named interesting is used to describe some path's agglomeration of ants to handle the balance between the convergent speed and the global solution searching ability. We throw the paths into different IL and the ants select their paths according to the paths' IL. At last, the viability of the approach has been tested with some typical travel salesman problems and encouraging results have been obtained.

Keywords: ant colony, pheromone, interesting level, agglomeration.

1 Introduction

Metaheuristics are well adapted to solve combinatorial optimization problems in practical situations. They allow an interesting compromise between the solution quality and computation time. Ant colony optimization (ACO), as a metaheuristic and relatively new search mechanism based on actual ant behavior, introduced by Marco Dorigo in[1], was found efficient in solving many problems since its introduction in the 90s.The first ACO algorithm, called ant system (AS)[1–2], initially was for solving the traveling salesman problem(TSP).

In recent years, the ACO have also been applied to other problems such as Vehicle Routing Problems(VRP), Graph Coloring, Scheduling, Partitioning Problems and Telecommunication Networks. The ACO is able to beat greedy randomized adaptive search procedure (GRASP) in all cases. An overview of ant algorithms can be found in[4]. In ACO algorithms, artificial ants act as computational agents which transmit information in some way and they are able to find good solutions to shortest path problems between a food source and their home colony[5].

The ACO is one of the most powerful in solving combinatorial optimization problems, but it has the contradictory between convergence speed and stagnation behavior like other metaheuristic algorithm. Many lectures try to solve this problem but the results give little help. In this paper, we design a self adaptive dynamic ACO based on interesting level (IL). This system can dynamically adjust the route pheromone updating strategy with the key parameters: interesting level. Many examples' results show that this strategy deals well with the contradictory between convergence speed and stagnation behavior.

D. Jin and S. Lin (Eds.): Advances in ECWAC, Vol. 2, AISC 149, pp. 117–122.
springerlink.com   © Springer-Verlag Berlin Heidelberg 2012

2 Ant Colony System: A Review

Ant algorithms are based on the real world phenomena that ants are able to find their ways to a food source and back to their nest Reference [1] was one in which is discussed what about it happens when an ant comes across an obstacle and it has to decide the best route to take around the obstacle. Initially, there is equal probability as to which way the ant will turn in order to negotiate the obstacle. If we assume that one route around the obstacle is shorter than the alternative route then the ants taking the shorter route will arrive at appoint on the other side of the obstacle before the ants which take the longer route. If we now consider other ants coming in the opposite direction, when they come across the same obstacle they are also faced with the same decision as to which way to turn. However, as ants walk they deposit a pheromone trail.

The ants which have already taken the shorter route will have laid a trail on this route so ants arriving at the obstacle from the other direction are more likely to follow that route as it has a deposit of pheromone. Over a period of time, the shortest route will have high levels of pheromone so that all ants are more likely to follow this route. There is positive feedback which reinforces that behavior so that the more ants that follow a particular route, the more desirable it becomes.

In the following, we convert this idea to a search mechanism for TSP. A TSP can be represented by a complete weighted directed graph $G = (V, A, d)$ where $V = \{1,2,\ldots, n\}$ is set of nodes (cities), $A = \{(i,j) \mid (i,j) \in V \times V\}$ the set of routings, and $d : A \to$ IN a weight function associating a positive integer weight d_{ij} with every $routing(i,j)^2$. The aim is to find a route of minimal length visiting every city exactly once. For symmetric TSPs, the distances between nodes are independent of the direction, i.e. $d_{ij} \neq d_{ij}$ for every pair of nodes. In the more general asymmetric TSP (ATSP) at least for one pair of nodes we have $d_{ij} \neq d_{ij}$. The TSP is a NP-hard optimization problem which has many applications and is extensively studied in literature [9]. It also has become a standard testbed for algorithms that try to find near optimal solutions to NP-hard combinatorial optimization problems. This is one more reason to apply our extensions of Ant System to this problem class.

3 A Self Adaptive Dynamic Ant System – ACO Based on Interesting Level(ACO-IL)

(1) definition of interesting level
It is well known that the pheromone in cities decides ants' paths selections. So the parameters' values that affect the pheromone are the key. In ACO, there are many parameters such as α, β, ρ, Q and m (ant quantity). α affects the global best solutions to get and β does the convergence speed. ρ, Q and M affects both of the global best and convergence speed but are not discussed in this paper. In ACS, they are const and isolated in algorithm processing. Based on the research of reference [9], we put forward a method with IL to help ants to select their favorite paths.

Theorem 1 the 2nd Newton Law(2NL)

The 2NL describes the relation of the quality M, trapping F and acceleration a as following:

$$a = F / M \qquad (1)$$

Then the speed theorem describes the equation of running initial speed V_0, length S, time T and acceleration a:

$$S = V_0 T + \frac{1}{2} a T^2 \qquad (2)$$

We make a suppose that The action that the ants search the shortest route abides by the 2NL and the speed theorem, then we suppose that:

$$a_{ij(f)}(t+1) = F_{i\,j(f)}(t+1) / M_i(t+1) \qquad (3)$$

Where:
$$M_i(t+1) = \tau_{iJ}(t) \qquad (4)$$

$$F_{i\,j(f)}(t+1) = \tau_{ij(f)}(t) - \overline{\tau_{iJ}(t)} \qquad (5)$$

Where: J is the cities set $\{j(1),j(2),j(3),....j(r)\}$ joined with city i, and t is the iterate number.

$$\overline{\tau_{iJ}(t)} = \frac{1}{r}\sum_{f=1}^{r}\tau_{ij(f)}(t) \qquad (6)$$

$\tau_{ij(f)}(t)$: represent the intensity of trail edge $(i,j(f))$ at time t.

So, we improve the ACO as following.

(1) the visibility $\eta_{ij(f)}(t)$

$$\eta_{ij(f)}(t+1) = 1/T_{ij(f)}(t+1) \qquad (7)$$

$$T_{ij(f)}(t+1) = \sqrt{\frac{2d_{ij(f)}}{|a_{ij(f)}|}} \qquad (8)$$

$T_{ij(f)}(t+1)$: the time that ant_k pass trail edge $(i,j(f))$; $d_{ij(f)}$: the length of trail edge $(i,j(f))$.

(2) Transition probability $P^k_{ij(f)}(t+1)$

The ant_k can probabilistically decide the next city to move to, according to formula (13).

$$P_{ij(f)}^k(t+1) = \begin{cases} \dfrac{\tau_{ij(f)}^{\alpha(t+1)}(t+1) \bullet \eta_{ij(f)}^{\beta(t+1)}(t+1)}{\sum_{s \in allowed_k} \tau_{iJ}^{\alpha(t+1)}(t+1) \bullet \eta_{iJ}^{\beta(t+1)}(t+1)} & if \quad j(f) \in allowed_k \\ 0 & otherwise \end{cases} \quad (9)$$

Where ① $\alpha(t)$: the heuristic factor defined as formula(10)

$$\begin{cases} \alpha_{ij(f)}(t+1) = \alpha_0 - \dfrac{1}{2t} a_{ij(f)}(t) & if \ G_i(t) > \Delta G_2 \\ \alpha_{ij(f)}(t+1) = \alpha_0 + \dfrac{1}{2t} a_{ij(f)}(t) & if \ \Delta G_1 > G_i(t) \\ \alpha_{ij(f)}(t+1) = \alpha_0 & else \end{cases} \quad (10)$$

$G_i(t)$ is the agglomeration of city i. it is calculated by formula(11). $\alpha_0, \Delta G_1$ and ΔG_2 are the initial values.

$$G_i(t) = 1 - \frac{1}{S_i(t)} \sqrt{\frac{r \sum_{i=1}^{r} (S_i(t)/r - ant(i))^2}{r-1}} \quad (11)$$

Where: $S_i(t) = \sum_{f=1}^{r} ant(f) = M * t$ $where$ $ant(f)$is the ant quantity of city i after iterate t.

Note: $S_i(t) = M * t$ because every ant must travel very city in one iterate.

② $\beta(t)$: the expected heuristic factor defined as formula (12)

$$\begin{cases} \beta_{ij(f)}(t+1) = \beta_0 + \dfrac{1}{2t} a_{ij(f)}(t) & if \ G_i(t) > \Delta G_2 \\ \beta_{ij(f)}(t+1) = \beta_0 - \dfrac{1}{2t} a_{ij(f)}(t) & if \ \Delta G_1 > G_i(t) \\ \beta_{ij(f)}(t+1) = \beta_0 & else \end{cases} \quad (12)$$

β_0 is an initial value, and the parameters are defined as above mentioned.

(3) Local update rule

The pheromone trails update given by formula (2) is modified as formula (13).

$$\tau_{ij(f)}(t+1) = \rho_{ij(f)}(t+1)\tau_{ij(f)}(t) + \Delta\tau_{ij(f)}(t+1) \quad (13)$$

Where: ①
$$\begin{cases} \rho_{ij(f)}(t+1) = (1 - \dfrac{a_{ij(f)}(t)}{t})\rho_0 & \text{if } \Delta G_2 < G_i(t)\tau_{ij} \\[2mm] \rho_{ij(f)}(t+1) = (1 + \dfrac{a_{ij(f)}(t)}{t})\rho_0 & \text{if } \Delta G_1 > G_i(t) \\[2mm] \rho_{ij(f)}(t+1) = \rho_0 \quad else \end{cases} \tag{14}$$

The $\rho_0(0<\rho_0<1)$ is a user-defined parameter, and the other parameters are defined as above mentioned.

$$② \quad \Delta\tau_{ij(f)}(t+1) = \sum_{k=1}^{m} \Delta\tau_{ij(f)}^{k}(t+1) \tag{15}$$

$$\Delta\tau_{ij(f)}^{k}(t+1) = \begin{cases} Q_{ij(f)}(t+1)/T_k, & \text{if the } k-th \text{ ant walks along the trail} \\ & \text{edge (i, j(f)) in its tour during time interval}(t, t+n) \\ 0, & otherwise \end{cases} \tag{16}$$

$Q_{ij}(t+n)$ will be changed on the $Lev_{ij}(t)$ of routing (i,j), and given by formula (17).

$$\begin{cases} Q_{ij(f)}(t+1) = (1 - \dfrac{a_{ij(f)}(t)}{t})Q_0 & \text{if } \Delta G_2 < G_i(t)\tau_{ij} \\[2mm] Q_{ij(f)}(t+1) = (1 + \dfrac{a_{ij(f)}(t)}{t})Q_0 & \text{if } \Delta G_1 < G_i(t) \\[2mm] Q_{ij(f)}(t+1) = Q_0 \quad else \end{cases} \tag{17}$$

where Q_0 is a constant and T_k is the tour time of the ant_k..

4   Experimental Study and Results

We study ACO-IL different TSP problems and compare it with the basic ACO. The initial parameters are: $\Delta G_1=0.39$, $\Delta G_2=0.91$, $\alpha(0)=2.0$, $\beta(0)=4.0$, $\rho_0=0.7$, $Q_0=100$, $NC=200$. The results are given by table 1.

Table 1 shows the results that ACO-IL has the better solutions than the current best solution for EL51, D198, KroA100 and Att532, but worse for Eil75. And it is better than ACO in the best solution, the worst solutions, the average and standard deviation. The results show that a combination of the parameters of ACO-IL performances best.

5   Conclusion

This paper presented an improved Ant System, more specifically, the An Ant Colony Algorithm with the 2nd Newton Law(ACO-2NL). The ACO-2NL was combined with a new updating rule and selection probability from the literature in order to improve the performance of the algorithm, mainly when applied to large instances of the problem. Therefore, the contribution of this paper is the presentation of a modified algorithm which demonstrated robustness in solving large scale problems.

Table 1 The results of ACO-2NL and ACO for different TSP cases

TSP cases	Current best solution	Basic ACO				ACO-IL			
		Best solution	Worst solution	Average	Standard deviation	Best solution	Worst solution	Average	Standard deviation
E151	428.1	434	439.3	456	9.24	428	431	426.3	2.89
Eil75	535	546	556.8	569	12.17	537	547	532.3	4.64
D198	15974	16608	16832.3	17123	249.73	15864	16550	16013.7	76.58
KroA100	21282	21429	21980.4	22848	692.92	21272	21773	21310.2	193.35
Att532	28131	29064	30071.6	31927	1212.67	28118	29303	28135.3	155.38

References

1. Dorigo, M.: Optimization, learning and natural algorithm. Politecnico di Milano, Italy (1992)
2. Colorni, A., Dorigo, M., Maniezzo, V.: Distributed optimization by ant colonies. In: Verela, F., Bourgine, P. (eds.) Proceedings of the First European Conference on Artificial Life (ECAL 1991). MITPress, Cambridge (1992)
3. Dorigo, M., Maniezzo, V., Colorni, A.: The ant system: Optimization by a colony of cooperating agents. IEEE Transactions on Systems, Man and Cybernetics 1(26), 29–41 (1996)
4. Dorigo, M., Caro, G., Gambar della, L.: Ant algorithms for discrete optimization. Artificial Life 2(5), 137–172 (1999)
5. Stutzle, T., Hoos, H.: MAX-MZN Ant System and Local Search for the Traveling Salesman Problem. IEEE, 309–314 (1999)
6. Mei, H.-B., Xie, L.-Q., Zhou, C.-H.: An Ant Colony Algorithm with the 2nd Newton Law. In: 2010 Second IITA Conference on Geoscience and Remote Sensing, vol. (1), pp. 550–554 (2010)

A Cooperative Design Approach of Fault-Tolerant Controller and Observer for Nonlinear Networked Control Systems Based on T-S Model

Xuan Li[1], Xiaomao Huang[1,*], and Xiaobei Wu[2]

[1] College of Engineering, Huazhong Agricultural University, Wuhan 430070, China
[2] Institute of Automation, Nanjing University of Science and Technology,
Nanjing 210094, China
lx@mail.hzau.edu.cn, huangxiaomao@mail.hzau.edu.cn,
wuxb@mail.njust.edu.cn

Abstract. The problem of integrity against sensor failures for the nonlinear networked control systems in which the plant can be described by T-S fuzzy models is studied based on state observer. Assuming that the time-delay is less than one sampling period, the system is modeled as a discrete time T-S fuzzy model with parametrical uncertainties. Based on the model, the fuzzy state observe is designed and according to possible sensor failures, an augmented mathematic model for the systems based on the fuzzy state observer is developed. Then the sufficient stability condition of integrity against sensor failures is given and the designs for fuzzy fault-tolerant controller and fuzzy observer are presented by using the theory of fuzzy control and the approach of Matrix Inequalities.

Keywords: sensor failures, stability, control law, bilinear matrix inequalities, time-delays.

1 Introduction

Networked Control Systems (NCSs) is feedback control system wherein the control loops are closed through a real-time network [1-3]. At present, the investigation on network control systems is mainly focused on the linear part of it, in which the controlled object is assumed to be a linear model. Generally, the linear system is only a special expression form for the nonlinear system under certain conditions. In many industrial process control areas, such as aerospace, weapons systems, robotics and other fields, controlled objects are essentially linear. The research of the nonlinear networked control systems can provide an important theoretical basis for applications in industry.

In recent years, more and more attention has been attracted to the study on nonlinear networked control systems. However, the research on the fault-tolerant control for the nonlinear networked control systems is not very common both abroad

---

* Corresponding author.

D. Jin and S. Lin (Eds.): Advances in ECWAC, Vol. 2, AISC 149, pp. 123–131.
springerlink.com   © Springer-Verlag Berlin Heidelberg 2012

and at home. Aiming at the networked control systems composed by a class of nonlinear plant and the state feedback controller, the sufficient conditions for the system maintaining exponentially stability against actuator or sensor failures are given respectively based on Taylor linearization theory and Lyapunov stability analysis method , and the fault-tolerant controller is designed using Linear Quadratic Optimal control theory in [4]. For a class of nonlinear networked control systems with time-varying delays and data dropout, based on the integrity fault-tolerant control theory and the LMI approach, the state feed-back guaranteed cost fault-tolerant controller is designed in [5]. Considering both network-induce delay and packet dropout, based on a T-S fuzzy model, the delay-dependent sufficient condition of robust H∞ integrity for nonlinear networked control systems with failures of actuator and sensor is analyzed and the robust H fault-tolerant controller is designed in [6]. Among these references, the conclusions are based on state feedback. In actual systems, as the complexity of network, the all states of the system are not convenience enough to be obtained. Then the state observer is prone to be used. Therefore it is essential to do research on nonlinear networked control systems with state observer.

In this paper, we aim at solving the problem of integrity against sensor failures for the nonlinear networked control systems in which the plant can be described by T-S fuzzy models based on state observer. Assuming that the time-delay is less than one sampling period, the system is modeled as a discrete time T-S fuzzy model with parametrical uncertainties. Then the cooperative design approach of the fuzzy controller and the fuzzy observer is given and the existence conditions of fuzzy fault-tolerant control law are testified in terms of the Lyapunov stability theory combined with bilinear matrix inequalities.

2   Problem Formulations

Nonlinear networked control systems studied in this paper is shown in Fig.1. Network only exists between sensor and controller, so the system has output time delay τ_k^{sc} .

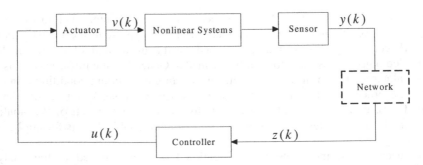

Fig. 1. Structure diagram of the nonlinear networked control systems

We assume:

(1) The sensor is driven by time, which is sampled with known fixed T and the controller and the actuator are event-driven;

(2) The output time delay τ_k^{sc} is time varying and $\tau_k^{sc} \in [0,T]$.

Consider the nonlinear system

$$\begin{cases} \dot{x}(t) = f(x(t), v(t)) \\ y(t) = g(x(t)) \end{cases} \tag{1}$$

where $x(t) \in R^n$, $v(t) \in R^p$ and $y(t) \in R^r$ are state vector, input vector and measure output vector respectively. $f(\cdot)$ is a nonlinear continuous function defined in compact space, as well as $g(\cdot)$ is also a nonlinear continuous function defined in compact space.

According to the reference [7], the system can be represented proximately by the T–S model given as the following:

Plant Rule i : IF $z_1(t)$ is M_1^i AND ... $z_r(t)$ is M_r^i THEN

$$\begin{cases} \dot{x}(t) = A_f^i x(t) + B_f^i v(t) \\ y(t) = C_f^i x(t) \end{cases} \qquad i = 1, 2, \cdots, g \tag{2}$$

where $M_1^i, \cdots, M_r^i$ are fuzzy sets, A_f^i, B_f^i, C_f^i are constant matrices of compatible dimensions, g is the number of IF-THEN rules, and $z_1(t), \ldots, z_r(t)$ is the premise variable vector. Considering the output time delay τ_k^{sc}, the discrete-time fuzzy model of the system can be obtained from (2) as follows

IF $z_1(k)$ is M_1^i AND ... $z_r(k)$ is M_r^i THEN

$$\begin{cases} x(k+1) = A_i x(k) + (B_i - \bar{B}_i(\tau_k^{sc}))u(k) + \bar{B}_i(\tau_k^{sc})u(k-1) \\ y(k) = C_i x(k) \end{cases} \tag{3}$$

where $A_i = e^{A_f^i T}$, $B_i = e^{B_f^i T}$, $\bar{B}_i(\tau_k^{sc}) = \int_{T-\tau_k^{sc}}^{T} e^{A_f^i s} ds B_f^i$, $u(k) \in R^p$ is control output vector.

According to the Ref. [8], we can obtain $\bar{B}_i(\tau_k^{sc}) = D_i F_i(\tau_k^{sc}) E_i$ and $F_i^T(\tau_k^{sc}) F_i(\tau_k^{sc}) < I$, where D_i and E_i are constant matrices determined by eigenvalues and eigenvectors of matrix A_f^i and B_f^i, $F_i^T(\tau_k^{sc})$ is an uncertain matrix satisfying $F_i^T(\tau_k^{sc}) F_i(\tau_k^{sc}) < I$ marked by F_i.

Given a pair of $(x(k), u(k))$, the final output of the fuzzy system is inferred as:

$$\begin{cases} x(k+1) = \sum_{i=1}^{g} \mu_i(z(k)) \left[A_i x(k) + (B_i - D_i F_i E_i) u(k) + D_i F_i E_i u(k-1) \right] \\ y(k) = \sum_{i=1}^{g} \mu_i(z(k)) C_i x(k) \end{cases} \tag{4}$$

where $\quad \mu_i(z(k)) = \dfrac{\omega_i(z(k))}{\sum_{i=1}^{g} \omega_i(z(k))}, \quad \omega_i(z(k)) = \prod_{j=1}^{l} H_j^i(z_j(k))$

with $H_j^i(z_j(k))$ representing the grade of $z_j(k)$ membership of in H_j^i. Then, it can be seen that

$$\omega_i(z(k)) \ge 0, \; i = 1, 2, \cdots, g \text{ and } \sum_{i=1}^{g} \omega_i(z(k)) > 0$$

for all k. Therefore, for all k, we have

$$\mu_i(z(k)) \ge 0, \; i = 1, 2, \cdots, g \text{ and } \sum_{i=1}^{g} \mu_i(z(k)) = 1$$

According to the system (4), the observer is designed as follow:

$$\hat{x}(k+1) = \sum_{i=1}^{g} \mu_i(z(k)) \left\{ A_i \hat{x}(k) + B_i u(k) + L_i \left[y(k) - C_i \hat{x}(k) \right] \right\} \tag{5}$$

where $L_i \in R^{n \times r}$ and $\hat{x}(k) \in R^n$ denote the fuzzy state observer gain matrix and the state estimate respectively.

Aiming at system (4), using state feedback control law based on observer as follows

$$u(k) = \sum_{i=1}^{g} \mu_i(z(k) K_i \hat{x}(k) \tag{6}$$

where $K_i \in R^{p \times n}$.

In order to formulate the possible sensor failure faults, the fault model must be established firstly. Considering possible sensor failure faults, we can introduce a switched matrix M to the system (4), and lay the matrix M before the measure output vector $y(k)$, where $M_i = diag(m_1, m_2, \cdots, m_r)$ and for $i = 1, 2, \cdots, r$

$$m_i = \begin{cases} 1 & \text{the } i\text{th sensor normal} \\ 0 & \text{the } i\text{th sensor failure} \end{cases}$$

Then the fuzzy observer equation becomes:

$$\hat{x}(k+1) = \sum_{i=1}^{g} \mu_i(z(k)) \left\{ A_i \hat{x}(k) + B_i u(k) + L_i \left[My(k) - C_i \hat{x}(k) \right] \right\} \tag{7}$$

Define the estimation error

$$e(k) = x(k) - \hat{x}(k) \tag{8}$$

From (4), (6), (7) and (8), we can obtain

$$e(k+1) = \sum_{i=1}^{g} \sum_{j=1}^{g} \mu_i(z(k))\mu_j(z(k)) \left[G_{ij}e(k) + H_{ij}x(k) - C_{ij}(e(k-1) - x(k-1)) \right] \quad (9)$$

where $G_{ij} = A_i + D_i F_i E_i K_j - L_i C_i$, $H_{ij} = L_i(I-M)C_i - D_i F_i(\tau_k^{sc})E_i K_j$,
$C_{ij} = D_i F_i E_i K_j$.

According to (4), (6) and (7), we have

$$x(k+1) = \sum_{i=1}^{g} \sum_{j=1}^{g} \mu_i(z(k))\mu_j(z(k)) \left[A_{ij}x(k) - \tilde{B}_{ij}e(k) + C_{ij}x(k-1) - C_{ij}e(k-1) \right] \quad (10)$$

where $A_{ij} = A_i + B_i K_j - D_i F_i E_i K_j$, $\tilde{B}_{ij} = (B_i - D_i F_i E_i)K_j, C_{ij} = D_i F_i E_i K_j$.

Let an augmented vector

$$\theta(k) = [x^T(k) \quad e^T(k)]^T \quad (11)$$

From (4), (10) and (11), the augmented system based on observer can be represented by:

$$\theta(k+1) = \sum_{i=1}^{g} \sum_{j=1}^{g} \mu_i(z(k))\mu_j(z(k)) \left\{ \begin{bmatrix} A_{ij} & -\tilde{B}_{ij} \\ H_{ij} & G_{ij} \end{bmatrix} \theta(k) + \begin{bmatrix} C_{ij} & -C_{ij} \\ C_{ij} & -C_{ij} \end{bmatrix} \theta(k-1) \right\} \quad (12)$$

Let

$$\hat{A}_{ij} = \begin{bmatrix} A_i + B_i K_j & -B_i K_j \\ L_i(I-M)C_i & A_i - L_i C_i \end{bmatrix}, \hat{D}_i = \begin{bmatrix} D_i \\ D_i \end{bmatrix}, \hat{E}_{ij} = [E_i K_j \quad -E_i K_j]$$

The system (12) can be rewritten as:

$$\theta(k+1) = \sum_{i=1}^{g} \sum_{j=1}^{g} \mu_i(z(k))\mu_j(z(k)) \left\{ (\hat{A}_{ij} - \hat{D}_i F_i \hat{E}_{ij})\theta(k) + \hat{D}_i F_i \hat{E}_{ij}\theta(k-1) \right\} \quad (13)$$

3 Main Results

Lemma 1[1]. If A , P and Q are finite-dimension constant matrices, then $Q = Q^T, P = P^T > 0$, we have

$$A^T PA + Q < 0 \Leftrightarrow \begin{bmatrix} Q & A^T \\ A & -P^{-1} \end{bmatrix} < 0 \quad or \quad \begin{bmatrix} -P^{-1} & A \\ A^T & Q \end{bmatrix} < 0$$

Lemma 2[2]. Given matrices W, M, N of appropriate dimensions and with W symmetric, then

$$W + N^T F^T(k)M^T + MF(k)N < 0$$

For all $F(k)$ satisfying $F^T(k)F(k) \leq I$, if and only if there exists a scalar $\varepsilon > 0$ such that

$$W + \varepsilon M M^T + \varepsilon^{-1} N^T N < 0$$

Lemma 3[9]. Given vectors a and b of appropriate dimensions then

$$2a^T b \le a^T X a + b^T X^{-1} b$$

where X is a arbitrarily positive-definite matrix of appropriate dimension.

Theorem 1: Consider the system (12), given the sensor failure pattern matrix $M \in \Omega$, if there exist symmetry positive-definite matrices $\bar{P}_1$, $\bar{P}_2$, scalars $\varepsilon_i > 0$, $\varepsilon_{ij} > 0$, $(i, j = 1, 2, ..., g)$ such that

$$\begin{bmatrix} \bar{P}_2 - \bar{P}_1 & 0 & \bar{P}_1 \hat{A}_{ii}^T & -\bar{P}_1 \hat{E}_{ii}^T \\ * & -\bar{P}_2 & 0 & \bar{P}_1 \hat{E}_{ii}^T \\ * & * & -\bar{P}_1 + \varepsilon_i \hat{D}_i \hat{D}_i^T & 0 \\ * & * & * & -\varepsilon_i I \end{bmatrix} < 0, \ 1 \le i \le g \tag{14}$$

$$\begin{bmatrix} 2\bar{P}_2 - 2\bar{P}_1 & 0 & \bar{P}_1(\hat{A}_{ij}^T + \hat{A}_{ji}^T) & -\bar{P}_1 \hat{E}_{ij}^T & -\bar{P}_1 \hat{E}_{ji}^T \\ * & -2\bar{P}_2 & 0 & \bar{P}_1 \hat{E}_{ij}^T & \bar{P}_1 \hat{E}_{ji}^T \\ * & * & -2\bar{P}_1 + \varepsilon_{ij} \hat{D}_i \hat{D}_i^T + \varepsilon_{ij} \hat{D}_j \hat{D}_j^T & 0 & 0 \\ * & * & * & -\varepsilon_{ij} I & 0 \\ * & * & * & * & -\varepsilon_{ij} I \end{bmatrix} < 0, \ 1 \le i < j \le g \tag{15}$$

then the system (12) is asymptotically stable and the equation (6) is a fault-tolerant control law based on the observer (5), where Ω is a set which consists of all possible sensor failure faults switch matrix M, "*" means the symmetric portion of the matrix.

Proof: Defining the following Lyapunov function

$$V(k) = \theta^T(k) P_1 \theta(k) + \theta^T(k-1) P_2 \theta(k-1)$$

where P_1 and P_2 are symmetry positive-definite matrices. Then

$$\Delta V(k) = \theta^T(k+1) P_1 \theta(k+1) + \theta^T(k) P_2 \theta(k) - \theta^T(k) P_1 \theta(k) - \theta^T(k-1) P_2 \theta(k-1)$$

Let

$$\bar{A}_{ij} = \begin{bmatrix} A_{ij} & -\tilde{B}_{ij} \\ H_{ij} & G_{ij} \end{bmatrix}, \ \bar{C}_{ij} = \begin{bmatrix} C_{ij} & -C_{ij} \\ C_{ij} & -C_{ij} \end{bmatrix},$$

$$a_{ij}^T = \mu_i(z(k)) \mu_j(z(k)) \begin{bmatrix} \theta(k) \\ \theta(k-1) \end{bmatrix}^T \begin{bmatrix} \bar{A}_{ij}^T \\ \bar{C}_{ij}^T \end{bmatrix} P_1$$

$$b_{\bar{m}\bar{l}} = \mu_{\bar{m}}(z(k)) \mu_{\bar{l}}(z(k)) \begin{bmatrix} \bar{A}_{\bar{m}\bar{l}} & \bar{C}_{\bar{m}\bar{l}} \end{bmatrix} \begin{bmatrix} \theta(k) \\ \theta(k-1) \end{bmatrix}$$

Based on *Lemma 3*, we have

$$a_{ij}^T b_{\bar{m}\bar{l}} \le \frac{1}{2} a_{ij}^T X a_{ij} + b_{\bar{m}\bar{l}}^T X^{-1} b_{\bar{m}\bar{l}} \tag{16}$$

where $X = P_1^{-1}$. For $0 \le \mu_i(z(k)) \le 1$, we have

$$\mu_i(z(k))\mu_j(z(k)) \begin{bmatrix} \theta(k) \\ \theta(k-1) \end{bmatrix}^T \begin{bmatrix} \bar{A}_{ij}^T \\ \bar{C}_{ij}^T \end{bmatrix} P_1 \mu_{\bar{m}}(z(k))\mu_{\bar{l}}(z(k)) \begin{bmatrix} \bar{A}_{\bar{m}\bar{l}} & \bar{C}_{\bar{m}\bar{l}} \end{bmatrix} \begin{bmatrix} \theta(k) \\ \theta(k-1) \end{bmatrix}$$

$$\le \frac{1}{2} \mu_i(z(k))\mu_j(z(k)) \begin{bmatrix} \theta(k) \\ \theta(k-1) \end{bmatrix}^T \begin{bmatrix} \bar{A}_{ij}^T P_1 \bar{A}_{ij} & \bar{A}_{ij}^T P_1 \bar{C}_{ij} \\ \bar{C}_{ij}^T P_1 \bar{A}_{ij} & \bar{C}_{ij}^T P_1 \bar{C}_{ij} \end{bmatrix} \begin{bmatrix} \theta(k) \\ \theta(k-1) \end{bmatrix} \rightarrow$$

$$\leftarrow + \frac{1}{2} \mu_{\bar{m}}(z(k))\mu_{\bar{l}}(z(k)) \begin{bmatrix} \theta(k) \\ \theta(k-1) \end{bmatrix}^T \begin{bmatrix} \bar{A}_{\bar{m}\bar{l}}^T P_1 \bar{A}_{\bar{m}\bar{l}} & \bar{A}_{\bar{m}\bar{l}}^T P_1 \bar{C}_{\bar{m}\bar{l}} \\ \bar{C}_{\bar{m}\bar{l}}^T P_1 \bar{A}_{\bar{m}\bar{l}} & \bar{C}_{\bar{m}\bar{l}}^T P_1 \bar{C}_{\bar{m}\bar{l}} \end{bmatrix} \begin{bmatrix} \theta(k) \\ \theta(k-1) \end{bmatrix}$$

Then we can obtain

$$\Delta V(k) \le \sum_{i=1}^g \mu_i^2(z(k)) \left\{ \begin{bmatrix} \theta(k) \\ \theta(k-1) \end{bmatrix}^T \begin{bmatrix} \bar{A}_{ii}^T P_1 \bar{A}_{ii} + P_2 - P_1 & \bar{A}_{ii}^T P_1 \bar{C}_{ii} \\ \bar{C}_{ii}^T P_1 \bar{A}_{ii} & \bar{C}_{ii}^T P_1 \bar{C}_{ii} - P_2 \end{bmatrix} \begin{bmatrix} \theta(k) \\ \theta(k-1) \end{bmatrix} \right\}$$

$$+ 2 \sum_{i<j}^g \mu_i(z(k))\mu_j(z(k)) \left\{ \begin{bmatrix} \theta(k) \\ \theta(k-1) \end{bmatrix}^T \begin{bmatrix} (\frac{\bar{A}_{ij}+\bar{A}_{ji}}{2})^T P_1 (\frac{\bar{A}_{ij}+\bar{A}_{ji}}{2}) + P_2 - P_1 & (\frac{\bar{A}_{ij}+\bar{A}_{ji}}{2})^T P_1 (\frac{\bar{C}_{ij}+\bar{C}_{ji}}{2}) \\ (\frac{\bar{C}_{ij}+\bar{C}_{ji}}{2})^T P_1 (\frac{\bar{A}_{ij}+\bar{A}_{ji}}{2}) & (\frac{\bar{C}_{ij}+\bar{C}_{ji}}{2})^T P_1 (\frac{\bar{C}_{ij}+\bar{C}_{ji}}{2}) - P_2 \end{bmatrix} \begin{bmatrix} \theta(k) \\ \theta(k-1) \end{bmatrix} \right\}$$

For $\mu_i^2(z(k)) > 0$ and $\mu_i(z(k))\mu_j(z(k)) > 0$, $\Delta V(k) < 0$ is equivalent to:

$$\begin{bmatrix} \bar{A}_{ii}^T P_1 \bar{A}_{ii} + P_2 - P_1 & \bar{A}_{ii}^T P_1 \bar{C}_{ii} \\ \bar{C}_{ii}^T P_1 \bar{A}_{ii} & \bar{C}_{ii}^T P_1 \bar{C}_{ii} - P_2 \end{bmatrix} < 0, \quad 1 \le i \le g \tag{17}$$

$$\begin{bmatrix} (\frac{\bar{A}_{ij}+\bar{A}_{ji}}{2})^T P_1 (\frac{\bar{A}_{ij}+\bar{A}_{ji}}{2}) + P_2 - P_1 & (\frac{\bar{A}_{ij}+\bar{A}_{ji}}{2})^T P_1 (\frac{\bar{C}_{ij}+\bar{C}_{ji}}{2}) \\ (\frac{\bar{C}_{ij}+\bar{C}_{ji}}{2})^T P_1 (\frac{\bar{A}_{ij}+\bar{A}_{ji}}{2}) & (\frac{\bar{C}_{ij}+\bar{C}_{ji}}{2})^T P_1 (\frac{\bar{C}_{ij}+\bar{C}_{ji}}{2}) - P_2 \end{bmatrix} < 0, \quad 1 \le i < j \le g \tag{18}$$

For $\bar{A}_{ij} = \hat{A}_{ij} - \hat{D}_i F_i \hat{E}_{ij}, \bar{C}_{ij} = \hat{D}_i F_i \hat{E}_{ij}$, In light of *Lemma 2* and using the *Lemma 1*, the inequality (18) can be equivalently rewritten as

$$\begin{bmatrix} P_2 - P_1 & 0 & \hat{A}_{ii}^T & -\hat{E}_{ii}^T \\ * & -P_2 & 0 & \hat{E}_{ii}^T \\ * & * & -P_1^{-1} + \varepsilon_i \hat{D}_i \hat{D}_i^T & 0 \\ * & * & * & -\varepsilon_i I \end{bmatrix} < 0 \tag{19}$$

Pre-and post-multiplying both sides of (19) with $diag(P_1^{-1}, P_1^{-1}, I, I)$ and its transpose, and let $\bar{P}_1 = P_1^{-1}, \bar{P}_2 = P_1^{-1}P_2P_1^{-1}$, we can obtain (14).

Similar to the proof process of (14), we can obtain (15).This completes the proof of theorem 1.

It is noted that the inequalities of theorem 1 are not linear matrix inequality. In order to obtain the controller gain K_i and the observer gain L_i, the inequalities of theorem 1 can be transformed to bilinear matrix inequality, which can be solved with MATLAB and PENBMI [10].

4 Conclusion

Focusing on nonlinear networked control systems which plant can be described by T-S fuzzy models with short time-delays, a control method based on state observer against sensor failures is investigated in this paper. Assuming that the network-induced delay is uncertain short time-varying delay, the system is modeled as a discrete time T-S fuzzy model with parametrical uncertainties. Then the designs for fuzzy fault-tolerant control law and fuzzy state observer are presented in terms of the Lyapunov stability theory combined with bilinear matrix inequalities.

Acknowledgments. This work was supported by the "Fundamental Research Funds for the Central Universities" (2010BQ008, 2011QC012, 2011QC010). Authors are thankful to the anonymous referees for the constructive comments in enhancing the content and structure of the manuscript.

References

1. Zhang, W., Branicky, M.S., Phillips, S.M.: Stability of networked control systems. IEEE Control Systems Magazine 21(1), 85–99 (2001)
2. Hespanha, J.P., Naghshtabrizi, P., Xu, Y.G.: A Survey of Recent Results in Networked Control Systems. Proceedings of the IEEE 95(1), 138–162 (2007)
3. Antsaklis, P., Baillieul, J.: Special Issue on Technology of Networked Control Systems. Proceedings of the IEEE 95(1), 5–8 (2007)
4. Zhang, Q.Y., Gong, D.W., Guo, Y.N.: Fault-tolerant control of a class of nonlinear networked control systems with time-varied delays. In: Proceedings of the 27th Chinese Control Conference, CCC, pp. 141–144 (2008)
5. Xie, N., Xia, B.: Guaranteed cost fault tolerant controller design for nonlinear networked control systems. In: Proceedings - 2010 3rd IEEE International Conference on Computer Science and Information Technology, ICCSIT 2010, vol. 7, pp. 379–382 (2010)
6. Li, W., Jiang, D.N.: Robust H∞ fault-tolerant control for nonlinear networked control system based on Takagi-Sugeno fuzzy model. Control and Decision 25(4), 598–604 (2010)

7. Takagi, T., Sugeno, M.: Fuzzy identification of systems and its applications to modeling and control. IEEE Transactions on Systems, Man and Cybernetics SMC-15(1), 116–132 (1985)
8. Xie, C.Y., Fan, W.H., Hu, W.L.: Modeling and control method of a class of networked control systems with short time-delay. J. Nanjing Univ. Sci. Technol. 33(2), 156–160 (2009)
9. Gao, H.J., Chen, T.W.: Stabilization of nonlinear systems under variable sampling: A fuzzy control approach. IEEE Transactions on Fuzzy Systems 15(5), 972–983 (2007)
10. Kocvara, M., Sting, M.: PENBMI. Version 2.0 (2004), Free Developer Version, http://www.penopt.com

The Sharing Methods of Traffic Data from EMME to TRANSCAD

Junyou Zhang, Juanjuan Liu, and Rui Tang

Shandong University of Technology, School of Traffic & Vehicle Engineering,
Traffic management and planning, China

Abstract. Through the analysis of the established process of Emme database, combined with the practical operation experience, find out the deficiency of the module Emme editor. In the light of the shortcoming raised the method of make use of the convenient edit of TransCAD and combined with strong Emme traffic assignment function, make traffic data input simplify and share.

Keywords: Emme, TransCAD, Traffic data.

1 Introduction

Many traffic prediction software were developed based on four steps at home and abroad. Emme which is the development of university of Montreal with its powerful distribution function is favoured by a lot of traffic engineers. But the size and number of Emme database file should be decided in advance , or will be complex to change.

This make a lot of inconvenience for the input traffic data.

As the base of the establishing various models , the collecting and processing of foundation of traffic data affect all stages of the late model development and results of model analysis. So the database establishment is particularly important. Meanwhile the basic network data occupied the main part in the traffic data . so the complete of Emme function which can be applied to the entire traffic prediction and research Emme traffic data based network data input data especially optimization problem is very practical , thereby, provide technical support to urban traffic planning in the application of the software for the convenient traffic prediction and Emme scientific decision-making urban infrastructure construction[1].

2 Emme Introduction and Problems Existing in the Application Process

Emme is professional software developed by Canada Montreal university for traffic prediction and traffic analysis. The run of Emme depends on the support of the database. The operation use the way of man-machine interactive. Meanwhile Emme also provides the user with a base interface, and calls to select a different program modules. The functional characteristics of Emme overall are very powerful. It can

D. Jin and S. Lin (Eds.): Advances in ECWAC, Vol. 2, AISC 149, pp. 133–138.

custom a powerful backend database, involves the all data are included, network and model can be convenient to convert, and execute automatic data consistency inspection during input process to ensure that the data is not lost. Interactive graphic editor can quickly make modification and renewal to database, and can be convenient to draw rivers, short line, charging streets and regional name. City information system integrae function of network and regional data, including traffic investigation, accident statistics, road surface characteristics, maintenance costs and user defined content. It has the analysis function to the present situation and the future area, calculate and display a cost function based on any of the minimum cost path. Emme can be defined means of transportation as many as 30 kinds, including cars, public transportation and auxiliary transportation means, while each bus way can include several vehicle types. This can make all the forms of transportation in the same connected network considered[2].

Emme function form is not limited, it can provide flow times, penalty function, bus travel times and demand function sets. Network and model has powerful calculation, and able to make interactive data calculation to sections, nodes, the bus routes, the bus route segment and the model. It has no fixed restrictions to demand model form but can convert demand model into a macro process, considering the needs of many network balance model. It has powerful traffic distribution fanction which could make separately distribution to cars or buses by balance method.

2.1   The Input of Emme Traffic Data

In the user interface of Emme, the user can be on different modules by clicking under section node button, link button, bus lines buttonon on the top of the interface to edit based network data, including mode (model name, types, description, etc.), node (common nodes and centroid), section (length, types, type of cars, lanes, function code and the user defined parameters, etc.), intersection, the bus route, etc.

The above input methods is intuitive and flexible. But the input of nodes need to edit and input to different types of nodes; And in the input section, must first add nodes, and to different types of sections the attribute should to be edited again. Therefore, this edit operation method is command trival, slow speed, data entry cycle long. Usually this method used only for the modification of existing Emme network, but not apply to enter basic network[3].

Emme support database established function, it can translate the contents of data into visual graphic elements. Emme batch of input document that can be read is IN files. So, as long as we editor the documents in the format of the Emme can read , then import into Emme through the command module can be called. The basic network file includes node number, coordinate, section of the beginning and the end, section length, type, lanes and so on. Emme can input IN documents in batch of input methods into database through the rapid control platform in tools menu, then form the visualization of graphics. Emme commands will input network as long as edit the node and sections attributes into IN file in the format that Emme readable format. So the creat of the network can be very simple as long as the IN file exist. This method is convenient and accurate.

Therefore, IN file become very imporant. The basic way to get IN file is to add attribute one by one in the excel, or add in CAD or other application software artificial statistical node coordinates, node number, the origin and the destination of section, length, lanes and road type and so on. But this method cost a lot of work[4].

3 The Solutions and Process of Realization

3.1 Sharing Date with Emme by Transcad

Through the above analysis shows that this method is used on the editor and modification of the network. Input by external file is fast and convenient when the IN file exist, butthe workload of input will be great if wo not have it.

The the well-known traffic planning software TransCAD of CALIPER company is the first geographic information system (GIS) esigned for traffic professional to store, display, manage and analyse traffic data . TransCAD combined GIS and transport model function into a single platform, on behalf of the most advanced GIS technology. It can be used to make and restructure map, establish and safeguard a geographic data set, or into the line of various means space analysis. TransCAD contain a variety of complex function of GIS, including polygon superposition, influence area analysis, geocoding, etc., and has an open system structure, support local network and the data on the wan share[5].

TransCAD support many kinds of file formats, such as BIN, DBF, XLS, SHP, DXF, etc. And DBF file can be edited through the excel, then save as IN files. The advantage of using TransCAD to edit network and derived is that the fanction of check connection of the worknet convenient display problems of sections, and section select function is convenient for batch processing section attribute. This greatly improve the efficiency of data input, and TransCAD is more universal and easy to operate than GIS.

According to the analysis above, we can combine the powerful network edit faction of TransCAD with the distribution functions of Emme to achieve the optimal result[6].

The process as shown in Fig.1:

Fig. 1. Technology process

3.2 Case Analysis

Input data with batch method need a specific format, each column have requirements to format and content.

Nodes data input formats such as table 1, the corresponding format type and describe such as table 2.

Table 1. Citing of data input formats

ID	Xi	Yi	Ui1	Ui2	Ui3
1257	110003143	10002031	0	0	0
1259	110003091	10002005	0	0	0
1258	110003100	10002010	0	0	0

Table 2. Node data input formats

row	content	type	description
1	a or a*	alpha	Update code (add, delete, modify) ,*Centroid indicator
2	i	integer	Node number
3	xi	real	X-coordinate of node
4	yi	real	Y-coordinate of node
5	ui1	real	User defined node data 1
6	ui2	real	User defined node data 2
7	ui3	real	User defined node data 3
8	lab	alpha	Optional node label

Input formats of section dates are shown in figure 3, the corresponding format type and description such as table 4.

Table 3. Data input formats

Inode	Jnode	Length	Modes	Type	Lanes	Vdf	Speed	Ul1	Ul2	Ul3
81	96	3.37	c	2	2	1	100.00	0	0	0
88	82	14.31	c	14	1	1	35.00	0	0	0
96	1148	2.44	c	14	1	1	35.00	0	0	0

Table 4. Sections data input formats

row	content	type	description
1	a or a=	alpha	Update code,= be two-way
2	i	integer	Node number of starting node (I-node) of link
3	j	integer	Node number of ending node (J-node) of link
4	len	real	Link length in length units
5	mod	alpha	List of modes allowed on this link.
6	typ	integer	Link type
7	lan	real	Number of lanes
8	vdf	integer	Volume-delay function number
9	ul1	real	User defined link data
10	ul2	real	User defined link data
11	ul3	real	User defined link data

Conclusion: section attributes of TransCAD traffic modeling need are ID, length, layer, capacity, speed, lanes, times and so on. But Emme also need the beginning and end of the node number of origin and destination of a section. Meanwhile lanes in Emme is the unidirectional number but TransCAD is in two-way. This needs to be transform in the TransCAD.

Input DXF file into TransCAD to generated MAP file, and check connectivity of the map network, then modify the nodes and sections which have problems. Classify the road through the survey, and edit attribute of capacity, speed, and lanes, times and so on. Add origin and destination of sections through the Fill and Formula tools , then convert the lanes into one-way lane number. Save network file as DBF files, then edit the IN file according to the requirement in excel[7]. Call the corresponding module of Emme to read the contents of the documents at last.

4 Conclusions

This text points out the shortcomings in theinput of the net work of Emme and the superiority in dealing with traffic data of TransCAD, accroding to the deep analysis of Emme and TransCAD traffic planning further analysis software, and combining with actual operating experience. Through the researching of data input formats of Emme module 2 and TransCAD data to find the relation. It is proposed that using the GIS platform of the TransCAD to deal with traffic data, then combine Emme and TransCAD to achieve the goal of simplified and sharing of the traffic date.

References

1. Lu, H.: Traffic planning theory and method, vol. 12, pp. 15–18. Tsinghua University Press, Beijing (2006)
2. Pei, Y.: Rode networks planing, pp. 18–20. China Communications Press, Beijing (2004)
3. Li, J.: Traffic Engineering, pp. 9–18. China Communications Press, Beijing (2002)
4. Sha, B., Yuan, Z., Miu, J., Cao, S.: Traffic demand prediction research based on TransCAD. Shanxi Technology (1), 24–26 (2006)
5. Ma, J., Pei, Y.: The development and application of TransCAD in urban transportation planning. Journal of Harbin University of Civil Engineering and Architecture 35(5) (2002)
6. Zhuang, Y., Lv, S.: The research of urban road impedance model based on TransCAD. Communications Standardization (146) (2005)
7. Zhang, L.: The application in traffic impact analysis of Emme / 2 traffic planning software. Northern Communications (2) (2010)

A Design of Practical Electronic Voting Protocol

ShiJie Guan*

Shenyang Ligong University
39960476@qq.com

Abstract. voting behavior in modern democratic society is a frequently occurring behaviors, electronic voting is always used. However there are some defects in existed electronic voting protocols. This paper is aim to introduce a few basic properties that the secure electronic voting protocol should have, then design a based on public key (double key),blind signature and ensure the password technology to realize the electronic voting protocol, the protocol has high security ,it is very suitable for large-scale election system.

Keywords: cryptography, electronic voting, blind signature, bit assurance.

1 Introduction

Nowadays, voting behavior in modern democratic society is a frequently occurring behaviors, such as the levels of Party Congress, NPC, CPPCC elections and accreditation vote etc. all of these activities are closely related to voting behavior. During the election, the traditional approach is to use a locked box which has a hold on the topside thus one can be placed into the slit in every election votes, votes from the election committee is special movement. At the time of the election, the bidder will fill out the ballot sheets into the ballot box, so as ensuring the ballot anonymity. To vote and then open the ballot box open roll. In order to ensure the election to be just, fair and open, from production to the whole process of votes counted. on the other hand, this action would cost a lot of labor force and material resources, to take various measures to prevent interference election normal damage events and the incidence of fraud.

However, the traditional manual ballot problems are significantly appeared: firstly, the artificial polling need a long time to created, secondly, the serious waste of human resources. Voters went to the designated polling station to vote, which undoubtedly increased the voting cost, as the human resources spending. Therefore the above reasons, voting methods need to be improved.

2 The History and Present Situation of Electronic Voting

Electronic voting is more efficient, more convenient and more flexible than the traditional manual vote. In early 1884, the great inventor Tomas Edison invented a kind of electronic voting device, he wanted to the legislature dep artment of

---

* ShiJie Guan, male, born in 1977 in Shenyang, Polytecnic School of Shenyang Ligong University, PhD Candidate, research direction: computer software and theory.

Massachusetts city testing the electronic voting in the real bidding systems, but he failed. However, the scientists of electronic voting have never stopped researching. With the development of the computer, people gradually began to research the electronic voting by the high technology, as the experience increasing they have found some electronic voting system.

The first electronic voting scheme is provided by the Chaum[1] in 1981. He used the public key cryptosystems, and digital signature to hide the voter's identity, but not an unconditional guarantee the voter's identity was tracking. Then, Demillao[2] and others have proposed in 1982 a more detailed plan. In these plan, voters must be encrypted information passed, until they receive the final voting results, but, all voters must cooperate to complete the voting, if a voter improper operation, it will disrupt the entire voting activities.

Thereafter, the development of electronic voting in two directions, one is based on homomorphic encryption technology electronic voting scheme, the technique can obscure the ballot content .the other is based on the anonymous channel technology electronic voting scheme, the technique can mask the identity of the voter.

Benaloh [3], Iverson [4],Sako [5] and others were proposed based on homomorphic encryption technology electronic voting scheme, each of these options have advantage and disadvantage factors. For an example, Iverson imitate electronic monetary agreement is going to try to solve this problem, but its main drawback is, if all the institutions collusion, the voters of secrets. More importantly, these regimens using high-order residual encryption technology, the need for large-scale transmission and calculation, is not suitable for large-scale vote.

On the other hand, there are some anonymous electronic voting schemes based on channel [6]. Anonymous channel is not tracked by e-mail system and billboard can hide information source channel. Chaum [7] in the anonymous channel technology pioneered untraceable e-mail system, therefore this system could be at least one reliable cover, the reliable sources of information can obscure. Then Nurmi [8] proposed one kind based on the ANDOS protocol of electronic voting scheme, this scheme is: in order to get on the ballot, thus voters must communicate with each other, Nurmi and Saloman other programmes also have such defects. Of course there are some good programs, such as Iverson [9] presented with zero knowledge proof protocol as the core electronic voting scheme, Sako, Kilian [10] proposed a typical multi election center program, Niemi, Renvall [11] proposed a voting centre project.

But the scheme is one of the greatest maladies which practical not strong, do not adapt to the large election. The first practical and suitable for large-scale voting scheme which composed of Fujioka, Okamoto and Ohta [12] it's proposed in 1992 by FOO scheme, FOO scheme, received greater attention, it is considered to be a better implementation of secure voting electronic voting protocol. Many universities and research institutions are improved in this paper, the developed corresponding electronic voting system. Such as the famous Massachusetts Institute of Technology (MIT) EVOX system (Washington), Washington University in Sensus system. So far, many scholars on the FOO protocol presents a problem in a variety of different solution [13, 14, 15], Xie Jinbao, Liu Huibo [16] put forward " based on the group signature blind, and secret sharing scheme of new electronic security voter model". Liu Shengli, Yang Bo, Wang Yumin [17] put forward " based on the elliptic curve cryptosystem voting protocol".

Through some literature [18, 19, 20, 21], we see the electronic voting system has reached a certain security, in addition many countries have already started to implement electronic voting, including America, Britain, France and other countries, I believe in that the near future our country will be through the electronic ballot election.

3 Requirements of Secure Electronic Voting Protocol

Research of electronic voting protocol scholar Cranor [22] for different vote on electronic voting protocol developed to achieve different requirements. This article made reference to his request. We will realize the secure voting agreement shall meet the requirements as follows:

(1) Complete -- if the agreement the parties are honest, so the election results are credible. All valid votes have been the correct statistics.

(2) Accuracy -- dishonest protocol participants cannot disrupt the election. To prevent forgery, modify other votes on election results have implications.

(3) Secret -- all votes are kept secret, no one can be a vote with a voter corresponding up and certain voters cast ballots in the content.

(4) No reusability -- each voter can only vote once, but cannot vote for.

(5) Legitimacy -- only authorized people to vote.

(6) Fairness -- nothing can influence the outcome of the election. In the election process should not reveal the election of intermediate results, thus the public vote and voting trends affect mood, so as to affect the final results of elections.

(7) Verification -- the election results are all concerned about the election results were verified is correct.

(8) Flexibility -- on the turnout and place should not limit, each voter 's voting activity independently of each other, unaffected, not together at the same time to vote, to vote the content without restriction.

(9) Convenience -- on voters to vote for, the necessary knowledge and skills should not be too high requirement, should be simple, convenient.

(10) Efficiency -- should track the password technology trends, will better safe efficient cryptographic technique applied in the agreement, to ensure that the protocol can efficiently.

The article (1) ~ (7) to realize the secure electronic voting protocol to safety requirements, any electronic voting protocol must satisfy the 7 safety requirements. Article (8) ~ (9) so that the electronic voting protocol is more practical. Article (10) will increase the efficiency of electronic voting protocol, and can improve the practicability of the agreement.

4 Design of New Electronic Voting Protocol

4.1 Create a New Protocol Environment

Included in the protocol management mechanism, candidates, voters tally, four participants. The agreement of some entity and symbolic representation of information as shown in table 1.

Table 1. Protocol in some entity and related information symbols

	identification number	public key	secret information	signature information
voters V_i, (i=1, 2···, m)	$V_1 {}^\sim V_m$	$e_{v1} {}^\sim e_{vm}$	decryption bit that random number: k_i, Blind number: r_i private key factor; $d_{v1} {}^\sim d_{vm}$	$S_{v1} {}^\sim S_{vm}$
Candidates C_j, (j=1, 2···, n)	$C_1 {}^\sim C_n$	$e_{c1} {}^\sim e_{cn}$	private key : $d_{c1} {}^\sim d_{cn}$	$S_{c1} {}^\sim S_{cn}$
management mechanism	A	e_a	private key : d_a	S_a

4.2 New Protocol Information Flow

Protocol information flow is shown in figure 1.

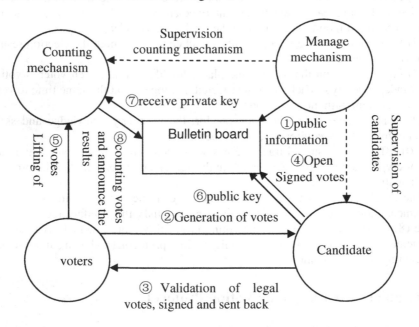

Fig. 1. Protocol information flow

Step 1: preparation stage

Management mechanism of A identifies:

(1) marked the vote a unique identifier bit string sign;

(2) candidates $C_1 \sim C_n$;

(3) counting mechanism T;

(4) the voting time arrangements and other public information;

(5) regulations began voting time t_0, end time t_1;

(6) of the vote in the legitimate voters $V_1 \sim V_m$, and voters need face-to-face registered;

The need to face to face, because through the network access to voter identity information can not ensure the true, the same voters may simultaneously at different status register.

The registration process is generally composed of the following steps:

(1) voters fill out voter information table (including the voting public key information);

(2) the governing body of the voter information into the database.

(3) the governing body of the candidate information into the database.

Then the above information is released to the public:

$\langle$sign, $(ID_1, C_1, e_{c1}) \sim (ID_n, C_n, e_{cn})$, T, $V_1 \sim V_m$, $(ID_1, e_{v1}) \sim (ID_m, e_{vm})\rangle$

Step 2: the registration phase

Voters in V_i, first determine their votes V_i, then calculate

(1) to V_i b_i=CON(sign, $H(v_i, k_i)$);

(2) the Bi X x_i =X(b_i, r_i);

(3) the x_i signature s_{vi} =$S_{vi}(x_i)$, and then the three tuple$\langle ID_i, x_i, s_{vi}\rangle$to every candidate C_j （j=1, 2$\cdots$n）, request signature.

Candidate C_j, received votes after voters sent out the following operations:

(1) to verify the signature for V_i, namely, verify that x_i, is equal to $e_{vi}(s_{vi})$, if not equal, refused to sign, if equal, then steps (2).

(2) verify whether voters V_i has applied for signature, has applied for a signature, then refuse to sign. If the voter V_i have not applied for signature, the signature, calculated s_{cjvi}=$S_{cj}(x_i)$. Candidate C_j $\langle ID_i, x_i, s_{cjvi}\rangle$ database and publish, proved V_i ask him to x_i signature, and the signature information is sent to the V_i s_{cjvi}

Step 3: voting phase

Voters received s_{cjvi} V_i signature information, proceed as follows:

(1) x_i=$e_{cj}(s_{cjvi})$ is established;

(2) according to each candidate blind signature calculation blind inverse transform R function y_{cjvi}=R(s_{cjvi});

(3) use of each candidate public computing: f_i=$e_{c1}(e_{c2}(e_{c3}\cdots(e_{cn}(v_i, k_i))\cdots))$;

(4) $\langle b_i, f_i, (C_1, y_{c1vi}), \cdots, (C_n, y_{cnvi}), S_{vi}(b_i, f_i, (C_1, y_{c1vi}), \cdots, (C_n, y_{cnvi}))\rangle$ send to T counting mechanism.

Step 4: the ticket stage

(1) counting mechanism T test

$\langle b_i, f_i, (C_1, y_{c1vi}), \cdots, (C_n, y_{cnvi}), S_{vi}(b_i, f_i, (C_1, y_{c1vi}), \cdots, (C_n, y_{cnvi}))\rangle$ is a first;

(2) counting mechanism T test $b_i=e_{cj}(y_{cjvi})$ $(j=1\cdots n)$ is established, if they are all equal to $\langle b_i, f_i, (C_1, \quad y_{c1vi}), \cdots, (C_n, y_{cnvi}), S_{vi}(b_i, f_i, (C_1, y_{c1vi}), \cdots, (C_n, y_{cnvi}))\rangle$ through billboard published.

Step 5: counting stage

After a time $t\,(t=t_1-t_0)$, the voting process is over, candidates announce key: $d_{c1}\tilde{\ }d_{cn},$

(1) counting mechanism T verification b_i_CON$(\texttt{sign}, \text{H}(d_{c1}(d_{c2}(d_{c3}\cdots (d_{cn}(f_i))\cdots))))$ is established, if established, then VI is a legitimate votes, otherwise illegal votes;

(2) Tally published in T $\langle b_i, v_i, k_i\rangle$, so that the public can be validated, statistical legal ballot number and published.

5 New Protocol and Its Performance Analysis

(1) The complete

If the agreement the parties are honest, the election result will be credible, apparently satisfied complete agreement.

(2) Accuracy

Voters, it is the only way to disrupt the election is constantly sending invalid votes, but we can find these acts of interference in the registration phase, because the voters have to before the vote to authority registration authority, then the administration will release all the voters $\langle ID_i, e_i\rangle$ information. Voters put $\langle ID_i, x_i, s_{vi}\rangle$ send candidates for signatures, candidates validated whether the ID_i is the first application of signature, if not, can only be considered voters vote for many voters, are malicious interference election. If the voter V_i is the first application of signature, then the candidate to verify voter V_i signature, which verify that x_i is equal to $e_{vi}(s_{vi})$, if the range that someone posing as voters voting V_i

(3) The secret of

Because before voting voters to register management mechanism, management mechanism to disclose all the voters' < IDi, EI > information, so the others can not be voter identification of IDi and Vi linking voters. Voters will vote to send candidates for signature, because of the use of blind signature technology, candidates can only see the voter identification IDi and encrypted ballot information, not to see the votes for the real content, and so there is no way to voter identification IDi and votes VI link, thus meet the secret.

(4) No reusability

If one person to vote two times, he must have a two effective ID, the normal procedure, each legal voter can only get a valid ID, so if anyone wants to vote two times, he must forge ID, while ID was voted by people and management mechanism for face to face, so the same individuals apply multiple ID is not possible, so the protocol can not meet the reusability.

(5) Legitimacy

If the illegal votes will be counted in the results, so the voters must be candidate signature, thus the illegal personnel think posing as legitimate workers vote, one must be able to decipher the digital signature in the candidates, we assume the existence of secure digital signature situations, this is not possible, and the protocol meets the legitimacy.

(6) Fairness

The agreement will be voting and counting the work into two stages. In the voting phase, voters sent not his fill the original true votes but after assurance processed vote b_i and use $e_{c1} \tilde{} e_c$ encrypted ballot counting mechanism f_i, receiving the votes, the bulletin board to their publication, so that the public inquiry, validation. In the counting phase, candidates announce private kcy $d_{c1} \tilde{} d_{cn}$, tally with the private key to unlock the f_i, and $\langle b_i, v_i, k_i \rangle$ together on the notice board released, so that the public inquiry, validation. It avoids the tally in the counting phase prior to the start of the intermediate results, revealing the election, affect the public's vote mood and trends, affect the final results of elections, thus the protocol satisfy the fairness.

(7) Verification

In the voting phase accordingly has published some necessary information, through these published information, the public can check the authenticity of a message. For validation of the required information is disclosed, including a variety of cryptographic algorithm, digital signature authentication key and the assurance of decrypted random number, so the public can be any of the voting process supervision, check whether the cheating behavior in examination results are correct, counting. Verify only need according to the corresponding information, instead of the voting process, such as how information is processed, does not need to know each vote's secret information.

6 Conclusion

This paper select a new method to improve the voting system security, designed a practical electronic voting protocol. The new agreement simplified the voter's work, solves the problem of inconvenient settlement agreement; when voters sent cannot open ballot assurance invalid key, cannot be distinguished from dishonest voter and dishonest counting people problems, namely solve security problems. In addition, due to the new deal put the candidate as a verification mechanism, thus, limiting the rights management mechanism, so that the possibility of cheating is smaller, increases the security of protocol.

References

1. Chaum, D.: Untraceable electronic mail, return address, and digital pseudonyms. Communications of the ACM (24), 84–88 (1981)
2. Demillo, R., Merritt, M.: Protocols for Data Security. Computer 02, 39 (1983)
3. Benaloh, J., Tuinstra, D.: Receipt-free Secret-ballot Elections. In: Proccedings of the Twentysixth Annual ACM Symposium on the Theory of Computing, pp. 544–553. ACM Press, New York (1994)
4. Iverson, K.R.: A cryptographic scheme for computerized general elections. LNC S, vol. 330, pp. 405–419 (1991)
5. Sako, K.: Electronic voting system with objection to the center. In: Proccedings of the Symposium on Cryptography and Information Security, Tateshina, pp. 92–130 (1992)

6. Chen, X., Wang, Y.: Based on anonymous communication channel secure electronic voting scheme. Chinese Journal of Electronics 31(3), 390–393 (2003)
7. Chaum, D.: Elections with Unconditionally-Secret Ballots and Disruption Equivalent to Breaking RSA. In: Günther, C.G. (ed.) EUROCRYPT 1988. LNCS, vol. 330, pp. 177–182. Springer, Heidelberg (1988)
8. Nurmi, H., Salomaa, A., Santean, L.: Secret Ballot Elections in Computer Networks. Computers & Security 36(10), 553 (1991)
9. Iversen, K.R.: A Cryptographic Scheme for Computerized General Elections. In: Feigenbaum, J. (ed.) CRYPTO 1991. LNCS, vol. 576, pp. 405–419. Springer, Heidelberg (1992)
10. Sako, K., Kilian, J.: Secure Voting Using Partially Compatible Homomorphisms. In: Desmedt, Y.G. (ed.) CRYPTO 1994. LNCS, vol. 839, pp. 411–424. Springer, Heidelberg (1994)
11. Niemi, V., Renvall, A.: How to Prevent Buying of Votes in Computer Elections. In: Safavi-Naini, R., Pieprzyk, J.P. (eds.) ASIACRYPT 1994. LNCS, vol. 917, pp. 141–148. Springer, Heidelberg (1995)
12. Fnjioka, A., Okaloma, T., Ohla, K.: A Pracaic:a Secret voting Scaieme for Large Scale Elections. Leture Notes in Computer Science 718, 244–251 (1992)
13. Thuringiensis, Lu, N., Zhu, Y.: No one needs a central mechanism of electronic voting protocol. Computer Engineering 30(11), 96–97 (2004)
14. Gao, H., Wang, Y., Wang, J.: A net electronic voting scheme based on Mix. Chinese Journal of Electronics 32(6), 1047–1049 (2004)
15. Chen, X., Wang, J., Wang, Y.: Based on receipt-free electronic voting. Chinese Journal of Computers 26(5), 557–562 (2003)
16. Xie, J., Liu, H.: Based blind group signature secret sharing, and the new safety electronic election model. Microcomputer and Application (9), 38–42 (2000)
17. Shengli, L., Bo, Y., Yumin, W.: Based on elliptic curve cryptosystem voting protocol. Journal of Electronics 22(1), 84–89 (2000)
18. Qi, D., Shuling, S.: The electronic voting research. Computer Application 18(4), 23–25 (1998)
19. Zhang, M.: Online voting technology security. The Computer Age (3), 38–39 (2004)
20. Cheng, C., Cheng, W., Kamfai: The electronic voting system security problem (English). Computer Engineering 25(special issue), 61–64 (1999)
21. Liu, J., Fu, X., Cheng, G.: Electronic voting security and application prospects. Computer Security 12, 24–26 (2004)
22. Cranor, L.F., Cytron, R.K.: Design and implementation of a security-conscious electronic polling system[R]. Washington University: Technical Report WUCS-96-02 (1996)

Design and Implementation of Simplified TCP/IP Stack Based on Embedded Network Interface

Lei Cheng and Jingchun Hu

School of Information Engineering, Nanchang Hangkong University,
Nanchang, China
450670770@qq.com, cyhj5102@yahoo.com.cn

Abstract. According to the characteristics of embedded system and it's network interface, we designed a simplified TCP/IP stack based on standard TCP/IP stack. This stack can works on its own or as a task of an embedded system. Based on 8-bit micro controller, we transplante the stack to embedded real-time operating system uCOS-II. Also, we did some tests and contrast the stack with existing protocol stack uIP. This stack has been proved practical.

Keywords: TCP/IP Stack, Embedded Network Interface, uCOS-II, uIP.

1 Introduction

Network is a universal feature of modern electronic devices, embedded systems are no exception. Make embedded devices access networks, not only widens the range of communication equipment, but also allows the operator more convenient for control the device. However, embedded systems have limited processing power, low storage resources, signal application place and other characteristics, standard TCP/IP protocol stack obviously can not be directly applied to 8-bit micro-control system. In this paper, we designed a tailored, streamlined TCP/IP protocol stack, mainly including ARP, ICMP, IP, UDP and other protocols.

2 TCP/IP Protocol Design

Some commonly used protocols in TCP/IP layered model as shown in Figure 1.

According to the characteristics of TCP/IP stack, from the following three aspects to write code: First, as each layer of the stack is composed of head and data part, and head is formed by a number of items, so we wrapped each head into the form of structure. Second, when a layer receives data, it is either passed up or processed in the layer, which need to determine the type of packet. For example, when the network interface receives a packet, if it is an IP packet, pass it up, if ARP packet, just call the corresponding code to analyse the data packet. Third, when the application has data to send, the data from the upper to the lower layer, the layers of packaging, and finally sent by the hardware interface. This requires the function of each layer encapsulation.

D. Jin and S. Lin (Eds.): Advances in ECWAC, Vol. 2, AISC 149, pp. 147–153.
springerlink.com   © Springer-Verlag Berlin Heidelberg 2012

What is left is the data unpack and the design of network chip driver. Because we do not design the TCP protocol, so the data unpack is relatively easy and can be completed in a single function. According to the type of network chip, there are three functions must be done to do actual hardware operation: chip initialization, send data, receive data.

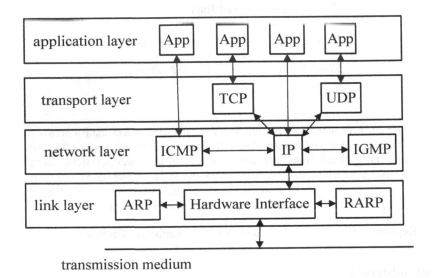

Fig. 1. TCP/IP layered model

2.1 ARP Protocol

In Ethernet, if two peers want to communicate with each other, in addition to know each other's IP address, but also need to know the MAC address[1]. ARP is used to obtain MAC address. In the general-purpose computer system, ARP cache is generally designed in the form of two-way data link, so that the entire cache can dynamically increase or decrease. But this non-linear list is time-consuming during the lookup of table entry matching, so this form does not apply to embedded systems. Therefore, we use a linear array, it is a continuous storage area, goes the fast speed of look-up. Due to the embedded network communication nodes are generally small, the ARP cache does not require large capacity, so they can be designed as fixed-size cache.

2.2 IP Protocol

IP protocol provides unreliable connectionless datagram delivery service. TCP, UDP, ICMP and other upper layer packets must be packed into IP datagram for transmitting. Standard IP protocol includes segmentation, reassembly and other features, it is complicated. In embedded control systems, data are mostly small and independent, so segmentation and reassembly is not required.

The two aspects to simplify the IP protocol: First, determine whether the received datagram is sent to me, if not, then discard the packet. Otherwise calculate IP checksum, remove IP head, submit data to upper layer. Second, only pack UDP, ICMP packet and pass the packet to the data link layer instantly.

2.3 ICMP Protocol

ICMP is used for control, testing and manage IP network. There are types of ICMP message, but most of them are not necessary in embedded systems. In order to enable users to understand whether the device is up to, as long as they can recognize the echo request from other customers and send echo reply to it, that is enough.

2.4 UDP Protocol

We uses UDP as the transport layer protocol. In theory, TCP's reliability is based on a number of complex measures and the resulting increase in overhead cost in return. TCP provides for connection-oriented, reliable service, while UDP is a connectionless-oriented. Because UDP has no reliable mechanism, it can give full play to the speed of the physical communications equipment, carried out at full speed data communications. And because UDP has no point to point access requirements, can have a "one to many", "many to many" of data transmission. UDP's overhead is very small, higher transfer rate than TCP, and real time. Therefore, it's wise for embedded TCP/IP protocol using UDP as transport layer protocol. Under some circumstances, embedded systems may also need high reliability data transmission. As the UDP protocol has no timeout retransmission, flow control, answer confirmation, emergency data acceleration mechanism, so in order to compensate for its shortcomings we can add some appropriate measures in application layer, such as packets with sequence identity, time to wait, retransmission mechanism, response confirmation mechanism and other complementary measures to ensure secure transmission of data correctly.

From the application point of view, this stack is mainly used in household appliances online. For the temperature, smoke and humidity sensors, once per second for centralized monitoring, the packet is small and its sending is frequent, it is enough for the device just broadcast its real-time status data to the network, so the choice of UDP is more appropriate.

3 Stack Processing

Stack processing flow shown in Figure 2.

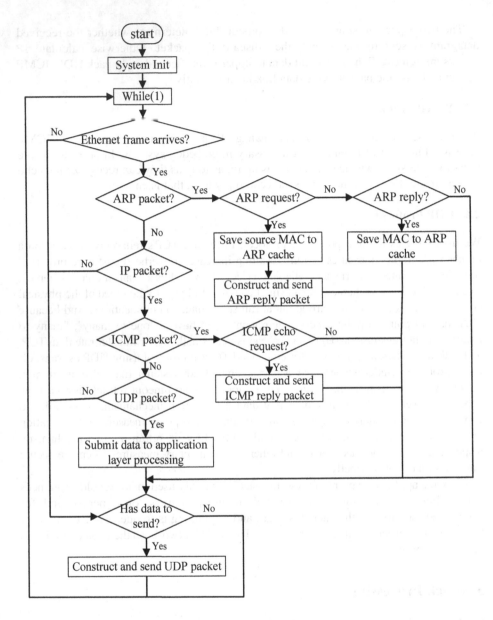

Fig. 2. Stack processing

First, when a frame arrives, the network interface control program read it into the buffer, and returns its length. Secondly, the main program determine the type of received packets (IP or ARP packet) and call appropriate unpacking code for further processing. If it is an ARP request packet, update ARP cache and construct ARP response packet, if it's an ARP replay packet, just update ARP cache. If IP packet, we should to further determine if it's a ICMP or UDP packet. If ICMP echo request packet, construct the echo response packet to it. If UDP packet, parse out the data and deliver it to application layer. If an unrecognized packets, discarded it.

4 Stack Tests and Applications

The stack test platform configured as follows: STC12C5A60S2 microcontroller, 62256 external RAM memory, RTL8019AS network chip, 12M crystal, a LAN computer. This protocol stack can be easily ported to various embedded operating systems, such as UCOSII, RT-Thread, as a task or thread to manage data transmission of network interface.

4.1 Data Transmission Tests[3]

First, test connectivity of the stack using the "ping" command. The results shown in Figure 3. It shows that this stack meets the time requirements of embedded systems and has more or less the same round-trip time with uip.

```
C:\Documents and Settings\Administrator>ping 192.168.1.4

Pinging 192.168.1.4 with 32 bytes of data:

Reply from 192.168.1.4: bytes=32 time=6ms TTL=128
Reply from 192.168.1.4: bytes=32 time=6ms TTL=128
Reply from 192.168.1.4: bytes=32 time=6ms TTL=128
Reply from 192.168.1.4: bytes=32 time=6ms TTL=128

Ping statistics for 192.168.1.4:
    Packets: Sent = 4, Received = 4, Lost = 0 (0% loss),
Approximate round trip times in milli-seconds:
    Minimum = 6ms, Maximum = 6ms, Average = 6ms
```

(a) The stack runs on its own. with the round-trip time 6ms, 0 packet lost.

```
C:\Documents and Settings\Administrator>ping 192.168.1.4

Pinging 192.168.1.4 with 32 bytes of data:

Reply from 192.168.1.4: bytes=32 time=7ms TTL=128
Reply from 192.168.1.4: bytes=32 time=7ms TTL=128
Reply from 192.168.1.4: bytes=32 time=7ms TTL=128
Reply from 192.168.1.4: bytes=32 time=7ms TTL=128

Ping statistics for 192.168.1.4:
    Packets: Sent = 4, Received = 4, Lost = 0 (0% loss),
Approximate round trip times in milli-seconds:
    Minimum = 7ms, Maximum = 7ms, Average = 7ms
```

(b) The stack runs as a task of UCOSII with the round-trip time 7ms, 0 packet lost[2].

Fig. 3. "ping" command test

```
C:\Documents and Settings\Administrator>ping 192.168.1.4

Pinging 192.168.1.4 with 32 bytes of data:

Reply from 192.168.1.4: bytes=32 time=4ms TTL=128
Reply from 192.168.1.4: bytes=32 time=4ms TTL=128
Reply from 192.168.1.4: bytes=32 time=4ms TTL=128
Reply from 192.168.1.4: bytes=32 time=4ms TTL=128

Ping statistics for 192.168.1.4:
     Packets: Sent = 4, Received = 4, Lost = 0 (0% loss),
Approximate round trip times in milli-seconds:
     Minimum = 4ms, Maximum = 4ms, Average = 4ms
```

(c) uIP1.0 runs on its own with the round-trip time 4ms, 0 packet lost.

Fig. 3. (*continued*)

Second, do a data transmission test, as shown in Figure 4. In the top of the figure is a network sniffer software MiniSniffer, bottom left is the serial debugging assistant, lower right is a network debugging assistant.

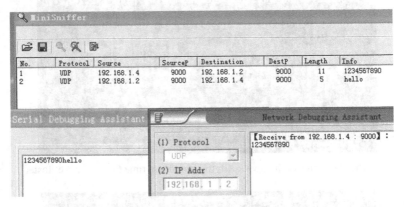

Fig. 4. Data transmission test. First, send data "0-9" to test platform through serial debugging assistant, test platform forward the data to the host software network debugging assistant. Second, network debugging assistant send "hello" to test platform, then test platform get the data and forward it to serial debugging assistant. Both sending and receiving data test are recorded by MiniSniffer.

4.2 Packet Loss Test[4]

Next, test the reliability of the stack. Test method is as follows: every 1 second, 0.1 second, 0.01 second send a 100 bytes UDP packet, send 1000 packets and do this ten times to calculate average packet loss. The relsults shown in Table 1.

From the table, when the data transmission rate is low, it is almost no packet loss. However, with the data transmission rate increases, packet loss is becoming worse. In embedded control systems, although the data sent frequently, but does not require high speed and large packets. In addition, if add application layer confirmation and retransmission mechanisms, even if there are packets loss, data can be safely delivered. This stack has been actually used in Ethernet to serial project.

Table 1. Packet loss test

Test environment	Data flow	1s/p	0.1s/p	0.01s/p
As a task of	send	0%	97.9%	96.6%
UCOSII	receive	0%	93.4%	91.1%
Without	send	0%	97.6%	96.2%
operate system	receive	0%	92.5%	90.8%

As the stack design with clear ideas, the codes are short (1.7K RAM, 7K ROM), so it is easy to modify and transplantation. This paper explored how to write a streamlined TCP/IP protocol stack in embedded control systems, not go into memory management and the affection of operate system., there needs to be improved and perfected.

References

1. ARP, http://baike.baidu.com/view/32698.htm
2. Labrosse, J.J., Beibei, S.: MicroC/OS-II The Real-Time Kernel. Beijing aerospace university press, Beijing (2003)
3. Liu, B., Feng, C.: The Hardware Realization of Embedded Web Server Using RTL8019AS. Journal of Yangtze University, 75–78 (2008)
4. Liao, L., Liu, X., Shi, Y.: A New Implement Method of Embedded Ethernet Communication Based on uc/os-II. Computer Measurement & Control, 1142–1147 (2010)

Research on RFID Application in Home Appliance PLM Information Management

Qinghua Zhang, Ni Tang, Zhuan Wang, and Guoquan Cheng

Department of Logistics Engineering, University of Science and Technology Beijing,
Beijing, China
{zqhustb,tangni1113}@163.com,
{wz64,cgq60}@vip.sina.com

Abstract. Different participants had different information requirements in home appliance PLM. End-users often use information related to the product such as maintenance service. But in many cases it is difficult to access complete relevant information. This situation was due to the separation of data and products or information recorded on the product nameplate was too little. This paper analyzed the participant's information requirements of PLM stages from production manufacture, inventory, sale and maintenance to recycle. RFID tag is used to collect and store product data. The related product information is written into RFID tag in the manufacturing and maintenance process. The information can be read at the distribution center, dealer, service center and the recycle station through RFID system.

Keywords: RFID, Home Appliance, PLM.

1 Introduction

There are two main problems in home appliance product information management. One problem is that the information obtained directly from product is limited. End-users often need some information related to the product, for example, maintenance service. But in many cases the product itself can not give enough relevant information. A great deal of information is lost as a product passes through its lifecycle from production manufacture, inventory, sale and maintenance to recycle. Generally, some information is lost when the product moves from the producer to the retail outlet, particularly details of product operation and function, but the greatest information loss occurs when the product moves from the retail outlet to the consumer. Over the time that a consumer owns a home appliance, the lost information includes the date and place of original purchase, the instruction manual, manufacturer contact information, and records on repairs. In addition, due to long-term use wear, it may also appear the lack of information about production date, producing area what were marked on product nameplate. Fig.1 is schematic of product information loss.

The other problem is that the information is not sharing enough in the entire product lifecycle of home appliance. And some of the information can not be got

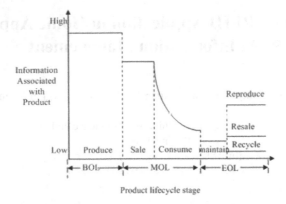

Fig. 1. Schematic of product information loss

accurately and in time. Though the bar code is widely used and can hold some data, its information capacity is limited and cannot be rewritten to add new message. In addition, bar code tag wears and tears easily after a long-term use, which leads to low efficiency of information collection, high error rate, even the data cannot be read. Home appliance PLM must use some advanced information technology to ensure information sharing timely and accurately between the various nodes of PLM.

Radio Frequency Identification (RFID) is a new wireless communication technology which has been applied to various areas such as distribution, transportation, tracing and tracking, etc. RFID is a non-contact technology that identifies objects attached with tags. It can also store some information in attached memory. RFID can be used to solve the problems mentioned above and strengthen the product lifecycle management. And there are some research works in this area [1], [2], [3], [4], [5] which used RFID to improve the effectiveness of data management.

2 Problems Analysis

Different participants of PLM have different information requirements in home appliance product lifecycle process. As to manufacturers, they mainly concerned about the inventory information, market share and sales, and usage feed back as well as products which parts prone to problems, maintenance rate, recycling information. As to the dealers, they need to know the potential customer distribution, product sales and inventory information, etc. Service centers would like to know the manufacturer of products, key parts, alternative parts and other information. As recycle agencies, they need to know the manufacturer, parts and manufacturing materials, and consumer purchase information. Consumers are more concerned about the upgrade of products and manufacturer information, where the nearest service center is and other information. Therefore, data collection and sharing are composed of basic product information and parts information that generated in the production processes, customer information and warranty card, and the maintenance and replacement parts information generated in service center, disassembly and recycle information, etc. Fig.2 is the information model diagram.

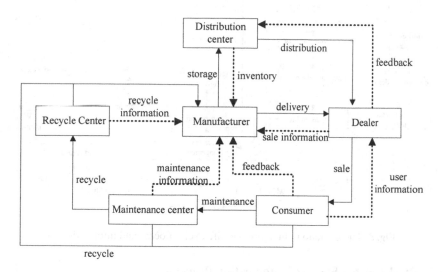

note： Solid line represents the actual logistics
Dash line represents information flow

Fig. 2. Information model diagram

3 Solution

3.1 Information Demand Analysis of PLM Stage

Product lifecycle consists of three main phases, beginning of life (BOL), middle-of-life (MOL) and end-of-life (EOL). BOL phase is a stage of generating product basic information, including parts attributes what should be used in the process of maintaining. The product information is written into RFID tag at this phase. The RFID tags will bind with the product to avoid product information and product separation in the BOL phase. In MOL phase, RFID is mainly used to provide information support reliability, availability and maintainability. The change of the product after maintenance is also written into RFID tag. In EOL, product recovery operations can be support by the information stored in RFID tag. The information stored in RFID tag is mainly used in MOL and EOL which incomplete product information often occurs in. Fig.3 shows overall product lifecycle process and information flows.

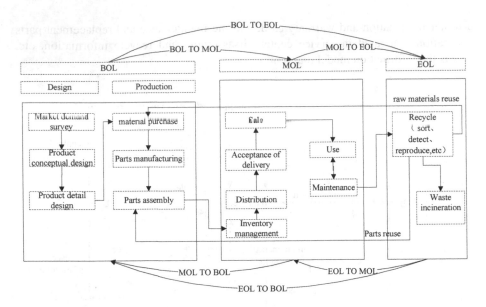

Fig. 3. The operation of the product life cycle process and information flow

3.2 PLM-Related Stages of Information Processing

The related stages of PLM include production processes, distribution center, sales, maintenance and recycle, and each of them is different in dealing with information using RFID.

Production Processes. Through the manufacturing process, the products getting off production line and qualified with RFID tags attached. In this process, the specific format codes will be written into RFID tags, the products and RFID tags will be then bond together. In this stage, the production information was written into RFID tag including product numbers, product name, color, model, production date, service life, manufacturer, key parts, and at the same time. This information was recorded into the database to ensure the product information tracking. This information will also be stored in database of information center which is convenient for the future management and authentication security. After information processing in the following several stages, manufacturing enterprises can real-time inquiry the sales information of products, customers' feelings when using the products, maintenance rate, recovery condition and other information, and finally realize the information sharing.

Distribution Center. Distribution center management which is based on the information stored in RFID can enhance the accuracy and quickness in operation, greatly improve the service quality and reduce operating costs, and save the labor of reading code. The typical application of RFID in distribution center is storage and outbound. The fixed RFID-Reader can be installed in the entrance of the warehouse, for example, when products with RFID tags enter the warehouse, the information in RFID tags can be obtained by reader, thus it can quickly handle inventory storage.

Sales. When selling the product, the product information stored in RFID tags is read, and the customer's purchase information is written into the RFID tags, such as buyer and warranty card number. Sales management based on RFID can supervise products more systematically and reliably, moreover, it can greatly reduce labor costs, error rate and improve efficiency. When dealers receive the products from enterprises, they judge the delivery is correct or not by reading the relevant product information stored in RFID tags, including manufacturers, production date, product information, the information of former intermediate processes and other information. Once miscarriage, dealers can return the products to manufacturing enterprises.

For the returned products, where and when the product is sold can be inquired by reading the relevant product information stored in RFID tags. In addition, it can quickly inquire the suppliers of these products, and inform them returned products name, model, quantity and other information. What's more, it can statistically analyze the sales information in a period and feedback consumers' complaints and advices to manufacturing enterprises.

Maintenance. The product information stored in RFID tags can be used to support maintenance, and the maintenance information can be written into the RFID tags while product is in maintenance. The maintenance history information stored in system is very important which can be used to analyze whether manufacturers has made mistakes in production, therefore, it's convenient for manufacturers to make progress in the next batch of products, and also make clear the responsibility to specific raw material suppliers [6]. It can statistically analyze the maintenance information in a period and feedback maintenance information of home appliance and statistical analysis results to enterprises, dealers and so on.

Recycling. With RFID support, product recovery can be taken into consideration while making the production and sales plans. The parts and materials will be disassembled by the information stored in RFID tags in recycling station, which is nearly impossible in traditional PLM information management of home appliance.

4 Conclusion

This paper analyzes the current issue of separation of information and products in home appliance PLM. The information missing and lack of sharing in home appliance PLM are the two main problems in information management. The RFID tag is used in this paper which can store the necessary information which will be used in PLM. In BOL phase, the product basic information is written into the RFID tag, and then will be used in MOL and EOL phase. And if the product has changed in maintenance, the replacement parts information can also be written to the RFID tag. Therefore, the information separation with product is resolved which increases the efficiency and reduces the cost of PLM.

References

1. Brewer, A., Sloan, N., Landers, T.L.: Intelligent tracking in manufacturing. J. Intell. Mfg. 10, 245–250 (1999)
2. McFarlane, D., Sarma, S., Chirn, J.L., Wong, C.Y., Ashton, K.: Auto ID systems and intelligent manufacturing control. Engng. Applic. Artific. Intell. 16, 365–376 (2003)
3. Djurdjanovic, D., Lee, J., Ni, J.. Watchdog agent: an infotronics-based prognostics approach for product performance degradation assessment and prediction. Advd. Engng. Inf. 17, 109–125 (2003)
4. Klausner, M., Grimm, W.M.: Sensor-based data recording of use conditions for product take back. In: Proceedings of the 1998 IEEE International Symposium on Electronics and the Environment, Oak Brook, IL, USA, pp. 138–143 (1998)
5. Saar, S., Thomas, V.: Toward trash that thinks: Product tags for environmental management. J. Ind. Ecol. 6, 133–146 (2003)
6. Jun, H.-B., Shin, J.-H., Kim, Y.-S., Kiritsis, D., Xirouchakis, P.: A framework for RFID applications in product lifecycle management. International Journal of Computer Integrated Manufacturing 22(7), 595–615 (2009)

Support Vector Machine Wavelet Blind Equalization Algorithm Based on Improved Genetic Algorithm

Yecai Guo, Baoge Li, and Kang Fan

College of Electronic & Information Engineering,
Nanjing University of Information Science and Technology, Nanjing, 210044

Abstract. In order to greatly overcome the disadvantages of low convergence rate, large mean square error and the local convergence of Wavelet Transform Constant Modulus blind equalization Algorithm(WTCMA), support vector machine wavelet transform constant modulus blind equalization algorithm based on improved genetic algorithm is proposed. This proposed algorithm can make full use of the global optimization ability of genetic algorithm to choose better parameters of Support Vector Machine(SVM) and uses SVM to initialize weight vector via a short initial data segment. When the weight vector of the equalizer is initialized by SVM, the proposed algorithm will carry out the WTCMA. The simulation result with underwater acoustic channel shows that the proposed algorithm outperforms the CMA and WTCMA in the convergent rate and mean square error.

Keywords: CMA, genetic algorithm, support vector machine, orthogonal wavelet transform.

1 Introduction

The underwater acoustic channel is a complicated time-varying multi-path channel and can cause serious inter-symbol interference(ISI). It is very necessary to employ blind equalization algorithms without training sequences for overcoming ISI. Constant modulus blind equalization algorithm(CMA) has good robustness and relatively simple structure, but it needs a large number of data samples and may converge to local minimum or diverge[1]. Orthogonal wavelet transform based CMA has faster convergence speed, but it also may be failed in local minimum[2][3]. Support Vector Machine(SVM) based CMA has good small sample learning ability and the feature of the global optimal solution, but the parameters in the SVM have great influence on the blind equalization performance. Reasonable parameters can make SVM have higher accuracy and better generalization ability [4][5].

In this paper, SVM and genetic algorithm are introduced into WTCMA, support vector machine wavelet transform constant modulus blind equalization algorithm based on improved genetic algorithm is proposed. In this proposed algorithm, after the parameters of SVM are optimized by genetic algorithm, the weight vector of equalizer are initialized by SVM via a short initial data segment. The simulation result with underwater acoustic channel shows the performance of the proposed algorithm.

D. Jin and S. Lin (Eds.): Advances in ECWAC, Vol. 2, AISC 149, pp. 161–166.

2 Orthogonal Wavelet Transform Based CMA

In orthogonal Wavelet Transform Constant Modulus blind equalization Algorithm (WTCMA), $a(n)$ is the original signal launched by signal source, $c(n)$ is channel impulse response vector, $w(n)$ is an additive Gaussian white noise vector in channel output, $y(n)$ is sent to the equalizer before orthogonal wavelet transform, $f(n)$ is the equalizer weight vector, $R(n)$ is the output signal after orthogonal wavelet transform, $z(n)$ is the output signal of equalizer. $e(n)$ denotes error signal. The relationship among variables are given as follows.

$$R(n) = Qy(n) . \tag{1}$$

$$z(n) = f^T(n)R(n) . \tag{2}$$

$$e(n) = z(n)[R - |z(n)|^2] . \tag{3}$$

$$f(n+1) = f(n) - \mu \hat{R}^{-1}(n)e(n)R^*(n) . \tag{4}$$

where, Q is orthogonal wavelet transform matrix, $R = E(|a(n)|^4)/E(|a(n)|^2)$, μ is step-size, $\hat{R}^{-1}(n) = \mathrm{diag}[\sigma_{j,0}^2(n), \sigma_{j,1}^2(n), ..., \sigma_{j,k_j}^2(n), \sigma_{J+1,0}^2(n), \sigma_{J+1,0}^2(n), \cdots, \sigma_{J+1,k_j}^2(n)]$, $\mathrm{diag}[\cdot]$ denotes diagonal matrix, $\sigma_{j,k_j}^2(n)$ is the average power estimation of $r_{j,k}(n)$ and $\sigma_{J+1,k_j}^2(n)$ is the average power estimation of $s_{J,k}(n)$. They may be estimated by following equations

$$\begin{cases} \hat{\sigma}_{j,k}^2(n+1) = \beta\hat{\sigma}_{j,k}^2(n) + (1-\beta)|r_{j,k}(n)|^2 \\ \hat{\sigma}_{J+1,k}^2(n+1) = \beta\hat{\sigma}_{J+1,k}^2(n) + (1-\beta)|s_{J,k}(n)|^2 \end{cases} . \tag{5}$$

where, β is the smoothing factor, and $0 < \beta < 1$. $r_{j,k}(n)$ is the coefficients of wavelet transform and $s_{J,k}(n)$ is the scale transform coefficient.

3 Support Vector Machine Based WTCMA

According to constant modulus signals, equalizer weight vector $f(n)$ can be initialized by SVM[6], the following cost function of Support Vector Machine Based WTCMA(SVMWTCMA) is minimized to obtain the updating equation of weight vector.

$$J(f) = \frac{1}{2}f^T(n) \cdot f(n) + C(Nv\varepsilon + \sum_{n=1}^{N}(\xi(n) + \tilde{\xi}(n))) . \tag{6}$$

The constraint conditions are written as

$$s.t. \begin{cases} (f^T y(n))^2 - 1 \le \varepsilon + \xi(n) \\ 1 - (f^T y(n))^2 \le \varepsilon + \breve{\xi}(n) \cdot \\ \xi(n), \breve{\xi}(n) \ge 0 \end{cases} \tag{7}$$

where, $C > 0$ is the penalty variable, parameter ε determines the insensitive zone width and the number of support vectors. $\xi(n), \breve{\xi}(n)$ are used to denote slack variable, C and ε are defined as[7]

$$\begin{cases} C = \bar{g}_n + 3\sigma_g \\ \varepsilon = 3\sqrt{\sigma_g^2 \dfrac{\ln N}{N}} \end{cases} \tag{8}$$

where, $\bar{g}_n$ is the expectation of $g(n)$ and $g(n) = \| x(n) \|^2$, σ_g is standard variance of $g(n)$. According to Iterative Reweighted Quadratic Programming(IRWQP)[8], equation (7) is turned into the following equation

$$L_p = \frac{1}{2} \| f(n) \|^2 + C(N v \varepsilon + \sum_{n=1}^{N} (\xi(n) + \breve{\xi}(n))) - \sum_{n=1}^{N} \alpha(n)[1 - z(n)(f^T(n)y(n)) + \varepsilon + \xi(n)]$$
$$- \sum_{n=1}^{N} (\mu(n)\xi(n) + \bar{\mu}(n)\breve{\xi}(n)) - \sum_{n=1}^{N} \tilde{\alpha}(n)[z(n)(f^T(n)y(n)) - 1 + \varepsilon + \breve{\xi}(n)] \cdot \tag{9}$$

where, $n = 1, 2, ..., N$, $\alpha(n) \ge 0, \tilde{\alpha}(n) \ge 0, \mu(n) \ge 0, \bar{\mu}(n) \ge 0$. The initial weight vector of equalizer can be solved by equation (10).

$$f(n) = \sum_{n=1}^{N} (\alpha(n) - \tilde{\alpha}(n))z(n)y(n) \cdot \tag{10}$$

If the following conditional equation is satisfied, then the updating of weight vector will stop, and this algorithm will switch to the wavelet constant modulus blind equalization algorithm

$$\begin{cases} AME(n) = \frac{1}{N} \sum_{n=1}^{N} (| z(n) |^2 - |\min(R1)|) \\ AME(n) - AME(n-1) \le \eta \end{cases} \tag{11}$$

Where, AME(n) is the mean modulation error, R1 is the distance of input point to each convergence point.

4 Improved Genetic Algorithm

Genetic algorithm including natural selection and evolution mechanism is a global probability search algorithm based on Darwin's theory of evolution. However, traditional genetic algorithm has low computational efficiency and may fail in local minimum. This paper the uses improved crossover operator and mutation operator to improve the computational efficiency and convergence speed. Improved genetic algorithm includes the following process: (1) Coding method. It is that putting a problem's feasible solution into the searching space of genetic algorithm. Generally, binary coding method can produce a mapping error, so floating point coding method is used in this paper. In this method, the value of each gene is expressed by a floating point in a certain range and individual coding length is equal to the number of decision variables.(2) Selection operator. It is that putting strong vitality individual into a new group and used to do the operation of survival of the fittest. According to the fitness value of each individual, we can know the higher fitness and the larger probability of inheriting from one generation to the next. This paper adopts the roulette wheel selection.(3) Crossover operator. It is to select two individuals from the group according to a greater probability, and exchanging their one or some place, and can produce new individuals, and determines the global search ability of genetic algorithm. But the difference of individuals is large in the initial stage and smaller in the later stage. If we choose a single crossover operator, it's easy to become premature convergence problem, so the adaptive crossover operator is adopted and written as

$$P_c(G+1) = P_c(G) - \frac{1-P_c(1)}{G_{\max}}.$$ (12)

where, G is the genetic generation, $G_{\max}$ denotes the largest genetic generation, $P_c(1)$ is the first crossover probability, and $P_c(G)$ is the Gth generation crossover probability. (4) Mutation operator. In mutation process, after one allele is replaced with another, a new genetic structure is produced. Mutation operation is an auxiliary method of producing new individuals and determines the local search ability of genetic algorithm. Traditional genetic algorithm will decline a class of genes in specific genes and will lead to gene defect. Gauss mutation operator is employed to overcome the defects of the traditional genetic algorithm. so we can use a random number to replace the existing genetic value, its mean value is P_m, and its variance value is P_m^2. According to the Gauss distribution, we know that some local area near the individual is the focus area of searching space. Accordingly, the local search ability of genetic algorithm can be improved.

5   Improved Genetic Algorithm Based SVMWTCMA

When the improved genetic algorithm is introduced into SVMWTCMA to optimize the parameters of SVM, SVMWTCMA will be easy to get the global optimal solution. The process of the optimization is shown in Fig.1.The steps of optimization are as follows:

Step 1: Initialization of the parameters of SVM and floating point coding operation. The number of group is 100, the parameters C and g need to be optimized and RBF(Radial Basis Function) is selected as kernel function.

Step 2: Calculation of fitness value. SVM is used to train the initial group and the individual's fitness is computed to improve the classification accuracy of SVM.

Step 3: Operation of GA. Selection operator chooses roulette wheel selection, crossover operator chooses adaptive crossover operator, mutation operator uses Gauss mutation operator. We can get new group through genetic manipulation.

Step 4: Determine whether the stopping criteria are satisfied. When the rate of correct classification achieving the requirement or having the same individual fitness for some continuous generations, we know stopping criteria are satisfied. Through the GA, we get the best parameters of SVM.

Step 5: SVM is used to initialize weight vector via using a short initial data segment, if equation (11) is satisfied, turn to the WTCMA.

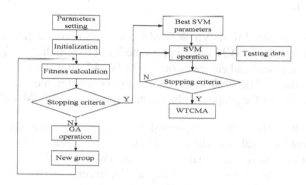

Fig. 1. Principle of IGASVMWTCMA

6 Algorithm Simulation

In order to test the performance of IGASVMWTCMA, computer simulation was carried out and compared with WTCMA and CMA. The underwater acoustic channel $c = [0.3132, -0.104, 0.8908, 0.3134]$, the transmitted signals were 16QAM,the signal-to-noise ratio was 20dB, the length of equalizer weight vector was 16. The first 100 points of input data were used to initialize weight vector via using SVM, the switching threshold ζ was set to 10^{-5}, the step-size λ was 0.9. The weight vector coefficients of CMA were initially set to 0 except that the 3rd tap was set to 1, the weight vector coefficients of WCMA were initially set to 0 except that the 3rd tap was set to 1, the weight vector coefficients of IGASVMWTCMA were initially set to 0 except that the 6th tap was set to 1. For CMA,WTCMA, and IGASVMWTCMA, the step-size was 0.000009,0.000018,and 0.000015,respectively. The simulation results were shown in Fig.2.

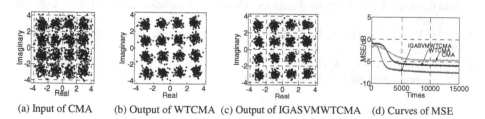

(a) Input of CMA   (b) Output of WTCMA (c) Output of IGASVMWTCMA   (d) Curves of MSE

Fig. 2. Simulation results

Fig.2(a)~(c) show the output constellations of IGASVMWTCMA is the clearest in all algorithm. Fig.2(d) shows that IGASVMWTCMA has an improvement of about 2000 steps for convergence rate and 8000 steps comparison with WCMA and CMA. IGASVMWTCMA has a drop of about 2dB and 3dB for mean square error(MSE) comparison with WCMA and CMA after convergence.

7 Conclusion

In this paper, the proposed IGASVMWTCMA uses the improved genetic algorithm to optimize the parameters of SVM and employs the SVM to initialize weight vector via a short initial data segment and can carry out the WTCMA after the weight vector of the equalizer is initialized. The simulation result with underwater acoustic channel shows that the performance of the proposed algorithm outperforms WTCMA and CMA.

Acknowledgment. This paper is supported by Specialized Fund for the Author of National Excellent Doctoral Dissertation of China (200753), Natural Science Foundation of Higher Education Institution of Jiangsu Province (08KJB510010) and "the peak of six major talent" cultivate projects of Jiangsu Province(2008026), Natural Science Foundation of Jiangsu Province(BK2009410), Natural Science Foundation of Higher Education Institution of Anhui Province (KJ2010A096), Jiangsu Preponderant Discipline "Sensing Net- works and Modern Meteorological Equipment Acknowledgment.

References

1. Guo, Y.: Communication signal analysis and processing. Press of Hefei University of Technology (2009) (in Chinese)
2. Cooklev, T.: An Efficient Architecture for Orthogonal Wavelet Transforms. IEEE Signal Processing Letters 13(2), 77–79 (2006)
3. Han, Y.: Wavelet Transform Based Blind Equalization Design and Algorithm Simulation. Anhui University of Science and Technology (2007) (in Chinese)
4. Deng, N., Tian, Y.: Support Vector Machine: Theory, Algorithm and Development. Science Press, Beijing (2009)
5. Chen, L., Chen, J.: An Improved P-SVM Method Used to Deal with Imbalanced Data Sets. In: IEEE International Conference on Intelligent Computing and Intelligent Systems, pp. 118–122. IEEE Press, New York (2009) (in Chinese)
6. Li, J., Zhao, J., Lu, J.: Simulation of Constant Modulus Blind Equalization Algorithm Initialized by Support Vector Machines. Computer Simulation 25(1), 84–87 (2008) (in Chinese)
7. Li, M., Kou, J., Lin, D.: The Basic Theory of Genetic Algorithm and Its Application. Science Press, Beijing (2002)
8. Kim, H.S., Cho, S.B.: Application of Interactive Genetic Algorithm to Fashion Design. Engineering Application of Artificial Intelligence 13(6), 635–644 (2000)

The H_∞ Control for Time-Discrete Stochastic Systems Driven by Martingales

Rui Zhang

College of Information and Science, Shandong University of Science and Technology,
Qingdao 266510, P.R. China
Z.R.2008@163.com

Abstract. This paper discusses the state feedback H_∞ control problem for the time-discrete stochastic systems driven by martingales. By completing square method, we obtain the H_∞ control for such systems.

Keywords: H_∞ control, time-discrete stochastic system, nonlinear stochastic system, Hamilton-Jacobi equation.

1 Introduction

In practice, the main purpose of $H\infty$ control design is to find the law to efficiently eliminate the effect of the exogenous disturbance. Theoretically, study of $H\infty$ control starts from the deterministic systems [6]. Recently, the robust problem, especially $H\infty$ controller designs, on stochastic Ito-type systems have received a great deal of attention and many researches about $H\infty$ problem have been given for such systems. Hinrichsen and Pritchard [1] developed an $H\infty$-type theory for a sort of linear Ito stochastic systems and proved the bounded real lemma for such systems. Applying a general Hamilton-Jacobi equation(HJE), Zhang and Chen [2] give an $H\infty$ control design approach for a kind of nonlinear Ito stochastic systems. Since then, many works and results are given, for further reference we prefer to [3], [4], [5], [7] and their references.

Generally speaking, for stochastic systems, it's difficult to describe the trajectory of the control and the state. A useful method is to study the corresponding discretization of the systems and simulate the system by computer means. In this paper, we study the $H\infty$ control problem for the discretization of the stochastic systems discussed by Zhang and Chen in [2]. The systems discussed in this paper is driven by time-discrete martingales. Since the $H\infty$ control for discrete systems is not easy to obtain, we introduce the quasi-$H\infty$ control instead. By the HJE, we obtain the quasi-$H\infty$ controller for such systems. This paper is organized as following: In section 2, we give the basic theories about the discrete systems driven by martingales and prove the discrete inequalities for such systems. In section 3, we discuss the $H\infty$ control designs for such systems and give the $H\infty$ control by solving a Hamilton-Jacobi equation.

D. Jin and S. Lin (Eds.): Advances in ECWAC, Vol. 2, AISC 149, pp. 167–172.
springerlink.com   © Springer-Verlag Berlin Heidelberg 2012

2 Preliminaries

Let $(\Omega; F; P)$ be a complete probability space equipped with be a complete probability space equipped with filtration $F = (F_i^n)$,and $M^n = (M_i^n)_{i=0,1,\cdots n}$ (with $M_0^n = 0$,and n can take ∞) be a sequence of square integral martingale corresponding to F and $\langle M^n \rangle = (\langle M^n \rangle_i)_{i=0,\cdots n}$ (with $\langle M^n \rangle_0 = 0$) is the compensator of M^n is describe as

$$x_{i+1}^n = x_i^n + f_i^n(x_i^n)\delta_i + \sigma_i^n(x_i^n)\Delta M_{i+1}^n, x_0 \in R^{nx} \qquad (2.1)$$

Where $\delta_i = \langle M^n \rangle_{i+1} - \langle M^n \rangle_i$, and f_i^n, σ_i^n satisfy Lipshitzian conditions with coefficient C and $E\sum_{i=0}^n [\|f_i^n(0)\|^2 + \|\sigma_i^n(0)\|^2] < \infty$ uniformly for every n . It's easy to see, for given $x_0 \in R^{nx}$ equation (2.1) exists unique solution in $l^2(n, R^{n_x})$. For convenience, denote t_i^n as $\langle M^n \rangle_i$. Suppose $V(t,x) \in C^{1,2}([0,\infty), R^{n_x})$, then by Taylor's expansion, we can have the Itò type formula for (2.1):

$$V(t_{i+1}^n, x_{i+1}^n) = V(t_i^n, x_i^n) + \frac{\partial V}{\partial t}(t_i^n, x_i^n)\delta_i + \frac{\partial V}{\partial x}'(t_i^n, x_i^n)f_i^n\delta_i$$
$$+ \frac{1}{2}\sigma_i^n{}' \frac{\partial^2 V}{\partial x^2}(t_i^n, x_i^n)\sigma_i^n\delta_i + \frac{\partial V}{\partial x}'(t_i^n, x_i^n)\sigma_i^n\Delta M_{i+1}^n + \alpha_i^n. \qquad (2.2)$$

Where $E|\alpha_i^n|$ is the higher infinitesimal. The following lemma can be gained by the usual method and the proof is omitted.

Lemma 2.1. Suppose $x^n = (x_i^n)_{i=1,2,\cdots,n}$ is the solution for discrete SDE (2.1) with initial $x_0^n = x_0 \in R^{nx}$ Then we have

$$E \sup_{0 \le i \le n} \|x_i^n\|^2 \le c_{x_0M}\left[\|x_0\|^2 + (1+\delta)E\sum_{i=0}^{n-1}\|f_i^n(0)\|^2\delta_i + E\sum_{i=0}^{n-1}\|\sigma_i^n(0)\|^2\delta_i\right], \quad (2.3)$$

$$E\sum_{i=0}^{n-1}\|\sigma_i^n(x_i^n)\|^2\delta_i \le c'\left[E\|x_n^n\|^2 + E\sum_{i=0}^{n-1}\|f_i^n(0)\|^2\delta_i\right], \qquad (2.4)$$

where $\delta = \max_{0 \le \delta_i \le n-1} \delta_i$, $c_{x0.M}$ and c' are positive constants.

3 H_∞ Control for Discrete Stochastic Systems

In this section, we let $\{M_t\}_{t \geq 0}$ be square integrable martingale and Mk is the time-discrete martingale at time $t_k = kh$, i.e. $M_k = M_{kh}$,where h is the mesh of partition of interval $[0, +\infty)$. Now we consider the time-discrete nonlinear stochastic system

$$
\begin{cases}
x_{i+1} = x_i + (f_i^n(x_i^n) + g_i^n(x_i^n)u_{i+1} + k_i^n(x_i^n)d_{i+1})\delta_i \\
\qquad\qquad\qquad + (h_i^n(x_i^n) + l_i^n(x_i^n)d_{i+1})\Delta M_{i+1}^n, \\
z_i = \begin{pmatrix} m_i^n(x_i^n) \\ u_i \end{pmatrix}, m_i^n(0) = f_i^n(0) = h_i^n(0) = 0, i = 0, 1, \cdots, n-1;
\end{cases}
\tag{3.1}
$$

Where $d = (d_i)_{k=0,1,\cdots n}$ (with $d_0 = 0$) stands for the exogenous disturbance, which is predictable with respect to $F_{ti}, \delta_i = \langle M \rangle_{ti} - \langle M \rangle_{ti-1}$. For given admissible control $u = (u_i)_{k=0,1,\cdots,n}$, Let $x(u, d, x_0) = (x_i(u, d, x_0))_{i=0,1,\cdots,n}$ be the solution of system(3.1) with initial state $x_0 \in \Re^{nx}$,then output $z = (z_i)_{k=0,1,\cdots,n}$ can be written as $z_i = z(x_i(u, d, x_0))$. Given $\gamma > 0$,we want to find an admissible control $u^{*,n}$ such that for any $d \in l_n^2(\Re^{nd})$, when $x_0 = 0$,the following inequality holds:

$$
E\sum_{i=0}^{n} \|z_i\|^2 \delta_i \leq \gamma^2 E \sum_{i=0}^{n} \|d_i\|^2 \delta_i, d \neq 0,
\tag{3.2}
$$

then we call $u^{*,n}$ the H_∞ control of (3.1).Moreover, if we define the perturbation operator $\tilde{L}_{zd} : l_n^2(\Re^{nd}) \to l_n^2(\Re^{nz})$ as

$$
\tilde{L}_{zd}(d) = z(x(t, u^{*,n}, d, 0, 0)), t \geq 0, d \in l_n^2(\Re^{nd}),
$$

And the norm of $\tilde{L}_{zd}$ as

$$
\left\| \tilde{L}_{zd} \right\|_\infty = \sup_{x(0)=0, d\neq 0, d\in l^2(R^{nd})} \frac{\|z\|_{l_n^2(R^{nz})}}{\|d\|_{l_n^2(R^{nd})}},
$$

Then the H_∞ control of (3.1) satisfies $\left\| \tilde{L}_{zd} \right\|_\infty \leq \gamma$,However, for thim-discrete nonlinear system (3.1), the following example shows that the control u satisfying inequality (3.2) is not very easy to find even in determent system.

Example 3.1. Assume the time-continuous system is given by

$$dx = (-x + u + v)dt,$$

$$z = \begin{pmatrix} x \\ u \end{pmatrix}$$

Then, for $\gamma = 1, u^* = -x$ is the H_∞ control for the system satisfying

$$\int_0^\infty \|z\|^2 \, dt \le \int_0^\infty \|d\|^2 \, dt.$$

However, the corresponding discretization of (3.3) with mesh h is

$$x_{i+1} = x_i + (-x_i + u_{i+1} + d_{i+1})h,$$

and the control $u = -x$ only have

$$E\sum_{i=0}^{n} \|z_i\|^2 h \le \gamma^2 E\sum_{i=0}^{n} \|d_i\|^2 h + \frac{1}{2}\sum_{i=0}^{n}(d_{i+1} - 2x_i)^2 h^2,$$

and the second term of the right side is higher order infinitesimal of h, which can be omitted when h is small enough.

In order to solve such dilemma, we give a weaker definition about H_∞ control and call it quasi-H_∞ control.

Definition 3.2. Given $\gamma > 0$, if an admissible control $u^{*,n} = (u_i, i = 1, 2, \cdots, N)$, $u_0 = 0$ for system (3.1) such that for any $d \in l_n^2(R^{n_d})$, when initial $x(t_0) = 0$, the following in-equality holds:

$$E\sum_{i=0}^{n} \|z_i\|^2 \delta_i \le \gamma^2 E\sum_{i=0}^{n} \|d_i\|^2 \delta_i + \alpha, d \neq 0,$$

Where α is infinitesimal when $h \to 0$, then we call $u^{*,n}$ the quasi-H_∞ control of system (3.1).

Similar to the results of the time-continuous case, the following result is also true.

Theorem 3.3. Suppose there exists a nonnegative solution $V \in C^2(\mathfrak{R}^{nx})$ to the HJE

$$
\begin{cases}
H_\infty^1(V(x)) := \dfrac{\partial V}{\partial x}' f + \dfrac{1}{2}m'm + \dfrac{1}{2}(\dfrac{\partial V}{\partial x}'k + h'\dfrac{\partial^2 V}{\partial x^2}l)(\gamma^2 I - l'\dfrac{\partial^2 V}{\partial x^2}l)^{-1}(k'\dfrac{\partial V}{\partial x} + l'\dfrac{\partial^2 V}{\partial x^2}h) \\
\qquad\qquad\qquad \dfrac{1}{2}h'\dfrac{\partial^2 V}{\partial x^2}h - \dfrac{1}{2}\dfrac{\partial V}{\partial x}' gg'\dfrac{\partial V}{\partial x} = 0, \\
\gamma^2 I - l'\dfrac{\partial^2 V}{\partial x^2}l > 0, V(0) = 0;
\end{cases}
\tag{3.3}
$$

Then

$$u^*_{i+1} = -g''_i\left(x^n_i\right)' \frac{\partial V}{\partial x}(t_i, x_i), i = 0, 1, \cdots, N-1$$

is an quasi-H_∞ control for system (3.1).

Proof. By Taylor formula, we have

$$V(x_{i+1}) - V(x_i) = [\frac{\partial V}{\partial x}'(t_i, x_i)(f_i + g_i u_{i+1} + k_i d_{i+1})$$

$$+ \frac{1}{2}(h_i + l_i d_{i+1})' \frac{\partial^2 V}{\partial x^2}(t_i, x_i)(h_i + l_i d_{i+1})]\delta$$

$$+ \frac{\partial V}{\partial x}'(t_i, x_i)(h_i + l_i d_{i+1})\Delta M_{i+1} + \alpha^n_i.$$

Complete the square together, we have, for any $T > 0$,

$$EV(x_n) - V(x_0) = \left[\frac{1}{2}E\sum_{i=0}^{n-1}\left\|u_{i+1} + g'_i \frac{\partial V}{\partial x}\right\|^2 + 2H^1_\infty(V) - \|z_i\|^2 + \gamma^2\|d_{i+1}\|^2\right.$$

$$\left. - \left\|d_{i+1} - (\gamma^2 I - l'_i \frac{\partial^2 V}{\partial x^2} l_i)^{-1}(l'_i \frac{\partial^2 V}{\partial x^2}h_i + k'_i \frac{\partial V}{\partial x})\right\|^2_{\gamma^2 I - l'_i \frac{\partial^2 V}{\partial x^2} l_i}\right]\delta + \alpha^n,$$

where $\alpha^n = \sum_{i=0}^{n}\alpha^n_i$. Let $u_{i+1} = u^*_{i+1}, x_0 = 0$, we can prove the results.

Remark 3.4. We can also see that theorem 3.3 still holds if HJE (3.3) is replaced by the Hamilton-Jacobi inequality. Here, the infinitesimal presents the difference between H_∞ control and quasi-H_∞ control, which is coming from the discretization of the system and corresponding Ito type formula.

Example 3.5. Let M be the standard Brownian motion taking values at the point $k\delta$, $k = 0, 1, 2, \cdots$ and $\delta > 0$ be the step of $[0, \infty)$. Then $\langle M, M\rangle_t = t$, $\delta_i = \delta$. The system (3.1) is given by

$$\begin{cases} x_{i+1} = x_i + (-2x_i - u_{i+1} - d_{i+1})\delta + 2x_i \Delta M_{i+1}, \\ z_i = \begin{pmatrix} 2x_i \\ u_i \end{pmatrix}, i = 0, 1, \cdots, n-1, x_0 \in \Re; \end{cases}$$

and $\gamma = 1$. To solve the corresponding HJE (3.3), we obtain the solution $V(x) = x^2$ and the H_∞ control u = 2x. Moreover, we also have

$$E\sum_{i=0}^{n}\|z_i\|^2 \delta_i \leq \gamma^2 E\sum_{i=0}^{n}\|d_i\|^2 \delta_i + \alpha,$$

where $\alpha = \sum_{i=0}^{n}(4x_i + d_i)^2 \delta^2 \to 0$ when $\delta \to 0+$.

4 Concluding Remarks

We have discussed the state feedback H_∞ control problem for the time-discrete stochastic systems driven by martingales. In some sense, we give an approach to simulate the trajectories of state and H_∞ control for the stochastic system discussed in [2]. Based on discretization, we use a quasi- H_∞ control to replace and obtain the solution for H_∞ control problems. However, since such systems are discretized from the time-continuous Ito systems, there are errors for such systems comparing with usual Ito stochastic systems, and how the errors affect the H_∞ ontrol still remains for further study.

References

1. Hinrichsen, D., Pritchard, A.J.: Stochastic H_∞. SIAM J. Control Optim. 36, 1504–1538 (1998)
2. Zhang, W., Chen, B.: State feedback H_∞ control for a class of nonlinear stochastic systems. SIAM J. Control Optim. 44, 1973–1991 (2006)
3. Zhang, W., Chen, B.: Stochastic H_2 / H_∞ control with (x; u; v)-dependent noise: Finite horizon case. Automatica 42, 1891–1898 (2006)
4. Boukas, E.K., Liu, Z.K.: Robust H_∞ control of discrete time Markovian jump linear systems with mode-dependent time-delays. IEEE Trans. Automat. Control 46, 1918–1924 (2001)
5. Chen, B.S., Zhang, W.: Stochastic H_2 / H_∞ control with state-dependent noise. IEEE Trans. Automat. Control 49, 45–57 (2004)
6. Zames, G.: Feedback and optimal sensitivity: Mode reference transformation, multiplicative semi-norms and approximative inverses. IEEE Trans. Automat. Control 26, 301–320 (1981)
7. Lin, X., Zhang, R.: H_∞ Control for Stochastic Systems with Poisson jumps. J. Syst. Sci. Complex 24, 683–700 (2011)
8. Lin, X., Zhang, R., Wang, X.: A Sort of Discrete Backward Stochastic Differential Equations and Its Convergence. In: 2009 International Colloquium on Computing, Communication, Control, and Management, pp. 431–434. IEEE Press, New York (2009)
9. Lin, X., Zhang, R., Wang, X.: Stability of Solutions of Discrete Reflected BSDEs. In: 2008 IEEE International Conference on Granular Computing, pp. 428–431. IEEE Press, New York (2008)
10. Lin, X., Sun, Q., Wang, X.: The robustness of fully coupled forward-backward stochastic differential equations. J. Phys.: Conf. Ser. 96, 1–5 (2008)

Outside Hospital Rehabilitation System for COPD Patients Based on GPRS Network

LiFeng Wei, ShaoChun Chen, and Shanshan Tang

College of Information Engineering, Shenyang University of Chemical Technology,
Shenyang, Liaoning, China
weilifeng62@sina.com, huxi2014@126.com, tss7114331@126.com

Abstract. The long time oxygen therapy (LTOT) and home mechanical ventilation(HMV) are lack of effective monitor and unified management in the treatment of COPD which can not give patients real-time monitoring and treatment. To overcome these shortcomings, GPRS network is applied to the rehabilitation therapy system which includes oxygen therapy, mechanical ventilation and the GPRS mobile network.

Keywords: COPD, GPRS, CC430.

1 Introduction

Chronic obstructive pulmonary disease is a chronic progressive respiratory disease which ranks the first of respiratory disease. Long time oxygen therapy (LTOT) and home mechanical ventilation(HMV) treatment is an important way to increase survival rate of COPD patients which the most can influence the measure of COPD prognosis [1, 2]. But long time oxygen therapy and home mechanical ventilation treatment are lack of effective monitor and management, COPD patients are poor compliance and the treatment is not standardized. Especially in acute exacerbations , the doctor can not give patients the real-time monitor and treatment. So improvement the compliance and normative of oxygen therapy and mechanical ventilation, real-time monitoring for COPD patients is an urgent problem to solve.

With the development of the network, especially the wireless mobile network technology continues to mature; it provides an effective technical means for COPD patients with low load of community telemedicine monitor. The COPD diagnosis and treatment experience combining with the self-developed medical oxygen generator and noninvasive positive pressure ventilator are used in this paper. And through the seamless integration of the wireless network and Internet, wireless remote rehabilitation system of oxygen therapy and noninvasive positive pressure ventilation to the COPD patient group are realized in community.

2 The Structure of the System

Physical rehabilitation system for COPD patients is based on wireless GPRS network which proposes a new scalable multi-level network system structure and a implementing

D. Jin and S. Lin (Eds.): Advances in ECWAC, Vol. 2, AISC 149, pp. 173–177.
springerlink.com   © Springer-Verlag Berlin Heidelberg 2012

method. It consists of treatment devices, terminal modules of the wireless mobile network and the remote data center. Wireless mobile network terminal module is equipped to terminal treatment equipments: oxygen machine and ventilator. Using the GPRS module, Wireless GPRS network module sends the data to the remote data server center after receiving the data of terminal equipments. The data is processed through the remote data server center to extract the pathological characteristics, and then transmitted to a remote medical treatment center, hospitals or communities through the Internet network. At last medical professionals statistics and observe the data to provide the necessary services. Specific content is shown in figure 1.

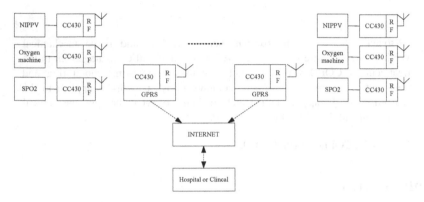

Fig. 1. The structure of the system

In the design, the wireless mobile network node includes the noninvasive ventilator node, medical oxygen machine node and some body sensors of physiological parameters, for example: oxygen saturation. Medical equipment status or treatment process can be monitored by mobile nodes. Network node increases the number of nodes according to the different needs which has great flexibility and scalability. At the same time the system accesses to the Internet network which can form a large community medical and hospital network.

3   The System Design

The system comprises wireless sensor network, wireless receiver and PC software.

3.1   The Design of Wireless Sensor Network

The design of wireless sensor network includes the physiological signal sensor or terminal equipments of the built-in wireless transmission module and wireless receivers. Wireless transmission module are respectively equipped in the body or installed in the oxygen machine and noninvasive positive pressure ventilation, as shown in Figure1.Each wireless terminal module is a wireless sensor node. The signal wireless transmission is completed by wireless transmitting module. Central control module collects the data from various wireless modules and then the data is packaged to transmit to the Internet through the wireless communication (GPRS).

3.1.1 Core Components of the Wireless Sensor Network

The processor is the heart of the system. The system is based on CC430 single chip microcomputer with powerful processing function, high processing speed and low power consumption. In the clock operating conditions of 1MHz, the current will be 200-400uA.When the clock is shutdown mode, the current is only 0.1uA [3].

CC430 chip has built-in wireless module CC1101. It is designed for low power wireless applications. This design is set in the 433MHz frequency band. RF transceiver is integrated with a highly configurable modem. This modem supports different modulation formats, the data transfer rate can up to 500kbps, and software may modify the baud rate. Most transmission distance reaches 300-500 meters with wireless waked-up function, sensitivity to 110dBm, high reliability. It can be widely used in various short distance wireless communication field.

3.1.2 Body Sensor Network Protocol

TI's SimpliciTI protocol is adopted which is a simple RF protocol for the low frequency network with low power consumption. It supports low power communication, store-and-forward mechanism. The amount of code is small, low cost, maximum use 8K Byte flash and 1K byte RAM [4].

The SimpliciTI protocol packages the network into several types of API functions. The point to point communication can be realized through directly calling the API function. Simplici TI can run at low cost hardware resources. In addition to flash required by the program and RAM used by random variable, Simplici TI does not require any other resource; it doesn't even need a timer.

3.1.3 The Terminal Node

The terminal device comprises the oxygen moudle, noninvasive ventilator and oxygen machine which are shown in figure 2.

A high sensitive finger clip sensor is used in oxygen saturation module. Its internal photodiode converts the light signal to the electrical signal. The signal is a mixed and weak volume with red light signal and infrared light signal which must be amplified。 And then successfully get red light and infrared light through separated circuit. The two signals are sent to low-pass filter circuit and high-pass filter circuits to get DC components and AC components, components are sent into AD conversion, and then calculate the oxygen saturation value.

The RMS medical BPAP20 of noninvasive ventilator is used; CC430 and BPAP20 are connected through RS232. Pressure, flow rate, real-time tidal volume, respiratory rate, minute ventilation and machine setting parameters are sent by built-in RF transceiver of CC430 in the ventilator work Similarly, The RMS medical AO10 of oxygen generator is used, CC430 sends the real-time flow rate or pulse volume, oxygen concentration and machine parameters out.

Fig. 2. The hardware of the system

3.2 The Wireless Central Receiver

The wireless receiver microprocessor also adopted the CC430 to receive the data from the terminal device. Each device data is repackaged after data correlation with array pointers. And then the data is sent out through GPRS module. The hub is achieved by using SIM900 Module. Schematic diagram of CC430 and SIM900 are shown in Figure 3.

The SIM900 wireless modules are the smaller double tri-band modules on the market today [5]. This module covers all of the GSM/GPRS networks that exist across the world - and subsequently enable you to exchange any data in specific format. GPRS enabled mobile SIM card to be placed in the adapter provided in Modem. The embedded client implements AT Commands [6] to communicate to the mobile devices. The fixed IP address is configured in the modem and after establishing communication with the internet continuous data will be transferred to that particular IP address by this modem.

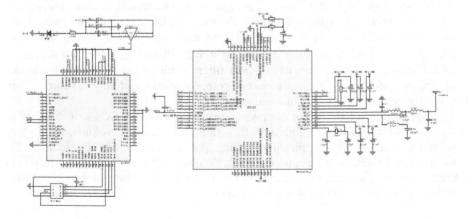

Fig. 3. The schematic diagram of CC430 and SIM900

3.3 PC Software

Currently software environment of web application development is lot, we use VC +
+, it includes all-around support of the web application development, for example: the
Csocket class for the WinSock. The system adopts C / S model. Due to the short
connection to connect, disconnect, will take up a lot of system resources. The bottom
of the system communication module uses the long connection. Member functions of
bind( ), listen( ), send( ) and recv( ) are called through the USM SSocket object to
realize the operation of binding, interception, sending data and receiving data.

4 Summary

The physical rehabilitation system based on wireless sensor network technology is
presented for the COPD patients which realize standardized treatment of the long time
oxygen therapy and noninvasive positive pressure ventilation, early warning during
acute exacerbation, real time monitoring management for patient group by individual
practitioner .It improves the quality of life of patients and scientific and effective use
of existing resources of medical purpose.

Acknowledgment. This work is supported by Shenyang Science and Technology
Program (No.F11-009-2-00).

References

1. Wedzicha, J.A., Bestall, J.C., Garrod, R., et al.: Randomized controlled trial of pulmonary
   rehabilitation in severe chronic obstructive pulmonary disease patients, stratified with MRC
   dyspnoea scale. Eur. Respir. J. 12, 363–369 (1998)
2. Wang, B., He, P.: Telemedicine—Technology, application, evaluation and Prospect. Space
   Medicine and Medical Engineering 12(4), 311 (1999)
3. TI company. CC430f5135 User Manual (2010)
4. TI company. SimpliciTI Sample Application User's Guide (2009)
5. Krishnamachari, B.: The Grid: Networking Wireless Sensors. Cambridge University Press,
   New York (2005)
6. Application Note for SIM100 TCP/IP AT Commands

Moving Track Control System of Agriculture Oriented Mobile Robot

Qunfeng Zhu, Dengxiang Yang, and Guangyao Zhu

Department of Electrical Engineering, Shaoyang University, Shao Yang, China
Xichang Electric Power Bureau, Xi Chang Si Chuan, China
Department of Electrical Engineering, College of Hunan Mechanical & Electrical Polytechnic,
Chang Sha Hu Nan, China
smallbeeseagull@163.com, tomgreen6@163.com,
zhuguangyao@gmail.com

Abstract. Aiming at moving track control of mobile robot, this paper studied the practical implementation of linear control algorithm in moving track control of robot, expounded and proved its feasibility and analyzed stability constraint condition of the algorithm, and then analyzed effects of control parameters on moving track through simulation. Finally, the paper applies the algorithm on actual control of robot, and designed gyroscope and photo-electricity encoder-based control system of mobile robot for green house application.

Keywords: robot, modeling, track control, gyroscope, photo-electricity encoder.

1 Introduction

All along the moving track control is key research of mobile robot, which is key technique for its high intellectualization and complete autonomy. Moving in 3-D environment, robot must collect data through sensors, analyze and fuse data with a algorithm to build up mathematic model of outer environment, which can reflect the characteristics of outer environment comprehensively and correctly, so that correct basis could be provided for track control strategy. The movement control of mobile robot mainly depends on perception of sensor, fusion of information, tracking of intelligent control algorithm and path planning.

2 Movement Control Basis of Mobile Robot

Under the reference coordinates system $\{XR, YR, \theta\}$ where robot stays, supposing actual gesture error vector $e = {}^R[x_2, y_2, \theta_2]^T$, where x_2, y_2, θ_2 is the parameters of robot target gesture. And the task to design track controller is to find a control matrix

$$K = \begin{bmatrix} k_{11}, k_{12}, k_{13} \\ k_{21}, k_{22}, k_{23} \end{bmatrix} \qquad (1)$$

D. Jin and S. Lin (Eds.): Advances in ECWAC, Vol. 2, AISC 149, pp. 179–184.
springerlink.com   © Springer-Verlag Berlin Heidelberg 2012

which satisfies equation (2)

$$\begin{bmatrix} v(t) \\ \omega(t) \end{bmatrix} = Ke = K \begin{bmatrix} x_2 \\ y_2 \\ \theta_2 \end{bmatrix}^R \tag{2}$$

$$\text{where } \lim_{x \to \infty} e(t) = 0 \tag{3}$$

Generally, we suppose target position is the origin of inertia frame, as shown in figure 1

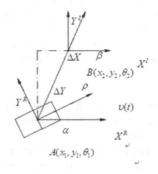

Fig. 1. Moving model of robot

In reference coordinate system {XI, YI, θ }, the robot motion is described as follow

$$\begin{bmatrix} \dot{x} \\ \dot{y} \\ \dot{\theta} \end{bmatrix} = \begin{bmatrix} \cos\theta & 0 \\ \sin\theta & 0 \\ 0 & 1 \end{bmatrix} \begin{bmatrix} v \\ \omega \end{bmatrix} \tag{4}$$

where $\dot{x}$ and $\dot{y}$ is the linear speed under reference coordinate system.

Let α denote angle between reference vector XR and vector $\hat{x}$ which connects robot moving center and final position, then if $\alpha \in I_1$,

$$I_1 = \left(-\pi/2, \pi/2\right] \tag{5}$$

Converting coordinates into polar coordinates where target position is set as origin, then we get

$$\rho = \sqrt{\Delta x^2 + \Delta y^2} \ ; \ \beta = -\theta - \alpha \tag{6}$$

Hence we get description for system in the new polar coordinates as follows below.

$$\begin{bmatrix} \dot{\rho} \\ \dot{\alpha} \\ \dot{\beta} \end{bmatrix} = \begin{bmatrix} -\cos a & 0 \\ \dfrac{\sin\alpha}{\rho} & -1 \\ -\dfrac{\sin\alpha}{\rho} & 0 \end{bmatrix} \begin{bmatrix} v \\ \omega \end{bmatrix} \tag{7}$$

where ρ denotes distance between robot center and target position, θ the angle between reference frame X_R and XI which is related to final position, and v and ω respectively tangential and angle speed.

3 Control System Design of Mobile Robot

In this system gyroscope and photo-electricity encoder are used for robot's track navigation. When the mobile robot moves forward or rotates, slipping friction between robot and ground is inevitable which makes system output control output can not be reflected into control system correctly. The angular velocity is measured by gyroscope, while linear velocity photo-electricity encoder. And for navigation system using gyroscope, the real-time accumulated robot position should be accurate enough. For this reason, two driven pulleys, whose motion amount can reflect the real course of robot, are installed right and left respectively at chassis, and accordingly two photo-electricity encoders are added as well. Hence the equivalent linear and angular velocity of robot can be calculated.from the instantaneous velocity of the two driven pulleys. To ensure close contact between driven pulleys and ground, coils are planned to be installed on the driven pulleys. The movement model of robot is shown in figure 1.

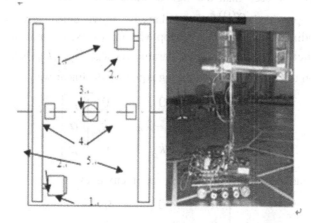

1. Encoder for driving motor 2. Driving motor 3. Gyroscope 4. Driven pulleys 5. caterpillar

Fig. 2. Schematic diagram of driving chassis and appearance of robot

The robot system hardware is composed of microprocessor LPC2138 of ARM7, angle measuring gyroscope, photo-electricity encoder, DC motor and motor driving circuit. Gyroscope uses uniaxial type of GQ-3, and consists of angle-measuring modules which is based on packaging of micro-mechanic gyroscope chips. The real-time angular data is output through its serial port with Baud rate of 9600bps. As for data format, 8 bits for data, 1 bit for stop, no bit for parity check.

4 Control Algorithm of Robot Control System

The linear velocity v and angular velocity ω of robot are control signal for track navigation, and for control algorithm control laws for v and ω need to be designed so that the robot can be drived to target location from current position.

In the track controlling system of mobile robot, linear control algorithm is used as follows

$$v = k_\rho \rho \; ; \quad \omega = k_\alpha \alpha + k_\beta \beta \tag{8}$$

From equation (7) we can get

$$\begin{bmatrix} \dot{\rho} \\ \dot{\alpha} \\ \dot{\beta} \end{bmatrix} = \begin{bmatrix} -k_\rho \rho \cos\alpha \\ k_\rho \sin\alpha - k_\alpha \alpha - k_\beta \beta \\ -k_\rho \sin\alpha \end{bmatrix} \tag{9}$$

There is no singularity in control system when $\rho = 0$, and a unique balanced point when $(\rho, \alpha, \beta) = (0,0,0)$, therefore it can drive robot to target location. And from equation (8) it can be seen that, the sign of angular velocity v is constant, and it remains positive provided $\alpha(0) \in I_1$; Otherwise move conversely, namely robot always keeps unidirectional from start to stop without reverse movement.

In actual application, combination of parameters k_ρ, k_α, k_β should meet local stability condition of control system. Making equation (9) linear we get

$$\begin{bmatrix} \dot{\rho} \\ \dot{\alpha} \\ \dot{\beta} \end{bmatrix} = \begin{bmatrix} -k_\rho & 0 & 0 \\ 0 & -(k_\alpha - k_\beta) & -k_\beta \\ 0 & -k_\rho & 0 \end{bmatrix} \begin{bmatrix} \rho \\ \alpha \\ \beta \end{bmatrix} \tag{10}$$

To make it local-indexed stable, then all matrix eigenvalues

$$A = \begin{bmatrix} -k_\rho & 0 & 0 \\ 0 & -(k_\alpha - k_\beta) & -k_\beta \\ 0 & -k_\rho & 0 \end{bmatrix} \tag{11}$$

must have negative real part. Namely all roots of characteristic polynomial of matrix A

$$(\lambda + k_\rho)(\lambda^2 + \lambda(k_\alpha + k_\rho) - k_\rho k_\beta) \tag{12}$$

have negative real part. Hence parameters k_ρ, k_α, k_β should meet

$$k_\rho > 0; k_\beta < 0; k_\alpha - k_\rho > 0 \tag{13}$$

5 Algorithm Simulation

To further analyze effect of k_ρ, k_α, k_β on moving track, simulation model is built up under Simulink environment, where the robot is driven to target position $(40, 30, 0^0)$ from starting point $(0, 0, 90^0)$ by control signal $\omega(t)$ and $v(t)$.

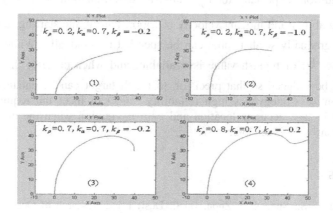

Fig. 3. Sketch map of simulation curve

Simulation results is shown in figure 3 from which it can be seen the system shows great difference under different parameter values, while all curves remain smooth. And only when the presented control algorithm-based control system meet the stability condition described in equation (13), namely $k_\rho > 0; k_\beta < 0; k_\alpha > k_\rho$, could the robot reach target position steadily. From comparison of the above-mentioned 4 simulation curves conclusion can be reached as follows:

(1) For sub-graph (4), the stability condition isn't met, and the curve can't stay at target point steadily. While for sub-graph (3), $k_\alpha = k_\rho$, right a critical condition, and the curve can stop at target. So, when the algorithm is actually implemented, the stability condition must be met.

(2) Comparing sub-graph (1) and (3) we can know the smaller the parameter k_ρ, namely the smaller the linear velocity of robot, the better the curve form, and the shorter the moving path. Practically, however, too short path brings about significant increasement of moving time. Hence in practical applications, on premise that hardwares permits, k_ρ can be increased to some extent during the middle course of robot movement so that moving time could be shortened, while at the final stage the speed must be reduced so that stability could be enhanced.

(3) Through comparison of sub-graph (1) and (2) it can be found that system performance with k_β of biggish absolute value is also be reduced. As change of k_β does nothing to linear velocity of robot, thus k_β value can be smallish.

6 Conclusion

In practical control of mobile robot for greenhouse application, owing to limit of software and hardware, selection of k_ρ, k_α, k_β can be different from ideal circumstance, and we must determine proper parameter values through trial according to actual condition. Especially for k_μ value, its determination is directly related to motor driving ability and moving velocity of robot. During the starting stage k_ρ can be increased gradually so that robot can be speeded up gradually. While during the middle course, k_ρ of biggish value is available, and when close to the target, k_ρ value should be reduced so that precision to reach target can be enhanced. In one word, the control strategy has high reliability, and can meet the requirements of mobile robot in fast, accurate location and path navigation when the mechanism is flexible and hardware is reliable.

References

1. Xu, H., Fu, L.-S., Sun, X.-J., Zhou, D.-S.: Design and Research of Open Construction Controller of Agricultural Robot Based on ARM. Journal of Agricultural Mechanization Research 5, 124–126 (2007)
2. Liu, Y., Dai, X., Liu, S.: Optimized Fuzzy Control System for Motion Trajectory of Robot Manipulator. Computer Engineering and Applications, 43–47 (2007)
3. Liang, K., Wang, B., Cai, D.-G.: A Study on and Implementation for A Practical Positioning System of Mobile Robot. Electro-Mechanical Engineering 4, 61–64 (2006)
4. Zhang, Y., Luo, Y., Zheng, T.: Mobile robot technology and it's applications. Publishing House of Electronics Industry, Beijing (2007)
5. Chen, Z., Xu, W., Liu, W.: Intelligent robot cutting system based on motion control card. Control & Automation, 4–2 (2008)

Project Supported the Fund

Hu Nan province Department of Education project (09C889).

A New Optimized Generator Maintenance Model and Solution

Xiangping Meng[1,3], Ye Yue[2,3], Zhongyu Shen[2,3], and Minmin Li[2,3]

[1] Changchun Institute of Technology, Changchun, Jilin Province, China, 130012
[2] Northeast Dianli University, School of Information Engineering, Jilin,
Jilin Province, China, 132012
[3] Jilin Province University Distribution Automation Engineering Research Center,
Jilin Province, China, 130012
yueye0321@sina.com.cn, shenzyu@foxmail.com,
liminmin2011@gmail.com

Abstract. Based on the background of traditional generator maintenance scheduling optimization model, the thesis has established one kind of new generator maintenance scheduling optimization model. As the fact that ant colony algorithm has the characteristic of slow convergence and easily falling into local optimum, by using fuzzy control rules, the thesis has improved the ant colony algorithm. This improved algorithm is applied to the proposed generator maintenance scheduling optimization model. As a result, simulation algorithms and models can achieve good results.

Keywords: Generator maintenance, Ant colony algorithm, Fuzzy control rules.

1 Introduction

With the quick development of economic, electricity demand increasing rapidly in the past few years, in order to get reliable and affordable electricity supply, reasonable electricity maintenance plan arrangement becomes very important. The purpose of Electrical repair electrical equipment is to extend the life cycle, the next failure or delay the average time, in addition, an effective maintenance can reduce the number of maintenance tasks of the interrupt, and the caused adverse effects [1]. In recent years, many studies have focused on the optimized methods of power plant maintenance scheduling, both the traditional mathematical programming methods, including dynamic programming, integer programming and mixed - integer programming, and the evolutionary algorithms (EAs), including genetic algorithms (GAs), simulated annealing (SA), the search algorithm (TS) [2] and ant colony algorithm (ACO) [3], have been widely used in maintenance scheduling optimization problems. Compared with other optimization algorithms, when solving a series of project portfolio issues, ant colony algorithm will show a higher computing speed and quality, such as the traveling salesman problem (TSP), DeJong test functions, etc. [4].

D. Jin and S. Lin (Eds.): Advances in ECWAC, Vol. 2, AISC 149, pp. 185–190.
springerlink.com © Springer-Verlag Berlin Heidelberg 2012

2 Maintenance Scheduling Optimization Model

2.1 Mathematical Model

The Objective Function
It is said that power plant maintenance scheduling optimization unit is belonged to optimize the safety operation of power systems research, however, it is a multi-objective and multi-constraint combinatorial optimization problems. In this thesis, using minimized operating costs of all equipment as the objective function, expressed as:

$$\min f(x) = \sum_{t=1}^{T}\sum_{i=1}^{N}(C_{it}+PP_{it}) + m\sum_{i=1}^{N}|R_{i0}-R_i| \tag{1}$$

Where: C_{it} is the maintenance costs of device i at time of t ; P is the average price for the period t; P_{it} is the output power of device i at time t; m is the excess cost of changing in each period, compared with the reported initial period; R_{i0} to declare the beginning period of the reporting; R_i is the initial period after optimizing; N as the number of devices; T as the total number of periods. Here, if you take a week as a period, then the one-year maintenance cycle will consist of 52 periods, you also can take one day, a month or a ten-day as one period, it is based on the actual situation. In order to calculate in a easy way, a week for one period is adopted in this thesis.

Formulation
For the constraints specified in PPMSO problems, it includes some common constraints: maintenance windows, available resources, load, continuity, complete availability, priority and reliability, as discussed below:

$$T_n = \{t \in T_{plan} : e_n \le t \le l_n - dur_n + 1\}, d_n \in D , \tag{2}$$

e_n implies the allowed earliest start time of maintenance for the device, l_n is the allowed latest ending time for the device, dur_n stands for the repair time for the device, d_n is the NO. n devices waiting for maintenance.

$$\sum_{d_n \in D_t}\sum_{k \in S_{n,t}} R_{n,k}^r \le R\,Avai_t^r, t \in T_{plan}\ \ r \in R \tag{3}$$

Where, $R_{n,k}^r$ is the required resource of d_n at time k. $R\,Avai_t^r$ is all the relevant resources that can be provided at time k.

$$T_2 = \{t \in T_{plan} : l_2 - dur_2 + 1 > t > s_1 + dur_1 - 1\}\ \ t \in T_{plan} \tag{4}$$

Where, S_n is the chosen starting time of d_n .

$$\sum_{d_n \in D} P_{n,t} - \sum_{d_n \in D_t} \sum_{k \in S_{n,t}} X_{n,k} P_n \geq L_t + f.L_t \quad , t \in T_{plan} \tag{5}$$

Where, L_t is the expected load at time t; P_n is the power generation of d_n ; f is the factor of load reserve demand [5].

2.2 Optimization Model

The optimization model establishing as follow : N units, corresponding allowed sequence of scheduled maintenance period, each node corresponding to a date during the scheduled period, the node contains three attributes, namely the day power generation, maintenance costs, maintenance cycle, ants will export series sequence of nodes based on the determined objective function, when the optimization process is complete. In this way, it can achieve, a group of time series economical, reliable unit maintenance date. Figure 1.

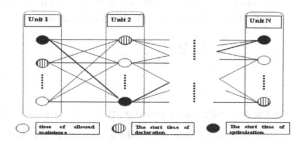

Fig. 1. Optimized model chart

Intuitively, PPMSO is that when knowing the initial reporting date, the output power and maintenance costs for each unit at different times, the additional cost caused by changing the initial period, find a set of optimal initial date of maintenance $F = \{f_1, f_2, ,,, f_n\}$, it belongs to the combinatorial optimization problems with constraints. Optimization process applied in this thesis as follows: Start finding optimization, placing an ant in each unit on the initial reporting date based on the starting date of reporting, ant moving randomly among the N units, when passing through all units that waiting for maintenance, there will form one path, then the equivalent cost to of this current path will be outputted. For the same units, we assume that the reference date of declaration nodes, each node changing (a date change), will cost m units, this cost should be included in the objective function, as the impact factor. In this way, after M times of iterations, and finally outputting the whole optimal path, that is, the best N units overhaul start date, according to each unit's maintenance cycle, the corresponding optimal maintenance time can be derived.

3 Fuzzy Control Theory

Fuzzy interface changes "precise state of the process" generated in the process of control to "blur", generally, it can be represented by fuzzy sets. After conversion, the generated "fuzzy quantity" will pass through the core of fuzzy logic control steps "fuzzy reasoning", "fuzzy reasoning" to simulate human reasoning, thinking, mainly based on the relationship in fuzzy logic and some fuzzy rules. "Ambiguity" is the process corresponding to the "ambiguity", it reduces the blur generated by fuzzy reasoning to the human understanding control actions. "Expert knowledge" includes "controlled process" knowledge and experience in the related field and the objectives that achieved in the control process, also made of the data base and fuzzy control rules.

4 ACO

4.1 Basic Concept

Ant colony algorithm is inspired by the behavior of real ants, particularly on how they find food. For the AS, the probability for an ant at time t to choose city j as the next target instead of current city i is given by the following formula:

$$P_{ij}^{k}(t) = \begin{cases} \dfrac{\left[\tau_{ij}(t)\right]^{\alpha}\left[\eta_{ij}\right]^{\beta}}{\sum\limits_{l \in N_{i}^{k}}\left[\tau_{il}(t)\right]^{\alpha}\left[\eta_{il}\right]^{\beta}}, & j \in N_{i}^{k} \\ 0, else \end{cases} \tag{6}$$

Where $\eta_{ij} = 1/d_{ij}$ is the heuristic value moving from city i to city j move, N_{i}^{k} is the collection of the cities that ants have not visited, parameter α and β controls the pheromone level and the relative importance of inspired value, $\tau_{ij}(t)$ is the pheromone between node i and node j. Once ant completes one iteration, the pheromone will update as below:

$$\tau_{ij}(t+1) = (1-\rho).\tau_{ij}(t) + \Delta\tau_{ij}(t) \tag{7}$$

$$\Delta\tau_{ij}(t) = \begin{cases} g(t)/C, walked \\ 0, else \end{cases} \tag{8}$$

Where $\rho \in (0,1)$ is a parameter, here $1 - \rho$ is the evaporation coefficient, g(t) is the pheromone intensity, C is the length of the path that ants traversed, it can be replaced by f(t) of formula (1).

4.2 Improved Ant Colony Algorithm

Basic ACO algorithms prone to have the problem of stagnation and falling into a local optimum, to solve this problem, we make some improvements of formula (7) to create a new pheromone update rule with one additional factor:

$$\tau_{ij}(t+1) = \begin{cases} \tau_{min}, \tau_{ij}(t+1) < \tau_{min} \\ \tau_{ij}(t+1), \tau_{min} < \tau_{ij}(t+1) < \tau_{max} \\ \tau_{max}, \tau_{ij}(t+1) > \tau_{max} \end{cases} \tag{9}$$

In this way, we can control the amount of pheromone between nodes in the range of $[\tau_{min}, \tau_{max}]$, and thus solve the problem of stagnation and easily falling into the local optimum.

Based on the randomness and fuzziness of the two parameters, fuzzy reasoning system to determine the fuzzy control rules is used in this thesis, this rules also helps controlling g(t) and ρ values , the rules is as follow:

IF there is no new results of continuously evolved S-generation, **THEN** selecting the smaller g(t) and larger ρ .

IF there is new results, **THEN** select the larger g(t) and smaller ρ .

5 Simulations

Here, the following declaration of 10 generating units maintenance data will be optimized by using the proposed units maintenance plan optimization model and the improved ant colony algorithm by fuzzy control rules:

Unit number	Power generation /MW	Scheduled maintenance period	The initial reporting period	The maintenance cycle	Maintenance costs
1	250	1,8	2	6	100
2	300	3,12	6	6	150
3	150	2,6	3	2	100
4	300	5,18	8	5	150
5	300	11,19	12	5	200
6	200	13,18	15	5	200
7	200	8,14	10	3	150
8	150	15,21	17	3	150
9	200	12,20	14	5	150
10	200	10,16	12	5	100

For this problem, different results generated by the basic ant colony optimization algorithm and the proposed model will be compared, model parameters as follows: $\alpha = 4$ $\beta = 1$, according to the specificity of unit maintenance problem, we assume $\eta_{ij} = 1$, it means that the assumed "distance" in the optimization searching process will not play a role. The results are compared:

Traditional ACO		The improved ACO	
Unit	Began to maintenance time And cyde	Unit	Began to maintenance time And cyde
2	5 (6)	2	4 (6)
5	13 (5)	6	14 (5)
8	16 (3)	7	11 (3)
		10	13 (5)
Cost/yuan	75624	Cost/yuan	74853
Time/s	256	Time/s	261

Fig. 2. The comparsion chart

6 Conclusion

The proposed new generating units maintenance scheduling optimization model and improved ant colony algorithm by fuzzy control rules, reducing the unit maintenance costs, at the same time improved the ant colony algorithm overcomes drawbacks that traditional ant colony algorithm easily falling into local optimum and slow convergence. It also enhances the algorithm's overall search capability, and improves the convergence speed. Finally, through simulation analysis to verify the effectiveness and practicality of maintenance model and the improved algorithm.

Acknowledgments. This work is supported by the National Natural Science Foundation of China under Grant No. 60974055 and Office of Science and Technology of Jilin Province (No.20100307).

References

1. Endrenyi, J., et al.: The Present Status of Maintenance Strategies and the Impact of Maintenance on Reliability. IEEE Transactions on Power Systems 16(4), 636–646 (2001)
2. Satoh, T., Nara, K.: Maintenance Scheduling By Using Simulated Annealing Method. IEEE Transactions on Power Systems 6(2), 850–857 (1991)
3. El-Amin, I., Duffuaa, S., Abbas, M.: A Tabu Search Algorithm for Maintenance Scheduling of Generating Units. Electric Power Systems Research 54, 91–99 (2000)
4. Wodrich, M., Bilchev, G.: Cooperative Distributed Search: The Ant's Way. Journal of Control and Cybernetics 26(3) (1996)
5. Ant colony optimization for power plant maintenance scheduling optimization—a five-station hydropower system, pp. 434–436 (2007)

Research on Evaluation Model of Harmonious Enterprise Based on AHP

Jinting Yang[1], Junqing Huo[2], and Tao Wang[2]

[1] Handan College
hdxyyjt@163.com
[2] School of Economics and Management Hebei University of Engineering,
Handan, China
huojunqing530@163.com, wt2009a@163.com

Abstract. Building a socialist harmonious society is the theme of Chinese social development, it is an important and complex system which needs all aspects of society involved in. As a cell of social development, Enterprise plays a vital role in the harmonious society. Only when enterprise becomes harmonious, economic can be harmonious, social can be harmonious. In this paper, a harmonious enterprise as a starting point, Harmony will be applied to the operation of an enterprise development. Win-win business outside of the macro-operation mechanism, Win-win business outside of the micro-operation mechanism, main win-win situation Internal operating mechanism of Internal operating mechanism are first grade indexes, extending the corresponding second index to build a harmonious enterprise's operational mechanism. Then use AHP to determine the weight of each index, and came to comprehensive evaluation. It is useful for researching enterprise's harmony.

Keywords: AHP, harmonious enterprise, harmonious index system.

A socialist harmonious society, is the social development goals which was raised in the party's recent Sixth Plenary Session of the "building a socialist harmonious society and a number of important issues". Many discordant phenomenon is the fact that we have to face, from the reform and opening up in 1978 to 2010, China's GDP growth from $ 147.3 billion to $ 5.8786 trillion, an average annual increase of 9.8% , and in 2010, surpassing Japan, ranked second in the world economic output. People are proud of these achievements, but with the rapid economic growth, a lot of problems have surfaced, the gap is too large, unemployment, environmental pollution, corporate credibility. To solve these problems, enterprise is the key. As the economic cell of society, enterprise is the driving force in building a harmonious society.

1 Research Statuses

1.1 Abroad Research Status

First stage: the formation of traditional experience, emphasizing the human factor, emphasizing the interests of business and co-workers.

D. Jin and S. Lin (Eds.): Advances in ECWAC, Vol. 2, AISC 149, pp. 191–196.
springerlink.com   © Springer-Verlag Berlin Heidelberg 2012

Second stage: formed in the classical management, which stresses the importance of employee sentiment, improves employee satisfaction, but not related to the whole business inside and outside.

Third stage: It was formed in the modern management, theory of the harmonious enterprise shape was proposed, which Involved in the production planning and the control of the collective work. Complex operating was figured out by diagram. Ensure that a large number of operations carried out in harmony.

1.2 Domestic Research at Present

Harmony Enterprises is a new theory, the harmonious thinking applied to business management began in recent years. In 1987, Xi you min professor of Xi'an Jiao Tong University proposed the theory of harmonious enterprise first in his doctoral thesis. Then there are also some scholars research the harmonious enterprise, but because of China's business development has been in the impact of political parties, change is relatively fast, harmonious enterprise process has just started, so lack of in-depth exploration.

2   AHP Process of Establishing

2.1   Establish a Hierarchical Structure Model and Constructed All Levels' Comparison Matrix

When applied AHP, first make the problem structured, hierarchical, construct a hierarchical structure model.

To compare the n-factor $X= \{ x_1 \ldots x_n \}$ on the impact of the size of a factor Z, that is, once taking two factors x_i and x_j . a_{ij} indicates the impact of the size of x_i and x_j of Z. All results of the comparison matrix $A= (a_{ij})_{N*N}$, A is called pairwise comparison between the judge Z-X matrix (comparison matrix).

2.2   Single-Level Sequencing and Consistency Examination

Determine the consistency of the test matrix as follows:

①Consistency index calculation CI : CI=$\dfrac{\lambda_{max} - n}{n-1}$.

②Find the corresponding average random consistency index RI. For n = 1, ..., 9, Satyr gives the value of RI, RI value is thus obtained, using the method to construct 500 random samples of the matrix: random from 1 to 9 and the last drawn numbers are each constructed anti-matrix, find the maximum eigenvalue λ_{max} del average ,

and define RI=$\dfrac{\lambda_{max}^{'} - n}{n-1}$. Then calculate the consistency ratio CR : CR=$\dfrac{CI}{RI}$.

3 The Basic Framework of the Operational Mechanism to Create a Harmonious Enterprise System

3.1 Creating a Win-Win Enterprise External the Macro-system Operating Mechanism of a Harmonious

Enterprise is not only the main element of social, economic, culture, but also the basic unit to promote social, economic and natural progress and harmony. Enterprise in the human and social, economic, cultural, natural and harmonious development plays cohesion, promoting the role. Based on this, enterprise need to operate on the basis of the law to fulfill their social responsibility to promote social welfare, improving community building, better for business development services for enterprises to create a relaxed social environment.

3.2 Creating a Win-Win Enterprise External the Micro-system Operating Mechanism of a Harmonious

Enterprises should be in the social, economic, cultural and natural environment to survive and develop. Need in the industry, suppliers, customers and other stakeholders posed by the micro-environment of growth and external growth. Therefore, the need to create and build a fair competition with industry rivals, win-win cooperation and harmonious development of operational mechanisms. Created orderly competition with suppliers, multifaceted cooperation and seek common development of strategic partnerships and harmonious development mechanism. Created a fair treatment, with stakeholders just to get along, take the initiative to be responsible, efficient and harmonious development of co-operation mechanism.

3.3 Created the Main Body of the Internal Operation Mechanism System

The main body of the internal operation mechanism system contains the decision of the operating mechanism, harmonious enterprise operating system and institutional mechanism system, harmonious enterprise operational mechanism of internal and external security of the system, harmonious enterprise self-organization of the operating mechanism system, harmonious enterprise system of spiritual culture of the operating mechanism, the assessment of a harmonious enterprise operational mechanism.

4 AHP in a Harmonious Enterprise Operational Mechanism in the Use of

4.1 Establish a Hierarchy

The establishment of an indicator: harmonious enterprise evaluation model A; Two indicators: external macro-win-win system of health and harmonious operational mechanism of B1, external micro-enterprise operational mechanism of harmony and win-win system of B2, the main body of the operating mechanism the enterprise internal system B3; three indicators: in harmony with social operating mechanisms C1, in harmony with the national economy, the operational mechanism of C2, and social and cultural harmony operating mechanism C3, in harmony with the natural operating mechanism C4, suppliers C5, the industry competitors C6, customers C7, other stakeholders C8, decision-making operational mechanism of C9, harmonious enterprise operational system and institutional mechanism for system C10, a harmonious internal and external security of the enterprise operational mechanism system C11, a harmonious enterprise operational mechanism of self-organization system C12, the spirit of cultural harmony and enterprise operational mechanism system C13 , assessing harmonious enterprise operational mechanism C14.

4.2   Build a Matrix to Determine the Levels of All

Determine the matrix include:

1. (B1, B2 B3) importance inter-comparison to A
2. (C1 C2 C3 C4) importance inter-comparison to B1
3. (C5 C6 C7 C8) importance inter-comparison to B2
4. (C9 C10 C11 C12 C13 C14) importance inter-comparison to B3

Table 1. AHP inter-comparison scale

relative importance	definition	annotation
1	Equally important	Two factors i, j is equally important
3	Slightly important	Experience to determine the factors that i slightly important than factor j
5	Very important	Experience to determine the factors i important than factor j
7	Obviously important	Important than factor j factor i, and has been proven in practice
9	Absolutely essential	Strongly feel the factor i comparing factor j to have an absolute predominance position
2,4,6,8	Midway between two adjacent states to determine	Expressed the need to take a compromise between the two values to determine
Reciprocal	$a_{ij} = 1/a_{ji}$	Factor i and factor j compared to determine the value obtained

4.3   Application of Eigenvalue Solution Techniques Seek to Determine the Maximum Eigenvalue Matrix and Eigenvector, Then Consistency Test

Judge the R of matrix among them. Be worth the following form:

Table 2. The R.I. value of judgment matrix

order	1	2	3	4	5	6	7	8	9
R.I.	0.00	0.00	0.58	0.90	1.12	1.24	1.32	1.41	1.45

Only CR <0.01 was considered to determine the matrix has satisfactory consistency.

The results are:

Table 3. Judgment matrix and weight with the code of Harmonious enterprise

A	B1	B2	B3	weight	maximum eigenvalue
B1	1	1/3	1/7	0.091	
B2	3	1	1/3	0.231	3.007
B3	7	3	1	0.677	

The form 3 gets C.R=0.006<0.01. Upon examination, comparison matrix has satisfactory consistency.

Table 4. Judgment matrix and weight with the code of external macro view

B1	C1	C2	C3	C4	weight	maximum eigenvalue
C1	1	3	5	2	0.492	
C2	1/3	1	2	1/2	0.154	4.0145
C3	1/5	1/2	1	1/3	0.091	
C4	1/2	2	3	1	0.261	

The form 4 gets C.R=0.0053<0.01. Upon examination, comparison matrix has satisfactory consistency.

Table 5. Judgment matrix and weight with the code of external tiny view

B2	C5	C6	C7	C8	weight	maximum eigenvalue
C5	1	3	1/2	5	0.283	
C6	1/3	1	1/7	2	0.087	4.0165
C7	2	7	1	9	0.570	
C8	1/5	1/2	1/9	1	0.059	

The form 5 gets C.R=0.0061<0.01. Upon examination, comparison matrix has satisfactory consistency.

Table 6. Judgment matrix and weight with the code of enterprise movement

B3	C9	C10	C11	C12	C13	C14	Weight	maximum eigenvalue
C9	1	2	2	5	4	9	0.390	
C10	1/2	1	1	2	2	5	0.192	
C11	1/2	1	1	2	2	4	0.190	6.0097
C12	1/5	1/2	1/2	1	1	2	0.087	
C13	1/4	1/2	1/2	1	1	2	0.095	
C14	1/9	1/5	1/4	1/2	1/2	1	0.043	

The form 6 gets C.R=0.0016<0.01. Upon examination, comparison matrix has satisfactory consistency.

Table 7. Influence on synthesis weight by the bottom hierarchical factor

	B1 0.091	B2 0.231	B3 0.677	Sort the results of the total layer C	Sequence
C1	0.492			0.045	8
C2	0.154			0.014	13
C3	0.091			0.008	14
C4	0.261			0.024	10
C5		0.283		0.065	5
C6		0.087		0.020	11
C7		0.570		0.132	2
C8		0.059		0.014	12
C9			0.390	0.264	1
C10			0.192	0.130	3
C11			0.190	0.129	4
C12			0.087	0.059	7
C13			0.095	0.064	6
C14			0.043	0.029	9

Available from the table, in the establishment of a harmonious enterprise operational mechanism, the various factors on the impact of a harmonious enterprise is not the same. C9, C7, C10, C11 weight greater than 0.1, summing the weights get 0.655, is key factor to build a harmonious enterprise, Should focus on it. C5, C13, C12, C1, C14 weight index greater than or close to 0.05, Although it is a secondary factor, it should be in the process of building a harmonious enterprise attention.

In summary, , through the AHP we analyzed the establishment of a harmonious enterprise in the weight of the various subsystems, for the establishment and evaluation of a enterprise laid the foundation of harmony and harmonious social development, enterprise creation is the process of enterprise development. Therefore, in the enterprise development process, to coordinate the various factors for corporate harmony and social harmony, Social enterprise should play a positive role in the economic.

References

1. Yang, J.: According to establishing of science development view harmonious enterprise operation mechanism system thesis. Hebei University of Engineering Journal 9 (2008)
2. Yang, J.: Constuct the methodology of harmonious culture. The Enterprise Manages 12 (2008)
3. Zhao, H., Xu, S., Jin, S.: AHP- a kind of simple decision method. Science Press (1996)

Smart Acoustic Control System

Rui Chen[1], Fangfang Peng[1], and Mohammad Reza Asharif[2]

[1] School of Computer and Information Engineering,
Central South University of Forestry & Technology, Changsha, Hunan, China
[2] Department of Information Engineering, Faculty of Engineering,
University of the Ryukyus, Okinawa, Japan
k038656@yahoo.com.cn, lanyl0000@126.com

Abstract. In this paper, a new algorithm of smart acoustic control (SAC) system is presented. That is, the acoustic response between two (or more) points could be controlled smartly. By control, we mean to have a well estimation of the acoustic path between two points and then to make the appropriate signal to cancel an unwanted noise or to emphasis to a desired signal (speech or music). In a sense this SAC system works as like as conventional ANC. This is because, we make the control system to be imposed acoustically. The computer simulation results support the theoretical findings and verify the robustness of the proposed algorithm.

Keywords: Acoustic room impulse response, Smart Acoustic Control (SAC) system, Active noise control (ANC).

1 Introduction

Study of the room acoustic is an important topic in all kinds of speech processing and audio systems. In hand free telephony or in teleconferencing system, acoustic echo canceller (AEC) [1] is used to remove the echo signal from speech. Here, echo is generated due to acoustic couplage between loudspeaker and microphone in a room. The echo degrades the intelligibility of the communication. Therefore, AEC tries to estimate the room acoustic response and make a replica of the echo signal and remove it. Acoustic noise control (ANC) [2] system is another example to reduce acoustic noise in a location of the room. Here, the acoustic noise is propagated through room acoustic and ANC tries to estimate this acoustic path to generate an opposite signal similar to noise and reduce it appropriately.

In all kinds of above-mentioned examples, we need to estimate and control the room acoustic response between two locations. Nevertheless, this control could be imposed electrically (AEC) or acoustically (ANC), the adaptive digital filter (ADF) is used to perform this job with an appropriate algorithm.

In this paper, a new algorithm of smart acoustic control (SAC) system is presented. That is, the acoustic responses could be controlled smartly. By control, we mean to have a good estimation of the acoustic path from the speakers to the microphone, and then to make the appropriate signal to cancel the unwanted signals.

D. Jin and S. Lin (Eds.): Advances in ECWAC, Vol. 2, AISC 149, pp. 197–202.

2 Application of Smart Acoustic Control System

As shown as Figure 1, suppose that we want to listen to a Jazz music in one portion of a room and at the same time other fellow wants to listen to a classic one in the other side of the room. Also, we do not want to use headphone as it totally isolate the person from surrounding. Other example is in a conference room or big hall, that we have two kinds of audiences. In one section, audiences want to listen in Japanese while in other section international audiences are seated and they want to listen to the speech in English. Again we do not want to use headphone as here is very costly to manage the system for each person and the Hall should be designed for that or we need transceiver, which is also costly. But if we design the acoustic response such that Japanese loudspeaker covers the desired location while English loudspeaker covers the other part, just by seating in the right place one can hear to desired language. There are much more applications of SAR system. ANC is an especial case of SAR, because in a room we want to reduce the noise source propagation to a location. In more general case, we can define acoustic channels similar as radio or TV channels. Imagine you want to change the channel of TV by using a remote control. The same is possible to be performed for acoustic channel. But, the difference here is location dependency of the remote control. That is, depending on place of the remote control, one can push a bottom to listen to a specific program that be propagated to that place only. If we move the remote control to other location in the room, we can select another program and set the acoustic path to listen only to specified program. Therefore, in SAR we require to change and control acoustic impulse response of the room, as we desire [4],[5],[6].

Of course, sound propagation through acoustic channel from one loudspeaker could cause perturbation for the other one. This is because in contrast to electromagnetic propagation and frequency division multiplexing (by using proper modulation technique) is not possible in acoustic wave propagation. Therefore, by using a powerful algorithm in adaptive digital filter, one can make the null point (zero point) of an acoustic source to be set in specific location and/or move it to any other location.

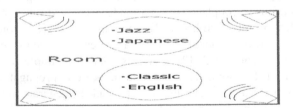

Fig. 1. Smart Acoustic Room (SAR) system

3 Smart Acoustic Control System

In this paper, we challenge to control the acoustic responses between two points as shown in Figure.2. That is by using two speakers and one microphone to make an acoustic null point at the microphone position. And also the Eriksson's method [7] is used for online secondary path modeling. Consider Eriksson's method is shown in Figure 3.

The output signal of the control filter y(n) can be expressed as:

$$y(n) = w^T(n)x_L(n) \qquad (1)$$

Where $w(n) = [w_0(n), w_1(n), \cdots w_{L-1}(n)$ is the tap-weight length L vector, $x_L(n) = [x(n), x(n-1), \cdots x(n-L+1)]$ is the L-samples reference signal vector, and $x(n)$ is the input signal from the reference microphone. A white Gaussian noise signal wqith zero-mean, $v(n)$ is injected at the output $y(n)$ of the control filter. The error signal $e(n)$ is given as:

$$e(n) = d(n) - y'(n) + v'(n) \qquad (2)$$

Where $d(n) = p(n) * x(n)$ is the primary disturbance signal at the error microphone, $y'(n) = s(n) * y(n)$ is the secondary canceling signal, $v'(n) = s(n) * v(n)$ is the modeling signal, * denotes the convolution operation, and $p(n)$ and $s(n)$ are impulse responses of the primary path and secondary path, respectively. The residual noise signal $e(n)$ is used as an error signal for $w(n)$, and as a desired response for $\hat{s}(n)$, i.e., $g(n0 = d(n) = e(n)$. Thus the error signals for w(n) and $\hat{s}(n)$ are, respectively, given as:

$$g(n) = [d(n) - y'(n)] + v'(n), \qquad (3)$$

$$f(n) = [d(n) - y'(n)] + [v'(n) - v'\hat{(n)}] \qquad (4)$$

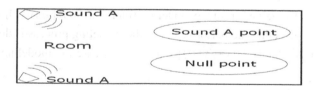

Fig. 2. Two speakers SAR system

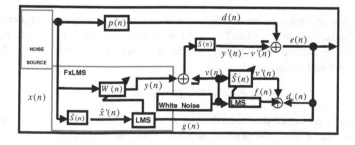

Fig. 3. Eriksson's method for online secondary path modeling

The coefficients of the control filter w(n) are updated by the FxLMS algorithm:

$$w(n+1) = w(n) + \mu_\omega g(n) \hat{x}'(n)$$
$$= w(n) + \mu_\omega \hat{x}'(n)[d(n) - y'(n)] + \mu_\omega \hat{x}'(n) \hat{v}'(n)$$

(5)

where μ_ω is the step size for the control process, $\hat{x}(n) = [\hat{x}(n), \hat{x}(n-1); \cdots \hat{x}(n-L+1)]$, and $\hat{x}'(n)$ is the reference signal $x(n)$ filtered through the modeling filter $\hat{s}(n)$. We see that the control process is perturbed by an undesired term $\mu_\omega \hat{x}'(n) \hat{v}'(n)$.

Assuming that $\hat{s}(n)$ is represented by an FIR filter of tap-weight length M, the filtered-reference signal is $\hat{x}'(n)$ is obtained as:

$$\hat{x}'(n) = \hat{s}^T(n)x_M(n)$$

(6)

Where $\hat{s}(n) = [\hat{s}_0(n), \hat{s}_2(n) \cdots \hat{s}_{M-1}(n)]^T$ is the impulse response of the modeling filter $\hat{s}(n)$, and $x_M(n) = [x(n), x(n-1) \cdots x(n-M+1)]^T$ is the M-sample reference signal vector.

The LMS update equation for $\hat{s}(n)$ is given as

$$\hat{s}(n+1) = \hat{s}(n) + \mu_s f(n)v(n)$$
$$= \hat{s}(n) + \mu_s v(n)[v'(n) - \hat{v}'(n)] + \mu_s v(n)[d(n) - y'(n)]$$

(7)

Where μ_s is the step size of the modeling process, $\hat{v}'(n) = \hat{s}'(n) * v(n)$ is an estimate of $v'(n)$ obtained from the modeling filter, and $v(n) = [v(n), v(n-1) \cdots v(n-M+1)]^T$.

Equation(7) shows that the performance of the modeling process is degraded by an undesired term $\mu_s v(n)[d(n) - y(n)]$ and in worst case the modeling process may diverge.

4 Simulation Results

In this section in order to verify the robustness of the proposed algorithm, some computer simulations were done.The acoustic paths $p(n)$ and $s(n)$ of the room were assumed to have exponential decaying shape that decreases to –60dB after M sample, which were defined as follows:

$$w_{1,2}(i) = Randn[\exp(-8i/M)]$$

(8)

where Randn is a normal distributed random number between +1,-1 with zero mean and unit variance. $P(n)$ is the primary acoustic path and $s(n)$ is the secondary acoustic path.

To measure the performance of the algorithm, the performance comparison is done on the basis of relative modeling error, Δs (dB), being defined as

$$
\Delta s = 10 \, \log \left\{ \frac{\sum_{i=0}^{M-1} \left[s_i(n) - \hat{s}_i(n) \right]^2}{\sum_{i=0}^{M-1} \left[s_i(n) \right]^2} \right\}
\tag{9}
$$

The input signal x(n) is the voice of a woman at her 20's and pronounced as "Good morning and welcome to…" in English, which.is shown in Figure 4. If we have a good estimation of secendary acoustic path s(n), the signal speech $d(n)$ wiil be canceled at the microphone position. The error signal from the microphone $e(n)$ is shown in figure 5. And the ratio of distance between secondary acoustic path $s(n)$ and estimated acoustic path $\hat{s}(n)$ is shown in figure 6. Δs converges to -38dB at the iteration 5000.

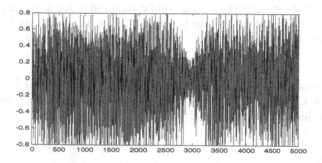

Fig. 4. Waveform of input signal x(n)

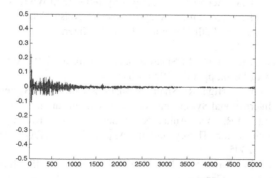

Fig. 5. Waveform of error signal

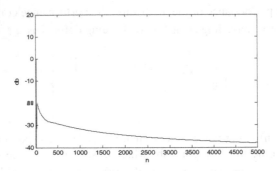

Fig. 6. Ratio of distance between secondary and estimated path

5 Conclusion

In this paper, a new smart acoustic control (SRC) system algorithm based on Eriksson's Method is presented. That is, the acoustic responses could be controlled smartly. By control, we mean to have a good estimation of the acoustic path from the speakers to the microphone, and then to make the appropriate signal to cancel the unwanted background signal (speech or music). The computer simulation results support the theoretical findings and verify the robustness of the proposed algorithm.

Acknowledgments. This paper is supported by Hunan Provincial Natural Science Foundation of China (No.10JJ5062) and Hunan Science and Technology Agency science and technology projects(No.2010GK3003).

References

1. Haykin, S.: Adaptive Filter Theory, 3rd edn. Prentice Hall (1996)
2. Kuo, S.M., Morgan, D.R.: Active Noise Control Systems. John Wiley & Sons, Inc. (1996)
3. Ohana, Y., Kohna, T.: Direct Fully: Adaptive Active Noise Control Algorithms Without Identification of Secondary Path Dynamics. In: IEEE International Conference on Control Application, Scotland (2002)
4. Alsharif, M.R., Higa, R., Chen, R.: Smart acoustic room. In: 2004 Autumn Meeting of the Acoustical Society of Japan, pp. 601–602 (2004)
5. Alsharif, M.R., Chen, R., Higa, R.: Smart Acoustic Room (SAR) System by Using Virtual Microphone. In: International Symposium on Telecommunications (2003)
6. Chen, R., Alsharif, M.R., Yamashita, K.: Smart Acoustic Room (SAR) System. In: Symposium on Information Theory and its Applications (SITA) Conference, Okinawa, Japan, pp. 913–916 (2005)
7. Akhtar, M.Y., Abe, M., Kawamata, M.: Modified Filtered-x LMS Algorithm based Cross-updating Active Noise Control System with Online Secondary-path modeling. In: International Technical Conference on Circuits Systems, Computers and Communications, Japan (2004)

A Study on the Technology of Intelligent Electric Lead Sealing Management System Based on RFID

Yong Pan[1], Kaihua Liu[1], and Yi Gao[2]

[1] School of Electronic Information Engineering
Tianjin University TianJin, China
panyong@126.com
[2] Research and Development Department
Qicheng Science and Technology Co., Ltd.TianJin, China

Abstract. Radio frequency identification is a key technology in the Internet of Things. This paper designs hand-held radio frequency identification terminal of intelligent electric lead sealing tags based on dual processors. This terminal uses STM32F103VET6 as its master controller, and Mifare CL RC632 as its radio frequency identification module. It complies with ISO14443 TYPEA/B and ISO15693 international standards and can read and write intelligent electric lead sealing tags within 10 cm. The actual measurement of intelligent electric lead sealing tags shows that this terminal is highly stable and comparable in 13.56M high frequency radio identification.

Keywords: Radio frequency identification, RFID, Intelligent electric Sealing, STM3, Antenna.

1 Introduction

The Internet of Things is an important part of the new generation of information technology. The concept "Internet of Things" is defined as follows: all things are connected via internet. According to the agreed protocol, any things will be connected to the Internet though sensing devices such as Radio Frequency Identification (RFID), Infrared Sensors, Global Positioning System (GPS), Laser Scanners to achieve information exchange and communications.

RFID (Radio Frequency Identification), also called electric tags, a key technology in the identification field of the Internet of Things, is a kind of communication technology. A complete set of RFID system is made up of readers, transponders and software system. Nowadays, a large amount of researches have been done on intelligent electric lead sealing management system and some results have been achieved. [1-6]With the scientific development, source-based RFID technology [7-8] will be increasingly be applied in the logistic field.

The old fashioned mechanic electric lead sealing is venerable to corrosion and manmade changes and short in lifetime. According to these shortcomings, we integrate a semiconductor chip with common lead sealing to form a safe and anti-counterfeit radio frequency identification terminal of intelligent electric lead sealing based on RFID. Based on the popularity and widespread application of computers, lead sealing can achieve electronic and information management via intelligent electric sealing management system, which will improve the daily supervision and inspection about lead sealing by the metrological service department.

2   The Hardware System Design

The system in this paper includes two parts, intelligent lead seal labels and handheld terminal. This system attempts to realize all levels of signature verification and safety management of intelligent lead sealing and also tries to regulate and manage the operators' operations. The system can not only run on a single computer but also supports network operation, and it can also be embedded in the management system as a subsystem. The framework of the system structure is shown in Figure 1.

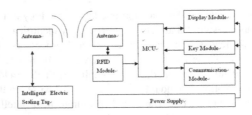

Fig. 1. The Structure System of Intelligent Electric Lead Sealing

2.1   The Handheld Terminal Hardware System Design

Intelligent electric lead sealing labels directly use the intelligent identification chip, Mifare RFID, introduced by NXP. This terminal can identify, read, write and encrypt HF RFID tags on 13.56 M band and reserves a secondary development interface for the identification of UHF RFID labels on 860~960M band.

The Master Controller Module: The terminal uses the low-power embedded control system, STM32F103VET6 ARM Cortex-M3, as its master controller module, which, with great performance and low energy consumption, ensures the powerful data processing abilities of the terminal and provides the hardware basis for the complicated applications. The main controller module is responsible for LCD, the keyboard and communication and it exchanges data and commands through a serial port and a sub-controller.

The Radio Frequency Identification Module: The radio frequency identification module uses CL RC632 chip from NXP. This chip has a highly integrated analog circuit for demodulation and decoding of responses. Its buffer output driver applies a minimum number of external components connected to the antenna. The longest distance in a close control can reach 10 cm and it supports MIFARE dual interface card IC and ISO14443 TYPEA/B and ISO15693 standards. The internal oscillator buffer connects 13.56MHz crystal and it jitters at a low phase and its power consumption is low.

The LCD Module-TFT LCD: The intelligent robot needs good human-computer interaction systems. This design adopts a 2.8-inch TFT LCD as the display module, which is used to display the information read and written by labels and the relevant information display of the human-computer interaction. The specific plan uses a 2.8-inch DST2001PH LCD screen driven by ILI9320.

2.2 The Antenna Design

As the frequency of MFRC632, 13.56 MHz, are short-wave band, a small loop antenna can be used. Small loop antenna can be square, round, oval and triangular, etc. This system uses square antenna.

There is no strict boundary between the maximum geometric dimension and operating wavelength and it is generally defined as:

$$L / \lambda \leq 1 / 2\pi \tag{1}$$

In the above formula, L is the maximum size of the antenna and λ is operating wavelength. For 13.56 MHz system, the maximum size of the antenna is around 50 cm. In the antenna design, the quality factor Q is a very important parameter. For PCB antenna of the inductively coupled radio frequency identification system, high quality factor leads to stronger current in the antenna coil, which, on the other hand, can improve the power transmission to PICC. The quality factor is calculated as:

$$Q=(2\pi f0 \bullet Lcoil) / Rcoil \tag{2}$$

In the above formula, $2\pi f0$ is the operating frequency, Lcoil is the antenna inductance and Rcoil is the antenna resistance. The transmission bandwidth of the antenna can be easily calculated through the quality factor:

$$B=f0 / Q \tag{3}$$

It can be seen from formula (3) that there is an inversely proportional relationship between the antenna bandwidth and the quality factor. Therefore, too high quality factor will lead to reduced transmission bandwidth, thereby weakening the modulation sidebands of PCB, causing PCB unable to communicate with the card. The best quality factor for a general system is $10\sim30$ and it cannot exceed 60. Considering the above factors and the convenience to achieve the goal, a way is adopted, that is using square wires to wound the outside of the PCB board to form the antenna. The circuit is shown in Figure 2 and the antenna PCB circuit is shown in Figure 3.

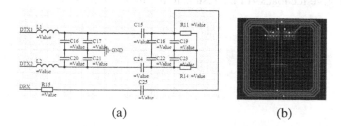

(a)   (b)

Fig. 2. (a) The Diagram of the Radio Frequency Antenna. (b)The PCB Diagram of the Radio Frequency Antenna.

3 The Software System Design

In the software system, the hand-held terminal can identify, read, write and encrypt the intelligent lead sealing tags. The master controller STM32F103VET6 gets and further processes the results from the sub-controller and drive TFT LCD, keyboard and the communication module to communicate with computers.

3.1 The Flowchart of the Master Controller

Firstly, the main program initializes peripheral equipments such as LCD, keyboard and the communication module, then, through the serial port, it sends commands and data to the sub-controller, which, according to commands, will make some relevant operations and preprocess the data, then send the data back to the master controller. The master controller will display the data again, send them to the computer's management system and then send the next command and data.

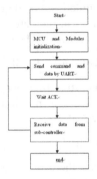

Fig. 3. The Main Flowchart

3.2 The Flowchart of the Sub-controller

The sub-controller program first initializes the sub-controller and the RF module and then waits for commands and data from the master controller. The RF module, controlled by these commands and data, reads and writes intelligent lead sealing labels, then give feedbacks of the results to the master controller.

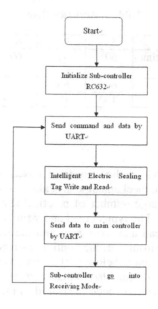

Fig. 4. The Flowchart of the Sub-controller

4 Testing and Analysis

Based on the design in the paper, the hand-held terminal is illustrated in Figure 5.

Fig. 5. The terminal

To test the system, first keep a distance of 5 cm, read and write labels repeatedly, observe the results displayed by the terminal LCD, and verify data accuracy. Then the distance will gradually be extended to 10 cm, and read and write labels repeatedly again. Some results are shown in Table 1. The measurement suggests that within 7 cm, the accuracy rate is 100%; beyond 7 cm, errors occur according to practical environment; beyond 10 cm, the read and write situation will see a sharp deterioration. This result is mainly caused by the performance of the chip itself.

Table 1. Accuracy Rate

distance (cm)	5	7	8	10	11	12
Total read and write times (times/min)	60	60	60	60	60	60
The number of correct times	60	60	60	59	37	12
Accuracy rate (%)	100	100	100	98.3	60.1	20

5 Conclusion

The intelligent electric lead sealing system based on RFID overcomes the shortcomings of the traditional mechanical lead sealing, it's good in encryption, long in life and easy to manage. A large number of practical tests show that this terminal is highly stable and comparable. The typical read and write distance is 10 cm. There is no dead zone within effective read and write area and the read and write operations are reliable. In the laboratory environment, the continuous read and write error rate of the card in the effective sensing area is below 0.01%. Its compatibility shows not only in that it can read the standard card, but also in that it can read cards that are far from the standard. At present, the system has been measured in the field and its operations are stable.

References

1. Liu, Y., Song, G., Wu, J., Lu, Y., Liu, M.: Fixed eccentric wheel device based electronic sealing synchronization control for continuous automatic F/F/S packaging machines. In: 2009 IEEE Student Conference on Research and Development (SCOReD), pp. 422–425 (2009)
2. Campos, M.L.B.: Evaluation of the sealing system and of the electric performance of the distribution lightning arresters. In: 14th International Conference and Exhibition on Electricity Distribution. Part 1: Contributions. CIRED, vol. 3, pp. 30/1–30/5 (1997)
3. Zhao, X.-F.: Design of electronic seal system based on PKI. In: 2011 International Conference on E -Business and E -Government (ICEE), pp. 1–3 (2011)
4. Ma, B., Xu, J.: Electronic seal system based on RSA algorithm and public-key infrastructure. In: 2010 2nd International Conference on Networking and Digital Society (ICNDS), vol. 1, pp. 636–639 (2010)
5. Xu, J., Ma, B.: Study of an electronic seal system based on elliptic curve cryptography and public-key infrastructure. In: 2010 2nd International Conference on Future Computer and Communication (ICFCC), vol. 2, pp. V2-760 – V2-763 (2010)
6. Park, J.-S., Oh, S., Cheong, T., Lee, Y.: Freight Container Yard Management System with Electronic Seal Technology. In: 2006 IEEE International Conference on Industrial Informatics, pp. 67–72 (2006)
7. Ritamaki, M., Ruhanen, A.: Embedded passive UHF RFID seal tag for metallic returnable transit items. In: 2010 IEEE International Conference on RFID, pp. 152–157 (2010)
8. Chin, L.-P., Wu, C.-L.: The Role of Electronic Container Seal (E-Seal) with RFID Technology in the Container Security Initiatives. In: 2004 International Conference on MEMS, NANO and Smart Systems, ICMENS 2004. Proceedings, pp. 116–120 (2004)

The Realization of Information Hiding Algorithm Based on Logistic Mapping

JunYing Sun and JiNing Feng

Electronics Department, Vocational And Technical Education College,
Hebei Teachers University Shijiazhuang, China
xxmmwang@163.com

Abstract. Information hiding technology is an advanced researching field, which recently spring up in international academic area. This article utilizes chaotic system, which possesses optimum pseudorandom characteristics, sensitivities to initial situations and such unpredictable characteristics and it applies LSB(Least Significant Bit),to produce embedded sequence and embed secret information to certain bit on pixel of a picture. It applies minimum pixel to alter optimum strategy. Experiment results show that crypticity of embedded secret information is fine; room for secret key is ample. This calculation can stand noise and other interference; authorized users can get the secret information easily and quickly, while illegal ones cannot get the right secret information anyway.

Keywords: logistic mapping, information hiding, space field algorithm.

1 Introduction

Nowadays, Internet technology booms so that information security becomes a hot research field. At present, information can be easily transferred and copied, which makes intellectual property protections harder; during the transmittance of information, people want to secretly transmit information and avoid perceiving by the third party. The development and prevalence of computer and internet communication technology, the transmittance and transaction of digital audio & video products and other digital publication is becoming more and more convenient, unauthorized copying and violation pirate is becoming severe day by day. In this situation, information hiding technology is more and more crucial. This article discusses a space field calculation on picture based on Logistic and Arnold chaotic mapping. As Arnold transformation is one by one mapping, embedded secret information won't bring conflict on position. And it applies minimum pixel changed strategy when embedding bit of secret information, making the information possesses optimum invisibility when it embeds information that includes huge content. Compared with other similar chaotic calculation, it possesses strong ability as standing illegal getting, insensibility is fine, similarity is fine, and it fits hiding of any size grey BMP pictures.

D. Jin and S. Lin (Eds.): Advances in ECWAC, Vol. 2, AISC 149, pp. 209–214.
springerlink.com   © Springer-Verlag Berlin Heidelberg 2012

2 Producing of Chaotic Sequence

2.1 Logistic Mapping

Logistic mapping namely insect bite model , is a crucial chaotic model. When certain breed of insect generates, the number of offspring if far more than the number of its predecessors. In this way, we can assume that the number of predecessors is negligible after the generation of offspring. Suppose Xn is the number of a certain insect on year n, and this number has something to do with year, n is equal to integer merely, the number is xn+1 on year n+1,and the relationship between them can be described by a function below:

$$Z_{n+1} = \lambda Z_n (1 - Z_n) \qquad Z_n \in (0,1) \ \lambda \in (0,4] \tag{1}$$

After thoroughly researching, Logistic system will present different characters for different values of ,and Logistic system bears times cycle divaricate, and becomes chaotic finally. Figure 1 is the Logistic chaotic sequence when =3.6543.We can see that sequence generated by being chaotic possesses optimum random statistic character and uniform distribution character.

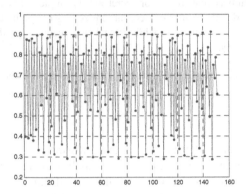

Fig. 1. Chaotic Sequence for λ =3.6543

2.2 Arnold Transform

Arnold transform was a kind of nonlinear mapping which introduced by V.J.Arnold who researched ergodic theorem. There are two dimensional, three dimensional or even N dimensional Arnold transform. This article applies two dimensional Arnold transform, and it is called cat mapping as usually do the experiment by a cat face.
    Function of cat mapping is:

$$\begin{cases} x_{n+1} = (x_n + y_n) \bmod 1, \\ y_{n+1} = (x_n + 2y_n) \bmod 1. \end{cases} \tag{2}$$

In this function, mod 1 means that it contains decimal fraction merely, namely x mod1=x-[x].Thus, space position of (x_n, y_n) is limited in unit cube [0,1]×[0,1],transform function(2) to matrix:

$$\begin{bmatrix} x_{n+1} \\ y_{n+1} \end{bmatrix} = \begin{bmatrix} 1 & 1 \\ 1 & 2 \end{bmatrix} \begin{bmatrix} x_n \\ y_n \end{bmatrix} = C \begin{bmatrix} x_n \\ y_n \end{bmatrix} \bmod 1. \tag{3}$$

Function (3) defined matrix C, when determinant |C|=1,cat mapping is an area mapping(there's no attractor).At the same time cat mapping is one by one mapping, every single point in unit matrix uniquely transformed to another point in this matrix, by this character, it makes secret information in different position gets different embedded position. Cat mapping possesses quite typical factors that can produce chaotic movement: stretch (multiply matrix C to enlarge x, y) and fold (absolute x and y to make them fold to single matrix). Actually, cat mapping is chaotic mapping. Extending gamma space as {0,1,2,...,N-1}×{0,1,2,...,N-1}, namely use the positive integer range from 0 to N-1; then extending the functions as quite common two dimensional reversible area functions:

$$\begin{bmatrix} x_{n+1} \\ y_{n+1} \end{bmatrix} = \begin{bmatrix} a & b \\ c & d \end{bmatrix} \begin{bmatrix} x_n \\ y_n \end{bmatrix} = C \begin{bmatrix} x_n \\ y_n \end{bmatrix} \bmod N \tag{4}$$

In this function a, b and c are positive integers, and its area should be |C|=ad-bc=1.In such limiting conditions, only three factors among four are independent, then how to make a,b and c independent and d is defined by area. Cat mapping defined like this is generalized cat mapping.

3 LSB Calculation Theory Based on Chaotic Notion

3.1 Chaotic and Positions for Embedded Information

Apply generalized cat mapping function (4) to produce spatial positions of every pixel that is embedded to carrier picture from secret picture: make coordination of every pixel on secret picture as initial value, then make three independent parameters and iteration times k of coefficient matrix C as secret key to produce iteration results (x ',y ') as the space position from (x,y),where processes secret picture to carrier picture. As chaotic character of mapping, when iteration times are huge enough, as for any two adjacent pixels of a secret picture, the positions they embeds into on the carrier picture will largely separate. And as this mapping is one by one mapping, the embedded positions iterates by different secret picture coordination will be different, thus, embedded information won't conflict on positions.

3.2 Algorithm on Embedding and Extracting

(1) Embedding algorithm

a. Designate independent parameters a,b,c of cat mapping, iterative times n and sequence initial value Z1 of Logistic mapping, makes Logistic mapping pre-iterate five hundred times.

b. Take a pixel w(i,j) from secret information, make x0=i,y0=j, iterate it for n times by formula (3-4), then get values of x and y (make sure $1 \leq x,y \leq N$).

c. Logistic mapping iterates once will get Zn, by which to get k, when Zn<0.25, k=2, when 0.25<Zn<0.55, k=3; when $0.55 \leq Zn<0.85$, k=4, when $0.85 \leq Zn<0.98$, k=5; when 0.98<Zn, k=6. And calculate number k bit bk of f(x,y).

d. If w(i,j)=bk, then take f '(x,y)=f(x,y); or execute the sub-processes as follow:

①t =1 ;

②Take f1 '(x,y)=f(x,y)+t or f2 '(x,y)=f(x,y)-t, and calculate number k bit bk11 and bk12 from f1 '(x,y) and f2 '(x,y) respectively.

③If bk11=w(i,j), then f '(x,y)=f1 '(x,y), and transfer to e; or if bk12=w(i,j), then f '(x,y)=f2 '(x,y), and transfer to e ; else, transfer to process④ ;

④t=t+1, repeat processes ②and③.

e. Take next pixel from secret information, repeat b~d of this calculation until all the pixels of secret information are embedded.

(2) Extract algorithm

We'd know secret key parameters of a,b,c,n and Z1 and the length of secret information when extract secret information. Apply w(i,j) stand for extracted pixel from secret information, the calculation can be expressed in detail as follow:

A. Designate independent parameters a,b,c of cat mapping, iterative times n and sequence initial value Z1 of Logistic mapping, makes Logistic mapping pre-iterate five hundred times.

B. For ready to be extracted pixel from secret information w(i,j), make x0=i, y0=j, iterate for n times by formula (3-4), to get x,y (make sure $1 \leq x,y \leq N$)

C. Logistic mapping iterates once will get Zn, when Zn<0.25, k=2, when $0.25 \leq Zn<0.55$, k=3; when $0.55 \leq Zn<0.85$, k=4, when $0.85 \leq Zn<0.98$, k=5; when 0.98<Zn, k=6.

D. Calculate number k bit bk2 of f ' (x,y), get w2(i,j)=bk2.

E. Repeat processes b~d of this calculation, until all the pixels from secret information w2(i,j)(i=1,2,…,M;j=1,2,…,M) are extracted.

4 Experiment Simulation and Analysis

4.1 Experiment Conditions

This experiment embeds a secret picture with 40×40 resolution to a carrier picture with 256×256 resolution, the three independent parameters a=1, b=33, c=3, iterative times k=50 of Arnold mapping, initial value of Logistic chaotic iterative sequence Z_1=0.40001.Result of experiment as follow:

Original Image   Original Carrier Image

Carrier Image Embeded
with Secret Information   Extracted Image

Fig. 2. Original Image, Carrier Image and Extracted Image

4.2 Analysis on Experiment Results

The veracity of extracting information can be described by original secret information and the similarity of secret information that being extracted, in order to precisely describe the similarity of those two pictures, we use relevant coefficient γ to evaluate this index. The relevant coefficient γ of the two pictures is defined as below:

$$\gamma = \frac{\sum_i \sum_j \left(A_{ij} - \overline{A}\right)\left(B_{ij} - \overline{B}\right)}{\sqrt{\left(\sum_i \sum_j \left(A_{ij} - \overline{A}\right)^2\right)\left(\sum_i \sum_j \left(B_{ij} - \overline{B}\right)^2\right)}}$$

We can conclude by Table 1 that relevant coefficient γ will increase with the increase of t, that is to say the similarity of secret information being extracted is getting better.

Table 1. Values of γ when t varies

t	8	16	32	64
γ	0.1152	0.1968	0.4000	0.9836

5 Conclusion

As the new technology in information security field, information hiding technology plays a more and more crucial role in ensuring the security of secret information and the high efficiency of transmitting secret information. This article mainly applies the space area information hiding brand new calculation, which combines with chaotic theory, based on researching in space area information hiding calculation and chaotic theory. Compared with similar calculations, this calculation possesses such characters: content being hiding is huge, insensibility is fine, strong anti-illegal extracting, can process any pictures with any sizes.

Acknowledgement. This work is supported by Jia Cheng Foundation of Hebei Teachers University, Education Department Foundation of He Bei province, National Science Foundation of He Bei province.(NO. F2011205023)

References

1. Lu, D., Liao, X., Han, J.: Picture Scrambling Algorithm Based on Logistic Mapping and Sort Transformation. Computer Technology and Development, 27–29 (2007)
2. Li, P., Zhang, X.-F., Tian, D.-P.: Digital image hiding method based on bit operation and chaotic sequence. Computer Engineering and Applications, 175–176 (2008)
3. Yang, H., Zhang, W., Wei, P.-C., Huang, S.: An Image Encryption Scheme Based on Random Chaotic Sequence. Computer Science, 205–208 (2006)
4. Zhang, D., Kang, B.: Digital Image Scrambling Based On Chaotic Sequence And ZF-02 Block Cipher Scheme. Computer Applications and Software, 21–29 (2007)
5. Feng, M.-K., Qiu, S.-S., Liu, X.-Y., Jin, J.-X.: Analysis on the randorrr like property of discrete chaotic sequences. System Engineering and Electronics (2008)

Color Image Watermarking Algorithm
Based on Angular Point Detection

Ying Tian and Xuesong Peng

College of Software, University of Science and Technology Liaoning, Anshan Liaoning China
{astianying,t_tianying}@126.com

Abstract. This paper presents a watermarking algorithm based on the Angular point detection in the wavelet domain. Vector graphics through the corners on the analysis of characteristics to obtain the outline of the image area. The host image by four wavelet transform to obtain frequency sub-image at all levels of the coefficient matrix. It treated the original watermarking signals using a periodic Arnold scrambling, then the scrambled signals no longer had the periodicity during the recovery. Scrambled watermark encrypted by a chaos sequence generated by Lorenz chaotic system, embedded into the host image. The experiments show that this paper presented algorithm in various attacks against the image has good robustness.

Keywords: Watermarking, Image corner feature, Arnold scrambling, Wavelet transform.

1 Introduction

In recent years, with the rapid development of digital technology, multimedia, image exchange of information reached a certain depth and breadth, However, accompanied by the copyright protection of digital products. Therefore, copyright protection of digital products has become an urgent need to address the research topic..Digital watermarking technology came into being. Over the years researchers have made a lot of digital watermarking algorithms embedded in the classic, in which the robustness of digital watermarking and hidden digital watermarking technology is the key.

In the past Image Watermarking Algorithm can be divided into two categories: digital watermarking algorithm and transform spatial domain digital watermarking algorithm. The former is a direct pixel by changing the value of the original image to embed watermark, the advantage is to achieve simple, easy, and can achieve the watermark invisibility[1]. However,the image against the attacks is weak. Another watermarking algorithm is carried out in the transform domain: the discrete Fourier transform (DFT), discrete cosine transform (DCT)[2], discrete wavelet transform (DWT)[3] and so on. Discrete wavelet transform is applied to the idea of digital watermarking for digital images using wavelet multi-resolution decomposition, into different space, different frequency sub-images, multi-resolution analysis of wavelet features can be well controlled in the vector image watermark distribution, and to adapt to human visual system (HVS), the watermark is better hidden. In recent years, digital watermarking based on wavelet transform technique has very good development, also is the digital watermarking research focus.

For existing digital watermarking algorithm in wavelet domain, its main purpose is to enhance the ability of resistance to geometric attacks, mainly on the embedding location and complexity of the algorithm up to consider, some algorithm is simple and easy to implement, but it will affect the visual watermark.Corner-based image analysis[4], digital watermarking can better reflect the characteristics of the image, the image based on wavelet analysis, image-based digital watermarking algorithm corners resist geometric attack ability. Most are based on the wavelet analysis of the image is decomposed into three sub-image, and corner-based watermarking methods are up in the 256-color grayscale watermark [5]. The method inherits the wavelet decomposition ability against geometric attacks and embedding the watermark in the choice of location area and avoided the false positive watermark is a robust copyright protection for digital watermarking methods.

2   Algorithm

2.1 Proposed Algorithm

The proposed watermark embedding scheme is four-dimensional color images in the wavelet transform analysis technique based on corner watermark embedded into the image wavelet coefficients. Corner is the target contour curvature local maximum points, the outline of the characteristics of mastery goal decisive role, once you find the outline of characteristics of the target will largely control the shape of the target. It is generally believed that two-dimensional corner point of rapid change image brightness, or image edge curve maximum curvature points, these points are important characteristics of image and graphics while retaining the same time, the information can effectively reduce the amount of data to improve the information content, calculation speed is conducive to a reliable image matching, making real-time processing possible.

First, the watermark image W with the Aronld scrambling for vector images obtained in four discrete wavelet transform coefficient matrix frequently on vector images by Harris corner detection operator for corner detection, and then the corner points are detected mapped to the vector images in all four sub-band decomposed image. Lorenz chaotic system[6] using the watermark is encrypted, the encrypted watermark information embedded into the image at all levels of the corner frequency matrix corresponding to the wavelet coefficients. This increases the amount of watermark embedding and watermark visibility and robustness are not.

2.2 Image Preprocessing

Image preprocessing is as follows:

- The watermark image $W(m \times m)$ do dimensional Aronld transform coefficients by scrambling to determine the watermark image to achieve the best scrambling to ensure the robustness of the watermark.

- Vector image of the Harris corner detection:

(1) Partial autocorrelation function of image intensity can be expressed as a Taylor polynomial:

$$E(x, y) = \sum_{u,v} |I_{x+u,y+v} - I_{u,v}|^2 = \sum_{u,v} w_{u,v}[xX + yY + O(x^2, y^2)]^2 \tag{1}$$

Where, I representative of the image pixel value.

(2) The image intensity of a point of extreme curvature of the autocorrelation function of matrix M can be approximated by the eigenvalues. In order not to find the eigenvalues of M, the use of Eq.2 and Eq.3:

$$Tr(M) = A + B \tag{2}$$

$$Det(M) = AB - C^2 \tag{3}$$

(3) Using the following formula to calculate the Harris corner algorithm functions:

$$R(x, y) = Det(M) - cTr(M)^2 \tag{4}$$

Just a point R(x, y) exceeds a certain threshold, the point is that the corner points. In this paper, the following method to determine in advance the image of the corner points, take the appropriate threshold number of points equal to the angle to get the angle you want to get points.

2.3 Watermark Embedding

Embedding algorithm is described below:

- Transforming the color space of true color host image of H, and the RGB color space conversion to LAB color space.
- LAB color space of the host image using wavelet analysis, selecting the four wavelet decomposition.
- After four wavelet decomposition, at all levels of the corresponding frequency component of matrix at the corner of embedded watermark, conor[i][j] is the carrier image decomposition of the corner location.
- The line of i and the column of j of watermarked image binary encrypted pixel b[i][j] embedded as follows:

$$If\,(\text{conor}[i][j]=1)\ \ M'_{ij} = M_{ij} \oplus b[i][j]$$

- The watermarked image of M'_i wavelet reconstruction, the reconstructed image is the watermarked image H_1.
- The image H_1 of space transformation from LAB to RGB color space. The resulting image of H' contains the watermark.

2.4 Watermark Extraction

Watermark extraction process is the inverse process of watermark embedding. Extraction process is as follows:

- True color images H' with watermark on the color space conversion. Be the image H_1.
- Four wavelet decomposition of images H_1, resulting in frequency components at all levels of the watermark image matrix. $M'_{ij}(i, j = 1,2,3,4)$
- The pixel value matrix of M'_i and $M_i(i, j = 1,2,3,4)$ compared to extract the watermark.
- Extracted watermark image section i line, j column binary encrypted the pixel $b'[i][j]$ extracted as follows:

$$if \ (\text{conor}[i][j]=1) \quad b'[i][j] = M_{ij} \oplus M'_i$$

- The extracted watermark information $b'[i][j]$ decrypted and reconstructed back 256-color image W'.
- Arnold the inverse transformation of scrambling of the watermark W', the watermark image W.

3 Experimental Results and Analysis

The true color 256×176 image with the original character image. Since the more corner points detected the time longer needed, but the greater the amount of embedded watermark. After several tests, taking into account the amount of elapsed time and the watermark. The watermark is a 32×32 256-color image.

Let M, N is the image height and width of the image after treatment compared to the original image I, the peak signal to noise ratio PSNR is defined as:

$$PSNR = 10 * \log_{10} \frac{3 * 255^2 * M * N}{\sum_{i}^{M} \sum_{j}^{N} \left[\left(R'_{i,j} - R_{i,j} \right)^2 + \left(G'_{i,j} - G_{i,j} \right)^2 + \left(B'_{i,j} - B_{i,j} \right)^2 \right]} \qquad (5)$$

Let L is the length of the watermark, the watermark embedding ω and extraction of the watermark ω', similarity NC measure is defined as:

$$NC = \frac{\sum_{i=1}^{L} \omega(i)\omega'(i)}{\sqrt{\sum_{i=1}^{L} \omega^2(i)} \sqrt{\sum_{i=1}^{L} \omega'^2(i)}} \qquad (6)$$

Fig. 1 shows the original image is not watermarked, watermarked image:

(a) the original image.   (b) the watermarked image.

Fig. 1. Comparison Chart watermark embedding

Fig. 2 shows the watermarked image, the extracted watermark image.

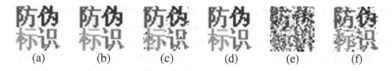

(a)   (b)   (c)   (d)   (e)   (f)

Fig. 2. Watermarked image and extracted watermark image

Figure 2 Comparison Chart watermark, (a) watermark image, (b) extracted watermark, (c) the watermark extracted after adding noise, (d) the watermark extracted 2 times magnification,(e) extracted watermark rotate 15 degrees, (f) cut 25% of the extracted watermark.From the chart we can see the true color watermarked image degraded the human eye can not detect the peak signal to noise ratio PSNR to 33.1db. PSNR value is greater than 30.00db. Indeed, the watermark information is hidden in the contour feature vector of image target area. The algorithm is fully reversible, in the absence of an attack can completely correct when the watermark information.

True color images with a watermark (256 176) carefully studied the various images of the algorithm's ability to resist attack. Test results include the following six areas: Gaussian smoothing, salt and pepper noise, shear testing, image compression, rotation test, mean filter. The test methods and parameters as shown in Table 1:

Table 1. The test methods and parameters

Testing methods	Parameters
Median filter	Filtering window is [5,5]
Gaussian smoothing	Filtering window is [3,3]
Shear test	25% cut image
Image compression	2 times magnification
Rotation test	Reverse rotation 15°

The results are shown in Table 2:

Table 2. Test results

Testing methods	Median filter	Gaussian smoothing	Salt noise	Shear test	Image compression	Rotation test
PSNR/db	24.3207	27.0574	19.3106	20.1293	33.1138	41.3931
NC	0.8585	0.8391	0.9822	0.8598	0.9998	0.9869

The experimental results show that the proposed watermarking algorithm have good robustness for various image attacks, especially for image compression, image rotation has good robustness.

4 Summary

Proposed a corner detection based on wavelet domain after the watermark embedding algorithm. The algorithm is based on the characteristics of the image contours using the idea of the 256-color image adaptive watermark sequence embedded into the wavelet transform the frequency coefficients. Using the detected corners, respectively, mapped to the IF at all levels of wavelet transform sub-images, so the image resist geometric attack and watermark embedding strength and resistance to compression than traditional algorithms. Experimental results show that: the algorithm against noise, attack, image compression, rotation attacks, attacks have cut robustness.

References

1. Wu, Q., Niu, X.-X., Yang, Y.(trans.), Katzenbeisser, S.: Petitcolasfap.-Information Hiding Steganography and digital watermarking. People's Posts and Telecommunications Press, Beijing (2001)
2. Cox, I.J., et al.: A Secure Robust Watermark for Multimedia. In: Anderson, R., et al. (eds.) IH 1996. LNCS, vol. 1174, pp. 185–206. Springer, Heidelberg (1996)
3. Ni, R., Ruan, Q.: Based on iteration mapping and image content adaptive watermarking algorithm. Communications 25(5), 182–189 (2004)
4. Zhang, X., Lei, M., Yang, D.: Multi-scale Curvature Product for Robust image Corner Detection in Curvature Scale Space. Pattern Recognition Letters 28(5), 545–554 (2007)
5. Voyatzis, G., Pitas, I.: Chaotic Watermarks for Embedding in the Spatial Digital Image Domain. In: IEEE ICIP, Chicago, Illinois, USA, October 4-7 (1998)
6. Ma, Z., Qiu, S.: One based on general cat map image encryption system. Communications 24(2), 51–57 (2003)

The Modeling and Simulation of First-Order Inverted Pendulum Control System

Zhiming Dong*, Lepeng Song, and Huanhuan Chen

Chongqing University of Science and Technology, Chongqing, China
zhmdong@yahoo.cn, slphq@163.com, chenhuanhuan@163.com

Abstract. In this paper, it give a research about the control problem of the balance of the first-order inverted pendulum. And this paper presents two methods of first-order inverted pendulum using PID, fuzzy controller and constructs the mathematical model of pendulum. The satisfied results are achieved with two schems. The simulation results show that the two schemes is satisfying. And the two schemes have their own characteristics.

Keywords: Inverted pendulum, Fuzzy control, Simulation.

1 Introduction

The inverted pendulum problem is one of the most important problems in control theory and has been studied excessively in control literatures. It is well established benchmark problem that provides many challenging problems to control design. The system is nonlinear, unstable, no minimum phase and under actuated. Because of their nonlinear nature pendulums have maintained their usefulness and they are now used to illustrate many of the ideas emerging in the field of nonlinear control [1]. The challenges of control made the inverted pendulum systems a classic tool in control laboratories.

PID control is a very typical control method in the field of automatic control. Its features is that it is very easy to operate and suitable for industrial field. Fuzzy logic controller, which is the logic on which fuzzy control is based, is much closer in spirit to human thinking and natural language than the traditional logical systems. In this paper, we will first build the Mathematical model of the inverted pendulum in section 2. And in section 3 and section 4 is construction of PID controller and fuzzy controller. At last it is simulation.

2 The Model of Inverted Pendulum System

Ignoring the air flow and all kinds of friction, we can abstracted inverted pendulum system into a car, uniformity and quality of blocks rod system. The ultimate goal of this application is to adjust the car position to make an inverted pendulum in the inverted vertical position[2]. The principle structure is shown Fig.1.

---

* The Natural Foundation of Chongqing City (cstc2011jjA1152).

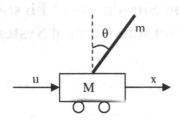

Fig. 1. Principle of single inverted pendulum structure

3 Inverted Pendulum Mathematical Modeling

The state space equations of system are[3-7]:

$$\dot{x} = AX + Bu$$
$$y = CX + Dn \tag{1}$$

Equations (1.1) on the solution of algebraic equations, by solving the following:

$$\dot{x} = \dot{x}$$

$$\ddot{x} = \frac{-(I + ml^2)b}{I(M + m) + Mml^2}\dot{x} + \frac{m^2 gl^2}{I(M + m) + Mml^2}\phi + \frac{(I + ml^2)}{I(M + m) + Mml^2}u$$

$$\dot{\phi} = \dot{\phi} \tag{2}$$

$$\ddot{\phi} = \frac{-mlb}{I(M + m) + Mml^2}\dot{x} + \frac{mgl(M + m)}{I(M + m) + Mml^2}\phi + \frac{ml}{I(M + m) + Mml^2}u$$

The last system state space equation are:

$$
\begin{bmatrix} \dot{x} \\ \ddot{x} \\ \dot{\phi} \\ \ddot{\phi} \end{bmatrix} =
\begin{bmatrix}
0 & 1 & 0 & 0 \\
0 & \dfrac{-(I + ml^2)b}{I(M + m) + Mml^2} & \dfrac{m^2 gl^2}{I(M + m)Mml^2} & 0 \\
0 & 0 & 0 & 1 \\
0 & \dfrac{-mlb}{I(M + m) + Mml^2} & \dfrac{mgl(M + m)}{I(M + m) + Mml^2} & 0
\end{bmatrix}
\begin{bmatrix} x \\ \dot{x} \\ \phi \\ \dot{\phi} \end{bmatrix} +
\begin{bmatrix}
0 \\ \dfrac{I + ml^2}{-I(M + m) + Mml^2} \\ 0 \\ \dfrac{ml}{I(M + m) + Mml^2}
\end{bmatrix} u
\tag{3}
$$

M the car quality   1.096Kg;
m pendulum mass   0.109 Kg;
B the frition coefficient of car   0.1N/m/sec;
l the length of pendulum   0.25m;
I pendulum inertia   0.0034 kg * m * m.

$$
y = \begin{bmatrix} x \\ \phi \end{bmatrix} =
\begin{bmatrix} 1 & 0 & 0 & 0 \\ 0 & 0 & 0 & 0 \end{bmatrix}
\begin{bmatrix} x \\ \dot{x} \\ \phi \\ \dot{\phi} \end{bmatrix} +
\begin{bmatrix} 0 \\ 0 \end{bmatrix} u
\tag{4}
$$

as external forces for input the system state equations are:

$$
\begin{bmatrix} \dot{x} \\ \ddot{x} \\ \dot{\phi} \\ \ddot{\phi} \end{bmatrix} = \begin{bmatrix} 0 & 1 & 0 & 0 \\ 0 & -0.0883167 & 0.629317 & 0 \\ 0 & 0 & 0 & 1 \\ 0 & -0.225655 & 27.8285 & 0 \end{bmatrix} \begin{bmatrix} x \\ \dot{x} \\ \phi \\ \dot{\phi} \end{bmatrix} + \begin{bmatrix} 0 \\ 0.883167 \\ 0 \\ 2.35655 \end{bmatrix} \dot{u}
$$

(5)

$$
y = \begin{bmatrix} x \\ \phi \end{bmatrix} = \begin{bmatrix} 1 & 0 & 0 & 0 \\ 0 & 0 & 1 & 0 \end{bmatrix} \begin{bmatrix} x \\ \dot{x} \\ \phi \\ \dot{\phi} \end{bmatrix} + \begin{bmatrix} 0 \\ 0 \end{bmatrix} \dot{u}
$$

When the input is pendulum angle and the output is cart 's displacement, the transfer function are:

$$
\frac{\Phi(s)}{X(s)} = \frac{0.0275s^2}{0.0102125s^2 - 0.26705}
$$

(6)

When the input is pendulum angle and the output is external forces, the transfer function are:

$$
\frac{\Phi(s)}{U(s)} = \frac{2.35655s}{s^3 + 0.0883167s^2 - 27.9169s - 2.30942}
$$

(7)

4 Fuzzy Controller

Inverted pendulum system has four input variables, if we use a fuzzy controller, each input variable has five linguistic values, so the rules may be up to more than 100. It has so many rules that we are difficult to design, therefore, we design two fuzzy cotrollers which contain the angle fuzzy controller and the displacement controller, controlling the angle and displacement respectively[7]. Two fuzzy controllers will be designed in parallel or in series to control inverted pendulum, fuzzy controller structure was changed in order to achieve the lower the dimension of the controller.

According to the force U1 and of the motor output are relation to the displacement x of car and the the rate of change displacement x', we can establish follow fuzzy logic control rules[8,9].

Table 1. Rules between displacement and speed

U_1		The displacement of car x						
		PL	PM	PS	O	NS	NM	NL
The speed of car x`	PL	NL	NL	NL	NS	NS	NM	PL
	PM	NL	NL	NS	NS	NM	O	PL
	PS	NL	NS	NS	NM	O	NM	PL
	O	NL	NS	NM	O	PM	PS	PL
	NS	NL	NM	O	PM	PS	PS	PL
	NM	NL	O	PM	PS	PS	PL	PL
	NL	NL	PM	PS	PS	PL	PL	PL

Table 2. Rules between displacement and angular

U_2	The angle of the pendulum θ						
	PL	PM	PS	O	NS	NM	NL
angular velocity of pendulum θ' PL	PL	PL	PL	PS	PM	O	NL
PM	PL	PL	PS	PM	O	NM	NL
O	PL	PS	PM	O	NM	NS	NL
NS	PL	PM	O	NM	NS	NL	NL
NL	NL	PM	PS	PS	PL	PL	PL

According to the force U2 and of the motor output are relation to t the angle of the pendulum θ and the the angular velocity of the pendulum θ'[8], we can establish follow fuzzy logic control rules

The design in matlab / simulink simulation platform is shown in figure 2

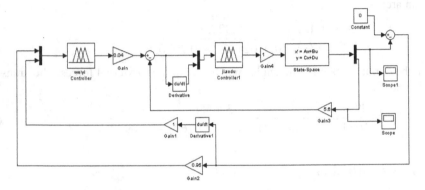

Fig. 2. The simulation model of inverted pendulum

5 Simulation Results

Simulation results with fuzzy control and PID control for inverted pendulum are put in the same scope, We can get the angle-curve in figure 3 and the displacement -curve in figure 4.

Through figure 10 and figure 11of the inverted pendulum control, the fuzzy control was significantly better than the PID control, fuzzy control 's overshoot is very small, and the adjust-time also is very small, it only has 0.8 seconds. The adjust-time of PID contro is 9 seconds, the overshoot is also larger relative to fuzzy control.

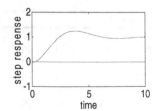

Fig. 3. Angle-curve

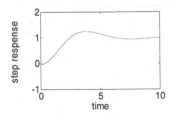

Fig. 4. Displacement -curve

6 Conclusion

This paper we established a mathematical model of the first-order inverted pendulum at first, and then designed PID controller and fuzzy controller for first-order inverted pendulum. We obtain good simulation curves from the two kinds of control, so it show that the angle of the pendulum and the the displacement of the car are controlled well by the PID-controller and fuzzy-controller. But compare to PID controller, the adjust-time of fuzzy controller is shorter, overshoot is smaller and the control accuracy is also higher.

References

1. Sun, L., Kong, H., Liu, C., Bi, L.: Inverted pendulum and research. Machine Tools and Hydraulic 104(7), 79–83 (2008)
2. Wang, J.Y., Liu, M., Li, H., Miao, Z.: The same value of fuzzy Controller output for two sufficient conditions. Control Theory and Applications 6(14), 324–327 (2009)
3. Li, H.: Fuzzy control system modeling. Chinese Science 32(9), 772–781 (2002)
4. Yang, W.: The first order based on MATLAB inverted pendulum control system modeling and simulation. Application of Electronic Components 1, 29–31 (2007)
5. Yang, S., Xu, L., Wang, P.: Single PID control of inverted pendulum. Control Engineering 6(14), 26–30 (2007)
6. Zhang, W., Wang, Y., Hong, Q., Miao, G.: Controlled inverted pendulum system based on principles of experimental design. Modern Educational Equipment 15(39), 35–39 (2010)
7. Li, H., Song, W., Yuan, X., Li, Y.: Based on Fuzzy reasoning. The time-varying system modeling systems. Science and Mathematics 8(13), 82–86 (2009)
8. Li, H.X., Yuan, X., Wang, J., Li, Y.: Fuzzy Systems and Fuzzy norm of the classification system. Chinese Science: Information Science 12(5), 135–140 (2010)
9. Ren, B., Li, Z.: Inverted pendulum system based on fuzzy human-simulated intelligent control. Science, Technology and Engineering 15(44), 167–171 (2010)

Based on Fuzzy PID Control of AC Induction Motor Vector Control System

Wang Yuhua and Miao Jianlin

College of Physics & Electronics Taizhou University Taizhou City Zhejiang Province 318000
Wangyh977@126.com

Abstract. AC induction motor for nonlinear, strong coupling, time-varying characteristics, combined with characteristics of the fuzzy control and the traditional PID control, in order to further improve performance of the AC induction motor vector control system, AC induction motor vector control system was designed based on Fuzzy-PID control . By matlab / simulink simulation experiments, we can see that the static and dynamic features of the nonlinear system was improved by this design method, allowing the system to obtain a better performance.

Keywords: AC induction motor, vector control system, Fuzzy PID, simulation.

1 Introduction

Since the emergence of vector control system [1], induction motor can be controlled like a DC motor as in industrial applications requiring high precision in the use of vector control system is to produce the same momentum as the rotating magnetic criteria, on the coordinate system in three-phase AC current through the stator three-phase / two-phase transformation can be equivalent to two-phase stationary coordinate system on the AC current through the synchronous rotation transformation is equivalent to DC. In recent years, in order to further improve the control effect of vector control system, many researchers have done a lot of research, such as fuzzy control of vector control systems: fuzzy control is a rule-based control, in the design of the controlled object does not require the establishment of precise mathematical model of [2-3]. Although fuzzy control system can obtain good dynamic performance, but the static error can not be eliminated. Conventional PID controller is the deviation of the entire control system adjusted so that the actual value of the controlled object and the predetermined value is consistent. But the main problem conventional PID controller is the PID parameter tuning, the control process will not be changed, so difficult to achieve the best control effect [4].

For ordinary fuzzy control and conventional PID controller problems, combined with conventional PID and the general advantages of fuzzy controller, this paper presents the use of fuzzy PID control AC induction motor vector control system model, the traditional PID and fuzzy PID control of induction motor vector control system for

D. Jin and S. Lin (Eds.): Advances in ECWAC, Vol. 2, AISC 149, pp. 227–232.

matlab / simulink simulation, can be seen from the simulation results, the fuzzy PID controller has better control performance.

2   AC Induction Motor Vector Control System

Asynchronous motor is a multivariable, strong coupling, higher-order nonlinear time-varying systems. The basic method of vector control is the use of three-phase / two-phase coordinate transformation. In this transformation, the Cartesian coordinates on the complex three-phase system is transformed into a two-phase plane coordinate system, with its two-phase dq axis coordinate system. Coordinates of the rotor shaft by the magnetic field orientation induction motor model can be derived under a model of the magnetic flux vector control system and coordinate transformation relations.

$$T_r \frac{|i_{mr}|}{dt} + |i_{mr}| = i_d \qquad (1)$$

$$\omega_{mr} = \omega_r + \frac{i_q}{T_r |i_{mr}|} \qquad (2)$$

$$\begin{bmatrix} i_d \\ i_q \end{bmatrix} = \sqrt{\frac{2}{3}} \begin{bmatrix} \cos \theta & \sin \theta \\ -\sin \theta & \cos \theta \end{bmatrix} \begin{bmatrix} 1 & -\frac{1}{2} & -\frac{1}{2} \\ 0 & \frac{\sqrt{3}}{2} & -\frac{\sqrt{3}}{2} \end{bmatrix} \begin{bmatrix} i_A \\ i_B \\ i_C \end{bmatrix} \qquad (3)$$

T_r - induction motor rotor time constant; i_{mr} - The rotor excitation current

i_d and i_q - the rotor shaft stator field oriented coordinates d and q-axis current component. θ - three-phase stationary coordinate system to any two-phase rotating coordinate system transformation, d-axis and the q-axis angle

i_A、 i_B、 i_C - Three-phase winding is the stator phase current zero line without a star connection, $i_C = -i_A - i_B$

ω_r - Rotor angular velocity ; ω_{mr} - Rotor flux angular velocity

Shaft with the rotor field-oriented coordinates given under the stator current reference flux models and replace the amount of the actual current and, and assuming control system without weakening, the rotor excitation current that is constant, you can remove the style (1) of the differential term, which can construct a type of indirect vector control system. Vector control system block diagram shown in Figure 1:

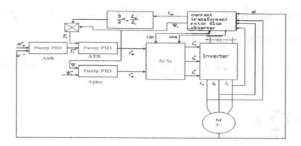

Fig. 1. Fuzzy PID vector control system diagram

Vector from the fuzzy PID control system structure to mimic the process of DC motor, flux regulator and speed regulator and control, respectively, in order to completely decouple the two subsystems, in addition to coordinate transformation, but with a fuzzy-PID control device, try to offset the rotor flux on the impact of the electromagnetic torque, this is mainly the speed regulator output signal divided by, when the controller coordinates the anti-conversion and motor coordinate transformation for the consumer, and this time, the inverter hysteresis can be ignored, then, the two systems in the fuzzy-PID, under the mediation can be seen as two completely independent subsystems.

3 Fuzzy PID Controller Design

Fuzzy PID controller based on fuzzy rules regulate the PID parameters of a real-time adaptive control system. Fuzzy rules is given in a different state of the PID parameters of real-time reasoning results.

3.1 Fuzzy PID Control Principle

This design of fuzzy PID controller based on the error e and error change rate ec as input. It can meet the different moments of e and ec self-tuning PID parameters required, the use of fuzzy control rules in real-time correction of the PID parameters can constitute a fuzzy PID controller.

Fuzzy PID controller PID tuning formula is:

$$\begin{cases} K_p(k) = K_p(k-1) + \gamma_p \Delta K_p \\ K_i(k) = K_i(k-1) + \gamma_i \Delta K_i \\ K_d(k) = K_d(k-1) + \gamma_d \Delta K_d \end{cases} \tag{4}$$

Which γ_p、 γ_i、 γ_d was correct speed, with the next formula can be seen with the current controller parameters can be fuzzy reasoning controller parameters and controller parameters derived from the weighted incremental and composition. Control the amount of time was the formula is:

$$u(k) = K_P(k)e(k) + K_i(k)x(k) + K_d(k)[e(k) - e(k-1)\}$$ (5)

Which : $x(k) = T \sum_{j=0}^{k} e(j)$;

3.2 The Establishment of Fuzzy Rules

The output K_P、K_i、K_d of fuzzy controller, as conventional PID controller proportional, integral, differential correction value input parameters, K_P、K_i、K_d, the fuzzy sets defined as {NB, NM, NS, ZO, PS, PM, PB}, will be mapped to their domain of [-3,3] on. Below is e, ec and kp, ki, kd membership function curve.

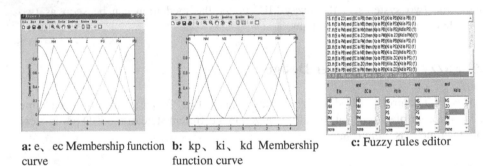

a: e、 ec Membership function curve   **b:** kp、 ki、 kd Membership function curve   **c:** Fuzzy rules editor

Fig. 2. Membership function curve and Fuzzy rules editor

In the fuzzy PID controller design, the core of fuzzy control rules, fuzzy control rules directly determine the effect of control of the controller.

In order to obtain good control results, K_P、K_i、K_d and e, ec are related as follows:

(1) when $|e|$ is large, K_P should be larger and K_i、K_d should be small, it will give the system response speed, and avoid excessive overshoot;

(2) $|e|$ as such, K_P should be smaller, so that overshoot smaller, K_i、K_d taked the appropriate value, then pay special attention to K_d greater impact on the system;

(3) When $|e|$ the small, K_P、 K_i should be larger, the system has good steady-state performance, and K_d should be appropriate to avoid a shock in the balance.

4   Simulation Experiment

Based on the above design, the fuzzy control, the traditional PID, Fuzzy PID simulation model shown in Figure 4, in which the simulation picture diagram of fuzzy PID and Figure b, respectively, a packaged composition.

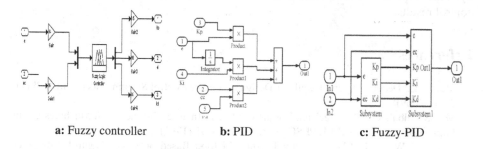

a: Fuzzy controller   b: PID   c: Fuzzy-PID

Fig. 3. Simulation model of controller

In this simulation, the selection of the motor parameters are: $P_N = 4kW$, $U_N = 380V$, $f_N = 50Hz$, $n_N = 1450r / \min$, the stator resistance $R_s = 0.435\Omega$, stator inductance $L_s = 0.071H$, rotor resistance, $R_r = 0.816\Omega$ rotor inductance $L_r = 0.071H$, mutual inductance $L_m = 0.069H$, rotational inertia. $J = 0.19kg \bullet m^2$. 0.7s add to the load. The speed of response curve:

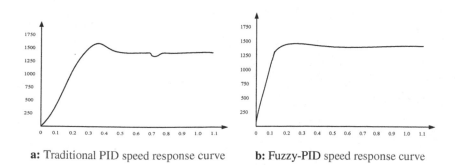

a: Traditional PID speed response curve   b: Fuzzy-PID speed response curve

Fig. 4. Speed response curve

5 Conclusion

This induction motor vector control system for the design of fuzzy PID controller, fuzzy controller and integrated the advantages of the traditional PID controller, not only to overcome the controlled object parameter variations and disturbances, in addition a fast response should be the best case dynamic parameters, so as to improve system performance objectives. Through the fuzzy PID controller and conventional PID controller simulation, the simulation results show that fuzzy PID control and conventional PID control system, although both contain integral control action, so the steady-state accuracy of the same, but with a fuzzy PID faster than the traditional PID dynamic response, smaller overshoot. So you can see in the fuzzy PID control, better control results.

References

1. Ma, X.-L.: Decoupling and Regulator Design of Vector Control System. Electricdrive 39(1), 3–10 (2009)
2. Mir, S., Elbuluk, M.E.: Precision Torque Control in Inverter-fed Induction Machines Using Fuzzy Logic. In: Proc. IEEE PESC 1995, pp. 396–401 (1995)
3. Dong, J., Wang, J., Ma, L.: Motor Loading Method Based on Fuzzy Control. Computer Measurement & Control 14(9), 1180–1182 (2006)
4. Yan, P., Tao, Z.-S., Zhao, Z.-H.: The PID Control System for Stepping Motor Based on the Improved Simplex Method. Drive and Control (8), 49–51 (2008)

Java Fault Emulation and a Preliminary Fault Injection Framework

Qiuhong Zheng

Computer Science and Information Technology School,
Zhejiang Wanli University, Ningbo, 315100, China

Abstract. Fault injection plays a critical role in the verification of fault-tolerant mechanism, software mutation testing and dependability benchmark for computer systems. Derived from the common java bug patterns from Findbugs, we propose a new scheme for emulation of software faults at the java bytecode level, and have developed a preliminary framework called JBugInjector to support fault injection application. Different from the traditional Orthogonal Defect Classification of faults, we analyze the peculiar bug patterns in java program development, and introduce these faults and bad practices into the target system to construct a scenario closer to reality. Through demonstration with simple examples, we find it possible to emulate many faults by mutation at bytecode level. With this method, fault injection without the source code of target application is enabled to assist dependability benchmark.

Keywords: Fault Injection, Software Fault Emulation, Bug Pattern, Findbugs.

1 Introduction

With the widespread application of computer systems, higher level of dependability is demanded. Many studies indicate that software faults led to more systems failures than hardware and take an increasing proportion as the software complexity grows [1].

Dependability benchmarking method has recently caught researchers' attention when examining dependability of computer systems and a research organization in Europe specially started the DBench [2] project to explore this method.

Fault injection is a technique for introducing faults into the target system according to certain fault models, and the behavior of the target is observed and recorded. Through this process, some qualitative or quantitative measures could be obtained for further analysis. Obviously, higher fault simulation accuracy is a necessity in development of fault injection tools to reproduce the real faults. So as to simulate software faults more accurately, common faults in industrial programs must be collected and classified, to establish fault models.

In this paper, based on the characteristics of software faults in java development, we analyze some common bug patterns detected by Findbugs [3], and propose a new scheme for java bug emulation, which apply mutation directly at the java bytecode level to inject faults into target system. To facilitate the use of the scheme, we design a preliminary framework and develop the corresponding tool called JBugInjector. Our

D. Jin and S. Lin (Eds.): Advances in ECWAC, Vol. 2, AISC 149, pp. 233–240.
springerlink.com   © Springer-Verlag Berlin Heidelberg 2012

scheme is specific to problems in OO development and java language features, and also introduces bad practices as potential faults. As reusability is emphasized in OO design and development, some bad practices and styles which are not faults originally might incur new faults after some kind of inheritance and reuse.

The remainder of this paper is organized as follows: Section 2 focuses on the related work about fault injection; In section 3, a new fault injection scheme is proposed with detailed analysis of common bug patterns; Section 4 consists of the design of JBugInjector and it functionality is demonstrated by working on demo programs; In Section 5, we conclude our research work and future works are given.

2 Related Works

Some fault injection tools [4,5,6,7] mainly deal with the common hardware faults. As the characteristic of hardware faults has been analyzed in great detail, and is relatively simple, the accuracy of simulation is out of the question.

Depending on the peculiarity of software, fault injection through mutation to emulate software faults is first proposed in [8], which is based on the analysis of over 500 high language faults from 9 common used programs, and a set of fault emulation operation is provided for IA32. It is showed that most of the mutations provide good accuracy in comparison with modifying source code directly. In [9], the researchers proposed a fault injection pattern system, but it is not perfect for software fault injection. JACA [10] implement the proposed pattern system based on Javassist, and injects faults by passing corrupted parameters.

However, features of Object-Oriented development are neglected in all these work. Regarding faults of OO development, Allen summed up 14 bug patterns according to his development experience [11]. Hovemeyer and Pugh proposed a bug pattern detecting technique by scanning target system [12], and implemented the famous Findbugs tool [3].

3 Scheme for Java Bug Pattern Emulation

To eliminate the spatial and time overhead, we propose to inject faults at java bytecode level by mutation, while achieving the same effect with modification at source code. The emulation of fault makes demands on deciding where and how to inject faults, which are solved by finding scan patterns and injection operation respectively. We analyze 28 detector patterns grouped in 10 classes, and propose 25 fault injectors. Fault emulation scheme for all the bug patterns above is shown in Table 1. In the following, we just show some typical patterns for illustration. The names of patterns are borrowed from [3].

3.1 CN Fault Class

CN_IDIOM_NO_SUPER_CALL: *Clone* method does not call *super.clone ().*

This non-final class defines a *clone()* method that does not call *super.clone()*. If this class ("*A*") is extended by a subclass ("*B*"), and the subclass *B* calls *super.clone()*, then it is likely that *B*'s *clone()* method will return an object of type *A*, which violates the standard contract for *clone()*. To emulate this fault, we should not only remove the invocation of *super.clone()*, but also create a new object as the return value.

3.2 DE Fault Class

DE_MIGHT_DROP: Method might drop exception. Sometimes in a try-catch block, there is no content in the catch statement, so the exception has no handler at all. Injector for this bug pattern involves deleting the exception handlers in the catch statements.

3.3 EC Fault Class

Because of the similarity of attributes names, programmers might be confused when referencing objects for comparison.

EC_ARRAY_AND_NONARRAY: This method invokes the *equals (Object o)* to compare an array and a reference that doesn't seem to be an array.

Through careful examine of the bug patterns, the common result is causing the *equals* method to constantly return false. To achieve the faulty condition, there are two possible ways: one is to change the value to be returned inside *equals* method, and the other is to find the target that stores the return value, and change the target variable to false.

3.4 EQ Fault Class

Java classes may override the *equals (Object)* method to define a predicate for object equality. Programmers sometimes mistakenly use the type of their class Foo as the type of the parameter to *equals( )*: *public boolean equals (Foo obj) {...}*

This covariant version of *equals ( )* does not override the version in the *Object* class, which may lead to unexpected behavior at runtime.

EQ_SELF_USE_OBJECT: The class defines some type of the *equals* method, but does not override the boolean *equals (Object)* method.

3.5 HE Fault Class

In order for Java objects to be stored in HashMaps and HashSets, they must implement both the *equals (Object)* and *hashCode ( )* methods. The important hidden rule is: Objects which compare as equal must have the same hashcode.

HE_EQUALS_NO_HASHCODE&HE_EQUALS_ USE_HASHCODE: This class overrides *equals (Object)*, but does not override hashCode(). Two ways are possible to simulate the bug: when a class defines both *equals* and *hashCode* methods, we can either delete the *hashcode* method or substitute it with *Object.hashCode*.

This type of bugs derives from the mismatch of equals and hashcode methods. As a result, any operation that breaks the correct relation of the two methods will introduce the fault.

3.6 IS2 & DC Fault Class

The common error to realize thread safety is to allow accessing or modifying shared state without synchronization. The reason for grouping IS2 and DC patterns together is they are both related with incorrect synchronization, and emulations of fault are the same.

DC_DOUBLECHECK: This method may contain an instance of double-checked locking. This idiom is not correct according to the semantics of the Java memory model.

IS2_INCONSISTENT_SYNC: The fields of this class appear to be accessed inconsistently with respect to synchronization. A typical bug matching this bug pattern is forgetting to synchronize one of the methods in a class intended to be thread-safe.

Emulation method is to cancel the synchronization mechanism by deleting the synchronized keyword.

Table 1. Scheme for bug patterns

No	Pattern Name	Scanning Method	Injecting Method
1	CN	1.Class implements cloneable interface 2. Class override the clone() method	Remove clone method
2		The same to the above	Remove the invocation of super.clone(), and create a new object as the return value
3	DE	1. Mmethod contains exception handle table attribute. 2. Get the exception handle code.	The exception handle code is made invalid.
4	EC	1. class correctly overrides equals(Object) 2. Get the Code attribute of the equals method	The original boolean equals(Object) method is made invalid. The return value of the new one is fixed with true/false.
5		1.The equals(Object) is invoked in the code. 2. Get the invocation point of the equals(Object)	Getting the invocation point and making the return invalid. The return value will be fixed with true or false.
6	EQ	Class correctly overrides equals(Object)	Modify the signature of the equals method, to violate the standard contact for boolean equals(Object).
7		The same to the above	The original equals(Object) method is made invalid. The new one inherits from the Object.equals(Object)
8		The same to the above	The same to the above
9		The same to the above	The same to the above
10		The same to the above	Remove hashCode() method.
11	HE	Class overrides hashCode()	The original int hashCode() method is made invalid. The new one inherits from Object.hashCode().
12		The same to the above	Remove equals(Object) method.
13		The same to the above	The original boolean equals(Object) method is made invalid. The new one inherits from Object.equals(Object).
14		The same to the above	The same to the above
15	IS2 DC	Existing Synchronization operation code	Remove the synchronized keyword
16	MS	Static private mutable fields are defined	The private keyword is replaced with public keyword.
17		Package static mutable fields are defined	The package keyword is replaced with public keyword.
18		Static final mutable fields are defined	The final keyword is removed.
19		Non-static private mutable fields are defined	The private keyword is replaced with public keyword.
20		The same to the above	The same to the above
21		The same to the above	The final keyword is removed
22	NP	The clone method has been defined	The original clone method is made invalid. A null value is returned by the new one.
23		The toString method has been defined	The same to the above
24		1. The equals method has been defined 2. The equals(Object) check for null as the arguments	Remove the statements of checking for the null argument.
25	OS	Existing the close invocation of IO stream	Remove the invocation of close method of IO stream

3.7 NP Fault Class

The fact that some commonly used methods might return a null pointer is neglected by most programmers, such as the *clone* and *toString* method.

NP_CLONE_COULD_RETURN_NULL: This *clone* method seems to return null in some circumstances, but clone is never allowed to return a null value. Recreating the condition is difficult, but it can be emulated by simply designating the return value of clone to be a null reference.

3.8 OS Fault Class

OS_OPEN_STREAM&OS_OPEN_STREAM_EXCEPTION_PATH: When a program opens an input or output stream, it is good practice to ensure that the stream is closed when it becomes unreachable. The emulation involves deleting the invocation of closing methods.

4 Design and Preliminary Validation of JBugInjector

4.1 JBugInjector Framework

In order to facilitate the application of the emulation, we develop a framework named JBugInjector, that supports for both dynamic and static application of fault injection.The interactions between modules are shown in Fig. 1, the arrows represent the dataflow.

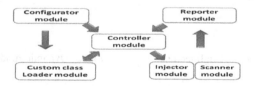

Fig. 1. Structure of JBugInjector

Configure Module. It reads configuration information from XML files, including target program, desired scanning patterns and emulations specified by the user.

Scanner Module. It provides a library of scanners. Different scanners are organized according to the structure information to find their scan patterns.

Injector Module. It provides a library of injectors. Once the scan patterns are ready, injectors are activated to emulate corresponding faults. After one injection, either a faulty version is generated, or the class loaded contains the specified fault.

Reporter Module. It mainly assists in receiving the information provided by scanner and injector, and generating reports in XML formats.

Classloader Module. To enable dynamic modification of java classes, a customized classloader is needed to get the target class once it is loaded, and notifies the controller to start injectors.

Controller Module. It is responsible for organizing and controlling the fault injection process. Depending on the settings, the controller could run in both static and dynamic modes.

In static mode, the controller first asks the configure module to provide the target project and the injection rules. Then the controller starts the corresponding scanners and injectors according to the rules. Once a scanning pattern is recognized, the fault is injected to form a new faulty version. This process is repeated until all the possible faults are injected. Finally, the scanning and injection report will be provided to record the fault record, and all the faulty versions are generated for later use. The whole procedure is depicted in Fig. 2.

In dynamic mode, information is extracted from the XML file, and the user is asked to select the desired locations and faults to be injected. After that, the class loader activates the target program, and if the loaded class is in the injection scope, the corresponding fault is injected. Once the target program completes, the controller will start a new round of loading and injecting. When all the specified faults have been processed, the controller finishes the experiment, and generates final reports. Fig. 3 shows the execution diagram.

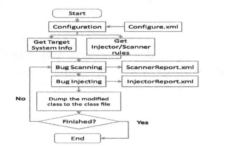

Fig. 2. Static mode of JBugInjector   **Fig. 3.** Dynamic mode of JbugInjector

4.2   Validation

To validate the efficacy of fault emulation, we apply JBugInjector to some demo programs, and compare the difference between the correct version and the modified class.

Two bug patterns, CN_IDIOM and HE_HASHCODE_NO_EQUALS are chosen to illustrate the effect of fault emulation. Experiments of the other patterns are similar.

Result of Fault Injector: CN_IDIOM (Table 2)
After fault emulation, the clone method is deleted from the class, thus successfully introduce the CN_IDIOM bug.

Result of Fault Injector: HE_HASHCODE_NO_EQUALS (Table 3)
For this bug pattern, we first encertain that both equals and hashCode are defined. According to our scheme, the equals method is deleted to emulation HE faults.

Table 2. Results of fault injection for CN_IDIOM

Information about original class	Modified class
public class HelloWorld extends ... implements java.lang.Cloneable{ public java.lang.String get(); public java.lang.Object clone() throws java.lang.CloneNotSupportedException; public boolean equals(java.lang.Object); }	public class HelloWorld extends ... implements java.lang.Cloneable {... public java.lang.String get(); public boolean equals(java.lang.Object); ...}

Table 3. Results of fault injection for HE_HASHCODE_NO_EQUALS

Information about original class	Modified class
public class HelloWorld extends ... implements java.lang.Cloneable{ ... public java.lang.Object clone() throws java.lang.CloneNotSupportedException; public boolean equals(java.lang.Object); public int hashCode(); ...}	public class HelloWorld extends ... implements java.lang.Cloneable {... public java.lang.Object clone() throws java.lang.CloneNotSupportedException; public int hashCode(); ...}

5 Conclusion

In this paper, we analyze several common bug patterns from Findbugs and propose the emulation of bug patterns at java bytecode level. A new preliminary framework called JBugInjector is developed to ease the application of fault injection.

Given the fact that it is the first try of our study, we mainly focus on the exploration of feasibility of the proposed method. Through the analysis and practice in this paper, we confirm that emulation at bytecode level provides a viable solution for fault injection in java programs. In the future, more common bug patterns will be processed with experimental evidence to support the efficacy of emulation scheme.

References

1. Chillarege, R., Bowen, N.: Understanding large systems failures - A fault injection experiment. In: Proc. of the 19th Int. Symp. on Fault-Tolerant Computing, Chicago, pp. 356–363 (1989)
2. Madeira, H., Kanoun, K., Arlat, J., et al.: Conceptual framework, deliverable CF2, preliminary dependability benchmark framework. DBench Project, IST 2000-25425 (2001)
3. Findbugs, http://findbugs.sourceforge.net/
4. Barton, J.H., Czeck, E.W., Segall, Z.Z., et al.: Fault Injection Experiments Using FIAT. IEEE Transactions on Computers 39(4), 575–582 (1990)
5. Kao, W.-L.: FINE: A fault injection and monitoring environment for tracing the UNIX system behavior under faults. IEEE Transactions on Software Engineering 19(11), 1105–1118 (1993)
6. Kanawati, G., Kanawati, N., Abraham, J.: FERRARI: A Tool for the Validation of System Dependability Properties. In: Proc. of the 22nd IEEE Int. Symp. on Fault Tolerant Computing, Boston, pp. 336–344 (1992)

7. Carreira, J., Madeira, H., Silva, J.G.: Xception: A Technique for the experimental evaluation of dependability in modern computers. IEEE Transactions on Software Engineering 24(2), 125–136 (1998)
8. Duraes, J., Madeira, H.: Definition of software fault emulation operators: A field data study. In: Proc. of Int. Conf. on Dependable Systems and Networks, San Francisco, pp. 105–114 (2003)
9. Leme, N.G.M., Martins, E., Rubira, C.M.F.: A Software Fault Injection Pattern System. Technical Report of Institute of Computing, Unicamp (2002)
10. Martins, E., Rubira, C.M.F., Leme, N.G.M.: Jaca: A Reflective Fault Injection Tool Based on Patterns. In: Proc. of International Conference on Dependable Systems and Networks (DSN 2002), pp. 483–487 (2002)
11. Allen, E.: Bug patterns in Java, 2nd edn. Apress (2002)
12. Hovemeyer, D., Pugh, W.: Finding bugs is easy. ACM SIGPLAN Notices 39(12), 92–106 (2004)

The Wireless Sensor Network Security Protocol Research in Internet of Things

John Lee[1], Hua Zhou[1], Zhihong Liang[1], Xiangcheng Wan[2], and Yuhong Chen[3]

[1] Yunnan Software Engineering Key Laboratory, Yunnan University, Kunming, China
[2] The 96224 Troops of the Chinese People's Liberation Army
[3] Yunnan Traffic Management Research Institute
Lijunlin8662@yahoo.com.cn, HZhou@ynu.edu.cn,
728500572@qq.com, wanghui2100@sina.com,
wanghui2100@sina.com, ynchyh@vip.tom.com

Abstract. Due to security protocols be Proven "safety" are few use in internet of things now, and design and analysis them is a very difficult thing, this paper research on the establishment of wireless sensor network security protocol in internet of things(IOT), the protocol provide a variety of services such as the mutual authentication of communication agents, the distribution extemporaneous public key, the encryption and decryption of dates and the sending or receiving of acknowledgement and so on in the environment of IOT, so security protocols must satisfy relational security properties, firstly, this paper analyses the architecture of internet of things and abstracts the communication agents satisfied the physical characteristics and the requirements of application of internet of things, and then put forward a security protocol(WSNIT) based on the requirements and characteristics of communication agents in internet of things, the lastly, testified this security protocol satisfied the SK security in AM model.

Keywords: Internet of things, wireless sensor network, security protocol, SK security.

1 Introduction

The technique of Internet of things is a new application technique, people often hot for development and the study of application, but neglected the security problem that is more important of wireless sensor network in internet of things [1]. Comparing wireless sensor network of internet of things (WSNIT) with traditional sensor network, the biggest difference is the gotten dates are used for local service as well as transmit to remote service by internet in wireless sensor network of internet of things [2], so, the architecture of wireless sensor network of internet of things can be abstracted base on as follow:

D. Jin and S. Lin (Eds.): Advances in ECWAC, Vol. 2, AISC 149, pp. 241–247.

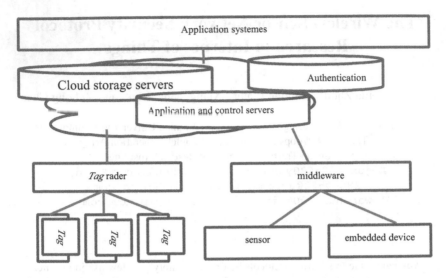

Fig. 1. Architecture of wireless sensor network of internet of things

In the figure 1, for the convenience of analysis, all servers in internet of things such as cloud storage servers, authentication servers, application and control servers and so on, all the servers are called "clouds servers", the communication and security protocol between these servers can use of the current mature equipment and techniques [3], and it isn't within the scoped of the present article.

The servers in the middle are middle equipment and the abstract of local servers between cloud and sensor equipment [4], they may consist of some servers, internet equipment, computing equipment or processing and transmission equipment, if cloud need authentication and date communication with Tag label[5], then the middle server in above figure is consist of read write device of Tag and corresponding equipment etc.

2   The Characteristics of WSNIT

Comparing with existing wireless sensor network, WSNIT has some characteristics as following [6]: (1). WSNIT need to establish integrated deal Centre, manage and control Centre and integrated remote storage Centre, which can deal with all kinds of dates integrated solid to satisfy the application requirement of different departments, industry, and users. The existing sensor network didn't form a uniform management mechanism and it often is limited to inner of company, even can't connected to the Internet, as well as users are limited to inner of company, and then composition more simple. (2). WSNIT need provide both remote and local service, while existing sensor network need only local service. (3). For there are many kinds of internet application in WSNIT, assume the internet among authentication servers, middle servers and sensor equipment is not safe and may existing attacker, then only assume the internet between sensor equipment and the equipment of receiving sensor signal are not safe while other parts use wired connections and secure technologies in existing wireless sensor internet.

3 Communication Agents of WSNIT

According to the analysis, the communication agents of WSNIT as shown in Figure 2:

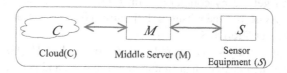

Fig. 2. Communication agents of WSNIT

In WSNIT, use a large of electronic labels and sensor equipment, and simplify the design of sensor, try to lengthen the using time and reduce production cost of sensor is an import fact, with security conditions, to design of sensor as simplify as possible. Base on thin consider, abstracted the communication agents of WSNIT.

In the above figure, assume that the way of transmit datagram between middle servers and sensor equipment is wireless communication channel, and attacker can get all messages of communication [7], while the transmit channel between cloud and middle servers is a secure and guaranteed channel that it has a mature of internet secure mechanism.

This network architecture is also fit for RFID system [8], C is equal to databases in the rear of cloud, M is s equal to the Tag that a read and write device, S is equal to label Tag.

4 Secure Protocol of WSNIT

In protocol, besides random number, all other messages were encrypt by one-way hash function based on K when it were transmitted, in the design of this protocol, set the way of encryption and transition of protocol as the base of design, lastly make CK verification to protocol.

Assume that cloud C sends inquired request to M and make verification to S firstly, according to the function, protocol is design as follow:

C products a secret random number Rc, and sends a inquired request to M and then sends Rc to M;

M sends Rc to S, after it receives request.

S computes $M1= Hk (Rc \oplus Rs \oplus IDs)$ and then sends $(M1, Rs)$ to M after it receives the request.

M products a random number Rm, and get $M2= Hm(Rc \oplus Rm \oplus IDm)$ by m that the shared key between M and C, then sends $(M1,M2,Rs,Rm)$ to C;

C needs to authenticate for each of M and S, when it authenticates M, authentication servers check for weather there is a $IDi(1 \leq i \leq h)$ and shared key k, the item h represents the number of sensors, making the equation $Hk(Rc \oplus Rm \oplus IDj)=M1$ satisfied, when it authenticates S, authentication servers check for weather there is a $IDi(1 \leq i \leq n)$ and shared key k, the item n represents the number of sensors, making the equation $Hk(Rc \oplus Rs \oplus IDi)=M1$ satisfied, if it has, then authentication is pass, and a new key Kc is generated and the Hm(IDj⊕Rm), Hm(IDj)⊕Kc, Hk(IDi)⊕Kc,Hk(IDi⊕Rs)⊕IDj are sent to M.

After M received it, computes itself with the key M shared with C and check for weather the result equal to Hm(IDm⊕Rm), if they are equal, then C authentication is passed, and then get Kc by computing Hm(IDm)⊕Kc, and get Hk(IDi⊕Rs) by the computing of XOR between itself's IDj with Hk(IDi⊕Rs)⊕IDj, and then sends Hk(IDi)⊕ Kc,Hk(IDI⊕Rs) to S.

S computes by its key k that shared with C to check if Hk(IDi⊕Rs) equal to Hk (IDs⊕ Rs), if they are equal, then the authentication pass, and gets Kc by the computing of Hk(IDs)⊕Kc, and sends the encrypted message M3= Kc⊕M is sent to C; if authentication fail, then keeps silence.

In the model, the design thought and the way of encryption also comes from protocol unit, we designed protocol unit by formalized method and according to the analyzed we know that protocol unit is safe in the model CK, while in WSNIT, the number of communication agents become to three, and the content and process of interactive information all change.

In the protocol, Rs that the response of the S to C changes randomly, which can prevents the position and parameter of S are obtained by the way that track the value of ID, the value of Rs changes every times by which jumping based on the random number stored in S, there stored sequence code Ni, which the number of Ni is L and the length of Ni is k bit in every S. Due to the values of Ni in every sensor are different, the random sequences are different that stored in every S, and it can sets itself ruler of jumping, when the length of Ni is k bit, there are L*2k kinds of output results, so it is not settled for attacker and very difficult to track.

In Step 4, m sent Hk(Rc⊕Rs⊕IDs) and Hm(Rc⊕Rm⊕IDm) two functions, aimed at that C authenticates M and S respectively, after the authentication of C successful, C sent Hm(IDj⊕Rm), Hm(IDj)⊕Kc, Hk(IDi)⊕Kc and Hk(IDi⊕Rs)⊕IDj to M, the aim is to make M and S authenticate C respectively, and distributes the secret key Kc to M and S.

In the last step, the way of encrypts changes from Hk(Kc) to Kc, the purpose is that make M and S don't know the shared key and identification number that stored in C each other.

The protocol that we get lastly as follow:

```
WSNIT:
Msg1 C→M: Rc
Msg2 M→S: Rc,
Msg3 S→M: Hk(Rc⊕Rs⊕IDs),Rs
Msg4 M→C: Hk(Rc⊕Rs⊕IDs),Hm(Rc⊕Rm⊕IDm),Rm,Rs
Msg5 C→M: Hm(IDj⊕Rm), Hm(IDj)⊕Kc,
Hk(IDi)⊕Kc,Hk(IDi⊕Rs)⊕IDj
Msg6 M→S: Hk(IDi)⊕Kc,Hk(IDi⊕Rs)
Msg7 S→M: Kc⊕M
```

5 The Proof of Protocol Security

Now begin a proof of the protocol security in above, Dolev-Yao model [9] is applied to the ability and action of attacker:

Definition 1. (AM model) Assume $\Sigma=(S, P, C, E,A)$ is a CK model , then in Σ:

$$\text{Intruder}_{AM}(X) = \text{lean? } m: P \rightarrow \text{Intruder}_{AM}(X \cup \{m\})\overline{}\text{say! } m: P \rightarrow \text{Intruder}_{AM}(X) \qquad (1)$$

The S represents communication agents, the P rcpresents the process monitored by attacker, $P \in S$, the X represents the set of knowledge owned by attacker, the C represents communication, the E represents event, the E represents action, the properties are:

$S=\{ S1,S2,...Sn, n\geq 2\}$
$P=\{P1,P2...Pm, m\geq 1\}$
$\exists Si \subset S, 1\leq i\leq n, Pk: Si , Pk \subset P$
$C=\{ C1,C2,...Cj, 1\leq j\leq \}, Ck: Si , Ck \subset C$

We can know from the above formula, the attacker is a kind of attacker that an ideal situation, it only can get messages from monitor communication channel or sends the messages that gotten by send communication channel.

Definition 2. (The security of SK) Assume $\Sigma=(S, P, C, E, A,I)$ is a CK model: (1). In the condition that communication agents didn't break through, $\exists Si, Sj \in S$, after they finish C that a conversation, they accepted the same secret key. (2). The probability that attacker can distinguish between the K that get from (1) and a random bit streams is $\leq 1/2+\delta$, the item δ represents a probability value that can't be ignored.

   Now begin the proof the protocol is secure in the model AM.

   Theorem 1 In the protocol of WSNIT, if the pseudo-random numbers that created by protocol is safe, then the authentication of C to S and M satisfy the safe of SK in the model AM.

Certificate:

In the Msg1--Msg4, assume the messages X={Rc,Rc,Hk(Rs⊕Rc⊕IDs),Rs, Hk(IDi⊕Rs), Hk(IDi)⊕Kc, Kc⊕M } are intercepted by attacker, in the condition that communication agents didn't break through, attacker cannot get the value of K and IDs, and for the every items are different in the X that a set of messages, any items cannot be integrated deduced by other items, and Kc cannot be integrated deduced by other items in the set of X, so protocol unit satisfy the definition 2 and properties (1).

   Now begin to proof the protocol unit satisfy definition 2 and properties (2).

   Assume Q0={ HkΩ (), HkΩ (), R, R}, Q1={ random(), random(), R, R}, Q0 and Q1are set. The R is a random number produced by attacker, the HkΩ () is a one-way hash function based on K. Which it is consistent with protocol and owned by attacker, random() is a random number, when attacker carry out attacks alternately, if attacker cannot gets the HkΩ (), it will uses random() substitute for HkΩ () continue attracting alternately.

The input of Ω that is an algorithm is a two-tuples (γ, γ, R, R), the probability that the two-tuples chooses for Q0 and Q1 is half, algorithm Ω use attacker A as a sub process, assume A can activate conversation in any interaction process. The description of algorithm Ω as follow: (1).Choses a random number, the random number is initialized when it came from factory in S, assume the number is L, and the length of Rs is h, then the kinds of pseudo-random numbers is L*2h. Assume the pseudo-random numbers of algorithm Ω is RΩ, R$\Omega \in \{1, ..., L*2h\}$. (2).Assume in AM model, the number of participators is n in the running protocol and the participators p1, p2, ... pn interacted each other, attacker randomly choose the first m (m$\leq$n/3) to attack alternately, then except the first m conversation that was activated by attacker, all other conversations shared kij, while the correspondence participators of conversations shared km. (3). When A activates pi that is a participator and establishes a new conversation or receives a new message, Ω represents the participator execute by the steps of protocol unit. (4). When the first m conversation (pi, pj, pk, Query(Rc))m is activated, Ω makes pi send Query(Rc) that a message to pj. (5). When pj receives the message Query(Rc), Ω makes pi send Rc that a message to pk. (6). When pk receives the message Rc, Ω makes pk send Hk(Rs$\oplus$Rc$\oplus$IDs),Rs that a message to pj. (7).When pj receives the message Rc, Ω makes pj send $(\gamma 1, \gamma 2, R1, R2)$ that substituted Hk(Rm$\oplus$Rs$\oplus$IDs), Hm(Rm$\oplus$Rs$\oplus$IDm),Rm,Rs to pi. (8).If conversation (pi, pj, pk, Query(Rc))m is chose for the test conversation to in qure, then Ω transmits the respond of inquiry γ to A. (9).If pi cannot finds k and IDs that match Rc and Rs and computes $\gamma 1, \gamma 2$ suitably, then the first m conversation (pi, pj, Query(Rc)) m is exposed, if attacker choose another conversation to test conversation by A or dose not choose test conversation, then the conversation will be stop, so Ω outputs the respond result of γ and then stops. (10). If A stop and output the respond result of γ, then Ω stop and output the result same as A. (11). From the algorithm described in above, the operation of A that activated by Ω and the normal operation of A against protocol signally are consistent.

In the first situation, when A choose the first m conversation as the test conversation, A gets the respondence γ. If the input of Ω is Q0, then the respondence is the true conversation key. If the input of Ω comes from Q1, then the respondence is a random number, while as indicated in above, the input probability value half is comes from Q0 or Q1 respectively. So, the respondence of the inquire to A by Ω and the description of test conversation are consistent, in this condition, the probability value that A guess the respondence is true conversation key, secret key or random number correctly is $1/2+\delta$, the δ could not be ignored. It is equal to that guess the input of Ω came from Q0 or Q1, by output b that same as A, the probability value that input came from Q0 or Q1 is $1/2+\delta$, the δ could not be ignored.

In the second situation, when A dose not choose (Hk(Rs$\oplus$IDs$\oplus$Rc), Rs)m, Ω outputs a random number, in this condition, it guesses correctly the probability of input comes from Q0 or Q1 correctly is also a half.

While the probability of the first condition is 1/ L*2h, the probability of the second condition is 1-1/ L*2h, so the probability that Ω guesses correctly is 0.5+δ)+0.5X(1-)=0.5+, that is could not be ignored and , so protocol unit satisfy definition 2 and properties (2).

6 Conclusion

This paper we analyses the architecture of internet of things and abstracts the communication agents satisfied the physical characteristics and the requirements of application of internet of things, and then establish a Secure protocol of WSNIT, it is based on the requirements and characteristics of communication agents in internet of things, the lastly, we testified this security protocol satisfied the SK security in AM model.

Acknowledgment. This work was supported by the Science and technology breakthrough plans foundation of Yunnan Province under Grant No.2005GG11-5 and by Information Security Professional Education System Constructions from Yunnan University.

References

1. Srivastava, L.: Pervasive, ambient, ubiquitous: the magic of radio. In: European Commission Conference From RFID to the Internet of Things, Bruxelles, Belgium (March 2006)
2. Michael, M.P.: ArchitcctnraL Solutions for Mobile RFID Services for the Internet of Things. In: IEEE Congress on Services (2008)
3. GonzaLez, G.R.: Early Infrastructure of an Internet of Things in Spaces for Learning. In: IEEE International Conferenec on Advance Learning Technologies (2008)
4. Weber, R.H.: Internet of things – Need for a new legal environment? Computer Law & Security Review 25, 522–527 (2009)
5. Broll, G., Rukzio, E., Paolucci, M., Wagner, M., Schmidt, A., Hussmann, H.: PERCI: pervasive service interaction with the internet of things. IEEE Internet Computing 13(6), 74–81 (2009)
6. Toma, I., Simperl, E., Hench, G.: A joint roadmap for semantic technologies and the internet of things. In: Proceedings of the Third STI Roadmapping Workshop, Crete, Greece (June 2009)
7. Atzori, L., Iera, A.: The Internet of Things: A survey. Computer Networks (2010)
8. Katasonov, A., Kaykova, O., Khriyenko, O., Nikitin, S., Terziyan, V.: Smart semantic middleware for the internet of things. In: Proceedings of the Fifth International Conference on Informatics in Control, Automation and Robotics, Funchal, Madeira, Portugal (May 2008)

NASL Based Vulnerability Detector
Design and Implementation for Remote Computer

Feng Xu[1] and Xiaoting Jin[2]

[1] School of Computer Science, Zhongyuan University of Technology
450007 Zhengzhou, China
[2] College of Public Administration, Henan University of Economics and Law
450002 Zhengzhou, China
flyjinxt@163.com

Abstract. This paper proposed a vulnerability detector design and implementation method, referred to as NRCVD, for probing given security holes of remote computers. After expounding the detector design idea, the work flow, plug-in definition and design method, we illustrated the detector architecture. All plug-ins for vulnerability detection within NRCVD are implemented by the scripting language of NASL. Experiment results and comparisons show that NRCVD is feasible, and it has the features of high security, fast-speed, and low resource-occupying ratio. Thus, it provides a good solution to cyber security testing and assessment.

Keywords: vulnerability detector, remote computer, NASL.

1 Introduction

Today, cyber security has become an increasingly important issue as the Internet has expanded, and once-private corporate networks conduct a great deal of business in public cyberspace over the Internet and wide area networks [1,2]. The danger of information being intercepted, corrupted, or misappropriated has escalated, as has the fear of hackers breaking into servers and altering source code published on Web sites or conducting Denial-of-Service attacks. As a result computer and network security procedures, practices, and technologies have developed. According to the recent statistical result of CERT /CC, vulnerabilities are one of the main causes of cyber security incidents [3, 4, 5].

Software vulnerabilities arise from deficiencies in the design of computer programs or mistakes in their implementation, and most vulnerabilities are a result of programming mistakes, in particular the misuse of unsafe and error-prone features of the programming language, such as Perl, C, C++, and etc. Although there are many plug-in based vulnerability detection methods, the following deficiencies of these techniques should take into account:

- The safety is poor.
- Most of the plug-in need to be compiled again before running.
- The working speed is slow.

D. Jin and S. Lin (Eds.): Advances in ECWAC, Vol. 2, AISC 149, pp. 249–254.

- It is difficult for plug-ins to upgrade.
- The resource-occupying ratio is high.

NASL is a scripting language designed for the Nessus security scanner. Its aim is to allow anyone to write a test for a given security loophole in a few minutes, to allow people to share their tests without having to worry about their operating system, and to grantee everyone that a NASL script can not do anything nasty except performing a given security test against a given target.

Inspired by the principles of NASL, this paper proposes a vulnerability detector design and implementation method, referred to as NRCVD, for probing given security holes of remote computer. The remaining of the paper is organized as follows. Section 2 discusses the design principles of NRCVD. In Section 3, theoretical analysis and experiment results are provided. Finally, Sections 4 contains our conclusion and the future work.

2 Detector Design

This section discusses the design idea, working flow, design methods of plug-ins, and architecture of NRCVD.

2.1 Design Idea

NRCVD are designed using by the network-based method, and it is implemented by simulating hacker attacks. Within the proposed method, all plug-ins are implemented using by the NASL scripting language in accordance with the specific characteristics of known vulnerabilities of the computer network and information system. Though building and sending tcp-ip packets for the object host (including server, workstation, switch, router, firewall, and etc.), vulnerabilities are obtained by analyzing the returned information from the destination according to the configuration information and established security strategies.

2.2 Work Flow

The working flow of NRCVD is illustrated in Fig. 1.

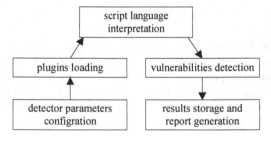

Fig. 1. Working flow of NRCVD

The functional modules in Fig. 1 are described as follows.

(1) Detector parameters configuration: setting the target host's IP address, number of threads, vulnerability category to be detected, and etc.

(2) Plug-ins loading: loading plug-ins according to the configuration parameters, and arranging the execution sequence of the loaded plug-ins.

(3) Script language interpretation: interpreting the function of the configuration parameters, and plug-ins syntax and semanteme.

(4) Vulnerabilities detection: simulating hacker attack, detecting the possible vulnerabilities of remote targets, and returns the detection results.

(5) Report storage and report generation: saving the detected vulnerability results, and generating testing report.

2.3 Design Method

2.3.1 Formal Definition

Within NRCVD, each plug-in can detect one or more known vulnerabilities. In order to describe the plug-ins design method, we let P represent the set of plug-ins, and its formal definition is given as follows.

$$P = \left\{ p \mid p = \left\langle \begin{array}{l} version, id, name, description, summary, \\ category, copyright, family, cveid, \\ requiredports, dependency, timeout, attacode \end{array} \right\rangle \right\}. \tag{1}$$

wnere $version$, id and $name$ represent the version, identification and name of the plug-in (p), respectively; $description$ denotes the detailed information of the plug-in p, including the function, the vulnerabilities' description, risk value, and the corresponding solution; $summary$ and $copyright$ represent the abstract and copyright information of p, respectively; $category$ is used to classify plug-ins of NRCVD (including host information collection, privileges obtaining, Dos attack, and port scan); $family$ represents the family which the plug-in belongs to, such as the back door, CGI misuse, Dos, and etc.; $cveid$ denotes the serial number CVE (Common Vulnerabilities & Exposures) of the vulnerability; $requiredports$ denotes the opened service or port of the target host which is required by the plug-in (p); $dependency$ expresses the other plug-ins which are plug-in dependency, which shows other plug-ins to be executed before the plug execution; $timeout$ represents maximum timeout interval of executing process; $attacode$ is attack code achieving the plug-in function, and the core of plug-in.

2.3.2 Design Method

In general, the vulnerabilities of computer network and information systems are interrelated with services. Therefore, the design method of the plug-ins for NRCVD is service oriented, and detailed design method is shown in Fig. 2.

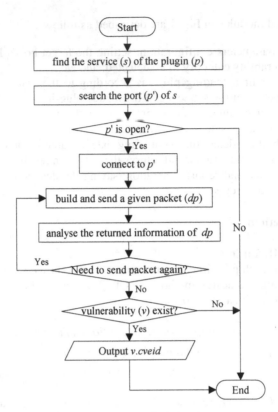

Fig. 2. Design method of plug-ins

2.4 Architecture

NRCVD is mainly composed of five modules: detecting parameter configuration, detection engine, plug-ins, detection knowledge database, result storage and report generation. The architecture of NRCVD is illustrated in Fig. 3.

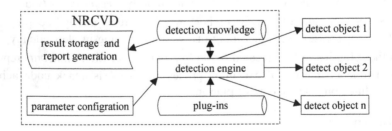

Fig. 2. Architecture of NRCVD

From Fig. 3 we see that the module of detection engine is the core of NRCVD detector, it is in charge of interpreting the configuration parameters, calling plug-ins to detect the known vulnerabilities of objects though simulating Hack attacks and using by detection knowledge database.

3 Experiment and Comparison

We realized the main programs of NRCVD with the language of VC 6.0, and all the attack scripts were implemented by NASL. In order to test the feasibility of NRCVD, security and efficiency, we did experiments in Zhengzhou key lab of network security assessment, and we used NRCVD in project of network security assessment system of *FoundLight*, which was supported in part by the grant of the Key Technologies R & D Program of Henan Province (no. 092102310038). For detecting vulnerability of remote computers, we carried out the experiments in the campus of Zhongyuan University of Technology.

The experimental results of the same and different campus show that NRCVD is feasible.

To test the security performance of NRCVD, we tested the security of NRCVD by the Tools of *beSTORM*, *Coverity Prevent* and *MU-4000*. The tested results show that the security of NRCVD is high.

The comprehensive comparison between NRCVD and the other similar techniques (tools) is shown in Table 1.

Table 1. Comparison between NRCVD and other similar techniques

	Nessus	Aurora RSAS	NRCVD
CPU utilization	low	high	low
memory occupancy rate	low	high	low
According to security level	no	no	yes
detection rate	fast	slow	fast

Experiment results and comparisons show that NRCVD is feasible, and it has the features of high security, fast-speed, and low resource-occupying ratio.

4 Conclusion

Computer system vulnerability assessment was initially used by the hacker attack technology development, and it is still an emerging research field. Experiment results and comparisons show that the NASL based vulnerability detector designed and implemented in this paper is feasible, and it has the merits of high security, fast-speed, low resource-occupying ratio.

In future work, we should perfect the plug-ins and detection knowledge databases of the presented method, so as to meet the requirement of system software and application software of the actual demand. Moreover, we should do more experiments to test the false positives (*FP*) and false negative (*FN*) of NRCVD.

References

1. Sun, F.: Gene-Certificate Based Model for User Authentication and Access Control. In: Wang, F.L., Gong, Z., Luo, X., Lei, J. (eds.) WISM 2010. LNCS, vol. 6318, pp. 228–235. Springer, Heidelberg (2010)
2. Sun, F.: Artificial Immune Danger Theory Based Model for Network Security Evaluation. J. Netw. 6, 162–255 (2011)
3. Sun, F., Kong, M., Wang, J.: An Immune Danger Theory Inspired Model for Network Security Threat Awareness. In: 2010 International Conference on Multimedia and Information Technology (MMIT 2010), vol. 2, pp. 93–95. IEEE Press, New York (2010)
4. Sun, F.: Practice Teaching New Model for Course of TCP/IP Principles and Applications. In: 2010 Third International Conference on Education Technology and Training, vol. 4, pp. 94–96. IEEE Press, New York (2010)
5. Sun, F., Wu, Z.: A New Risk Assessment Model for E-Government Network Security Based on Antibody Concentration. In: 2009 International Conference on E-Learning, E-Business, Enterprise Information Systems, and E-Government, pp. 119–121. IEEE Press, New York (2009)

A Filtering Algorithm for Point Cloud Data

Feng Zeng[1,*], Zhichu Zhong[1], and Jianan Ye[2]

[1] Computer School, JiaYing University, Meizhou, Guangdong, China, 514015
[2] School of Materials Science and Engineering, Dalian University of Technology, Dalian, Liaoning, China, 116024
zengfeng@jyu.edu.cn

Abstract. To against the demerits of point cloud processing of reverse engineering, a quantitative filtering and compacting algorithm is presented. Statistical method was used to found out the mathematical distribution of experimental data, and the confidence interval of mathematic expectation was calculated by the mathematical distribution. The valid data's mathematic expectation within confidence interval was output as the return value of data processing. The case studies justified the feasibility and stability of this method.

Keywords: reverse engineering, point cloud, data compacting, data filtering.

1 Introduction

The point cloud data measured through real entity are used to reconstruct the digitized models of the entity in reverse engineering. The redesign, analysis and the processing of the digitized models are realized[1-2], and it is a important technology under the background of digitized and intelligentized design and the quick response manufacturing. The reverse engineering is widely used in industry manufacturing, biological morphology, digital entertainment and heritage protection[3-4]. The high-accuracy collection and processing for point cloud data are the key technologies. The point cloud data are very huge and contain some noisy data. The standard Gauss algorithm and mean value method are usually used for denoising, and the effect is shown in Fig. 1. The standard Gauss algorithm can make the data keep the original appearance, but it can't process the singular point and bug points well. The averaging filter is used in mean value method which can eliminate bug points but loss the part of original appearance of the point cloud data.

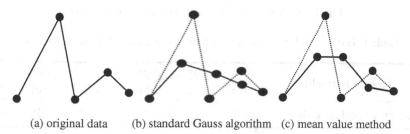

(a) original data   (b) standard Gauss algorithm   (c) mean value method

Fig. 1. Effect of two processing methods

---

* Corresponding author.

For the research of data reduction, Martin[5] suggested the orthogonal cross sections method to reduce the data which was widely used in median filtering for image processing. The advantage of this method is higher algorithm efficiency, while the disadvantage is insensitive to the shape of real entity. K.H.Lee[6-7] used the method of distance tolerance to reduce the point cloud data according to the property of laser scanning system, and it offered an basic algorithm which can be quantified.

To above mentioned situations, a filtering algorithm for point cloud data is presented in this paper. Statistical method is used to filter the point cloud data according to the property of laser measuring equipment. Statistical distribution of point cloud data is calculated in local neighborhood. And then the points outside the confidence interval are deleted according to confidence coefficient.

2 Mathematical Distribution of Point Cloud Data

The first step of the algorithm is to obtain the statistic distribution in local neighborhood of point data. Taking the self-design of reverse engineering laser measuring equipment for example, the laser probe was assembled 0.14m above the quiescent entity to collect the data. 10 group data were collected while the sampling frequency of system is 5000HZ and the number of samples was 548864. The range of the data was from 0.132m to 0.146m. The frequency distribution of one representative sample data is shown in Fig. 2. The data were processing by SPSS while the KS(Kolmogorov-Smimov) test method and the SW(Shapiro-Wilk) method were used to test the normal distribution of data, and the result is shown in Table 1.

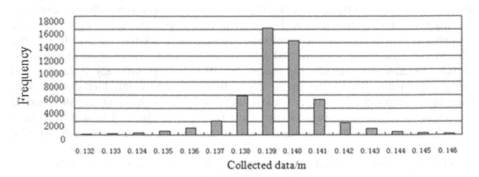

Fig. 2. The frequency of one group collected data in 0.14m

Table 1. Testing result by Kolmogorov-Smirnov method and Shapiro-Wilk method

Tests of normality	Kolmogorov-Smirnov			Shapiro-Wilk		
	Statistic	df	Sig.	Statistic	df	Sig.
One group collected data	.372	28	.000	.479	28	.000

The result of KS testing is that the sample data's significance probability(Sig) is smaller than 0.5, so the distance data collected by the laser sensor were not normal distributed. The KS testing also showed that the data were not the homogeneous distributed, poisson distributed and exponential distributed. The Sig of SW testing shown in Table 1 were also smaller than 0.5, and finally the whole collected data were not normal distributed.

We transform the sample data to the natural logarithm, and then, test the data after transformation using the same methods. The Sig is significantly greater then 0.5 by using KS testing method and SW method, as shown in Table 2. That means the sample data measured by the reverse engineering are logarithmic normal distribution.

Table 2. Testing result by Kolmogorov-Smirnov method and Shapiro-Wilk method for logarithm transformation with the sample data

Tests of Normality	Kolmogorov-Smirnov			Shapiro-Wilk		
	Statistic	df	Sig.	Statistic	df	Sig.
Data after logarithm transformation	.072	28	.202	.960	28	.259

The collected data of other distances were tested by KS method and SW method. The Sig value is all greater then 0.05, as shown in Fig. 3. Thus, the self-development reverse engineering laser measuring system is logarithmic normal distribution within the measuring range.

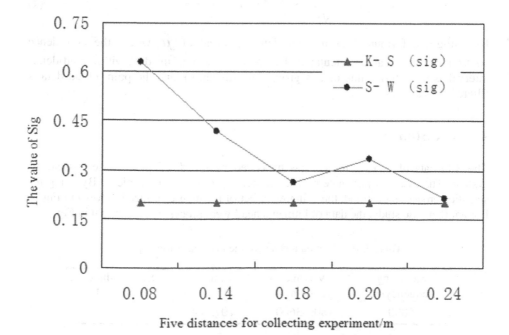

Fig. 3. The result tested by KS method and SW method to five distances

3 Statistically Analysis for Point Cloud Data

If a group of sample data X are normal distribution, as shown in formula (1), and, the confidence coefficient of certain given μ is 1-α, then, the confidence interval of μ is as shown in formula (2).

$$Z = \frac{\overline{X} - \mu}{\sigma/\sqrt{n}} \rightarrow N(0,1) \tag{1}$$

$$\frac{1}{n}\sum_{i=1}^{n} x_i - Z_{\alpha/2} \cdot \frac{\sigma}{\sqrt{n}} < \mu < \frac{1}{n}\sum_{i=1}^{n} x_i + Z_{\alpha/2} \cdot \frac{\sigma}{\sqrt{n}} \tag{2}$$

The 1-α = 0.95 where α = 0.05, and $Z_{\alpha/2}$ =1.96, we get the confidence interval of μ is:

$$\overline{X} - \frac{1.96\sigma}{\sqrt{n}} < \mu < \overline{X} + \frac{1.96\sigma}{\sqrt{n}} \tag{3}$$

The sample data were logarithmic normal distribution, so the formula (3) must transform to the formula (4):

$$\overline{\ln x} - \frac{1.96\sigma}{\sqrt{n}} < \mu < \overline{\ln x} + \frac{1.96\sigma}{\sqrt{n}} \tag{4}$$

By using the formula (4), the confidence interval of μ where the confidence coefficient is 0.95 can be calculated. The medium value of the data within confidence interval is the return value of this group data (distance), and the point cloud data is filtered by statistics.

4 Case Studies

The 3-D data of mouse surface was measured by self-development laser measuring system. The essential parameters of the system were shown in Table 3. By using the statistics processing method, the skinning effect of the mouse surface is shown in Fig. 4. For another case study, the data of human's head were processed and shown in Fig. 5.

Table 3. Essential parameters of the laser measuring system

Sampling frequency/Hz	Voltage range/mv	Number of sample data	Size of the data file/M
5000	1800-3600	10730000	24

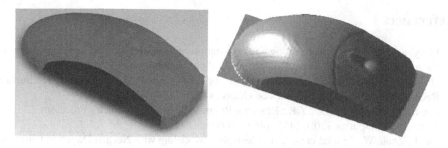

Fig. 4. Skinning effect of the mouse surface

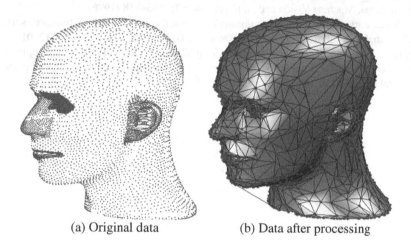

(a) Original data   (b) Data after processing

Fig. 5. Data of human's head

5   Conclusion

The filtering algorithm for point cloud data base on statistics is presented. The statistical method was used to found out the mathematical distribution of experimental data, and the confidence interval of mathematic expectation was calculated by the mathematical distribution. The medium value of the data within confidence interval is the return value of this group data. The results of case studies justified the feasibility and stability of this method.

Acknowledgments. This work was supported by Foundation for Distinguished Young Talents in Higher Education of Guangdong, China. NO.LYM10121.

References

1. Flory, S., Hofer, M.: Surface fitting and registration of point clouds using approximations of the unsigned distance function. Computer Aided Geometric Design 27(1), 60–77 (2010)
2. Hui, H., Dan, L., Hao, Z., et al.: Consolidation of Unorganized Point Clouds for Surface Reconstruction. ACM Transactions on Graphics 28(5), 176–183 (2009)
3. Hu, G.M., Yang, Y.L., Lai, Y K · Research Progress of Digital Geometry Processin. Chinese Journal of Computers 32(8), 1451–1469 (2009)
4. Ge, J., Zhou, W., Liu, L., et al.: Large-leaf-plant Modeling with Normal Direction Clusterin. Journal of Computer-Aided Design & Computer Graphic 22(11), 1940–1944 (2010)
5. Milroy, M.J., Bradley, C., Vickers, G.W.: Automated laser scanning based on orthogonal cross sections. Machine Vision and Applications 9(3), 106–118 (1996)
6. Lee, K.H., Park, H., Son, S.: A framework for laser scan planning of freeform surfaces. International Journal of Advanced Manufacturing Technology 17(3), 171–180 (2001)
7. Lee, K.H., Woo, H.: Use of Reverse Engineering Method for Rapid Product Development. In: 23rd International Conference on Computers and Industrial Engineering, Computers and Industrial Engineering, vol. 35(1), pp. 21–24 (1998)

Electronic Library Acceptance Research: Based on the Theory of Reasoned Action

DaQing Zheng

School of Information Management & Engineering,
Shanghai University of Finance & Economics, Shanghai, 200433

Abstract. The research explained Chinese digital library system acceptance through the framework of the theory of reasoned action. This study not only examined linear factors contributing to adoption of digital library, but also a non lineal interaction between these factors. The research results revealed that attitude and subjective norm have a significant effect on acceptance intension, and there also is negative interaction between attitude and subjective norm, which mean, the theory of reasoned action is powerful to explain Chinese digital library acceptance.

Keywords: digital library, subjective normal, attitude, intension.

1 Introduction

Research and development in digital libraries have grown rapidly in the 1990s. Whereas the traditional focus of digital library has been on the technological developments ranged from the "hard" side, such as information retrieval and access, and system integration[1], to the "soft" side, such as social informatics and copyright management issues[2]. There is now a call for user-focused research. Although millions of money has been spent on building "usable" systems, research on digital libraries has shown that potential users may not use the systems in spite of their availability. There is a need for research to identify the factors that determine users' adoption of digital library[3, 4]. This study investigates the factors that influence the user's adoption, based on the theory of reasoned action.

2 Background and Literature Review

Definitions of digital library are abundant, due to the interdisciplinary nature. From the research oriented perspective, we can view the digital library as a "convenient and familiar shorthand to refer to electronic collections and conveys a sense of richer content and fuller capabilities than do terms such as database or information retrieval system"[5]. Compared to the traditional library, the major advantages of digital library include the easier tracked resource, remote, fast and fair access, and increased flexibility and power techniques for users searching[6]. So, as the definition of the information system, the digital library is also the combination of machines and users. The acceptance of digital library is also instructed by cognitive process.

D. Jin and S. Lin (Eds.): Advances in ECWAC, Vol. 2, AISC 149, pp. 261–266.
springerlink.com   © Springer-Verlag Berlin Heidelberg 2012

The theory of reasoned action (TRA) is designed to explain virtually any human behavior, and should therefore be appropriate for studying the determinant of computer usage behaviors as a special case [7]. According to the TRA, adoption is determined by intention to use a particular system, which in turn is determinate by attitude and subjective norm. One key benefit of using TRA to understand system usage behavior is that provides a framework to analysis the effects of personal cognitive process. TRA explains how the "attitude" affects the "behavior" through the "intension". According to the theory of reasoned action, the "intention" is determined by both the "attitude", held by individual towards it, and the subjective norm. The subjective norm refers to the social pressure which individual perceives after taking some certain behavior[8]. The theory of planned behavior is widely applied in a variety of IS contexts, organizational contexts, and user populations, and has already been proved to have more explanations, because of that, many researchers have suggested and used TRA as a foundation to investigate individuals' IT usage behavior [9-11].

3 Research Hypotheses

Early in the 1980, the theory of reasoned action proposed by Ajzen and Fishbein had already verified the significant influence which the attitude has influence on intention within the field of consumer behavior[12]. This model has gotten a lot of supports from empirical results [13, 14]. As a subject of the adoption of the information retrieval system behavior, the users' attitude towards the use of the information retrieval platform has the positive or negative emotion. Therefore, we can infer that the attitude has a positive effect on the digital library user's behavior intension.

H1. Attitude has a positive effect on the behavior intention
The construct of subjective norm reveals that the individual will consider about the feeling of the people who have a close relationship with him [12]. From the perspective of the digital library, the subjective norm reflects the influence which the surrounding people have on those users. Thus, the hypotheses2 is as following.

H2. The subjective norm has a positive effect on the digital library user's behavior intention.

4 Methodology

The study explored the phenomenon of the acceptance of digital library in China, and we took digital library system of Shanghai University of Finance & Economic as the research technical objects. In China, many digital resources are configured in universities and academic institutes, and the majority of students do not start to utilize the digital library systematically until they have the chances to conduct the postgraduate researches. Therefore, the research chose the university teachers and postgraduate students of Shanghai University of Finance & Economic as the survey samples.

It took about 1 month in total to release and collect the sample data, divided into 3 batches, 162 questionnaires were sent out, and we mainly used the method of filling in and collecting the questionnaires under our surveillance to ensure the objects of the study meet the requirements of the sample, we received 145 valid questionnaires, with an average recovery rate of 87%. In all valid questionnaires, post graduates accounted for 81.38% of the total objects of the study, doctor graduate students and above accounted for 8.28%, others who did not complete their education background accounted for 10.34%.

To validate the constructions, we examined reliability, convergent validity, and discriminant validity. We adopted Cronbach α and composite reliability (CR) to test the consistency of each variable. According to the empirical value, the common coefficient of Cronbach α is required to be larger than 0.65, and larger than 0.7 under ideal condition. If α value is too small, then it means that the question of the variable is not a good measure of the variable. The CRs range from 0.855 to 0.931. All values of α and CRs are above 0.7, which are good for the requirements.

Convergent validity measures the correlation among item measures of a given construct using the average variance extracted (AVE) values, which should be above the threshold of 0.50. The AVE values are shown as diagonal elements in table1.

Table 1. Correlation between construction and AVE

	BA	UI	SN
BA	0.69176 (0.83172)		
UI	0.653919	0.662245 (0.81384)	
SN	0.326458	0.420429	0.870430 (0.93297)

Discriminant validity refers to the extent to which different constructs diverge from one another. We used the cross-loading method to assess discriminant validity of the scale. The value of PBC4 and PBC5 is low, so we omitted them. The final Factor loading matrix is shown as table2. From the chart we can see that, all the factor loading values are valid. Furthermore, Discriminant validity is also verified with the squared root of the average variance, which is listed in the brackets of the diagonal units in table1, and extracted for each construct higher than correlations between it and all other constructs (off-diagonal elements), suggesting that the items share more variance with their respective construct than with other constructs.

Table 2. Rotated factor loadings and cumulative variance contribution rate

	factor 1	factor 2
BA1	.856	-.180
BA2	.862	-.266
BA3	.820	-.220
BA4	.576	-.462
SN1	.546	.751
SN2	.590	.713

5 Analysis Results and Discussion

At first, we tested the influence which the subjective norm and the attitude have on the acceptance intention. The structure is shown as table3. It clearly shows that the attitude and the subjective norm do have a significant effect on the digital library acceptance intention, namely the research hypotheses H1 and H2 are verified.

Table 3. Results of structural model

	Behavior intension	**p<0.05;***p<0.01,
Behavior attitude	0.578***(7.281130)	T-statistics are shown in parentheses
Subjective norm	0.232**(2.652936)	
R2(%)	0.476	

In addition, in IS research and non IS area, consensus has been formed that there is a linear and a nonlinear relationship between the attitude and subjective norm, which have been discussed in some researches [13, 15-17]. When we look at the practical application of the digital library system, in general, individual usually uses the library information retrieval system out of the work and study, when the users have a strong inclination to use the digital library, the improvement of the environmental pressure may reduce the acceptance intention of the users, namely when the users hold a positive attitude, the increase of the subjective norm may possibly reduce the acceptance intention; when the favorable environmental atmosphere about using the digital library is created, the further increase of the users' attitude may reduce the users' acceptance intention, namely in the case of high subjective norm, the increase of the users' attitude may reduce the acceptance intention.

Compared to linear model, we performed a non linear model against attitude and subjective norm as well as their multiplication (Table4). With the multiplication term, the model showed a strong, significant interaction effect.

Table 4. The linear model and non linear model

	Linear model	Non Linear model
behavior	0.578***	1.556**
Subjective norm	0.232**	1.292*
Behavior × Subjective norm		-1.672*
R^2(%)	0.476	0.511

According to the result of the linear model, we can consider that the attitude and subjective norm have a positive influence on acceptance intention, which has already been proved by the major portion of researches. However, through the analysis of the correlative relationship between the attitude and subjective norm, we can see that, when the attitude variables remain constant, the subjective norm has a positive effect on acceptance intention; and when the subjective norm remains constant, the attitude has a positive effect on acceptance intention as well. Compared with the conclusion from the linear regression, this conclusion further illustrates the relationship between the acceptance intention, the subjective norm and attitude. The results which shows the negative correlative relationship between the attitude and subjective norm reveals that in the case of high subjective norms, the increase of the marginal attitude will not lead to the increase of the marginal acceptance intention; likewise, in the case of the high attitude, the increase of the marginal subjective norm will impair the increase of the acceptance intention, and the relationship between the subjective norm, attitude and acceptance intention proved by the linear model, belongs to the special case of the nonlinear model under some situations.

From the perspective of the practice of electronic library, we can know that the subjective norm variables are one of the external influential factors, and the attitude is the internal factor of the receptors. The results of the research illustrates that in order to impel customers to accept the electronic library, we should keep the two influential factors, which are attitude and subjective norm, in balance.

6 Conclusion

This research proves that under the circumstance in which the individual is not mandated to use the digital library system, or is able to decide whether to use the system according to individual needs by oneself, the subjective norm and attitude have a significantly positive influence on the utilization intention, that means, when the digital library information system utilization attitude of the users is high, if we further improve the users' subjective norm, it will not lead to the corresponding improvement of the acceptance intention, due to the fact that the high attitude will restrain the positive influence which high subjective norm has on the acceptance intention, so in the case of the high acceptance attitude, if we hope to further enhance the users' acceptance intention, we need to consider other aspects; by the same token, when the users' utilization of the e-library information system is highly affected by the subjective norm, if we increase the users' attitude of the e-library, it will not lead to the corresponding increase of the acceptance intention, since the high subjective norm will restrain the positive influence which high attitude has on acceptance intention.

Acknowledgement. This paper was supported by general project of Chinese Ministry of Education (Research on theoretical model and application of Chinese e-government adoption) 09YJC630147), Shanghai Nature Science Foundation (Research on Transfer, Measure and Control of IT Operational Risk--Cases on Banking Industry, 11ZR1411900) and Leading Academic Discipline Program, 211 Project for Shanghai University of Finance and Economics.

References

[1] Adam, N.R., et al.: SI in digital libraries. Communications of the ACM 43(6), 64–72 (2000)

[2] Anderson, L.C., Lostpiech, J.B.: Rights Management and Security in the Electronic Library. Bulletin of the American Society for Information Science 22(1), 21–23 (1995)

[3] Miller, J., Khora, O.: Digital Library Adoption and the Technology Acceptance Model: A Crosse-country Analysis. The Electronic Journal of Information Systems in Developing Countries 40(16), 1–19 (2010)

[4] Nov, O., Ye, C.: Resistance to change and the adoption of digital libraries: an integrative model. Journal of American Society for Information Science and Technology 60(8), 1702–1708 (2009)

[5] Borgman, C.L.: What are digital libraries? Competing vision. Informatin Processing and Management 35, 227–243 (1999)

[6] Hong, W., et al.: Determinants of User Acceptance of Digital Libraries: An Empirical Examination of Individual Differences and System Characteristics. Joural of Management Information System 18(3), 97–124 (2002)

[7] Davis, F.D., et al.: User Acceptance of Computer Technology: a Comparison of Two Theoretical Models. Management Science 35(8), 982–1003 (1989)

[8] Ajzen, I., Fishbein, M.: Attitudinal and normative variables as predictors of specific behavior. Journal of Personality and Social Psychology 27(1), 41–57 (1973)

[9] Jackson, C.M., et al.: Toward an Understanding of the Behavioral Intention to Use an Information System. Decision Sciences 28(2), 357–389 (1997)

[10] Taylor, S., Todd, P.A.: Understanding Information Technology Usage: A Test of Competing Models. Information Systems Journal 6(2), 144–176 (2001)

[11] Rehman, T., et al.: Identifying and understanding factors influencing the uptake of new technologies on dairy farms in SW England using the theory of reasoned action. Agricultural Systems 94(2), 281–293 (2007)

[12] Ajzen, I., Fishbein, M.: Understanding Attitudes and Predicting Social Behavior. Prentice-Hall, Englewood Cliffs (1980)

[13] Venkatesh, V., et al.: User Acceptance of Information Technology: Toward a Unified View. MIS Quarterly 27(3), 425–478 (2003)

[14] Hung, S.-Y., et al.: User acceptance of Intergovernmental services: An example of electronic document management system. Government Information Quarterly 26, 387–397 (2009)

[15] Titah, R., Barki, H.: Nonlinearities Between Attitude and Subjective Norms in Information Technology Acceptance: A Negative Synergy? MIS Quarterly 33(4), 827–844 (2009)

[16] Terry, D.J., et al.: Attitude Behavior Relations: The Role of In-Group Norms and Mode of Behavioural Decision-Making. British Journal of Social Psychology 39, 337–361 (2000)

[17] Bansal, H.S., Taylor, S.F.: Investigating Interactive Effects in the Theory of Planned Behavior in a Service-Provider Switching Context. Psychology and Marketing 19(5), 407–425 (2002)

Database Design and Information Display of Three-Dimensional Wind Turbine Demonstration System

Guangwei Yan, Yanhui Xing, and Fengjiao Dai

School of Control and Computer Engineering,
North China Electric Power University,
Beijing, China
yan_guang_wei@126.com, royalstudents@163.com,
daifengjiaosw@163.com

Abstract. In order to provide more data support to users of three-dimensional wind turbine demonstration system, an information database for wind turbine has been built up on the basis of three-dimensional simulation. ORACLE database system is used as the database server. The client uses OCCI interface to operate database. Some functions such as data storage and access have been realized in the system. After formatting, obtained information will be displayed on a fixed position of the screen, so that users can easily get information and data related to wind turbine while placing oneself in a three-dimensional scene. Thus it makes user-friendly browsing and access to information possible in the system.

Keywords: ORACLE database, wind turbine, three-dimensional simulation.

1 Introduction

As the rapid development [1] of China's wind power industry, there is an increasing need in wind power technology training for staffs. Three-dimensional graphical simulation technology is intuitive, convenient, practical and economic. With these characteristics it can efficiently support wind power staffs' training.

Research and application of three-dimensional simulation becomes more and more popular at present. For example diesel engine virtual examine and repair were studied by Dalian Jiaotong University [2],Zhang Zhaoyan has done some research on 3D transformer substation simulation[3]. However, using three-dimensional graphics technology for wind turbine simulation study is rare. This system uses 3DSMax and Pro / E to build a three-dimensional model for wind turbine. After that, three-dimensional simulation of wind turbine will is achieved by OSG. The introduction of a virtual person and the use of ORACLE database technology for acquisition and storage of wind turbine data make easier access to the scene related information available while users are in a three-dimensional scene.

D. Jin and S. Lin (Eds.): Advances in ECWAC, Vol. 2, AISC 149, pp. 267–272.
springerlink.com   © Springer-Verlag Berlin Heidelberg 2012

2   System Framework

This system consists of interface, control modules and database system. Fig.1 shows the system structure. Through user interface, you can see a virtual human in the course of an inspection tour, the process of which is: enter the tower, climb the ladder, inspect in the box, use wrench to tighten the screws and so on. Control module is responsible for concrete realization of functions: the organization of the scene, roaming route calculation, behavior control of the virtual human and acquisition of database information. This paper focuses on database design and information displaying of the system. Oracle database takes the responsibility for storing data.

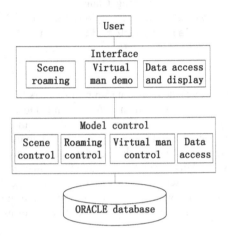

Fig. 1. System structure

3   Roaming Path

In this paper, roaming in wind turbine scene is achieved by fixed path roaming of OSG. Path is formed by points. The attributes of points are: location (x, y, z), Euler angles (a, b, c) and time. Insertion of points will be done at critical positions such as turnings in the path and places for device inspection.

Full path is generated by interpolation between points. Using 3DMax tools the coordinates of key points can be obtained. Time is determined by path order and the Euler angles can be calculated from location of the observing target. Associating roaming path operator with roaming path contributes to automatically update of transformations matrix of the current camera position which is conducted by roaming path operator in each frame. Thus as an input, this transformation matrix helps to adjust camera position and orientation when rendering the system.

4   Design of Database System

ORACLE is one of the most widely used databases in the world. It superiors to other databases in stability, development tools, database security measures and data storage capacity.

4.1   Data Flow Diagram

Fig. 2 is a system data flow diagram. It describes the data flow and information processing of wind turbine in the system from the perspective of information transmission and processing. In this figure, circles represent data processing and arrow lines represent data stream.

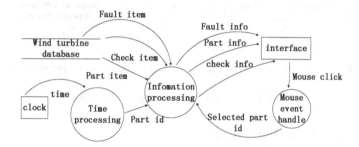

Fig. 2. Data flow diagram

Table 1 shows the specific description of each data processing.

Table 1. Database processing

Process name	Process
Mouse event handler	Find which part the mouse clicks on and return the part id.
Time processing	Return the id of the part whose information would be displayed based on the current time.
Information processing	Get the nodes' information from database and display it on the screen with right format.

This system produces display information based on the input clock and mouse click event. Fig. 3 (a) is a program flow chart of the input clock processing. Fig. 3 (b) shows the processing flow of a mouse click event.

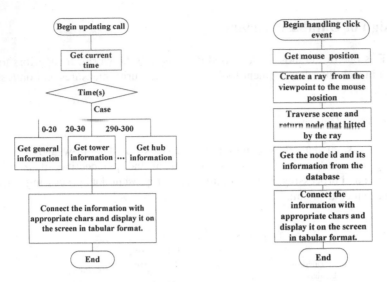

a. Time drive information display   b. Mouse event drive information display

Fig. 3. Information handing program flow chart

4.2 Conceptual Data Model Design

Based on the requirement analysis, an E-R diagram of the database is drawn, as shown in Fig.4. E–R diagrams do well in describing the relationship between system entities.

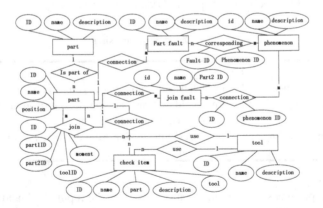

Fig. 4. E-R diagram

4.3 Data Model Design of the Database

Design table structure [4] of the wind turbine database, as follows:

1. Design a table for each entity. Every column in the table stands for an entity attribute. Set the key property and other constraints.

2. For many-to-many entity relationship, create a table which contains key property of each entity corresponding to it.

3. For one-to-many relationship between the entities, add the key property of the other entity to the entity which corresponding to "many" as a foreign key.

5 Database Operations

5.1 Selection of Database Operation Method

In Visual Studio environment there are several ways to connect ORACLE database. Among these, using the ORACLE Call Interface OCCI for data operations stands out for its fast speed, powerful data control, supporting of C/C++ language and cross-platform. As a result, OCCI is this system's choice.

5.2 Operation Steps

Use the following steps to operate database by OCCI:

1. Build the working environment object of database operation which called Environment. Use nonsupport model of tread mutual exclusion.

2. Generate a Connection object, and make a connection with database.

3. Create operating statement.

4. When reaching critical points in time or a mouse click event happens, system will re-read database records automatically and display them on the screen in tabular format.

5. Disconnect from the database.

5.3 Data Display

Three-dimensional graphics only display information on a fixed area of the screen. It will not affect scene display. The system uses OSG engine to realize functions.

Steps:

1. Create a camera node object of OSG. Set the projection rectangular window to be ortho2D (0, 1280, 0, 800).

2. Create a node object called Geode, two text objects and a picture object. One text displays system title and the other wind turbine information. The picture object is for fault information display.

3. Add the Geode node to the new generated Camera node.

4. Add the Camera node to the scene root node.

6 Experimental Result

Fig.5 is a screenshot of the system interface. In Fig.5 the virtual human is in the course of an inspection tour. Users can see related parts information on the screen. Using interactive mode such as mouse click can also get selected parts information which produced by the system automatically. Fig. 5 (a) shows wind turbine information automatically displayed by the system when inspection begins. Blade information is shown in Fig.5 (b) when the user clicks on a blade.

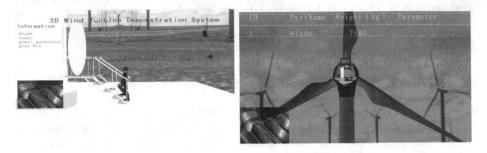

a. Enter the tower   b. User click the blade

Fig. 5. Pictures of the scene

7 Conclusion

This article discusses about database design and information display of the three-dimensional wind turbine demonstration system. Oracle is chosen as a database server and database access interface OCCI plays the role in manipulating database. While ensuring fast access to data, the system also guarantees not only security but also stability. It plays an important part in helping new staffs to master the professional knowledge quickly and vividly.

References

1. Li, J., Shi, P., Gao, H.: China Wind Power Outlook 2010, pp. 10–11. Hainan Press, Hainan (2010)
2. Jiang, D., Ma, T., Cheng, S.: Research on the Key Technique of Diesel Engine Virtual Inspection and Repair System. Diesel Locomotive (4), 7–10 (2005)
3. Zhang, Z., Duan, X., Wang, X.: Application of Visual Simulation System in Simulation of Substation. Computer Simulation 25(2), 252–256 (2008)
4. Sa, S., Wang, S.: Database System Introduction, pp. 229–232. Higher Education Press, Beijing (2000)

Collaborative WWW-Based CAI Software Achieved by Java

San-jun Liu[1] and Zheng-jiang Yi[2]

[1] Institute of Information Engineering of Jiaozuo University;
Jiaozuo, Henan province, China
[2] Institute of Economic Management of Jiaozuo University;
Jiaozuo, Henan province, China
liusanjun1975@163.com, yizhengjiang@163.com

Abstract. The WWW application model based on the internet as a convenient and rapid means of sharing information is popular with users. The users view multimedia teaching contents by accessing hypertext pages, and execute free explore learning by clicking hyperlinks. However, the teaching based only on the hypertext pages has many shortcomings. The computer-assisted teaching software described by this article overcomes the many shortcomings. The software is achieved by J + + and consists of the server section, the students section, the teaching section by the teachers, the courseware maintenance section. This software actually is a teaching framework system, and all teaching materials are prepared by the teachers, and the teachers and students go into their by visiting their roles by accessing their own URL addresses. The students execute the learning, discussion, exercises and testing in the on-site control of the teachers.

Keywords: real-time, data-source data, frame.

1 Function of the System

1.1 Learning of Hypertext Pages to Learn

The teachers unified guide the learning pages of the students. The teachers choose the learning pages, send the URL address of the learning pages to the students, and guide the students to the Web pages to learn. This process increases the pertinence of learning, and meets the organized and controlled teaching needs.

1.2 Real-Time Discussion

The real-time discussion is an essential feature of collaborative teaching software. The system provides two kinds of tools to support real-time online discussion. One tool is the text conversation. If the speaker does not select the users, the conversation is public speak, otherwise the conversation is privately speak and is only sent to the selected users. There is a text field in each user's discussion panel to display the speaker and the corresponding speech of each speak. The other tool is the shared whiteboard. The teachers and students can write contents on the whiteboard, and the contents on the whiteboard of all users are consistent. The shared whiteboard supports lines, circles, rectangles, text, graphics and other basic operations, and can choose the color of lines.

1.3 Interactive Exercises

The teachers send the guided message of practice, the students automatically enter the practice interface, and the system automatically transferred questions from the questions database for the students to do. In this process, the system compares the answers of the students with the standard answers to judge if the answers of the students is correct, and pops dialog boxes to give students the encourages information. The interactive exercises help students to farther understand the teaching contents.

1.4 Online Testing

The testing is a necessary teaching step in the process of teaching. The teachers can check the study of the students by the testing, and understand the learning results of the students to provide a basis for improving the teaching. The students can know their mastery of the contents to increase the interest in learning. Currently, each test paper is prepared by choosing questions from the test database and set the corresponding scores by the teachers. Tests are time-limited.

2 Overall Design of the System

To accomplish the functions from the system perspective, there are two ways of working: one ways is the client / server approach (such as students do questions), the client-side send an access application, the service-side gives the respondence.

The other ways is web-based collaboration applications (such as real-time discussion, etc.). In this way, any party could be a message sender, while the other party becomes the recipient of the message. The sender's action is caused by all kinds of event-driven of the user interface, and its timing is accidental. The receiver must always be in a listening state to receive.

Since Java's security restrictions, the socket connection between the two users' browsers can not be directly established. The communication between the two users' browsers must be transferred by the Java application program in a Web server. To sum up, in order to achieve the function of the system, the overall design of the system is shown in Fig. 1.

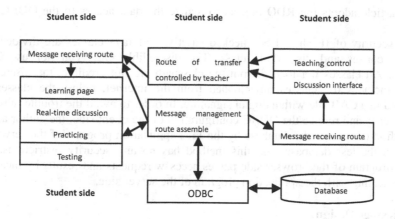

Fig. 1. Overall design of the system

Instructions:

a) The students and teachers, respectively, establish communication links with the Java application program in the Web server by means of client/server. The connection monitoring route of the server side is responsible for monitoring the Socket connection application from the clients. The server side should establish a message management route for each linked client, and the message management route is responsible for receiving messages from the corresponding client, and analyzing and managing the messages.

b) The students and teachers, respectively have a message receiving route which is responsible for receiving message from the serve side. The message receiving route reflects the requirement of the received messages on the user interface by calling the method in the object of the user interface. Therefore, during the process of designing the message receiving route, it is a basic method to achieve the exchange visits between the objects in Java that the objects of the user interface are set as a parameter passed to the objects in the message receiving route.

c) The process of accessing the test database is executed as follows: the students send the corresponding message to the serve, the serve side accesses ODBC-based database via the COM component RDO to access, and feedbacks the accessing results to the students.

3 Several Key Technology

3.1 Database Access

Visual J + + do a special extension for the Java class database according to the Microsoft operation system, and the users can access any software supporting COM in the Java programs. For accessing to the database, two methods of DAO(Date Access Object) and RDO (Remote Data Object) are provided in J + +. The software in this article adopts the RDO objects to realize the data access to the ODBC data source.

The security of IE should be checked strictly, and Java classes are divided into credible classes and discredible class. The discredible class can not use COM services. All classes not loaded from the classpath are not credible, the discredible classes includes the classes downloaded from the internet, unless the classes are packaged in a CAB file with a digital signature. In order to avoid the trouble raised by the security and reduce the ODBC configuration of the client side, the author adopts the method of visiting the database by the Java application program of the serve side to access the test database, and this method has no any security restrictions. The Applet program of the browser side passes accessw requests and results by means of communicating with the application program of the server side.

3.2 Message Design

The whole system is built based on the messages, the server side and client side both bear message management work. For a large number of messages based on collaborative applications, the serve side is only responsible for transferring such messages and such messages are managed by the client side. Each kind of messages is consisted of the message header and the content, and the message header is used for the system discrimination. The whole system mainly comprises six kinds of messages: chat messages; whiteboard messages; guide messages; users log in and log off messages; accessing to the test database messages, and other messages (such as boarding grade and checking grade).

3.3 Achieve of URL Guide

Inputting and transferring of URL The teachers click on "Page guide" button, the system accesses   learning content index and pop up a dialog box so that the teachers can select the chapters and problems to study. Then the system sends the corresponding URL address to the student client side via the Socket path, and a new browser displaying the page is pop up for the confirmation.

Displaying of URL The steps are as follows:

a)   using the URL class to form URL objects according the receiving character strings;

b) obtaining Applet Context of the small application program by means of get Applet Context of the Applet class;

c) displaying the content of URL by means of show Document of the Applet Context interface. The specific format and parameters of the method are described as follows:

Wherein, the target parameter is used to appoint the location of displaying of the file contents, and the specific value and the role of the target parameter are shown the Table 1:

Table 1.

Value	Location of file display
"_self"	Current frame
"_parent"	Parent frame
"_top"	Top frame
"_blank"	Displaying in a new open browser window
"Frame name"	Displaying in an appointed frame

In the teacher side, the author adopts the "-blank" method, which open another browser window to display the URL content for the teachers to confirm the content URL. In the student side, the author adopts the multi-frame method, which arranges the multi-frame in the HTML file, and one frame is used to display the Java Applet, the other frame displays the content of URL pages sent by the teachers. The size of the specified frame in the HTML file can be adjusted, so that the students can adjust the frame border to read the content more easily.

3.4 Conclusion

This article describes a specific implementation of WWW-based collaborative teaching software system. The software actually is a teaching framework system, and the specific teaching contents are arranged and designed by the teachers by using the maintenance parts of the system. Thus, the subjective initiative of the teachers can be well exerted. The author intends to make the following improvements to the system: a) providing the function of automatic test paper; b) introducing a more comprehensive model of student to estimate the learning level of the students; c) providing a variety of teaching methods, so that the system can be operated in both circumstances of the attendance and absence of the teachers to increase the intelligence of the CAI.

References

[1] Inmon, W. H., Wang, Z.-H. (trans.), et al.: Establishment of data warehouse. Mechanical Industry Press, Beijing (2000)
[2] Liu, F.: Design and optimization of database connection pool based on Java. Micro-Computer Applications (2008)
[3] Peng, C.-Y.: Development guide of JAVA application system. Mechanical Industry Press, Beijing (2004)
[4] Jaworski, J., Kang, C. (trans.): Java Development guide. China Water Conservancy and Hydropower Press, Beijing (1996)

Research of Network Mass Storage Mode Based on SAN

San-jun Liu[1] and Zheng-jiang Yi[2]

[1] Institute of Information Engineering of Jiaozuo University; Jiaozuo, Henan province, China
[2] Institute of Economic Management of Jiaozuo University; Jiaozuo, Henan province, China
liusanjun1975@163.com, yizhengjiang@163.com

Abstract. In the face of new needs of the current network storage, the traditional DAS, NAS and advanced SAN network storage technology are analyzed, and the technical differences between the three technologies are compared. The redundant technology and main performance indicator system are analyzed by using system reliability theory, and thus derive that the RAID acting as the key and core storage node of the storage management systems still has some unideal situations. After an in-depth study of the optimization strategy, the data reconstruction and other key technologies of "RAID 50, small write" in the RAID design, this article provides an ideal realization algorithm, and designs a high-performance disk array system program using the two-processor and CACHE with a large capacity. Thus, the network mass storage based on SAN is achieved.

Keywords: SAN(Storage Area Network), Mass storage, RAID, Small write.

1 Introduction

With the sustained, rapid growth of the large data applications, such as the simulation, modeling, BU browsing, multimedia, interactive process, e-commerce, data warehousing and data acquisition, the greater storage equipment capacity is needed. However, ad the continuous development and construction of the network technology, the huge network customers make the traditional file server as the bottleneck of the network service. Especially in the data storage, not only the data storage capacity is exponential growth, but also the storage equipment performance, expansibility, security, manageability and so on have further requirements. Driven by the huge storage market demand, the storage technology is developing from the traditional mainframe-centric storage structure to the network storage system. In the network storage technology, the information access and sharing services of the network information system is provided by the network storage equipment. Its main features are its super big storage capacity, large dta transger rate and high system availability and remote backup, and so on.

2 Introduction of Major Storage Network Technology

After the large-scale development, the storage technology has been completely beyond the underlying hardware technology of the hard disk level, and has risen to

D. Jin and S. Lin (Eds.): Advances in ECWAC, Vol. 2, AISC 149, pp. 279–284.

become one of the four pillars of IT side by side with the computer, network and communication. According to the connection mode of the host with the storage system, there are three main network storage technologies:

2.1 Direct Attached Storage DAS

The DAS (Direct Attached Storage) technology is the first network storage structure to be used. In the DAS, the data storage function is shared by the server and storage equipment. DAS itself is the hardware stack, and has no any storage management system. Therefore, the maintenance, management and basic I/O operation of the storage equipment are conducted through the server operating system. This occupies the CPU and memory of the server system. Thus, the greater the amount of data, the longer the backup and restore, and the greater the dependence and influence on the serve hardware.

2.2 Network Attached Storage NAS

The NAS (Network Attached Storage) technology integrates distributed storage equipments as a large data storage center for the centralized management. NAS directly connects the storage equipments to the internet, and shares memory resources or even data by adopting the Client/Server mode. NAS supports the file-lever data sharing. However, NAS and LAN are in the same physical network, and NAS depends on certain characteristics of LAN. NAS needs a very big network bandwidth and very high CPU processing power, and when the conditions mentioned above are not met, the network will be congested and the performance thereof will be reduced.

2.3 Storage Area Network SAN

The SAN (Storage Area Network) technology connects many storage equipments and servers by using switches, hubs and cables based on the optical fiber channel technology to form a "back-end" network. As shown in Fig.1, in the SAN, the exchange of the storage management information of data between the servers and storage equipments, and between the storage equipments is conducted adopting the communication mode of multipoint-to-multipoint. Meanwhile, the back-end technology feature of SAN make SAN can quickly and efficiently transfer data directly from a storage equipment to another storage equipment, rather than through the corporate LAN. As an independent back-end storage network, SAN has high transmission rate and high reliability. SAN become an important technology of enterprise-class storage solutions due to its good performance.

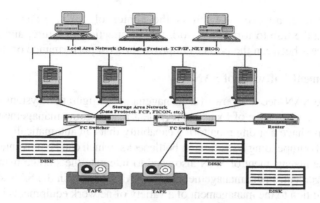

Fig. 1. Storage structure diagram of SAN

3   Main Components of SAN

3.1   Network Structure of SAN

The components of the network interconnection structure includes host bus adapters, bridges, routers, optical fiber, traps and switches. The communication between all components follows the link protocol. The configuration of the network interconnection structure is shown in Figure 2.

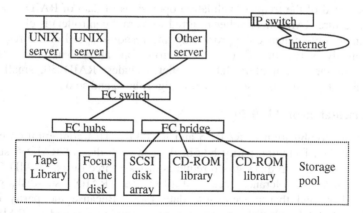

Fig. 2. Configuration of network interconnection structure of SAN

1) Host bus adapter: the host can be connected to the storage equipment by the host bus adapter.

2) Network bridge: FC, SCSI and ESCON are connected by the network bridge.

3) Router: Data can be transferred between the SCSI bus and FC port by the router.

4) Hub: The role of the hub in the SAN is similar to that in the traditional network. The server can access the storage network through one or more FC hub or switch. The feature of the hub is sharing a 100Mbps bandwidth, having less ports, usually 7-12, and having two nodes that can communicate with each other at the same time. The ports of the hubs are very cheap, and are very suitably used in small applications.

5) Switch: Logically, FC switch is the center of SAN. The switch can be combined with the hub to form a network interconnection structure, and allows a full bandwidth access between the ports to improve the cost performance of the system.

3.2 Management Software of SAN

The large-scale SAN needs a powerful management configuration system to manage it. As the need of a wide range of focus management resource, the management software of SAN should also have a strong monitoring capability that can automatically detect failures to be generated orimpending transmission bottlenecks, which can be prevented in advance. Meanwhile, the management of SAN also need to integrate the storage management and network management. In the management of the SAN, the data in the SAN should also be managed in addition to the management of a variety of network equipments in SAN.

4 Core RAID Technology in SAN System

RAID assembles many separate physical disks in different modes to form a logical disk to provide the function of fault tolerance and the storage performance higher than a single disk. RAID technology now has seven basic RAID levels from RAID 0 to 6 via continuous developments. In addition, there are some combination forms of basic RAID levels, such as RAID 10 (combination of RAID 0 and RAID 1), RAID 50 (combination of RAID 0 and RAID 5), and so on.

RAID is implemented by software RAID and hardware RAID. In order to improve the speed of the frequent validation operations of data of RAID system, the design of hardware RAID (hereinafter referred to as raid controller) is chosen. This program can not only enhance the processing and response speed of I/O request, but also can greatly solve the CPU from serious operation burdens. The key technologies of the design of RAID controller includes: RAID 50, small writing optimization strategy, and data reconstruction algorithms, and so on.

4.1 Implementation of RAID 50

RAID 50 is the combination of RAID 5 and RAID 0. This configuration executes the spin-off of data including parity bits of data in each disk of RAID 5 sub-disks group. Each RAID 5 sub-disks group requires at least three disks. RAID 50 has a better feature of failure tolerance because it allows each RAID 5 sub-disks group to have a bad disk, and that prevents the data loss. Because the parity bits are distributed in RAID 5 sub-disks group, the reconfiguration speed of RAID 50 is enhanced relative to RAID 5 with the same number of disks. The advantages of RAID 50: the higher fault tolerance; the RAID group downgraded increases the cost of the system, and affects the throughput, but due to the use of the parity checking redundant mode, the reconstruction time of RAID 50 is longer than that in the mirror mode, but shorter than that in RAID 5 having the same configuration.

4.2 Optimization of the Small Write

In order to improve the efficiency and throughput of the overall system, the effective optimization and consolidate of the small write operation are needed. There are the following primary methods to solve the problem of the small write:

1) By using the delayed write policy, on the one hand, the write request of the host is completed in writing CACHE, on the other hand, when the writing CACHE get to a high water level, the write request of the host is written to the disk by the RAID controller of DESTAGE, and is I/O consolidated by the CACHE module in the DESTAGE process.

2) When dealing with the operation of order small write, because of the continuous access to the logical block address LBA, the I/O consolidation is executed on the controller device side. By consolidating many continuous small writer into one big write, the number of disk access is reduced and the overall performance of the system is improved.

3) We know that according to the small write, two read disk requests and two write disk requests are needed in the lowercase flow, and three read disk requests and two write disk requests are needed in the capital flow. Thus, the operation of the small write can be identified and the lowercase flow of read-modify-write that can special solve the problem of the small write can be adopted to optimize the operation of the small write.

4.3 Achievement of Data Reconstruction Algorithm

The RAID controller should have the corresponding error recovery mechanism in addition to application and some fault tolerance, for recovering the system from a fault tolerance mode to ensure the system run normally. The sub-section level reconstruction algorithm which has less affect on the user's response time is final chosen by comparing the sub-section level reconstruction algorithm and the disk based reconstruction algorithm. However, the reconstruction speed of the sub-section level reconstruction algorithm is unsatisfactory, some improvements should be done to the sub-section level reconstruction algorithm:

1) raising the priority of the reconstruction locks to appropriately enhance the reconstruction speed.

2) I/O is redirected in the reconstruction process, and the pseudo codes of the redirection is as follows:

```
If the reconstruction is doing,
{
the sub-section of the data block is identified.
If the sub-section has been reconstructed,
the data block in the hot spare disk is directly
accessed.
else
{
The data reconstruction process is called to execute
the data reconstruction.
The result is available to the user process, and the
data block is written into the backup disk..
}
}
else
```

The reconstructed data block in the hot spare disk is directly accessed. If the sub-section including the data block is reconstructed, the redirecting operation should be adopted to directly execute the I/O operation in the hot spare disk. This needs only a I/O request, and does not need the online data reconstruction which costs a large number of I/O operations. Thus, the I/O redirection operation can effectively enhance the performance of the system and the response time to the user. The result shows that the response time to the user can be enhanced 10% to 20% by adopting the redirection process.

4.4   Increasing the Degree of the Reconstruction Process, That Is Many Sub-sections Are Reconstructed at the Same Time

The improved reconstruction algorithm of the sub-section level can take full advantage of the available bandwidth of the disk, and reduce the occupancy rate of the system resources in the reconstruction process. Furthermore, the reconstruction speed thereof is fast and can enhance the response speed to the user's request.

References

[1] Tao, Z.: Research of SAN storage resource management system-design and implementation of backup and recovery modules [MS Thesis]. Computer institute of Northwest industry university, Xian (2005)

[2] Clark, T.: IP-SANS A Guide to iSCSI, iFCP and FCIP Portoeol for Storage Area Netwokrs. Pearson Education (December 2005)

[3] Wei, Z.: Research and implementation of mass storage technology based on IP-SAN [PhD thesis]. Computer institute of Northwest industry university, Xian (2007)

[4] Wang, X.-N., Jiang, B.-S., Xu, J.: Research and implementation of disk array Cache based RAID5. Computer Engineering 29(3), 149–151 (2006)

[5] Feng, Y.-Y., Han, J.: Research and implementation of file storage technology based on SAN network. Micro-Computer Applicaiton 4(23), 45–50 (2008)

[6] Chong Jr., F.: Host bus adapter based scalable performance storage architecture. US, G06F 1314, 6684274 (1999)

The Degree Distribution of Random Attachment Networks

Qinggui Zhao[1,2]

[1] Department of Mathematics and Science, Hunan First Normal University,
Changsha, China
[2] School of Mathematics and Statistics, Chongqing University of Arts and Sciences
Chongqing, China
csupolaris@sina.com

Abstract. In this paper, we investigate networks grow in time according to an attachment rule that is random. Based on Markov chain theory, paper provides a rigorous proof for the existence of the steady-state degree distribution of the network generated by this model and gets its corresponding exact formulas. Moreover, we found this kind of networks don't obey the power-law distribution.

Keywords: Random attachment networks, Markov chain, Degree distribution, Barabási-Albert model.

1 Introduction

Many nature and manmade complex systems can be fruitfully represented and studied in terms of networks or graphs [1]. The last decade has witnessed the birth of a new movement of interest and research in the study of complex networks, i.e. Networks whose structure is irregular, complex and dynamically evolving in time, with the main focus moving from the analysis of small networks to that of systems with thousands or millions of nodes, and with a renewed attention to the properties of networks of dynamical units (see [2-10]). Research on fundamental properties and dynamical features of such complex networks has become overwhelming.

Since the ground-breaking papers by Barabási and Albert on scale-free networks [11] and by Watts and Strogatz on small-world networks [7], the interest on large, growing, and complex networks has soared. Consider a network consisting of N nodes, each having a number k of links to other nodes in the network. Since the network is large, k can be regarded as a random variable with a probability distribution $P(k)$. Barabási and Albert discovered [2,11] that many networks in nature appear to exhibit the scale-free feature in that the distribution $P(k)$ exhibits a power-law behavior in a range of k − values over several orders of magnitudes: $P(k) \sim k^{-\gamma}$. The mechanisms leading to the power-law distribution are argued to be growth and preferential attachment [2,4], where the former means that the size of the network keeps increasing with time and the latter underlies that the relative probability for an

D. Jin and S. Lin (Eds.): Advances in ECWAC, Vol. 2, AISC 149, pp. 285–289.

already heavily connected node to get new links is proportionally large. Growth and preferential attachments appear to be the fundamental organizing principle of the many complex networks.

A scale-free network, by its definition, permits a high degree of organization such as the existence of a set of nodes with great numbers of links. The scale-free situation is, however, idealized, for which the distribution $P(k)$ is strictly power-law. Indeed, the illuminating scale-free model proposed by Barabási et al [11] predicts a universal power-law scaling behavior with exponent $\gamma = 3$.

Dorogovtsev et al [12] introduced a more general model of growing networks and allowed multiple edges between vertices, where each new vertex has an initial attractiveness A . Simultaneously, m new directed edges coming out from non-specified vertices are introduced. The probability that a new link points to a given vertex is proportional to $k = A + q$ with q being in-degree of the vertex. They found that this has a power-law tail in the degree distribution.

Krapivsky et al [13] considered growing networks with $\Pi(k) \sim k^\alpha$, and found different behaviors arise for $\alpha < 1$, $\alpha > 1$ and $\alpha = 1$. For $\alpha < 1$, the number of vertices with k links, $P(k)$, varies as a stretched exponential. For $\alpha > 1$, a single vertex connects to nearly all other vertices. In the borderline case $\Pi(k) \sim k$, the power-law $P(k) \sim k^{-\gamma}$ is found.

In this paper, we study a comprehensive quantitative description of only random attachment network model, we call it as random attachment networks. Based on the Markov chain theory, we propose a new approach to provide a rigorous proof for the existence of steady-state degree distribution of this mode.

2 Model Description

The power-law scaling in the Barabási-Albert model indicates that growth and preferential attachment play important roles in network development. But are both of them necessary for the emergence of power-law scaling? To address this question, one limiting cases of the Barabási-Albert model has been investigated, which contain only one of these two mechanisms [4].

In this section, we will introduce the model.

(i) Growth: at t=0, starting with a small number m_0 of vertices, which has a total degree N_0 . At every time step, we add a new vertex with $m(\le m_0)$ edges into the network. The new vertex is linked with m different vertices already existing in the system.

(ii) Attachment: when choosing the vertices to which the new vertex connects, we assumed that the new vertex connects with equal probability to the vertices already present in the system, i.e., $\Pi(k_i) = 1/(m_0 + t)$, independent of k_i .

3 Degree Distribution

We define vertex i as the node added at time i and the degree $k_i(t)$ of vertex i at time t.
Let

$$P(k,i,t) = P\{k_i(t) = k\} \tag{1}$$

be the probability of vertex i having degree k at time t, and moreover let the network degree distribution be the average over all its vertices at time t [12],

$$P(k,t) = \frac{1}{t}\sum_{i=1}^{t} P(k,i,t) \tag{2}$$

and

$$P(k) = \lim_{t \to \infty} P(k,t) \tag{3}$$

Recall that $k_i(t)$ is a random variable for any fixed t and it is non-homogeneous Markov chain for variable t [14].
From (ii), the state-transition probability of this Markov chain is given by

$$P\{k_i(t+1) = l \mid k_i(t) = k\} = \begin{cases} 1 - \dfrac{m}{m_0 + t}, & l = k \\[2mm] \dfrac{m}{m_0 + t}, & l = k+1 \\[2mm] 0, & otherwise \end{cases} \tag{4}$$

Where $k = 1, 2, \cdots, m+t-1$ and $i = 1, 2, \cdots$

Lemma 3.1. $P(m)$ exists, moreover, $P(m) = \dfrac{1}{m+1}$.

Proof: From Eq.(2), it follows that

$$P(m,i,t+1) = (1 - \frac{m}{m_0 + t}) P(m,i,t) , \tag{5}$$

Since $P(m, t+1, t+1) = 1$, one has

$$P(m,t+1) = \frac{1}{t+1}\sum_{i=1}^{t+1} P(k,i,t+1) = \frac{1}{t+1}(1 - \frac{m}{m_0 + t}) P(m,t) + \frac{1}{t+1} \tag{6}$$

Then, by iteration and $P(m,1) = 1$,

$$P(m,t) = \frac{1}{t}\prod_{i=1}^{t-1}(1 - \frac{m}{m_0 + i}) \cdot [1 + \sum_{l=1}^{t-1}\prod_{j=1}^{l}(1 - \frac{m}{m_0 + j})^{-1}] \tag{7}$$

Next, let

$$x_t = 1 + \sum_{l=1}^{t-1} \prod_{j=1}^{l} [1 - \frac{m}{m_0 + j}]^{-1}, \quad y_t = t \prod_{i=1}^{t-1} [1 - \frac{m}{m_0 + i}]^{-1},$$

Using **Stolz Theorem**, we get

$$P(m) = \lim_{t \to \infty} P(m,t) = \lim_{t \to \infty} \frac{x_t}{y_t} = \lim_{t \to \infty} \frac{x_{t+1} - x_t}{y_{t+1} - y_t} = \lim_{t \to \infty} \frac{1}{1 + \dfrac{mt}{m_0 + t}} = \frac{1}{m+1} \tag{8}$$

Lemma 3.2. For $k > m$, if $P(k-1)$ exists, then $P(k)$ also exists:

$$P(k) = \frac{m}{m+1} P(k-1) \tag{9}$$

Proof: From the markovian property,

$$P(k,i,t+1) = (1 - \frac{m}{m_0 + t}) P(k,i,t) + \frac{m}{m_0 + t} P(k-1,i,t)$$

When $k > m$, $P(k,t+1,t+1) = 0$. We get

$$P(k,t+1) = \frac{t}{t+1}(1 - \frac{m}{m_0 + t}) P(k,t) + \frac{mt}{(m_0 + t)(t+1)} P(k-1,t)$$

the above difference equation has the following solution

$$P(k,t) = \frac{1}{t} \prod_{i=1}^{t-1}(1 - \frac{m}{m_0 + i}) \cdot \{ P(k,1) + lP(k-1,l) \sum_{l=1}^{t-1} \frac{m}{m_0 + l} \cdot \prod_{j=1}^{l}(1 - \frac{m}{m_0 + j})^{-1} \}$$

similar to **Lemma 3.1**, we have

$$P(k) = \frac{m}{m+1} P(k-1)$$

Then, (9) is proved.

Theorem 3.3. The steady-state degree distribution of random attachment networks exist, and is given by

$$P(k) = (\frac{m}{m+1})^{k-m} \frac{1}{m+1} \tag{10}$$

Proof: By induction, from **Lemma 3.1** and **Lemma 3.2**, we get **Theorem 3.3** easily.

From (10), we know this kind of networks don't obey the power-law distribution.

4   Conclusion

We have constructed a class of random attachment networks. Based on Markov chain theory, paper provides a rigorous proof for the existence of the steady-state degree distribution of the network generated by this model and gets its corresponding exact formulas.

Acknowledgements. This work is supported by Hunan First Normal University and the Project Foundation of Chongqing Municipal Education Committee (Grant No. KJ101210).

References

1. Bollobás, B.: The Grid: Modern Graph Theory. Springer, New York (1998)
2. Albert, R., Jeong, H., Barabási, A.-L.: The Diameter of the World-Wide Web. Nature 401, 130–131 (1999)
3. Newman, M.E.J.: Scientific collaboration networks. I. Network construction and fundamental results. Phys. Rev. E 64, 0161311–0161318 (2001)
4. Albert, R., Barabási, A.-L.: Statistical mechanics of complex networks. Rev. Mod. Phys. 74, 47–97 (2002)
5. Dorogovtsev, S.N., Mendes, J.F.F.: Evolution of networks. Adv. Phys. 51, 1079–1187 (2002)
6. Bollobás, B., Riordan, O.: Mathematical results on scale-free random graphs. In: Handbook of Graphs and Networks, pp. 99–121. Wiley-VCH, Berlin (2002)
7. Watts, D.J., Strogatz, S.H.: Collective dynamics of 'small-world' networks. Nature 393, 440–442 (1998)
8. Wagner, A.: The Yeast Protein Interaction Network Evolves Rapidly and Contains Few Redundant Duplicate Genes. Mol. Biol. Evol. 18, 1283–1292 (2001)
9. Wasserman, S., Faust, K.: Social network analysis: Methods and applications. Cambridge University Press, Cambridge (1994)
10. Garcia-Domingo, J.L., Juher, D., Saldanña: Degree correlations in growing networks with deletion of nodes. Physica D 237, 640–651 (2008)
11. Barabási, A.-L., Albert, R.: Emergence of Scaling in Random Networks. Science 286, 509–512 (1999)
12. Dorogovtsev, S.N., Mendes, J.F.F., Samukhin, A.N.: Structure of Growing Networks with Preferential Linking. Phys. Rev. Lett. 85, 4633–4636 (2000)
13. Krapivsky, P.L., Redner, S., Leyvraz, F.: Connectivity of Growing Random Networks. Phys. Rev. Lett. 85, 4629–4632 (2000)
14. Zhao, Q., et al.: The degree distribution of fixed act-size collaboration networks. Pramana-Journal of Physics 73, 955–959 (2009)

Acknowledgements This work is supported by Hunan First Normal University and the Project Foundation of ... Hunan Municipal Education Committee (Grant No. ...K100...).

References

1. ...
2. ...
3. ...
4. ...
5. ...
6. ...
7. ...
8. ...
9. ...
10. ...
11. ...
12. ...
13. ...
14. ...

Two-Degree-of-Freedom Control Scheme for Inverse Response Processes with Time Delay

Jinggang Zhang, Zhichneg Zhao, and Yuanyuan Qu

Department of Automation, Taiyuan University of Science and Technology,
Taiyuan, China 030024
jg_zhang65@163.com

Abstract. A two-degree-of-freedom (2DOF) control scheme is proposed for inverse response processes with time delay. The proposed control structure consists of reference model, feedforward controller and feedback controller. The feedforward controller is designed based on the inverse model of the plant, so that set-point tracking response can follow the one defined by a reference model. The feedback controller, designed using direct synthesis method based on disturbance rejection, can assure the excellent disturbance rejection performance on the premise of guaranteeing stability of the closed-loop system. Compared with existing control method, the proposed control structure can assure system of good set-point tracking performance and disturbance rejection performance simultaneously, and the controller parameters can be adjusted conveniently.

Keywords: inverse response processes, time delay, two-degree-of-freedom.

1 Introduction

Inverse response is the dynamic behavior of the system showing step response in the opposite direction initially to that of the steady-state direction. Such a behavior is exhibited by processes such as drum boilers in distillation columns. The reason for the inverse response is that the process transfer function has zeros in the open right half plane (RHP) [1]. This nonminimum phase (NMP) characteristic of the process affects the achievable closed-loop performance because the controller operates on wrong sign information at the beginning of the transient. Generally, these processes are particularly difficult to control and thus require special attention.

To overcome the difficulty mentioned above, many methods of process control are included in the design of controller for inverse response processes. There are two categories of control structures in classical methods. The first category of structure is PID controller with many kinds of tuning methods [1-6]. The second category of structure has an inverse response compensator [7-9]. These control methods are one-degree-of-freedom control structure, that can't make system to get good set-point tracking performance and disturbance rejection performance simultaneously.

In this paper, a two-degree-of-freedom control structure is proposed for the inverse response process with time delay. And the controller design method is given. Compared with existing control methods, the proposed method allows system to get good set-point tracking performance and disturbance rejection performance simultaneously. Control

D. Jin and S. Lin (Eds.): Advances in ECWAC, Vol. 2, AISC 149, pp. 291–296.
springerlink.com   © Springer-Verlag Berlin Heidelberg 2012

structure is simple and the controller parameters are easy to adjust. The effectiveness of the method is showed by the simulation results.

2   Two-Degree-of-Freedom Control Structure

The two-degree-of-freedom control structure proposed in this paper is presented as shown in figure 1. It consists of feedback controller, set-point feedforward controller and reference model. In the figure 1, r is the set-point input, y is the process output, d is the disturbance signal, G_P is the controlled process plant, G_C is the feedback controller, G_F is the feedforward controller and G_R is the reference model.

The feedback controller is used to ensure the stability of the system and to reject disturbance. The set-point tracking characteristics are implemented by the feedforward controller and reference model.

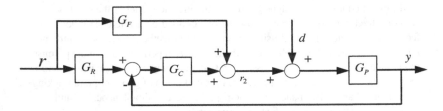

Fig. 1. Two-degree-of-freedom control structure

The closed-loop transfer function for set-point input is derived as

$$\frac{y}{r} = \frac{G_R(s)G_C(s)G_P(s)}{1+G_C(s)G_P(s)} + \frac{G_F(s)G_P(s)}{1+G_C(s)G_P(s)} \tag{1}$$

The closed-loop transfer function for rejection input is derived as

$$\frac{y}{d} = \frac{G_P(s)}{1+G_C(s)G_P(s)} \tag{2}$$

According to the definition of internal stability in feedback control system, when G_F and G_R is stable, internal stability is can be guaranteed by design G_C .

Obviously, if

$$G_F(s) = \frac{G_R(s)}{G_P(s)} \tag{3}$$

can be implemented, then equation (1) can be transformed as

$$\frac{y}{r} = G_R(s) \tag{4}$$

In this case, set-point tracking characteristics and disturbance rejection characteristics are decoupled and can be designed respectively. Obviously, when system model is perfect and without disturbance, the system shown in figure 1 is open-loop stable and

G_C have no control action. This means that function of closed loop feedback control is to reject the disturbance and model uncertainty.

In the nominal case, the process model $G_P(s)$ will be factored as:

$$G_P(s) = G_{PM}(s)G_{PA}(s) \tag{5}$$

where $G_{PM}(s)$ is reversible part of the process model, $G_{PA}(s)$ contains the right half plane zeros and time delays.

The set-point tracking characteristics are defined by $G_R(s)$. Without loss of generality, $G_R(s)$ can be designed as:

$$G_R(s) = \frac{G_{PA}(s)}{(\varepsilon s + 1)^n} \tag{6}$$

where n is the relative order of $G_{PM}(s)$, ε is an adjustable parameter. Then, equation (3) and (4) can be transformed as

$$G_F(s) = \frac{G_{PM}^{-1}(s)}{(\varepsilon s + 1)^n} \tag{7}$$

$$\frac{y}{r} = \frac{G_{PA}(s)}{(\varepsilon s + 1)^n} \tag{8}$$

In order to improve disturbance rejection characteristics, the feedback controller G_C is designed based on the direct synthesis method in reference [10].

According to equation (4), feedback controller $G_C(s)$ can be expressed as

$$G_C(s) = \frac{G_P(s) - (\frac{y}{d})}{(\frac{y}{d})G_P(s)} \tag{9}$$

Let the desired closed loop transfer function for disturbances be specified as $(y/d)_d$, and assume that a process model $G_P(s)$ is available. The design equation for the feedback controller $G_C(s)$ is given.

$$G_C(s) = \frac{G_P(s) - (\frac{y}{d})_d}{(\frac{y}{d})_d G_P(s)} \tag{10}$$

3 Controller Design

Consider the following typical inverse response process with time delay.

$$G_P(s) = \frac{K(-T_z s + 1)e^{-\tau s}}{(T_1 s + 1)(T_2 s + 1)} \tag{11}$$

where K is proportional gain, T_z is RHP zero resulting in inverse response, τ is time delay, and T_1, T_2 are time constant.

Using the first order Taylor approximation, the term $(-T_z s+1)$ can be seen as a pure time delay $e^{-T_z s}$. The transfer function of inverse response process with time delay is expressed as

$$G_P(s) = \frac{Ke^{-(\tau+T_z)s}}{(T_1 s+1)(T_2 s+1)} = \frac{Ke^{-\theta s}}{(T_1 s+1)(T_2 s+1)} \qquad (12)$$

where $\theta = \tau + T_z$.

According to equation (6) and (7), $G_R(s)$ and $G_F(s)$ can be given as respectively

$$G_R(s) = \frac{(-T_z s+1)e^{-\tau s}}{(\varepsilon s+1)^2} \qquad (13)$$

$$G_F(s) = \frac{(T_1 s+1)(T_2 s+1)}{K(\varepsilon s+1)^2} \qquad (14)$$

Using design method of reference [10], $G_c(s)$ can be designed as a PID controller, its parameters is expressed as

$$K_c = \frac{1}{K} \frac{(3\lambda+\theta)[(T_1+T_2)\theta+T_1 T_2]-\lambda^3-3\lambda^2\theta}{(\lambda+\theta)^3} \qquad (15)$$

$$T_i = \frac{(3\lambda+\theta)[(T_1+T_2)\theta+T_1 T_2]-\lambda^3-3\lambda^2\theta}{(T_1+T_2+\theta)\theta+T_1 T_2} \qquad (16)$$

$$T_d = \frac{3\lambda^2 T_1 T_2+T_1 T_2\theta(3\lambda+\theta)-(T_1+T_2+\theta)\lambda^3}{(3\lambda+\theta)[(T_1+T_2)\theta+T_1 T_2]-\lambda^3-3\lambda^2\theta} \qquad (17)$$

4   Simulation Results

In this section, the same numerical example as in that work by I-Lung Chien [3] will be used here to test the closed loop performance of the proposed control structure. The transfer function model of this example is

$$G_P(s) = \frac{(-0.2s+1)e^{-0.2s}}{(s+1)^2}$$

The parameters of the controller proposed in this paper are: $\varepsilon = 0.4$ $\lambda = 0.5$, the parameters of the controller $G_c(s)$ are: $T_i = 1.5281$; $K_c = 4.1084$; $T_d = 0.4040$. The parameters of the PID controller in reference [3] are: $T_i = 1$; $K_c = 0.83$; $T_d = 1$.

Suppose that the set-point is a unit step signal at $t=0$ and the disturbance is a negative unit step signal in the controlled variable at $t=8$. The controller parameters of methods above are given in Table 1, where the comparisons of some performance indices are also seen.

Table 1. Performance indices to comparison the simulation results

Tuning method	K_c	T_i	T_d		set-point tracking		disturbance rejection	
					IAE	ITAE	IAE	ITAE
reference[3]	0.83	1	1	nominal	1.2065	0.8609	1.2121	3.9744
				-20% mismatch	1.2115	0.9053	1.2123	3.7435
				20% mismatch	1.225	0.9272	1.212	3.7438
This paper	4.1084	1.5281	0.4040	nominal	1.2	0.8563	0.3799	0.7081
				-20% mismatch	1.2048	0.8969	0.3805	0.7146
				20% mismatch	1.224	0.8671	0.3904	0.7792

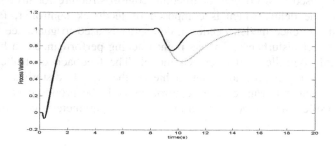

Fig. 2. Output response with perfect model

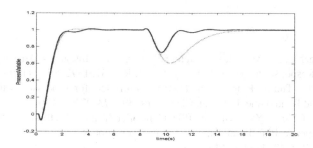

Fig. 3. Output response with +20% model mismatch

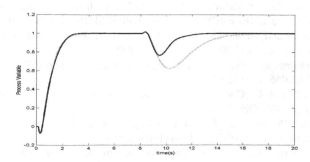

Fig. 4. Output response with -20% model mismatch

The closed loop set-point response and disturbance response of the nominal system for the two control methods are shown in Fig.2 (solid: present method; dash: reference [3] method). Suppose that there exists a +20% mismatch and -20% mismatch in model parameters, the closed loop set-point response and disturbance response of the perturbed system for the two control methods are shown in Fig.3 and Fig.4 (solid: present method; dash: reference [3] method). The results demonstrate that the proposed method shows the superior performance over the reference [3] method.

5   Conclusions

This paper proposed a two degree of freedom control structure for inverse response process with time delay, which is composed of feedback controller, feedforward controller and reference model. Feedback controller is used to guarantee the system stability and reject disturbance. The set-point tracking performance can be achieved by feedforward controller and reference model. The feedback controller designed using direct synthesis method can achieve the good disturbance rejection performance. Compared with existing control method, the proposed two-degree-of-freedom controller has simple structure and the parameters can be adjusted conveniently.

Acknowledgments. This work was partly supported by Shanxi (P. R. China) Natural Science Foundation under Grant No.2011011011-2.

References

1. Scali, C., Rachid, A.: Analytical Design of Proportional – Integral – Derivative Controllers for Inverse Response Processes. Ind. Eng. Chem. Res. 37, 1372–1379 (1998)
2. Luyben, W.L.: Tuning Proportional–Integral Controllers for Processes with both Inverse Response and Dead-Time. Ind. Eng. Chem. Res. 39, 973–976 (2000)
3. Chien, I.L., Chung, Y.C.: Simple PID Controller Tuning Method for Processes with Inverse Response Plus Dead-Time or Large Overshoot Response Plus Dead-Time. Ind. Eng. Chem. Res. 42, 4447–4461 (2003)
4. Sree, R.P., Chidambaram, M.: Simple Method of Tuning PI Controllers for Stable Inverse Response Systems. J. Indian Inst. Sci. 83, 73–85 (2003)
5. Chen, P.: A New Design Method of PID Controller for Inverse Response Processes with Dead Time. In: Proc. of IEEE-ICIT, pp. 1036–1039. IEEE Press, New York (2005)
6. Chen, P.: Design and Tuning Method of PID Controller for a Class of Inverse Response Processes. In: Proc. of ACC 2006, pp. 274–279 (2006)
7. Zhang, W.: Quantitative Performance Design for Inverse Response Processes. Ind. Eng. Chem. Res. 39, 2056–2061 (2000)
8. Alcántara, S., Pedret, C.: Control Configuration for Inverse Response Processes. In: Proc. of 16th MED, pp. 582–586 (2008)
9. Alcántara, S., Pedret, C.: Analytical Design for a Smith-type Inverse-response Compensator. In: Proc. of ACC, pp. 1604–1609 (2009)
10. Chen, D., Seborg, D.E.: PI/PID Controller Design Based on Direct Synthesis and Disturbance Rejection. Ind. Eng. Chem. Res. 41, 4807–4822 (2000)

A Novel Security Evaluation Model
for E-government Intranet Systems

Xiaoting Jin

College of Public Administration, Henan University of Economics and Law
450002 Zhengzhou, China
flyjinxt@163.com

Abstract. It is very important for administrators to ensure the security of E-government. Based on the principles of immunity, a novel security evaluation model for E-government intranet systems, referred to as SEME, was proposed in this paper. Utilizing "immunity" is a good solution for cyberspace security. In the presented model, artificial immune lymphocytes were used for network intrusions detection, and simultaneously, the risk of E-government intranet systems was evaluated from the threats of network attack and vulnerability. Experimental results show that the presented model is feasible, and it has the features of.

Keywords: E-government, intranet, security evaluation.

1 Introduction

In today's world, Information and Communication Technologies (ICT) is transforming the governmental processes in serving citizens (G2C), businesses (G2B) and governments (G2G) [1, 2, 3]. An E-government intranet is a private network used by government employees, and it facilitates collaboration, information sharing and coordination of activities within the civil service. Moreover, the electronical Government Intranet System (eGIS), which tries to connect individual government departmental networks into a secure intranet, is one of the most important components of the E-government infrastructure, and GINS is operational since the coming into operation of the Government Online Centre (GOC).

One of the main issues of trust in E-government implementation is security, such as user authentication, access control, and etc [4, 5]. Government has approved the implementation of ISO/IEC 17799 Security Standard across the Civil Service. However, these standards only provide a structured approach to address security within an organization. Therefore, designing and implementing more effective evaluation approaches, for securing eGIS is an important issue, because, the governmental information within eGIS is usually very sensitive. Furthermore, security has an important role in trust formation of citizens and their adoption of E-government.

The artificial immune danger theory based model for network security evaluation [6] introduced the concept of danger degree for network security risk evaluation, improved the formal definitions of the traditional immune detectors, described the model architecture, and gave the principles of network intrusion detection and security risk assessment. The presented model can evaluate the security risk, which is caused by network attacks. However, whether can it be used for eGIS has not mentioned.

Sun and Wu proposed a risk assessment model for E-government network security based on antibody concentration [7], which has the features of real time and self-adaptive. But, it only took the attack frequency and intensity into account, and the vulnerabilities was ignored.

In theoretically, all of the security approaches that are common in cyberspace are applicable to E-government. But E-government is a little different from E-commerce and other computer networks. Usually government networks can communicate to each other better than business networks, because, most of them are connected for transferring information, but businesses are competitors and they don't disclose their sensitive information.

This paper proposes a novel security evaluation method for E-government intranet system. The remaining of the paper is organized as follows. Section 2 summaries the concept of E-government. In Section 3, theoretical analysis and experiment results are provided. Finally, Sections 4 contains our conclusion and the future work.

2   E-government

The initiatives of government agencies to use ICT tools and applications, Internet and mobile devices to support good governance, strengthen existing relationships and build new partnerships within citizens, are known as E-government initiatives (see table 1). As with E-commerce, E-government represents the introduction of a great wave of technological innovation as well as government reinvention. It represents a tremendous impetus to move forward in the 21st century with higher quality, effective government services and a better relationship between citizens and government [8].

Table 1. Reinventing local governments and the E-government initiative [8]

Paradigm shifts in public service delivery		
	Bureaucratic paradigm	*E Government paradigm*
Orientation	Production cost-efficiency	User satisfaction and control, flexibility
Process organization	Functional rationality, departmentalization, vertical hierarchy of control.	Horizontal hierarchy, network organization, information sharing.
Management principle	Management by rule and mandate	Flexible management, interdepartmental team work with central coordination
Leadership style	Command and control	Facilitation and coordination, innovative entrepreneurship.
Internal communication	Top down, Hierarchical	Multidirectional network with central coordination, direct communication.
External communication	Centralized, formal, limited channels	Formal and informal direct and fast feedback, multiple channels
Mode of service delivery	Documentary mode and interpersonal interaction	Electronic exchange, non face to face interaction
Principles of service delivery	Standardization, impartiality, equity.	User customization, personalization

3 Proposed Method

3.1 Formal Definitions

In SEME, an *antigen* is a binary string, and it is extracted from the IP packets transferred in eGIS, and an *antibody* is used to recognize antigens. Therefore, the structure of *antigen* and *antibody* is same. Let *Bs*, *Ab*, and *Ag* represent the set of binary strings, antibodies, and antigens, respectively. So, we have $Ab \subset Bs, Ag \subset Bs$.

In SEME, an artificial *lymphocyte* is defined for evaluating the network attack risk. Let *N* and *L* represent the set of natural numbers and artificial lymphocytes, respectively, and formal definition of *L* is defined as follows.

$$L = \{s \mid s = \langle ab, age, num, con \rangle, ab \in Ab \ \& \ \&age, num \in N\}. \tag{1}$$

where *age*, *num*, and *con* denotes the age, antigen matched number, antibody concentration of *s* at time *t*, respectively. In order to describe the method of an antibody how to recognize an antigen, the matching function is defined as follows.

$$match(x, y) = \begin{cases} 1 & \begin{array}{l} \exists i, j \wedge i - j \ge r, 0 < j < i \le k, x.ab_m = y_m, \\ x \in L, y \in Ag, j \le m \le i \wedge i, j, k, m, r \in N \end{array} ; \\ 0 & otherwise. \end{cases} \tag{2}$$

3.2 Model Architecture and Principles

3.2.1 Model Architecture

The architecture of SEME is distributed, and it is illustrated in Fig. 1.

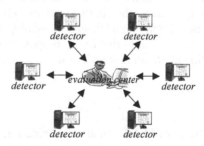

Fig. 1. Architecture of SEME

In Fig. 1, the *detector* is located in each host of eGIS and used to detect intrusions and vulnerabilities. The functions of the *evaluation center* include two aspects: bacterin distribution and network intrusion evaluation.

3.2.2 Model Principles

Fig. 2 illustrates the principles of security evaluation for network attacks of the *detector* within SEME.

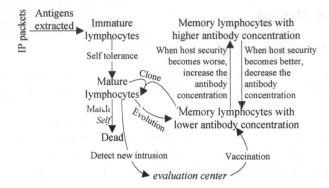

Fig. 2. Security evaluation principle for network attacks

In SEME, antigens (Ag) of the sensor are binary strings, having the characteristics of network activity and being extracted through the process of antigen presentation. Because of *Nonself* $\cup$ *Self* = *Ag* , the *Self* of *Ag* is used as self elements for the self tolerance of immature lymphocytes. The new immature lymphocytes of SEME have to experience a self tolerance period: the cells will be eliminated if it matches any self antigens (negative selection). The immature lymphocytes that survived in self tolerance period will evolve into mature ones, where the mature cells have a fixed lifecycle: the cells will be eliminated if they do not match enough antigens in their lifecycle; they will be activated if they get enough antigens. However, the activated cells will be eliminated if they do not receive co-stimulation.

In each detector, once an immune cell detects an intrusion (i.e., it has matched enough antigens), it will clone itself and create a lot of similar cells to protect the network against more intense intrusions, and the concentration of the immune cells will increase. On the contrary, when the intrusions are eliminated, the antibody concentration will decrease to zero immediately, indicating that the alert has been cleared. In a word, the network security risk in our proposed approach is evaluated by calculating the concentration of all kinds of antibodies.

4   Simulation

In order to verify the validity of SEME, simulation experiments were carried out. A total of 10 computers in a network were under surveillance, so as to test the ability of risk assessment for eGIS.

In the experiments, the antigen was extracted from source IP, destination IP, port number, protocol type. In order to simplify the algorithm of string matching, the author fixed the length of antigen (the length is 88). The match function adopts *r-contiguous-bits* matching rule (r=8). As limited by the computer capability, e.g., the size of memory, computation speed, etc., the number of lymphocytes in SEME was restrained, the proportional under 400, however, the more the lymphocytes, the better the system.

During the experiment, the author tested *smurf* and *teardrop* attacks, there was one nonself packet among 10 packets, and each experimental result is the average value.

200 IP packets were captured from network each time, and, they were transformed into antigen format to be processed.

The *smurf* and *teardrop* attack intensity of our simulations is listed in Table 2.

Table 2. Attack intensity (Packets/Minute)

Time(M)	1	2	3	4	5	6	7	8	9	10	11	12	13	14	15
Smurf	110	123	139	116	144	151	126	138	143	155	162	161	156	118	122
Teardrop	120	135	138	128	154	150	121	146	151	160	152	171	165	126	130
Time(M)	16	17	18	19	20	21	22	23	24	25	26	27	28	29	30
Smurf	135	154	143	149	133	134	161	167	165	154	145	154	159	134	156
Teardrop	125	170	165	158	144	149	174	163	170	158	161	170	170	165	174
Time(M)	31	32	33	34	35	36	37	38	39	40	41	42	43	44	45
Smurf	144	149	133	134	162	167	165	154	145	154	159	135	155	143	148
Teardrop	162	161	178	116	127	135	118	121	129	165	162	118	123	128	146
Time(M)	46	47	48	49	50	51	52	53	54	55	56	57	58	59	60
Smurf	133	135	160	166	165	154	146	154	159	134	154	143	149	132	137
Teardrop	144	142	158	127	155	121	143	132	138	155	147	145	165	166	145

Fig. 3 shows the risk assessment result for network security apperceived by SEME.

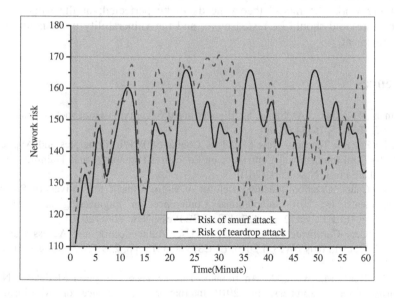

Fig. 3. Evaluated result

As can be seen in Fig. 3, SEME can aware the network security situation, which are caused by network attacks. Therefore, the proposed approach is valid.

We also found that the risk assessment result for each kind of network attacks is consistent with the real network attack intensity occurring in the network. At the same time, we find that when a network attack occurs and the attack intensity increases, our estimated attack intensity also increases and follows the trend of the real attack intensity. When the real attack intensity reaches the maximum, so does our estimated attack intensity. On the other hand, as the real attack intensity decreases, the estimated attack intensity decreases as well. However, the decreasing slope of the estimated attack intensity curve is smaller than that of the real attack intensity. The decreasing slope is related to the antibody concentration maintaining period. A small decreasing slope keeps the network at a high alert level against the same attack occurring again in a short period of time.

Moreover, in order to evaluate the vulnerabilities of eGIS, the author perfects SEME from the aspects of physical security, administrative security, data security, and technology security. Further more, the vulnerability assessment is in line with the requirements of CC, GB/T 20984-2007, GB/T 18336-2008, and GB/T 25058-2010.

5 Conclusion

Based on the principles of immunity and classified security criteria of information system, this paper presents a novel security evaluation model for E-government intranet systems, referred to as SEME. Simulation results show that the proposed model is feasible, and it can evaluate the risk of E-government intranet systems from the network attack and vulnerability in two aspects.

In future work, the model theory needs to be perfected; at the same time, the matching threshold should be tested more, and the vulnerability assessment program needs refinement.

References

1. Bahman, N., Akbar, J.A., Sahar, S.: E-government Security: A Honeynet Approach. Int. J. Adv. Sci. Tech. 5, 75–84 (2009)
2. Sun, F.: Practice Teaching New Model for Course of TCP/IP Principles and Applications. In: 2010 Third International Conference on Education Technology and Training, vol. 4, pp. 94–96. IEEE Press, New York (2010)
3. Jin, X.: Course Practice Teaching Reform of E-government for University Students of Managements. In: 2010 International Conference on Engineering and Educational Technology (EEET), vol. 2, pp. 437–439. IEEE Press, New York (2010)
4. Sun, F.: Gene-Certificate Based Model for User Authentication and Access Control. In: Wang, F.L., et al. (eds.) WISM 2010. LNCS, vol. 6318, pp. 228–235. Springer, Heidelberg (2010)
5. Sun, F., Kong, M., Wang, J.: An Immune Danger Theory Inspired Model for Network Security Threat Awareness. In: 2010 International Conference on Multimedia and Information Technology (MMIT 2010), vol. 2, pp. 93–95. IEEE Press, New York (2010)
6. Sun, F.: Artificial Immune Danger Theory Based Model for Network Security Evaluation. J. Netw. 6, 162–255 (2011)
7. Sun, F., Wu, Z.: A New Risk Assessment Model for E-government Network Security Based on Antibody Concentration. In: 2009 International Conference on E-Learning, E-Business, Enterprise Information Systems, and E-Government, pp. 119–121. IEEE Press, New York (2009)
8. Ndou, V.M.: E-government for Developing Countries: Opportunities and Challenges. EJISDC 18, 1–24 (2004)

Data Preprocessing for Web Data Mining

Wei Zhang[1] and Tinggui Chen[2]

[1] College of Computer Science & Information Engineering,
Zhejiang Gongshang University, Hangzhou, 310018, China
[2] Contemporary Business and Trade Research Center,
Zhejiang Gongshang University, Hangzhou, 310018, China
zhangwei19891210@126.com, ctgsimon@gmail.com

Abstract. Data preprocessing includes data cleaning, data integration, data transformation and data reduction. Data cleaning is aimed to remove unrelated or redundant items through two processes. Data integration includes three main problems and each of them can be solved by kinds of methods. Data transformation includes data generalization and property construction and standardization. Three algorithms can be used to normalize the data. The last step data reduction is used to compress the data in order to improve the quality of mining models. All these four steps are interrelated to each other and shouldn't be separated. They work together to improve the final result of data mining.

Keywords: Data preprocessing, cleaning, integration, transformation, reduction.

1 Introduction

Data preprocessing is the most important part of the entire data mining processes. The purpose of data preprocessing is to turn the Web logs into some reliable, complete and accurate sources to satisfy the need of data mining algorithms. Statistics show that in data mining processes, the process of data preprocessing accounts for 60% of the entire workload. Because the data from real world are often incomplete and inconsistent with the noise, the data preprocessing can improve the quality of the data for data mining. It not only can save a lot of time and space, but also are better able to play a role in decision making and forecasting. There are four steps of data preprocessing, including data cleaning, data integration, data transformation and data reduction.

2 Data Cleaning

The users' logs usually contain many unrelated or redundant items that will have a negative impact on data analysis and should be deleted. Data cleaning is often used to remove these items.

[1] Fill in the absent values:

A. Fill in the missing values with the average value of the samples in the same group. For example, the value of wage for some employees is absent, and then we can classify all the employees according to an attribute such as department and fill the absent values with the average value of wage of all employees in the same department.

B. Fill in the missing values with the average value of all samples. For example, the average income of all employees is 4000 and we can use it to fill the absent values.

C. Fill in the absent values with a global constant such as "unknown". But it may cause errors when the number of absent values is large.

[2] Smooth noisy data:

A. Binning: For example, A set of sorted data: 1, 4, 7, 9, 13, 19, 24, 26 and 36, Smooth by equal-frequency: 1, 4, 7; 9, 13, 19; 24, 26, 36. Smooth by bin median: 4, 4, 4; 13, 13, 13; 29, 29, 29. Smooth by bin boundaries: 1, 1, 7; 9, 9, 19; 24, 24, 36.

B. Clustering: Through this method, the similar values can be organized into groups, and then we can get some isolated data. The method can be briefly showed by figure1:

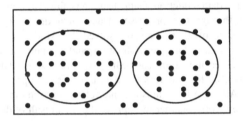

Fig. 1. Clustering

3   Data Integration

Data integration is to merge the data from multiple data sources together to build a consistent data storage form. The main problems of data integration are as follows:

The first problem is to compare two attributes from two different data sources. For example, you should determine whether an attribute called customer_id in one database is the same to the attribute called customer_number in the other database.

The second problem is reduction. If a property can be derived by one attribute or another group of attributes, the difference between the attributes' names may lead to the reduction. Some reduction can be examined by a variety of methods. For example, you can calculate the correlation coefficient of attribute A and attribute B, the formula is as follows:

$$r_{A,B} = \frac{\sum_{i=1}^{N} (a_i - \overline{A})(b_i - \overline{B})}{N \sigma_A \sigma_B} = \frac{\sum_{i=1}^{N} (a_i b_i) - N \overline{A} \overline{B}}{N \sigma_A \sigma_B} \tag{1}$$

N is the total number of tuples, a_i is the value of A in the tuples numbered i, b_i is the value of B in the tuples numbered i. $\overline{A}$ is the average value of A and $\overline{B}$ is the average value of B. σ_A is the standard deviation of A and σ_B is the standard deviation of B.

$-1 \leq r_{A,B} \leq 1$, if the result is larger than 0, it may indicates that A or B can be removed as reduction. If the result is 0, it indicates that A and B are unrelated. If the result is smaller than zero, it indicates that one attribute will prevent the existence of another attribute.

The last problem is conflict detection and data processing of the values. For the same entity from different data sources in the real world, the values may be different. For example, the currency in each country has different units and the prices of hotels in different cities may not only involve different currencies, but also involve different services.

Therefore, when try to match properties in one database with those in the other database, we should take the structure of data into consideration and ensure that the functional dependencies and referential constraint of the properties in the original system should in accordance with those in the target system.

4   Data Transformation

Data transformation is mainly to convert data to the forms which are suitable for data mining. It involves the following elements:

[1] Data generalization: You can use the method concept hierarchy, that is, you can replace the low-level or raw data with high-level concepts.

[2] Property construction and standardization: You can use the given property to construct a new property and add it to the attribute set. Then you can narrow or enlarge the data in proportion by normalizing to make the data into a specific range. Data normalization methods commonly include three ways:

A. Normalize the data by the largest value or the smallest value. Assume that m_A is the smallest value of attribute A and M_A is the largest value of attribute A, the formula is as follows:

$$v^{'} = \frac{v - m_A}{M_A - m_A}(new \_ M_A - new \_ m_A) + new \_ m_A \tag{2}$$

So v in A is mapped to $v^{'}$ in a specific range $[new\_m_A, new\_M_A]$. The method is used to change the form of raw data and maintain the links between the data.

B. Z-zero normalization: The method is used to normalize the value v of attribute A to $v^{'}$ by using the average value $\overline{A}$ and the standard deviation σ_A. The formula is as follows:

$$v^{'} = (v - \overline{A})/\sigma_A \tag{3}$$

C. Decimal scaling normalization: The method normalizes the data of property A by moving the position of decimal point. The median of the movement of the decimal point depends on the absolute maximum value, the formula is as follows:

$$v' = v/10^{j} \qquad (4)$$

For example, assume that attribute A has a range of 867 to 832. The absolute maximum value is 867, and then you can get the range -0.867 to 0.832 after normalization.

5   Data Reduction

It takes a long time to analyze the massive and complex data, so it is not feasible. In the premise of not affecting the mining results, you can use the methods such as numerical aggregation and removing data redundancy to compress the data in order to improve the quality of mining models and reduce the time complexity. The steps of data reduction are summarized as follows:

[1] Data cube gathering: Use some operations to gather the data in the data cube structure.

[2] The selection of attribute subset: You can detect irrelevant, weakly relevant or redundant attributes.

The method for selection of attribute subset can be shown as follows:

A. Selection by gradually moving forward: The process uses the empty set as a starting set, then determines the best attributes of the original attribute set, add them to the empty set and add the best attributes of the remaining original set to the collection in each step later.

B. Deletion by gradually moving backward: The process begins with the entire set of attributes and removes the worst properties of the attribute set at each step.

[3] Dimension reduction: You can use the encoding mechanism to reduce the size of data set and two effective dimension reduction methods are wavelet transform and principal component analysis.

6   Conclusion

Data pre-processing mainly consists of 4 steps and each step has a variety of methods. During the process, all the methods should combine with the professional knowledge and practical application. The whole process should combine human resources with computer technology and with particular emphasis on more exchanges between customers and experts. After preprocessing, if the differences between the data mining results and the desired outcome are very big, you should consider a second-time pre-processing after excluding the problems caused by initial data to correct errors introduced in the first-time pre-processing. If the result is still abnormal, you should think another method. Remember that after the end of a step, if the data of the results have the characteristics which the data of previous processing steps have, you should again return to the previous steps. For example, after data integration, you should conduct data cleaning again to detect and remove the redundant may cause by integration.

Acknowledgments. This research is supported by Research Fund for the Doctoral Program of Higher Education of China (Grant No. 20103326110001 and 20103326120001), Humanity and Sociology Foundation of Ministry of Education of China (Grant No. 11YJC630019), Zhejiang Provincial Natural Science Foundation of China (No. Z1091224 and Y7100673), Zhejiang Provincial Social Science Foundation of China (Grant No. 10JDSM03YB), the Scientific Research Fund of Zhejiang Province, China (Grand No. 2011C33G2050035), Research Project of Department of Education of Zhejiang Province (No. Y200907458), the Contemporary Business and Trade Research Center of Zhejiang Gongshang University (No. 1130KUSM09013 and 11JDSM02Z). We also gratefully acknowledge the support of Science and Technology Innovative project (No. 1130XJ1710215).

References

1. Dunham, M.H.: Data Mining Introductory and Advanced Topics. Tsinghua University Press, Beijing (2003)
2. Han, J., Camber, M.: Data Mining: Concepts and Techniques, vol. 3. China Machine Press, Beijing (2001)
3. Jiawei Han, Xiaofeng Meng, Jing Wang,, The Research of Web Mining(in Chinese)[J].The Research and Development of Computer.38(4):405-414(2001)
4. Chen, J.: Research on Individualized Information Services Based on Internet. Sci-Tech Information Development & Economy 15(3), 96–98 (2005) (in Chinese)
5. Chen, A., Chen, N., Zhou, L.: Data Mining Technologies and Applications, vol. 3. China Machine Press, Beijing (2006)
6. Liu, L., Song, H., Lu, Y.: Application of Web usage mining. Computer Science 9 (2003)
7. Zhu, H.: Data Preprocessing Algorithm of Web Log Mining. China Master's Theses Full-text Database (August 2010)

Robust H_∞ Control for Itô Stochastic Systems with Markovian Switching and Probabilistic Sampling

Hua Yang[1,2], Huisheng Shu[2,*], Yan Che[3], and Xiu Kan[3]

[1] College of Information Science and Engineering, Shanxi Agricultural University,
Taigu, Shanxi 030801, China
[2] Department of Applied Mathematics, Donghua University, Shanghai 200051, China
hsshu@dhu.edu.cn
[3] School of Information Science and Technology, Donghua University,
Shanghai 200051, China

Abstract. In this paper, the problem of robust H∞ control for sampled-data stochastic systems with Markovian switching and parameter uncertainties is investigated. By converting probabilistic sampling into time-varying delays, the concerned system is transformed into a time-varying delays stochastic system. Then, by constructing a new Lyapunov functional, a sufficient condition is derived to guarantee the exponential mean-square stability of the system as well as the H∞ norm is less than a given level. And then, the expression of desired controller is obtained in terms of the solution to certain linear matrix inequalities. Finally, a numerical simulation example is exploited to show the effectiveness of the theoretical results.

Keywords: Robust H_∞ control, stochastic systems, stochastic sampling, time-varying delay, Markovian switching.

1 Introduction

In the past few decades, the study of stochastic systems has received much attention, see e.g. [7,9,10] and the references therein. Recently, the problem of H_∞ control for stochastic systems has been considered. When both the Markovian jump parameters and time delays appearing in the stochastic systems, the H_∞ control problem has been studied in [4,5]. Sampled-data H_∞ control of systems has been studied in the last decades. Besides the approach of lifting technique and hybrid discrete/continuous model, in [1-3,6,8] the method of probabilistic input delay is adopted, that is sampled data H_∞ control problem is solved by converting probabilistic sampling into time-varying delays. To the best of the author's knowledge, the problem of robust H_∞ control via sampled-data state feedback in Markovian jump stochastic systems with parameter uncertainties is still open and remains unsolved.

In this paper, we are concerned with the problem of robust H_∞ control for Markovian jump stochastic systems with parameter uncertainties and probabilistic sampling. The objective is the design of state feedback controllers which guarantee

---

* All correspondences concerning this paper should be addressed.

D. Jin and S. Lin (Eds.): Advances in ECWAC, Vol. 2, AISC 149, pp. 309–316.
springerlink.com   © Springer-Verlag Berlin Heidelberg 2012

not only the robust mean-square exponential stability but also a prescribed disturbance attenuation level for the resulting closed-loop system. A sufficient condition for the solvability of this problem is obtained and the expression of desired controllers is presented. An illustrative example is given to show the potential of the techniques.

Notation. R^n and denote, respectively, the n dimensional Euclidean space and the set of all $n \times m$ real matrices, and $\| \cdot \|$ refers to the Euclidean norm in R^n , $L_2[0, \infty)$ is the space of square-integrable vector functions over $[0, \infty)$, represents the transpose of M , $\varepsilon\{x\}$ stands for the expectation of stochastic variable x .

2 Problem Formulation

Consider the following Itô stochastic system with Markovian switching and parameter uncertainties:

$$dx(t) = [(A(r(t)) + \Delta A(r(t)))x(t) + B(r(t))u(t) + E(r(t))v(t)]dt + G(r(t))x(t)dw(t),$$
$$y(t) = C(r(t))x(t) + D(r(t))u(t), \tag{1}$$

where $x(t) \in R^n, u(t) \in R^p, y(t) \in R^r$ and $v(t) \in L_2([0, \infty); R^q)$ are the state vector, control input, controlled output and exogenous disturbance, respectively. $w(t)$ represents a scalar Wiener process (Brownian motion) on a complete probability space that is independent of Markov chain $r(t)$.

In this paper, $\{r(t) \geq 0\}$ is a discrete-state Markov process taking values in a finite state space $S = \{1, 2, \cdots, N\}$ with the following transition probilities:

$$P\{r(t + \Delta t) = j \mid r(t) = i\} = \begin{cases} \gamma_{ij}\Delta t + o(\Delta t), & \text{if } i \neq j, \\ 1 + \gamma_{ii}\Delta t + o(\Delta t) & \text{if } i = j, \end{cases}$$

where $\Delta t > 0$ and $\lim_{\Delta t \to 0} o(\Delta t) / \Delta t = 0$ .Here $\gamma_{ij} \geq 0$ is the transition rate from mode i to mode j if $i \neq j$ while $\gamma_{ii} = \Sigma_{j \neq i}\gamma_{ij}$.

$A(r(t)), B(r(t)), C(r(t)), D(r(t)), E(r(t))$, and $G(r(t))$ are known constant matrices with appropriate dimensions, and $\Delta A(r(t)) = M(r(t))F(t)N(r(t))$, $F(\cdot) : R \to R^{k \times l}$ satisrying $F(t)^T F(t) \leq I, \forall t \geq 0$.

For state-feedback sampled-data control with zero-order hold (ZOH), the controller takes the following form:

$$u(t) = u_d(t_k) = K(r(t))x(t_k), \quad t_k \leq t < t_{k+1}, \quad k = 0, 1, 2, \cdots, \tag{2}$$

where u_d is a discrete-time signal, t_k denotes the sampling instant. Now, for $t_k \leq t < t_{k+1}$ we denote $t_k = t - (t - t_k) \triangleq t - d_k(t)$. Then, the system (1) can be transformed to the following continuous-time system with time-varying delay:

$$dx(t) = [(A(r(t)) + \Delta A(r(t)))x(t) + B(r(t))K(r(t))x(t - d_k(t)) + E(r(t))v(t)]dt$$
$$+ G(r(t))x(t)dw(t),$$
$$y(t) = C(r(t))x(t) + D(r(t))K(r(t))x(t - d_k(t)), \quad t_k \leq t < t_{k+1}. \tag{3}$$

In this paper, we assume that the sampling periods switch between c_1 and c_2 in a random way and that $0 < c_1 < c_2$. Such a phenomena is refereed to as the stochastic sampling, which can be represented by utilizing a random variable ρ with the probabilities $\text{Prob}\{\rho = c_1\} = \beta$ and $\text{Prob}\{\rho = c_2\} = 1 - \beta$, where $\beta \in [0 \ 1]$ is a known constant. Following the similar line in [6], the system (3) can be rewritten as

$$dx(t) = [(A(r(t)) + \Delta A(r(t)))x(t) + \alpha(t)B(r(t))K(r(t))x(t - \tau_1(t))$$
$$+ (1 - \alpha(t))B(r(t))K(r(t))x(t - \tau_2(t)) + E(r(t))v(t)]dt + G(r(t))x(t)dw(t),$$
$$y(t) = C(r(t))x(t) + \alpha(t)D(r(t))K(r(t))x(t - \tau_1(t))$$
$$+ (1 - \alpha(t))B(r(t))K(r(t))x(t - \tau_2(t)), \tag{4}$$

where

time-varying delays $\tau_1(t)$ and $\tau_2(t)$ satisfying $0 \leq \tau_1(t) < c1, 0 \leq \tau_2(t) < c2$ where $\alpha(t) = 1$ if $0 \leq d(t) < c1$, but $\alpha(t) = 0$ if $c1 \leq d(t) < c2$.

Definition 1. The closed-loop system (4) with $v(t) = 0$ is said to be exponentially stable in the mean square if there exist constants $\mu > 0$ and $\delta > 0$ such that

$$\mathcal{E}\{\| x(t) \|^2\} \leq \mu e^{-\delta t} \lim_{-2c_2 \leq \theta \leq 0} \mathcal{E}\{\| \phi(\theta) \|^2\}$$

where $x(t) = \phi(t), t \in [-2c2, \ 0]$, and $\phi(\cdot)$ is a continuous function.

Then, given a prescribed level of noise attenuation level $\gamma > 0$, the robust H_∞ control problem addressed in this paper can be formulated as follows:

a) The resulting closed-loop system (4) with $v(t) = 0$ is robustly exponentially mean-square stable;

b) Under the zero-initial condition, the controlled output $y(t)$ satisfies

$$\mathcal{E}\left\{\int_0^\infty \|y(t)\|^2 \, dt\right\} < \gamma^2 \int_0^\infty \|v(t)\|^2 \, dt$$

for all nonzero $v(t) \in L_2([0, \ \infty); R^q)$ and all admissible uncertainties.

3 Main Results

Theorem 1. Given controller gain matrices K_i, the sampled-data closed-loop system (4) with $v(t) = 0$ is robustly exponentially mean-square stable if there exist scalars ε_i and matrices $P_i > 0, Q_1 \geq 0, Q_2 \geq 0, R_1 \geq 0, R_2 \geq 0, S, U, V$, and W satisfying

$$\begin{bmatrix} L + \varepsilon_1 \Phi_1 \Phi_1^T & \Phi_2^T \\ * & -\varepsilon_1 I \end{bmatrix} < 0, \quad \forall i \in S \tag{5}$$

where

$$L = \begin{bmatrix} G_1 + G_2 + T_2^T & T_3 & L_1 & L_2 & L_3 & L_4 \\ * & L_1 & 0 & 0 & 0 & 0 \\ * & * & L_2 & 0 & 0 & 0 \\ * & * & * & L_3 & 0 & 0 \\ * & * & * & * & L_2 & 0 \\ * & * & * & * & * & L_3 \end{bmatrix}, T_1 = \begin{bmatrix} \Xi & 0 & aP_iB_iK_i & 0 & B_i\Sigma_i & P_iE_i \\ * & Q_2 - Q_1 & 0 & 0 & 0 & 0 \\ * & * & 0 & 0 & 0 & 0 \\ * & * & * & -Q_2 & 0 & 0 \\ * & * & * & * & 0 & 0 \\ * & * & * & * & * & 0 \end{bmatrix},$$

$\Xi = P_iA_i + A_i^TP_i + \sum_{j=1,\cdots,N}\gamma_{ij}P_j + Q_1 + G_i^T(P_i + F)G_i, \ \Sigma_i = (1 - \alpha)P_iK_i,$
$\Gamma_2 = \begin{bmatrix} S & V - U & U - S & -W & W - V & 0 \end{bmatrix}, \ T_3 = \begin{bmatrix} \sqrt{2c_1}S & \sqrt{2c_2}U & \sqrt{2}gV & \sqrt{2}gW \end{bmatrix},$
$L_1 = \begin{bmatrix} \sqrt{c_1}A_i & 0 & \alpha\sqrt{c_1}B_iK_i & 0 & \sqrt{c_1}\Sigma_i & \sqrt{c_1}E_i \end{bmatrix}^T, L_2 = \begin{bmatrix} gA_i & 0 & \alpha gB_iK_i & 0 & g\Sigma_i & gE_i \end{bmatrix}^T,$
$L_3 = \begin{bmatrix} 0 & 0 & f\sqrt{c_1}B_iK_i & 0 & -f\sqrt{c_1}B_iK_i & 0 \end{bmatrix}^T, L_4 = \begin{bmatrix} 0 & 0 & fgB_iK_i & 0 & -fgB_iK_i & 0 \end{bmatrix}^T,$
$\Lambda_1 = diag\{-R_1 \ -R_1 \ -R_2 \ -R_2\}, \Lambda_2 = -R_1^{-1}, \Lambda_3 = -R_2^{-1}, F = c_1R_1 + (c_2 - c_1)R_2,$
$\Phi_1 = \begin{bmatrix} N_i & 0 & 0 & 0 & 0 & 0 & 0 & 0 & 0 & 0 & 0 \end{bmatrix}^T, f = \sqrt{\alpha(1 - \alpha)}, g = \sqrt{c_2 - c_1},$
$\Phi_2 = \begin{bmatrix} M_i^T & 0 & 0 & 0 & 0 & 0 & 0 & \sqrt{c_1}M_i^T & gM_i^T & 0 & 0 \end{bmatrix}.$

Proof: Choose the following Lyapunov-Krasovskii functional as

$V(x,t,i) = V_1(x,t,i) + V_2(x,t,i) + V_3(x,t,i) + V_4(x,t,i), \ V_1(x,t,i) = x^T(t)P_ix(t),$

$V_2(x,t,i) = \int_{t-c1}^{t} x^T(s)Q_1x(s)ds + \int_{t-c2}^{t-c1} x^T(s)Q_2x(s)ds,$

$V_3(x,t,i) = \int_{t-c1}^{t}\int_{s}^{t} l^T(\theta)R_1l(\theta)d\theta ds + \int_{t-c2}^{t-c1}\int_{s}^{t} l^T(\theta)R_2l(\theta)d\theta ds$

$\qquad + \alpha(1 - \alpha)(\int_{t-c_1}^{t}\int_{s}^{t} h^T(\theta)R_1h(\theta)d\theta ds + \int_{t-c_2}^{t-c_1}\int_{s}^{t} h^T(\theta)R_2h(\theta)d\theta ds),$

$V_4(x,t,i) = \int_{t-c_1}^{t}\int_{s}^{t} g^T(\theta)R_1g(\theta)d\theta ds + \int_{t-c_2}^{t-c_1}\int_{s}^{t} g^T(\theta)R_2g(\theta)d\theta ds,$

where

$l(t) = (A(r(t)) + \Delta A(r(t)))x(t) + \alpha B(r(t))K(r(t))x(t - \tau_1(t))$
$\qquad + (1 - \alpha)B(r(t))K(r(t))x(t - \tau_2(t)) + E(r(t))v(t),$
$h(t) = B(r(t))K(r(t))x(t - \tau_1(t)) - B(r(t))K(r(t))x(t - \tau_2(t)), \ g(t) = G(r(t))x(t).$

It follows from (5) that $L + \varepsilon_1 \Phi_1 \Phi_1^T + \varepsilon_1^{-1} \Phi_2^T \Phi_2 < 0$, and it is easy to see that $\Phi_1 F^T(t) \Phi_2 + \Phi_2^T F(t) \Phi_1^T \leq \varepsilon_1 \Phi_1 \Phi_1^T + \varepsilon_1^{-1} \Phi_2^T \Phi_2$, then, by Itô lemma ,isometry and Schur complement , it follows that

$$\mathcal{E}\{\| x(t) \|^2\} \leq \frac{\rho}{\min_{i=1,\cdots,N}\{\lambda_{\min}(P_i)\}} e^{-\varepsilon T} \sup_{-2c2 \leq \theta \leq 0} \mathcal{E}\{\| \phi(\theta) \|^2\},$$

for $\forall T > 0$,therefore, by Defintion 1, the proof is complete.

Theorem 2. Given controller gain matrices K_i and a positive constant γ ,the sampled-data closed-loop system in (4) is robustly exponentially mean-square stable with an H_∞ disturbance attenuation level γ if there exist scalars $\varepsilon_1 > 0$ and matrices $P_i > 0, Q_1 \geq 0, Q_2 \geq 0, R_1 \geq 0, R_2 \geq 0, S, U, V$, and W satisfying

$$\begin{bmatrix} \bar{L} + \varepsilon_1 \Phi_1 \Phi_1^T & \Phi_2^T \\ * & -\varepsilon_1 I \end{bmatrix} < 0, \quad \forall i \in S \tag{6}$$

where $\bar{L}$ is obtained from adding T_6, T_7 in the first row and first column of L,

$T_6 = L_{10}^T L_{10} + L_{11}^T L_{11}, \quad L_{10} = \begin{bmatrix} C_i & 0 & \alpha D_i K_i & 0 & (1-\alpha)D_i K_i & 0 \end{bmatrix},$
$T_7 = diag\{0,0,0,0,0,-\gamma^2 I\}, \quad L_{11} = \begin{bmatrix} 0 & 0 & fD_i K_i & 0 & -fD_i K_i & 0 \end{bmatrix}.$

Proof: By the similar techniques as those in the above, we have $\mathcal{E}\{y^T(t)y(t)\} - \gamma^2 v^T(t)v(t) + \mathcal{E}\{LV(x,t,i)\} < 0$, suppose $x(0) = 0$, we have

$$\mathcal{E}\{\int_0^\infty \| y(t) \|^2 \, dt\} < \gamma^2 \int_0^\infty \| v(t) \|^2 \, dt,$$

for all nonzero $v(t) \in L_2([0, \infty); R^q)$ .The proof is complete.

4   Robust H_∞ Controller Design

Theorem 3. Given a positive constant γ, there exist state-feedback controller K_i such that the sampled-data closed-loop system in (4) is robustly exponentially stabilizable with an H_∞ disturbance attenuation level γ if there exist scalars $\varepsilon_1 > 0$ and matrices $P_i > 0, Q_1 \geq 0, Q_2 \geq 0, R_1 \geq 0, R_2 \geq 0, S, U, V$, and W satisfying the following LMI condition:

$$\begin{bmatrix} \Theta_1 & \pi_3 & \pi_5 & \pi_6 & \pi_8 & \pi_{10} & \pi_{11} & \pi_{13} & \pi_{15} \\ * & \pi_4 & 0 & 0 & 0 & 0 & 0 & 0 & 0 \\ * & * & \pi_7 & 0 & \pi_9 & 0 & 0 & 0 & 0 \\ * & * & * & \pi_7 & 0 & 0 & 0 & 0 & 0 \\ * & * & * & * & -\varepsilon_1 I & 0 & 0 & 0 & 0 \\ * & * & * & * & * & -\varepsilon_1 I & 0 & 0 & 0 \\ * & * & * & * & * & * & -\pi_{12} & 0 & 0 \\ * & * & * & * & * & * & * & \pi_{14} & 0 \\ * & * & * & * & * & * & * & * & \pi_{16} \end{bmatrix} < 0, \ \forall i \in S, \quad (7)$$

where

$$\Theta_1 = \pi_1 + \pi_2 + \pi_2^T + \Upsilon_7, \quad \Theta_2 = A_i \bar{P}_i + \bar{P} A_i + \bar{Q}_1 + \gamma_{ii} \bar{P}_i,$$

$$\pi_1 = \begin{bmatrix} \Theta_2 & 0 & \alpha B_i \bar{K}_i & 0 & (1-\alpha)B_i \bar{K}_i & E_i \\ * & \bar{Q}_2 - \bar{Q}_1 & 0 & 0 & 0 & 0 \\ * & * & 0 & 0 & 0 & 0 \\ * & * & * & -\bar{Q}_2 & 0 & 0 \\ * & * & * & * & 0 & 0 \\ * & * & * & * & * & 0 \end{bmatrix}$$

$$\pi_4 = diag\{\bar{R}_1 - 2\bar{P}_i, \ \bar{R}_1 - 2\bar{P}_i, \ \bar{R}_2 - 2\bar{P}_i, \ \bar{R}_2 - 2\bar{P}_i\},$$

$$\pi_7 = diag\{-\bar{R}_1, -\bar{R}_2\}, \quad \pi_{12} = diag\{\gamma_{i1}^{-1}\bar{P}_1, \cdots, \gamma_{ii-1}^{-1}\bar{P}_{i-1}, \gamma_{ii+1}^{-1}\bar{P}_{i+1}, \cdots, \gamma_{iN}^{-1}\bar{P}_{iN}\},$$

$$\pi_{14} = diag\{-I, -I\}, \quad \pi_{16} = -\bar{P}_i - c_1^{-1}\bar{R}_1 - (c_2 - c_1)^{-1}\bar{R}_2.$$

Moreover, desired controller gain matrices are given by

$$K_i = \bar{K}_i \bar{P}_i^{-1}, \quad i = 1, \cdots, N.$$

Proof: Define

$$J = diag\{J_1, J_2, I, I, I, \varepsilon_1, J_3, I, I, J_4\}, \quad J_1 = diag\{\bar{P}_i^{-1}, \bar{P}_i^{-1}, \bar{P}_i^{-1}, \bar{P}_i^{-1}, \bar{P}_i^{-1}, I\},$$

$$J_2 = diag\{\bar{P}_i^{-1}, \bar{P}_i^{-1}, \bar{P}_i^{-1}, \bar{P}_i^{-1}\}, J_3 = diag\{P_i^{-1}, \cdots, P_i^{-1}\}_{(N-1)n \times (N-1)n}, \quad J_4 = P_i^{-1}.$$

By Schur complement, performing a congruence transformation to the equivalent matrix of (6) by J, we obtain (7), and then the proof is completed.

5   Illustrative Example

The system data of (4) are assumed as follows:

the step size $\Delta = 0.005$; $c_1 = 0.02$; $c_2 = 0.04$; $\alpha = 0.8$.

$$\begin{bmatrix} \gamma_{11} & \gamma_{12} \\ \gamma_{21} & \gamma_{22} \end{bmatrix} = \begin{bmatrix} -2 & 2 \\ 1 & -1 \end{bmatrix}, \ A_1 = \begin{bmatrix} -1 & 0 \\ -1 & -0.5 \end{bmatrix}, \ A_2 = \begin{bmatrix} -5 & 1 \\ 0 & -2.5 \end{bmatrix}, \ B_1 = \begin{bmatrix} -2 \\ 1 \end{bmatrix}, \ B_2 = \begin{bmatrix} -0.5 \\ 0.5 \end{bmatrix},$$

$$C_1 = \begin{bmatrix} 1 & 0 \\ 0 & 1 \end{bmatrix}, C_2 = \begin{bmatrix} 0.5 & 0 \\ -0.2 & 0.3 \end{bmatrix}, D_1 = \begin{bmatrix} 1 \\ 0 \end{bmatrix}, D_2 = \begin{bmatrix} 0 \\ 1 \end{bmatrix}, E_1 = \begin{bmatrix} 0.1 \\ 0.1 \end{bmatrix}, E_2 = \begin{bmatrix} 0.2 \\ 0.3 \end{bmatrix}, G_1 = \begin{bmatrix} 0 & 0.3 \\ -0.3 & 0 \end{bmatrix},$$

$$G_2 = \begin{bmatrix} 0 & 0.3 \\ 0.2 & 0 \end{bmatrix}, M_1 = \begin{bmatrix} 0 & 0.3 \\ -0.3 & 0 \end{bmatrix}, M_2 = \begin{bmatrix} 1 & 0.2 \\ 0 & 0.3 \end{bmatrix}, N_1 = \begin{bmatrix} 0 & 1 \\ 2 & 1 \end{bmatrix}, N_2 = \begin{bmatrix} 1 & 0 \\ 0.5 & 1 \end{bmatrix}.$$

Using YALMIP3.0 and SeDuMi 1.1 to solve(7), we obtain

$$K_1 = \bar{K}_1 \bar{P}_1^{-1} = [0.9010 \quad -4.4538], \quad K_2 = \bar{K}_2 \bar{P}_2^{-1} = [0.4632 \quad -7.4803], \quad \gamma^* = 0.1480.$$

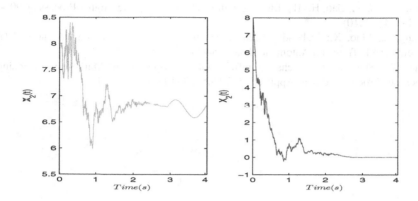

Fig. 1. The state evolution of $x_2(t)$ the uncontrolled and the closed-loop system

6 Conclusions

In this paper, we have investigated a robust H_∞ control problem for Itô's stochastic system with uncertainty, Markovian switching and probabilistic sampling.

Acknowledgments. This work is supported by the National Natural Science Foundation of P.R. China.(No.60974030)

References

1. Shen, B., Wang, Z., Liu, X.: A stochastic sampled-data approach to distributed H-infinity filtering in sensor networks. IEEE Transactions on Circuits and Systems - Part I (accepted for publication), ISSN 1057-7122
2. Fridmen, E., Shaked, U.: Input delay approach to robust sampled-data H_∞ control. In: Fridmen, E., Shaked, U. (eds.) 43rd IEEE Conference on Decision and Control, Atlantis, Paradise Island, Bahamas, December 14-17 (2004)
3. Fridmen, E., Shaked, U., Suplin, V.: Input/Output delay approach to rubost sampled-data H_∞ contrl. System & Control Letters 54, 271–282 (2005)
4. Wei, G., Wang, Z., Shu, H.: Robust H_∞ control of stochastic time-delay jumping systems with nonlinear disturbances. Optimal Control Applications & Methods 27(5), 255–271 (2006)
5. Wei, G., Wang, Z., Shu, H.: Nonlinear H_∞ control of stochastic time-delay systems with Markovian switching. Chaos, Solitons & Fractals 35, 442–451 (2008)

6. Gao, H., Wu, J., Shi, P.: Robust sampled-data H_∞ control with stochastic sampling. Automatica 45(7), 1729–1736 (2009)
7. Shu, H., Wei, G.: H_∞ analysis of nonlinear stochastic time-delay systems. Chaos, Solitons & Fractals 26(6), 37–47 (2005)
8. Wu, J., Chen, X., Gao, H.: H_∞ filtering with stochastic sampling. Signal Processing 90(4), 1131–1145 (2010)
9. Huang, L., Mao, X.: Delay-dependent exponential stability of neutral stochastic delay systems. IEEE Trans. on Automatic Control 54(1), 147–152 (2009)
10. Mao, X.: Stability of stochastic differential equations with Markovian switching. Stochastic Process and their Applications 79, 45–67 (1999)

Joint Channel and Synchronization Estimation Algorithm in FH-OFDM Systems

Yongmao Cheng, Jie Wang, Wenping Sheng, Yu Lin, Chunying Yang,
and Tingjun Li

Naval Aeronautical and Astronautical University, Yantai 264001, China
litingjun99@126.com

Abstract. A joint pilot-aided channel, carrier frequency offset, sampling frequency offset and timing offset algorithm is presented for frequency hopping orthogonal frequency division multiplexing (FH-OFDM) systems. Pilots of frequency domain are exploited to estimate Carrier Frequency Offset (CFO), Sampling Frequency Offset (SFO), and Channel's Impulse Response (CIR) coefficients are estimated. To improve system performance Timing Offset (TO) and Inter-Carrier Interference (ICI) are compensated in time domain. New ML scheme using training sequences is presented to get the initial value of CFO and SFO, which will accelerate convergence of Recursive Least Squares (RLS) estimation. Relationship of bandwidth efficiency and hop rate of system is expressed. Simulation results show our algorithm outperforms conventional algorithms.

Keywords: frequency hopping, orthogonal frequency division multiplexing, timing offset, carrier frequency offset, channel estimation.

1 Introduction

Orthogonal Frequency Division Multiplexing has received considerable attention as one of the key techniques in broadband communication. In military battle field communication systems, carrier frequencies should hop among wide RF range for anti-jam operation to provide operational robustness. RF carrier frequency hopping OFDM (FH-OFDM) is attractive for military communication. FH-OFDM is sensitive for timing offset (TO), sampling frequency offset (SFO) and carrier frequency offset (CFO) in receiver. These synchronization errors cause inter-carrier interference (ICI) and inter-symbol interference (ISI) in FH-OFDM system.

Joint SFO and ML channel estimation algorithm is discussed under the condition of no CFO in [1]. Differential modulation is exploited for FH-OFDM system in [2], avoiding channel estimation. Joint phase noise, CFO and CIR estimation approach based on expectation maximization (EM) is presented in [3]. CFO, SFO and CIR estimation approach based on recursive least square (RLS) with ideal synchronization is proposed in [4].

D. Jin and S. Lin (Eds.): Advances in ECWAC, Vol. 2, AISC 149, pp. 317–322.
springerlink.com   © Springer-Verlag Berlin Heidelberg 2012

2 FH-OFDM System Model for Aviation Mobile Communication

Fig. 1 shows FH-OFDM system applying the proposed algorithm. N is the total number of subcarriers. K is the number of used subcarriers. The k-th subcarrier of m-th symbol after cyclic prefix (CP) insertion can be represented as

$$x_m(t) - \sum_{k \in \Gamma} X_m(k) e^{j2\pi kt/(NT)}, \qquad -T_g \le t \le NT \tag{1}$$

Where T is sample interval, N_g is the length of CP, $T_g = N_g T$, $k \in \Gamma = \{-K/2,...,-1,1,...,K/2\}$. The symbol after transmitter front-end can be written as

$$s_m(t) = x_m(t) e^{j2\pi f_{FH} t} \tag{2}$$

Where f_{FH} is carrier frequency. CIR under assumption of wide sense stationary uncorrelated scattering (WSSUS) can be represented as:

$$h_l(t) = \sum_{i=0}^{L-1} \gamma_i \delta(t - \tau_i) \tag{3}$$

Where γ_i and τ_i are complex gain and time delay of i-th path separately, L is the total number of resolvable paths. The channel is assumed to be time-unvarying in one hop duration, i.e. $h=[h_0,h_1,...,h_{L-1}]$ is invariant during one hop. $N_g \ge L$ is assumed in order to prevent ISI.

Δf is the CFO caused by the Doppler shift and difference of oscillator frequencies between transmitter and receiver. Let $\varepsilon = \Delta f NT$. $\eta = \Delta T / T$ represents normalized SFO caused by sampling frequencies difference between transmitter and receiver. The sampling frequency in receiver is assumed to be $1/T'$, where $T' = (1+\eta)T = T + \Delta T$. TO in receiver is assumed to be $d + \Delta d$, d is the integer part, Δd is the fractional part. n-th sample of m-th symbol after cp removing in time domain of receiver can be expressed as:

$$Y_m(k) = \sum_{n=0}^{N-1} r_{m,n} e^{-j2\pi nk/N} = \sum_{i=-K/2}^{K/2-1} X_m(i) H(i) e^{j2\pi(\varepsilon_i N_m + kd + k\Delta d)/N} \delta_{ik} + W_m(k) \tag{4}$$

Where $N_m = N_g + (N+N_g)m$, $H(i) = \sum_{l=0}^{L-1} h_l e^{-j2\pi li/N}$, $\varepsilon_\eta = (\eta+1)\varepsilon$. $H(i)$ is the channel response at k-th subcarrier. $w_m(n)$ is additive white Gaussian noise (AWGN) with zero mean and variance of σ_w .

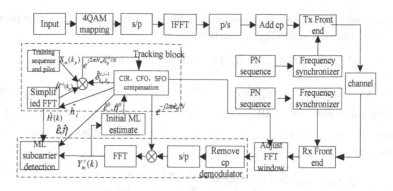

Fig. 1. FH-OFDM system model with proposed algorithm

3 Joint Synchronization and Channel Estimation Algorithm

3.1 ICI Mitigation in Time Domain

Signal of FH-OFDM in receiver can be expressed as below.

$$Y_m(k) = \sum_{l=0}^{N-1} r_{m,n} e^{-j2\pi nk/N} = \sum_{i=-K/2}^{K/2-1} X_m(i) H(i) e^{j2\pi \varepsilon_i N_m/N} \delta_{ik} + W_m(k) \tag{5}$$

Residual TO, CFO and SFO will induce ICI in frequency domain. Known from (5), time domain factor causing ICI in frequency domain is mainly composed of common factor $\exp(j2\pi\varepsilon_\eta n / N)$ and independent factor $\exp(j2\pi kn\eta / N)$. We compensate for the common factor as shown in Fig. 1. Frequency domain after compensation can be represented respectively as (6).

$$Y_m^c(k) = \sum_{i=-K/2}^{K/2-1} X_m(i) H(i) e^{j2\pi \varepsilon_i N_m/N} \delta_{ik}^c + W_m^c(k) \tag{6}$$

where $W_m^c(k) = \sum_{n=0}^{N-1} w_m(n + N_m) e^{-j2\pi n(k+\hat{\varepsilon}+\hat{\varepsilon}\hat{\eta})/N}$, $\hat{\varepsilon}_\eta = (1+\hat{\eta})\hat{\varepsilon}$, $\hat{\eta}$ and $\hat{\varepsilon}$ are estimation value of η and ε respectively.

3.2 Joint CFO, SFO and CIR Estimation Based on RLS

RLS can realize rapid capture of joint estimation with low stable error. We exploit pth frequency domain pilot in each hop of FH-OFDM signal in receiver to estimate CFO, SFO and CIR. The cost function is defined as:

$$C(\hat{\mathbf{h}}^i, \hat{\varepsilon}^i, \hat{\eta}^i) = \sum_{p=1}^{i} \lambda^{i-p} |e_{i,p}|^2 \tag{7}$$

where λ is the forgetting factor of the RLS algorithm, $\hat{\mathbf{h}}^i = [\hat{h}_0^i, \hat{h}_1^i, ..., \hat{h}_{L-1}^i]^T$, $e_{i,p} = Y_m^c(k_p) - X_m(k_p) \hat{H}^i(k_p) e^{j2\pi N_m \hat{\varepsilon}_{k_p}^i/N} \delta_{k_p,k_p}^{c,i}$, $\hat{H}^i(k_p) = \sum_{l=0}^{L-1} \hat{h}_l^i e^{-j2\pi k_p l/N}$, $\hat{\varepsilon}_{k_p}^i = k_p \hat{\eta}^i + (1+\hat{\eta}^i)\hat{\varepsilon}^i$, $\hat{\delta}_{k_p,k_p}^{c,i} = \frac{1}{N} \sum_{n=0}^{N-1} e^{j2\pi n[k_p \hat{\eta}^i + (1+\hat{\eta}^i)\hat{\varepsilon}^i - (1+\eta^c)\varepsilon^c]/N}$, η^c 和 ε^c are estimated value of CFO and SFO in previous symbol respectively.

To make use of linear RLS algorithm, we linearise the non-linear estimation error by first-order Taylor's series approximation.

$$e_{i,p} \approx Y_m^c(k_p) - \{f(X_m(k_p), \hat{\omega}^{i-1}) + \nabla f(X_m(k_p), \hat{\omega}^{i-1})(\hat{\omega}^i - \hat{\omega}^{i-1})\} \tag{8}$$

where $f(X_m(k_p), \hat{\omega}^{i-1}) = X_m(k_p)\hat{H}^i(k_p)e^{j2\pi N_m \ell_{k_p}^i/N}\hat{\delta}_{k_p,k_p}^{c,i}$, $\hat{\omega}^i = [\hat{\omega}_0^i, \hat{\omega}_1^i, ..., \hat{\omega}_{2L+1}^i]$ contains the estimated value of CIR, CFO and SFO. $\hat{\omega}_l^i = \text{Re}\{\hat{h}_l^i\}(l = 0,1,...,L-1)$, $\hat{\omega}_{l+L}^i = \text{Im}\{\hat{h}_l^i\}(l = 0,1,...,L-1)$, $\hat{\omega}_{2L}^i = \hat{\varepsilon}^i$, $\hat{\omega}_{2L+1}^i = \hat{\eta}^i$. The gradient vector can be determined by

$$\nabla f(X_m(k_p), \hat{\omega}^i) = [\frac{\partial f(X_m(k_p), \hat{\omega}^i)}{\partial \hat{\omega}_0^i}, \frac{\partial f(X_m(k_p), \hat{\omega}^i)}{\partial \hat{\omega}_1^i}, ..., \frac{\partial f(X_m(k_p), \hat{\omega}^i)}{\partial \hat{\omega}_{2L+1}^i}]^T \tag{9}$$

where $\frac{\partial f(X_m(k_p), \hat{\omega}^i)}{\partial \hat{\omega}_l^i} = X_m(k_p)e^{-j2\pi lk_p/N}e^{j2\pi N_m \ell_{k_p}^i/N}\hat{\delta}_{k_p,k_p}^{c,i}$, $\frac{\partial f(X_m(k_p), \hat{\omega}^i)}{\partial \hat{\omega}_{l+L}^i} = j\frac{\partial f(X_m(k_p), \hat{\omega}^i)}{\partial \hat{\omega}_l^i}$,

$\Xi_{i,p} = X_m(k_p)\hat{H}^i(k_p)e^{j2\pi N_m \ell_{k_p}^i/N}(j2\pi N_m \hat{\delta}_{k_p,k_p}^{c,i}/N + \frac{1}{N}\sum_{n=0}^{N-1}j2\pi n e^{j2\pi n(\ell_{k_p}^i - \varepsilon_\eta^c)/N})$.

Progress of the RLS based joint CFO, SFO and CIR estimation algorithm has the following steps:

Define $\mathbf{P}^0 = \delta^{-1}\mathbf{I}_{2L+2}$, where δ is the regularization parameter. Estimate the initial value of CFO, SFO and CIR parameters.

Update the parameter $\mathbf{P}^i$ and e^i at the ith iteration, then update the estimated value at the ith iteration.

3.3 ML Estimate Algorithm for Initial Value of Parameters

In order to improve the estimation accuracy, we present an improved joint ML estimation algorithm for ε and η.

Let probability density function as the ML cost function, we can get the the estimated ε and η

$$\hat{\varepsilon}, \hat{\eta} = \arg\max_{\varepsilon,\eta} p(\mathbf{Y}_1, \mathbf{Y}_2 | \varepsilon, \eta) = \arg\max_{\varepsilon,\eta} p(\mathbf{Y}_2 | \varepsilon, \eta, \mathbf{Y}_1)p(\mathbf{Y}_1 | \varepsilon, \eta,) \tag{10}$$

As the 2 training sequences in preamble are identical, i.e., $\mathbf{X}_1 = \mathbf{X}_2$, we can get

$$\mathbf{Y}_1 = \mathbf{\Phi}_1 \mathbf{X}_1 \mathbf{H} + \mathbf{V}_1 \tag{11}$$

$$\mathbf{Y}_2 = \mathbf{\Theta}(\varepsilon, \eta)\mathbf{Y}_1 + \mathbf{\Pi} \tag{12}$$

$\mathbf{\Pi} = -\mathbf{\Theta}(\varepsilon, \eta)\mathbf{V}_1 + \mathbf{V}_2$, $\mathbf{\Theta}(\varepsilon, \eta) = diag\{e^{j2\pi(N+N_g)(k\eta+\varepsilon_\eta)/N}\}$, $k \in \Gamma$.

Then the ML cost function can be expressed as:

$$\hat{\varepsilon}, \hat{\eta} = \arg\min_{\varepsilon,\eta} \sum_{k \in \Gamma} |Y_2(k) - e^{j2\pi(N+N_g)(k\eta+\varepsilon_\eta)/N}Y_1(k)|^2 \tag{13}$$

4 Simulation of System Performance

QPSK modulation is used in this FH-OFDM system with the total number of subcarrier N=64. It is assumed that $d + \Delta d = 5.2T$, $\varepsilon = 0.212$ and η=0.000112.

Fig. 2 and Fig. 3 show FH-OFDM system BER performance as a function of SNR and CFO. Curve E is our algorithm, curve C is algorithm in [4]. Curve A, B represent our algorithm without initial value estimate, ICI mitigation respectively. As shown of curve A in Fig. 2, algorithm performance will decrease intensively with high BER, if ignore the estimation of initial value. As shown of curve B, if we ignore the mitigation of ICI, ICI will be the main cause of high BER when SNR is high. As shown in Fig. 3, if we ignore initial CFO and SFO value estimation, curve E will degenerate into curve A, whose BER will increase intensively when CFO is high. In Fig. 3 our algorithm BER performance is better than curve B which ignored ICI.

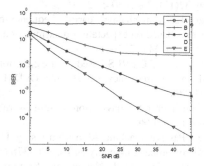

Fig. 2. BER performance versus SNR with different algorithm

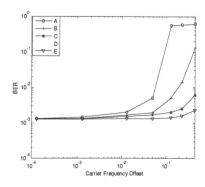

Fig. 3. BER performance versus CFO with different algorithm

5 Conclusion

We propose a low complexity joint pilot-aided channel, carrier frequency offset, sampling frequency offset and timing offset estimation algorithm for FH-OFDM system with TO, CFO and SFO. Simulation of FH-OFDM system with 1000 hop/s shows that our algorithm outperforms conventional algorithm.

References

1. Gault, S., Hachem, W.: Joint sampling clock offset and channel estimation for OFDM signals: Cramer-Rao bound and algorithms. IEEE Transactions on Signal Processing 54(9), 1875–1885 (2006)
2. Xiong, J.-Q., Gan, L.-C., Zhu, Y.-C.: Performance analysis of short-wave FH/OFDM system based on joint iterative demodulation and decoding algorithm. Journal of Electronics & Information Technology 32(12), 3041–3045 (2010)
3. Zhang, J., Mu, X., Hanzo, L.: Joint channel, carrier-frequency-offset and noise-variance estimation for OFDM systems based on expectation maximization. In: IEEE 71st Vehicular Technology Conference, pp. 1–5 (May 2010)
4. Nguyen-Le, H., Le-Ngoc, T., Ko, C.C.: RLS-based joint estimation and tracking of channel response, sampling, and carrier frequency offsets for OFDM. IEEE Transactions on Broadcasting 55(1), 84–94 (2009)
5. Li, T., Lin, X.: Research on intergrated navigation system by rubidium clock. Journal on Communication 27(8), 144–147 (2006)
6. Li, T.: Research on TDOA Passive Tracking Algorithm Using Three Satellites. Journal of Naval Aeronautical and Astronautical University 24(4), 376–378 (2009)
7. Li, T., Lin, X.: GPS/SINS Integrated Navigation System Based on Multi-scale Preprocessing. Journal of Wuhan University 36(1), 6–9 (2011)
8. Li, T.: The Phonetic Complex Data Based on FPGA. Key Engineering Materials 475, 1156–1160 (2011)

Electronic Truck Scale Wireless Remote Control Cheating Monitoring System Using the Voltage Signal

Yanjun Zhao and Yang Pan

Hebei United University, No 46 Xinhuaxi Road, Tangshan, Hebei, China
zhyj_ts@sohu.com, yjzhao@heuu.edu.cn

Abstract. The electronic truck scale wireless remote control cheating method can lead to large economic losses. Aimed at the wireless remote control cheating method, a new wireless remote control monitoring system using the output voltage signal of the weighing sensor is brought out. According to the change times of the sensor output voltage signal, whether the wireless remote control cheat method exists or not can be determined. Based on the monitoring principle, the wireless remote control monitoring system is designed. Experimental results show that the monitoring system can on-time detect the electronic truck scale wireless remote control cheat method.

Keywords: wireless remote control, cheating, monitoring system, Electronic truck scale, voltage signal.

1 Introduction

Electronic truck scale, as a convenient, fast, standard weighing instruments, are widely used in automotive transportation. In China, more and more electronic truck scales are used for the accounts. To seek illegal profits, lawbreakers make many kinds of electronic truck scale cheating system. The main cheating method is to reduce truck self-weight and increase the goods weight when weighing through a variety of methods. Many companies have brought significant economic losses.

The wireless remote control cheating method is very subtle. When the truck weighting, the lawbreaker sends a wireless remote control signal, the receiving device receives the cheating signal and change the measurement circuit parameters, so the vehicle weight is changed. The cheating method uses more convenient, so it brought more serious consequences[1-6]. In this paper, the electronic truck scale wireless remote control cheating monitoring system using the voltage signal is designed.

2 Principle of the Electronic Truck Scale

The electronic truck scale weighing system includes the scale platform, the pressure sensors, the signal conversion and the weighing meter. Electronic truck scale system is shown in Figure 1.

Four weighing sensors support the scale platform. When the truck is on the platform, the pressure sensors generate the deformation and output the mV signal

D. Jin and S. Lin (Eds.): Advances in ECWAC, Vol. 2, AISC 149, pp. 323–326.
springerlink.com   © Springer-Verlag Berlin Heidelberg 2012

which is proportional to the truck weight. The signal is transmitted to the meter after the signal conversion and the truck weight can be calculated. The difference between the full loaded weight and the empty weight of the truck is the truck goods weight.

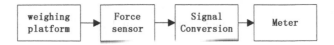

Fig. 1. The electronic truck scale weighing system

3 Principle of the Wireless Remote Control Cheating Method

The normal wiring of the weighing sensor is shown in Figure 2. The system includes five wires, two power line, two signal line and the shielded line. The wireless remote control cheating wiring is shown in Fig. 3. The lawbreaker peels the communication cable and cuts off the power line (E +) of the sensor. The cheating system strings into the power line of the sensors (E +) and does not affect the electronic truck scale normal work. When the remote control cheating signal is received by the remote control receiver, a resistor strings into the sensor power line. The Wheatstone bridge supply voltage is less than 12V, the weight of the truck is changed. The cheating method can generate several tons error, the action is very subtle and difficult to find.

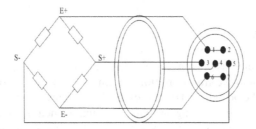

Fig. 2. The normal wiring diagram of Electronic truck scale weighing sensor

4 The Wireless Remote Control Cheating Monitoring System

Aimed at the wireless remote control cheating method, the cheating monitoring system is brought out. The monitoring system includes two parts: the voltage signal detection system and the data processing system. Take the double-axial truck for an example. The multi-axial car is similar with the double-axial truck. The normal weighing voltage signal is shown in the figure 4.

In the weighing voltage figure, when the front wheel of the double-axial truck is on the weighing platform, the pressure of the weighing sensor is bigger than the Initial state, so the output voltage signal is changed and gradually become larger. The voltage variation is proportional to the pressure variation of the sensor. The section a1 to a2 represents the case of the output voltage signal. When the front wheel is on the platform completely and the back wheel is not on the platform, the pressure of the sensor changes very small, so the output voltage of the sensor changes very small too. The section a2 to b1 is shown the case of the sensor output voltage signal.

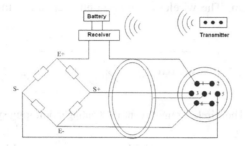

Fig. 3. The remote control cheating wiring diagram

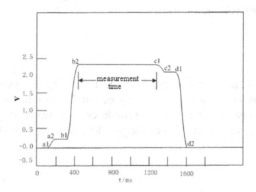

Fig. 4. The normal weighing voltage signal

In general, The goods is loaded in the rear of the truck, so the pressure of the truck back wheel is bigger than the pressure of the truck front wheel. When the truck back wheel is on the platform, the output voltage change rate of the weighing sensor is bigger than the output voltage change rate while the front wheel is on the platform, so the output voltage has a great change. The section b1 to b2 on the figure shows the voltage change while the back wheel is on the platform. While the all truck wheels are on the platform, the pressure of the truck is not changed, so the output voltage is not changed. The output voltage is proportional to the truck weight. When the truck wheels are out of the platform, the voltage signal is opposite to the voltage signal when the truck wheels are on the platform.

According to the sensor output voltage signal figure of the double-axial truck, whether the wireless remote control cheating method exists or not can be judged directly. If the change times of the output voltage signal which the truck is on the platform and out of the platform are equal to 2, there is no the wireless remote control cheat and can normal weighing. Otherwise, if the change times of the output voltage are greater than 2, there is the wireless remote control cheat. Refuse to weighing and alarming.

The wireless remote control cheating monitoring system is mainly consists of the voltage signal monitoring system, the enlarge unit, the A/D conversion unit and the data processing system. The wireless remote control cheating monitoring system is shown in figure 5.

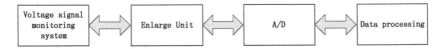

Fig. 5. The wireless remote control cheating monitoring system

The output voltage signal of the weighing sensor is received by the voltage signal monitoring system. The voltage signal is enlarged through the amplifier circuit and is changed into the digital signal through A/D. The voltage signal change times can be gotten from the output voltage figure.

5 Conclusion

The electronic truck scale wireless remote control cheating test system is established. The cheating monitoring system is designed and studied experimentally. The experiment results show that the wireless remote control monitoring system using the sensor output voltage signal can detect the wireless remote control cheating method on time.

References

1. Liu, J.Q.: The Technology Trends, Development Trends and Industry Direction of the Weighing Sensor. Weighing Instrument 34, 1–5 (2005)
2. Fu, Y.M., Liu, A., Liu, P.: The Anti-cheating Measures of the Electronic Truck Scale. Industrial Measurement, 11–13 (May 2002)
3. Zhang, F., Gao, Y., Li, M.: The Cheating and Anti-cheating Method of the Electronic Truck Scale. Weighing Instrument 32, 18–21 (2003)
4. Liu, X.M., Zhang, B., Yang, L.: The Photoelectrical Detecting System for Motorcycle Orbit. Optoelectronic Technology & Information 17, 55–57 (2004)
5. Huang, H.M.: Dynamical Compensating Method for Weighting Sensor Based on FLANN. Transducer and Micro System Technologies 25, 25–28 (2006)
6. Zhou, Q., Wang, S.G.: Property Contrast of Several Photoelectric Measure Circuits. Modern Electronic Technology, 87–89 (August 2008)
7. Pereira, J.M., Girao, P.M., Postolache, O.: Fitting Transducer Characteristics to Measured Data. IEEE Instrum. and Meas., 26–39 (December 2001)
8. Liu, Y.: Hardware Design of Remote Intelligent Monitoring System on Environmental Parameters. Journal of Anhui Agri. Sci. 37(6) (2009)
9. Huang, Z., Wei, Y.: Design and implementation of the real-time optical intensity watching system. Journal of Dongguan University of Technology 16(1) (2009)

Optimal Design Based on Computer Simulation for Improving Thermal Comfort in the Passage Space around Arcade Buildings in a Hot and Humid Climate

Jin Wei, Fangsi Yu, and Jiang He

Department of Architecture and Urban Planning, College of Civil Engineering and Architecture, Guangxi University, Nanning, Guangxi, China

Abstract. Arcade building is a unique architectural form in Lingnan Area of China. Its overhead space should provide a more comfortable passage space. However, during the stage of reconstruction design of Nanning Pedestrian Street, arcade buildings were completely retained in the structure and historical spirit as possible, whereas thermal comfort as a passage space was ignored. Through the analysis of field investigation, it was suggested that tree-planting, water-holding pavement and membrane structure would be effective cooling strategies. These passive strategies can improve thermal comfort in arcades and surrounding passage space. In order to optimize the thermal improving effect of these strategies, an optimal design was carried out based on computer simulation. Simulation results can be visualized in color images on a PC. From these images, a designer can easily understand where is cool, what degrees surface temperatures could be reduced in the passage space.

Keywords: numerical simulation, arcade building, passage space, thermal comfort, hot-humid climate.

1 Introduction

As climatic-oriented architecture, arcade buildings are popular in hot and humid climates such as in Lingnan Area of China. Arcades at the first floor of arcade buildings allow wind blow through, keep out rain, shade the sun, and provide spaces for shopping and walking. The layout form of an arcade building is close to out-corridor buildings and space attached to the arcade building is used as a commercial passage.

Orientation of an arcade building decides the direction of colonnades of its both-sides buildings, and greatly influences thermal comfort in the passage space. For example, North-South-oriented colonnade space is affected at the time of morning hours in the East-West-oriented Street; East-West-oriented colonnade space is affected during the period of afternoon hours in the North-South-oriented Street.

Arcade building in Lingnan Area has been studied by many researchers. Most of previous studies focused on the relationship between people and architecture. There is still lack of studies about thermal comfort in the passage space of arcade buildings.

D. Jin and S. Lin (Eds.): Advances in ECWAC, Vol. 2, AISC 149, pp. 327–334.
springerlink.com   © Springer-Verlag Berlin Heidelberg 2012

Furthermore, it was nearly not found the study about optimal design for improving thermal comfort around arcade buildings in Lingnan Area.

The subject of the present study is to clarify thermal environment problems in a pedestrian street which is located in a southern city called Nanning; propose effective thermal improving strategies; optimize the thermal improving effects of the proposed strategies based on computer simulation.

2 Description of the Pedestrian Street

2.1 Climatic Characteristics of the Location

The targeted pedestrian street is at the central downtown of Nanning City (N22' 13' ~ 23' 32', L107' 45' ~ 08' 51'). Nanning is located just south of the tropic of Cancer, where the summer season occurs between April and November and the highest daily air temperature exceeds 30℃. The rainy season occurs between May and August, at which time relative humidity averages over 80%. The monthly mean wind velocity is around 1m/s throughout the year. Therefore a lack of wind is another characteristic. Heating is required for only two months, December and January. The climate of Nanning can be summarized as follows: winter is very short at two months and the summer is long, hot and humid without breeze.

2.2 Characteristics of Nanning Pedestrian Street

Nanning Pedestrian Street consists of two roads: Xingning Road and Minsheng Road, as shown in Fig.1. It is an old commercial district in Nanning. Nanning Pedestrian Street may be characterized by arcade buildings. The arcade has a rational form with a former store and a latter living room, which means the first floor and passage space can be used for commerce, and the second floor and inside rooms can be used for living. At the beginning of 2000, Nanning Pedestrian Street was entirely rebuilt. Arcade buildings were completely retained in the structure and historical spirit as possible (see Fig.2), whereas consideration in thermal comfort was not enough. The next section will describe thermal environment problems found in the street.

2.3 Thermal Environment Problems in Nanning Pedestrian Street

This study focuses on the passage space around arcade buildings in Xingning Road with East-West orientation, and Minsheng Road with North-South orientation. Based on analysis results of field investigation in Nanning Pedestrian Street, thermal problems can be summarized as follows:

1. The passage colonnade space under the arcade buildings in the E-W-oriented road is influenced by solar radiation in the afternoon. This leads to uncomfortable thermal environment in the passage space.

Fig. 1. Plan of Nanning Pedestrian Street

Fig. 2. Photos of Nanning Pedestrian Street in 1932 and 2011 (*left and right*)

2. Because the passage space outside the colonnade space lacks of effective solar control strategies such as tall trees and shading shelters, most of the outside space around arcade buildings is exposed to the sun during daytime on a sunny day.

3. The ground is paved with light-color tiles, which is impermeable. These materials reflect more solar radiation to the passage space. Moreover, such material has high heat capacity and absorbs more solar heat. After sunset, the surroundings will be warmed by heat released from these pavement materials.

3   Thermal Improving Strategies

As described above, one of major causes to worsen thermal comfort in the arcade is solar radiation including direct and reflected solar radiation to the passage space. Therefore, solar control around arcade buildings would be the most important strategy. Another effective cooling strategy would be to lower surface temperatures around arcade buildings. According to the above analysis, the following passive cooling strategies can be considered:

1. Plant trees to the East-West-oriented street for shading solar radiation which directly irradiates the passage space.

2. Install membrane roofs to increase the sun-shaded space in the pedestrian street.

3. Apply water-holding (cool) pavement to the ground in the pedestrian street for keeping the ground surface cool by water evaporation.

Fig.3 illustrates a schematic description how to apply the above-mentioned strategies to the study street. To optimize the thermal improving effects of these strategies, an optimal design was carried out based on computer simulation. The simulation method and a case study will be described below.

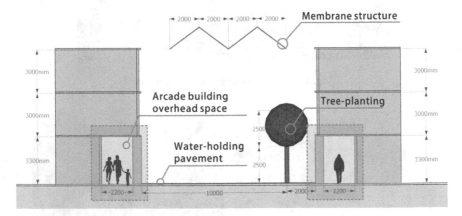

Fig. 3. Schematic description of the proposed cooling strategies in the E-W street

4   Computer Simulation

4.1   Description of the Simulation Tool

The simulation process is outlined in Fig.4. The simulation was performed using 3D-CAD models for buildings, trees and other structures in the area being analyzed. Three-dimensional spatial forms of the buildings, trees, etc., and two-dimensional ground surfaces were divided into mesh grids, and thermophysical data of construction materials, such as albedo and conductivity and solar transmittance, were assigned to the grids. A uniform mesh size (spatial resolution) of 0.4 m was used in the simulation. The external surface temperature for each mesh can be determined by solving a non-steady-state one-dimensional heat balance equation in the vertical direction of the surface. The non-steady-state one-dimensional heat conduction equation for each mesh was solved using the above-mentioned heat balance data as boundary conditions for external surfaces. Boundary conditions for internal surfaces are the indoor air temperature for the buildings and the underground temperature for the ground. As outputs of the simulation, temperatures of all external surfaces can be predicted and visualized on the 3D models.

4.2 Comparison Study

The 3D-CAD models for buildings, trees and other structures in the street were generated based on GIS data and field investigation data. Simulations for the following three cases were performed based on the 3D-CAD models for the street.

Case1: Present (without the application of the proposed strategies); Case2: Proposal1 (plant trees that are 5 m height as shown in Fig.3); Case3: Proposal2 (membrane structures are installed in the E-W-oriented road).

Fig.5 is a view of 3D-CAD models for Proposal2, looking from East to West in Minsheng Road. A membrane dome roof was installed at the position where the two roads cross. Another membrane roof was installed on the west side of the dome roof in Minsheng Road.

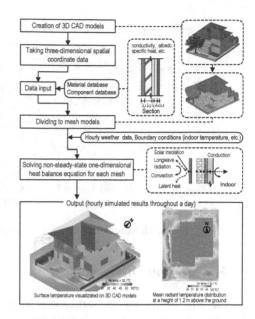

Fig. 4. Diagram of simulation process

Input weather data was hourly weather data of a typical sunny summer day for Nanning. The weather data is schematized in Fig.6. The maximum and minimum air temperatures are 33℃ and 25℃ respectively. The peak value of total horizontal solar radiation reached 800 W/m^2. Wind speeds were below 1m/s during the daytime.

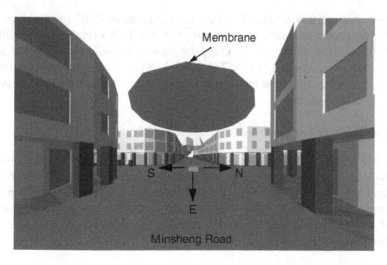

Fig. 5. A view of 3D-CAD models for the street (Proposal 2)

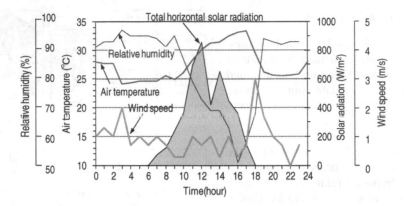

Fig. 6. Meteorological data for a typical sunny summer day in Nanning

5   Results and Discussion

Simulation results for the afternoon hour (15:00) were used for the following discussions. The reason for selecting the time is that the solar radiation to the vertical (wall) surface is peak at this afternoon time. A thermal sensation index called mean radiant temperature (MRT) was used to evaluate thermal comfort for a person in the pedestrian street. The mean radiant temperature at a point is defined as the uniform temperature of an imaginary enclosure in which radiant heat transfer from the human body or object equals to the radiant heat transfer in the actual non-uniform enclosure.

5.1   Surface Temperature Distribution

Fig.7 presents simulated results of surface temperatures at 15:00 for the three cases. These thermographs are surface temperatures in the E-W-oriented road (Minsheng Road), looking from West to East. From the figure, it can be seen that the ground surface temperature was reduced by 7℃ from 41°C to 34℃ after applying the cool pavement. It can be also found that surface temperatures of the sun-shaded pavement show 3-4℃ lower values.

5.2   Mean Radiant Temperature

Fig.8 is horizontal distributions of MRT at a height of 1.5 m above the ground at 15:00. As seen in the figure, MRTs in the arcade for Proposal 1 and 2 were about 1℃ lower than those for Present. This MRT reduction resulted from reduction of the ground surface temperature. As revealed by the right lower part of thermographs for Proposal 1 and 2, MRTs near arcade buildings in the S-N-oriented road were 2-3℃ lower than those for Present.

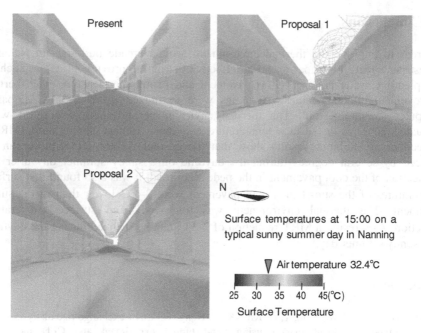

Fig. 7. Simulated surface temperature distributions in the E-W oriented street

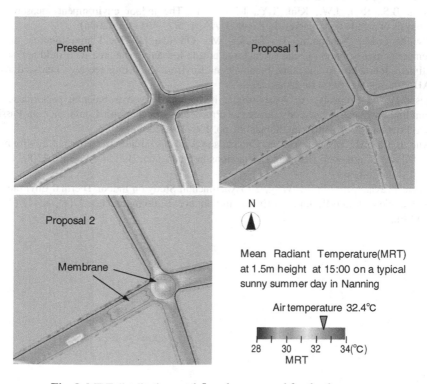

Fig. 8. MRT distributions at 1.5 m above ground for the three cases

6 Conclusions

In order to improve the thermal environment around arcade buildings in Nanning Pedestrian Street, this paper proposed effective thermal improving strategies such as tree-planting, water-holding (cool) pavement and membrane structure. The thermal improvement effects of these strategies were optimized using a 3D-CAD-based computer simulation tool. The thermal environment and thermal comfort were evaluated in terms of surface temperature and mean radiant temperature (MRT), respectively. Simulation results show that the ground surface temperature can be reduced by several degrees during the daytime on a sunny summer day after the application of the cool pavement in the pedestrian road. It was also found that surface temperatures of the sun-shaded cool pavement were 3-4 ℃ lower than in the sunlit pavement. The thermal environment was improved from surface temperature reduction. As a result, a MRT reduction of 2-3 ℃ can be realized during the daytime on a sunny summer day.

References

1. Kim, B.S., Roh, J.W., Kim, T.Y., Kim, K.H., Hong, G.P.: Air exchange rate analysis of the arcade-type traditional market using wind tunnel experiment and CFD model. J. AABE 5(1), 161–167 (2006)
2. Kim, B.S., Roh, J.W., Kim, T.Y., Kim, K.H.: The indoor environment measurement analysis of arcade-type markets in Korea. J. AABE 5(1), 191–198 (2006)
3. Tsujihara, M., Nakamura, T., Tanaka, M., Ohtsuka, J.: Field investigation on air temperature distribution inside an enclosed arcade located in the area with mild and sunny climate. Journal of Architecture, Planning and Environmental Engineering (Transactions of AIJ) 508, 43–50 (1998) (in Japanese)
4. Elseragy, A., Elnokaly, A.: Assessment criteria for form environmental performance of building envelope in hot arid climate. In: Proceedings of the 24th Conference on Passive and Low Energy Architecture, Singapore, pp. 156–162 (2007)
5. Xue, J., Ran, M., Wu, Y.: Test and Comparison Analysis on Summer Thermal Environment of Qilou Veranda with Different Orientations in Quanzhou. Building Science 27(8), 17–23 (2011) (in Chinese)
6. Chen, J., Wang, L., Liu, J., Wang, Y.: Experimental Study on Indoor Thermal Environment in Traditional Qilou Building in Haikou in Summer. Building Science 27(4), 42–47 (2011) (in Chinese)

Design of Controlling System for Linear Stepper Motor Based on DSP2407

HongMin Zheng[1] and WenFeng Cui[2]

[1] Engineering Training Center, Northwestern Polytechnical University, Xi'an, Shaanxi, China, 710072
zhenghongmin_nwpu@163.com
[2] School of Mechatronics, Northwestern Polytechnical University, Xi'an, Shaanxi, China

Abstract. According to the starting out-of-step and mechanical impact of stroke termination effectively on linear stepper motor, the control system based on DSP2407 was proposed. The paper introduced the system configuration, designed the micro-step control circuit, and control arithmetic was improved. The experiment results indicated that the linear stepper motor driven by this system can not only be started smoothly, restrained impact during the end of deceleration, but can also be accelerated and decelerated conveniently and simply.

Keywords: linear stepper motor, DSP2407, micro-step control.

1 Introduction

Linear stepper motor is a linear motion device, which can make electrical pulse signal into linear micro-step motion directly. In condition of open-loop, it can provide accurate and reliable straight-line displacement, velocity and acceleration control, and it can be used to constitute an open-loop system without feedback. However, due to its structural characteristics, there were some shortcomings as follows [1-2]:

(1) It can not automatic accelerate and decelerate smoothly as DC motor and AC motor, out-of-step or blockage will appear probably in circumstance of high speed and large variation.

(2) Its Whirling speed is not steady enough, step angle is overlarge, especially, and concussion phenomenon will appear occasionally.

Focus on shortcomings of linear stepper motor, this paper carry out two optimization for the application of its driver: control the process of acceleration and deceleration when speed (pulse frequency) of system fluctuate largely; Providing subdivision function for the system, which can promote positional accuracy and steady degree at lower speed; Theoretical analysis and experimental results show that system's dynamic and static performance has been improved noticeably by optimization design.

2 Control System Configuration

As the core of the system control unit TMS320LF2407A was used to achieve a signal generating.TMS320LF2407A was a powerful general-purpose I/O port and PWM

D. Jin and S. Lin (Eds.): Advances in ECWAC, Vol. 2, AISC 149, pp. 335–341.

output function, it can output a 16-way PWM waveform. PWM frequency change can be realized stepper motor speed and position control [4-5]

Control signal was used to set up the linear stepper motor operation mode and software integration of the motor control the start and stop, the positive/negative circumrotate, acceleration/deceleration and location and orientation. Display circuit with real-time display in the operation of the motor speed, and other parameters. The system schematic was shown as Figure 1.

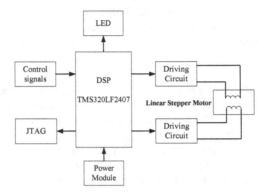

Fig. 1. System schematics

3 Acceleration/ Deceleration Control Arithmetic

Motor started from a low frequency, run at high speed after the uniform. When approaching of another terminal location, the stepper motor run slowdown, and then it could accurately stop in the terminal position. Starting vibration and mechanical impact of route end is restrained effectively. The process can be a variety of variable speed control mode time curve [6].

Line acceleration/deceleration control Curve was shown as Figure 2 (a). Plans in the segment AB, BC, CD, were on behal of acceleration, uniform and deceleration. The letter 'N' was behal of linear stepper motor current pulse frequency.

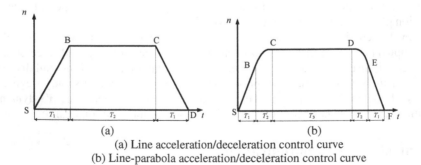

(a) Line acceleration/deceleration control curve
(b) Line-parabola acceleration/deceleration control curve

Fig. 2. Control curve of linear motor

Generally speaking, the use of linear speed of operation, there was oscillation in the process of acceleration/deceleration and uniform process, and could not transform smoothly. The function between acceleration and time $a(t) = dv(t)/dt$ was not consecutive, and the concussion phenomenon occurred. This will affect the life of the motor and mechanical systems, using Line-Parabola acceleration/deceleration control curve as it was shown in Figure 2 (b). First it accelerated by line-Parabola curve. Then it was uniform motion, at last it was decelerated by line-Parabola curve. The control arithmetic was improved, which eliminated starting out-of-step mechanical impact of distance termination effectively, and ensured the motor running smoothly.

The above curves were continuous, could not program, the paper used legitimate step to achieve speed control. It met the required operating frequency through the sub-level frequency jump gradually. When frequency increasing of the linear stepper motor the function between frequency and time was

$$f(t) = f_s + f_m(1 - e^{-t/\tau})$$ (1)

In the expression, f_s was starting frequency of line stepper motor, f_m was the highest frequency of line stepper motor. The letter τ was the drive system Time constant.

If the operating frequency was f_g, the total time of the Speed-up phase can be obtained from the expressions. it is

$$t_\tau = \tau[\ln f_m - \ln(f_m - f_g + f_s)]$$ (2)

Will speed up the dispersion of uniform for n, the type known from time to speed up, two adjacent speed change for the time interval. The acceleration paragraph was evenly divided into paragraphs n, from the above expression

$$\Delta t_\tau = \frac{t_\tau}{n}$$ (3)

In the expressions, N is the number or speed level.

The frequency of each level

$$f_k = f_s + f_m(1 - e^{-k\Delta t_\tau/\tau})$$ (4)

The step number in each level of speed N_k is

$$N_k = f_k \times \Delta t = \frac{f_k t \tau}{n}$$ (5)

When the program running, The speed of each step should take the calculated step number, Motor running every step subtracted by one. When the number reached zero, it explained that the steps of that speed level was end. Then, it was the next level, until end of the process of acceleration.

4   Micro-step Control

The stepper motor has the characteristics of fast start-up and shutdown, precision stepping, directly receive digital and non-cumulative error. Therefore, in digital control system it was extensively applied. However, because stepper motor had the problem of greater distance, not smooth Speed and low-speed vibration, it limited the application of high-precision occasions. In order to solve this problem, it is necessary to employ micro-step control [7].

The working principle of Stepper motor micro-step is: When each pulse inputting, instead of winding current input or removal all, but only a part of the corresponding winding was changed. Then mover of the motor only moved a part of one step. Here winding current is no longer a square wave, but a ladder wave, rated current was stepped inputting or removal. Current divided into a number of steps, the mover on the same number of steps to turn away from certain.

For the micro-step control of stepper motor, the better current waveform is discrete sine function. Constantly current vector constant amplitude uniform rotation method is employed. That is two-phase currents were changed at the same time, letting the current synthesis vector i_H circumrotate by constant amplitude and uniform speed. The micro-step control function was shown as

$$\begin{cases} i_A = i_M \cos\theta \\ i_B = i_M \sin\theta \end{cases} \tag{6}$$

4.1 Hardware Circuit Design

For simple structure flexible control two-phase stepper motor driver ASIC A39555SB was used in this paper. Power driver circuits made up of two pieces of A39555SB and external circuit, and then connect with the DSP2407 IO, drive circuit and DSP2407 IO connecting circuit was shown as in Figure 3.

Two pieces of A39555SB were received in the two-phase stepper motor windings in order to complete the stepper motor drive control. As the control core, DSP2407 achieved logic control for the A39555SB by software programming.

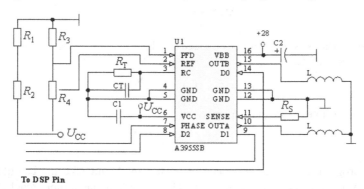

Fig. 3. Linear stepper motor micro-step drive circuit

One piece of A39555SB was used to drive one motor windings of the operating voltage went up to 50V and current up to 1.5A. 3 nonlinear internal DAC combination of internal PWM current control can realize the full stepper motor 1/2, 1/4, 1/8 mode. Internal PWM current control circuit with an external reference voltage determines the current mode (fast decay mode, slow decay mode and mixed-decay modes). At the same time, A39555SB also provided a perfect protective measures, including over-temperature protection, over-current protection and under voltage protection.

4.2 Software Design

There were three nonlinear DAC converters in A3955SB. The flowing motor windings maximum current I_{max} could be determined by setting pin2 V_{REF} and current sampling resistor R_S. $I_{max} = V_{REF}/3R_S$. The relation between input number D_2, D_0, D_1 and load current was shown as in table 1.

Table 1. DAC Output Data

DAC Input			Current ratio %	V_{REF}/V_S
D_2	D_1	D_0		
1	1	1	100	3.00
1	1	0	92.4	3.25
1	0	1	83.1	3.61
1	0	0	70.7	4.24
0	1	1	55.5	5.41
0	1	0	38.2	7.85
0	0	1	19.5	15.38
0	0	0	No output	

Based on Table 1 a micro-step control table should be founded for the motor. The core unit transmission control characters to I/O by looking up the table with software method. Two-phase hybrid stepping motor worked in the four kinds of status, the largest number of subdivision is 8, so the table included a total of 32 different data. Only under different micro-step state the offset was different, and for different rotational direction the lookup direction was different. Therefore only this can be a Table. Process flow chart was shown in Figure 4.

The key of programming is: (1) Under different micro-step state the offset was different (2) For different rotational direction the lookup direction was different. And rotational direction should be deal with correctly at the end of the table, so the data can be recycled viewed. Micro-step digital control characters of the core data sheet program were shown as follows

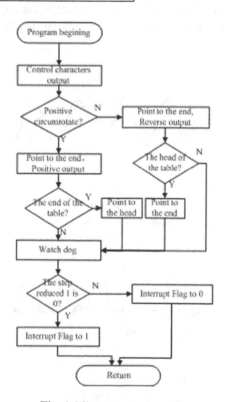

Fig. 4. Micro-step program flow

```
static const int run_Table[]=
{0x00CC,0x00BD,0x00AE,0x00009F,0x000F,0x001F,0x002E,0x003
D,0x004C,
0x005B,0x006A,0x0079,0x0070,0x0071,0x0062,0x0053,0x0044,0
x0035,0x0026,
0x0017,0x0007,0x0097,0x00A6,0x00B5,0x00C4,0x00D3,0x00E2,0
x00F1,0x00F0,
0x00F9,0x00FA,0x00DB};
```

5 Experimental Results

Serial connecting a small resistance to the winding of stepper motor, and observe the voltage drop of the resistance by oscilloscope, because the waveform of the resistance's voltage is equal to the waveform of winding's electric current. Figure 5.1, Figure 5.2 and Figure 5.3 show the waveform of winding's electric current on the condition of 8-subdivision, 4-subdivision and no subdivision in entire step respectively. Figures show that in the condition of subdivision, the waveform of motor winding's electric current approximate to stepped sine waveform, moreover, with the number of subdivision increases, the precision of approximation was improved, and the waveform of current become more smooth; When subdivision was not used, the waveform of motor winding's current approximate to square waveform. Figure 5.4 shows the diagram of control pulse waveform at end deceleration, we can see that the frequency of Stepping pulse control reduced significantly, which played a role in deceleration.

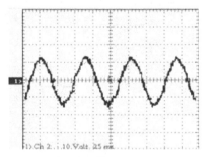

Fig. 5.1.
Current waveforms as 1/8 micro-step

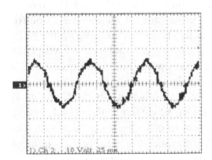

Fig. 5.2.
Current waveforms as 1/4 micro-step

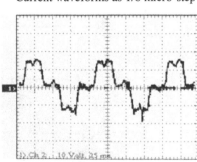

Fig. 5.3.
Current waveforms as no micro-step

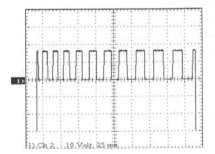

Fig. 5.4.
The waveform of deceleration process

Analysis above indicate that, in condition of subdivision, stepper motor was drove by current of sine waveform, and in condition of no subdivision, stepper motor was drove by electric current of square waveform; moreover, subdivision drive can enhance the stability of stepper motor, and reduce noise and vibration.

6 Conclusion

In this paper, improving drive performance of linear stepper motor is the start point, which adopt DSP2407 as control core, and introduce a design proposal with characteristic of acceleration, deceleration and micro-step control, especially, the realization of subdivision control based on A3955SB was introduced in detail, meanwhile, the control strategy of acceleration and deceleration was improved. Experimental results show that the linear motor droved by this system not only start smoothly, with small extremity impact of deceleration, but also display the advantages of convenient and easy on acceleration and deceleration control.

References

1. Yue, Y.: Linear motor technical manual. Machinery Industry Press, Beijing (2003)
2. Choi, J.-K., Jung, S., Baek, K.-R.: A Data Reorganization Algorithm to Improve ransmission Efficiency in CAN Networks. Intelligent Control and Automation 344, 438–443 (2006)
3. Zhou, L., Yang, S., Gao, X.: Modeling of stepper motor control system and running curve simulation. Electric Machines and Control 15(1), 20–25 (2011)
4. Wang, L., Cheng, C.-S.: DC excitation control of a wind induction synchronous generator using a micro-controller. IEEE Asia-Pacific Conference on Circuits and Systems, 1173–1176 (2004)
5. Cmosija, P., Kuzmanovic, B., Ajdukovic, S.: Microcomputer Implementation of Optimal Algorithms for Closed-Loop Control of Hybrid Stepper Motor Drive. IEEE Transaction Industrial Electronics 47(6), 1319–1325 (2000)
6. Zheng, X.: Research on restrain technique of ending impact of aviation linear stepping motor. Micro-motor 38(4), 53–55 (2005)
7. Song, S.: Design of 2 phase Hybrid Stepping Motor Driver. Electric Transmission 36(2), 59–64 (2006)

Controlling Study of D-STATCOM
Based on PSO-PID Algorithm

Qingxin Zhao[1,*], Hui Wang[1,2], Haijun Xu[2,3], and Liang Zhao[4]

[1] Changchun Institute of Technology, Changchun, Jilin province, China, 130012
[2] Jilin Province University Distribution Automation Engineering Research Center,
Jilin province, China, 130012
[3] Changchun University of Technology, Department of Electrical and Electronics Engineering,
Changchun, Jilin province, China, 130012
[4] Jilin Electric Power supply company, Siping power supply company, Siping,
Jilin province, China, 136000
zhaoqingxin1984@163.com, wanghui841013@163.com,
xuhaijun001@126.com, zilong6662002@163.com

Abstract. In distribution power system, with the increasing of pulsed loads and inductive load, these power quality problems was becoming more and more serious. Distribution Synchronous Static Compensator (DSTATCOM) was generally used to solve power quality problems in distribution systems. In the research processing of the device, the control strategies of DSTATCOM play an important role. In this article a novel adaptive control strategy was based on an improved Particle Swarm Optimization (PSO) algorithm PID. With this control method, the compensator can stabilize the PCC voltage around a prescribed value. A distributes power system is developed in MATLAB environment, the validity of the method proposed is verified.

Keywords: DSTATCOM, Intelligent Control, Particle Swarm Optimization, PID, Power quality.

1 Introduction

In a distribution system, the presence of unbalanced and non-linear loads can cause unbalanced and distortion in the system quantities. This would put the distribution feeder under greater burden. To prevent the unbalanced and distorted currents from being drawn from the distribution bus, Flexible ac transmission system (FACTS) controllers are proving to be very effective in using the full transmission capacity while increasing operational efficiency and maintaining reliability of power systems. These controllers are based on power electronic devices and have fast response time. Advanced FACTS controllers are based on voltage sourced converters and include: Static synchronous compensator (STATCOM), Distribution static compensator

---

* Qing-xin Zhao(1969-), female, Associate Researcher, Research Intelligent control theory, Control theory and Control engineering.

D. Jin and S. Lin (Eds.): Advances in ECWAC, Vol. 2, AISC 149, pp. 343–348.
springerlink.com   © Springer-Verlag Berlin Heidelberg 2012

(DSTATCOM), Static synchronous series compensator (SSSC), and Unified power flow controller (UPFC)[1][2].

DSTATCOM can be used to ensure that the current drawn from the distribution bus is balanced and sinusoidal in distribution system. In the open literature, conventional proportional-integral (PI) controller, and conventional proportional-integral-derivative controller (PID) controller, are some of the control schemes that have been proposed to control DSTATCOM. As it is well known, distribution system is typical nonlinear systems. In addition, the process of fine tuning of these controllers in such a highly nonlinear environment is a complex and challenging task. Also it should be mentioned that conventional PI and PID control schemes of DSTATCOM device which are used to maintain the bus bar voltage may not contribute significantly to improving system stability[3][4].

To overcome the previously mentioned drawbacks, this paper proposes an adaptive control strategy for a DSTATCOM based on PSO-PID Algorithm. The PSO-PID DSTATCOM controller exhibits adaptive system behaviors.

2   DSTATCOM Model

Fig. 1 shows the basic circuit diagram of the DSTATCOM system with lagging power factor loads connected to 1-phase 1-wire distribution system. DSTATCOM is connected in parallel with the system. The circuit model comprises a voltage source (with controllable magnitude and angle) connected to a power system in parallel through a series combination of resistance (R) and inductance (L). Here L represents the leakage reactance of the coupling transformer and R represents the equivalent resistance of the DSTATCOM system. In the distribution voltage level, the switching element is usually the integrated gate bipolar gate bipolar transistor (IGBT), due to its lower switching losses and reduced size. Moreover, the power rating of custom power devices is relatively low. Consequently, the output voltage control may be executed through the pulse width-modulation (PWM) switching method.

Under steady-state balanced three phase conditions, the total three phase real power and reactive power may be expressed in terms of $d - q$ quantities, where the $|v|$ and $|i|$ are the peak values of phase voltage and phase current, respectively, and β_i and β_v are the phase angels for phase voltage v_a and phase current i_a .

$$P = \frac{3}{2}|v\|i|\cos(\theta_v - \theta_i) = \frac{3}{2}(v_q i_q + v_d i_d) \tag{1}$$

$$Q = \frac{3}{2}|v\|i|\sin(\theta_v - \theta_i) = \frac{3}{2}(v_q i_d - v_d i_q) \tag{2}$$

Fig.1 depicts the active power flow and reactive power flow from the DSTATCOM to the distribution system through the PCC bus. The inverter output power $(P_e^* + jQ_e^*)$ must be equal to the sum of the power consumed by the

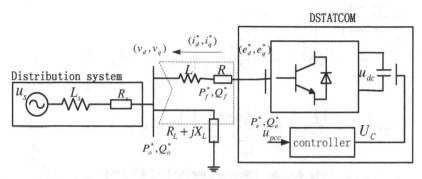

Fig. 1. Power-flow diagram of the DSTATCOM and power system

Coupling transformer and the filter $(P_f^* + jQ_f^*)$ and the power delivered to the distribution system and load $(P_o^* + jQ_o^*)$ which is defined as DSTATCOM output power. The power balance equations are as follows:

$$P_e^* = P_f^* + P_o^* \tag{3}$$

$$Q_e^* = Q_f^* + Q_o^* \tag{4}$$

Using (1) and (2), the inverter output active power and reactive power in $d-q$ axis are given by:

$$P_e^* = \frac{3}{2}(e_d^* i_d^* + e_q^* i_q^*) \tag{5}$$

$$Q_e^* = \frac{3}{2}(e_q^* i_d^* - e_d^* i_q^*) \tag{6}$$

Substitution of above formulas:

$$e_d^* = R i_d^* - \omega L i_q^* + |v| \tag{7}$$

$$e_q^* = R i_q^* + \omega L i_d^* \tag{8}$$

It is obvious from (7 and (8) that the output voltage commands of inverter, e_d^* and e_q^*, can be directly obtained from the current commands, the PCC bus voltage, and the coupling transformer and filter parameters [5].

3 Control Structure of DSTATCOM

Fig.2 shows the block diagram of the proposed direct output voltage PSO-PID control scheme. A reference voltage U_{pcc}^* is compared with actual measuring which

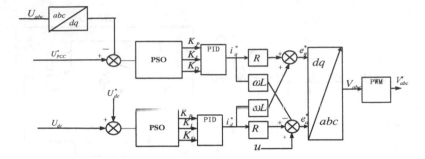

Fig. 2. Control Structure for the DSTATCOM

is expressed as U_{per}. This error signal, U_{per}, is processed in a PSO-PID controller and a reference reactive current i_q^* is obtained. Besides, a reference dc bus voltage U_{dc}^* is compared with sensed dc bus voltage U_{dc} of DSTATCOM, which results in a voltage error U_{der}. This error signal is also processed in a PSO-PID controller and a reference active current i_d^* is obtained. By the formulas (7) and (8), i_d^* and i_q^* are transmitted e_d^* and e_q^*, and then transform the $d-q$ voltage commands to abc voltage commands. After modulating the triangular carrier, which can generate pulse width modulation drive signal to control the intelligent power module action, and produce the required compensation voltage. Thus the voltage can maintain the dc capacitor voltage and stable points of common connection voltage.

4   PSO-PID Based Tuning of DSTATCOM

PSO shares many similarities with evolutionary computation techniques such as Genetic Algorithms (GA). The system is initialized with a population of random solutions and searches for optima by updating generations. However, unlike GA, PSO has no evolution operators such as crossover and mutation. In PSO, the potential solutions, called particles, fly through the problem space by following the current optimum particles.[6]

In Particle Swarm Optimization, Position of the particle represents the Results of problem. Each particle keeps track of its coordinates in the solution space which are associated with the best solution that has achieved so far by that particle. This value is called personal best, pbest. p_i, Another best value that is tracked by the PSO is the best value obtained so far by any particle in the neighborhood of that particle. This value is called gbest. p_g. Each individual is treated as a volume-less particle (a point) in the d-dimensional search space. The i th particle is presented as. $X_i(m) = (x_{i1}(m), x_{i2}(m), \cdots\cdots, x_{id}(m))$ The best previous position of the

i th particle is recorded and represented as $P_i = (p_{i1}, p_{i2}, \cdots, p_{id})$. The index of the best particle among all the particles in the population is represented by the symbol g. The rate of the position change for particle i is represented as $V_i = (v_{i1}, v_{i2}, \ldots, v_{id})$ .the particles are manipulated according to the following equation:

$$v_i(m+1) = \omega v_i(m) + c_1 r_1 (p_i(m) - x_i(m) + c_2 r_2 (p_g(m) - x_i(m)) \qquad (9)$$

$$x_i(m+1) = x_i(m) + v_i(m+1) \qquad (10)$$

Where c_1 and c_2 are two positive constants, and r_1 and r_2 are two random functions in the range [0,1].

The basic steps of PSO algorithm to solving this optimization are as follows:

Step 1: initialize the particle population randomly, there are three parameter k_p, k_i, k_d adopting real coding, ie the speed and position of all the particles in initial population. The value and error of parameter as velocity and position of the particle.

Step 2: Set PSO iterations, In each iteration process, evaluate the particles population according the fitness function. The fitness function is:

$$J = \min \sum_{j=1}^{N} \sum_{t=t0}^{T} \frac{1}{2} (|\Delta u(t)| + |\Delta u(t-1)|) * \Delta t \qquad (11)$$

Step 3: Update particle personal best p_i.

Step 4: Update particle gbest p_g.

Step5: Calculate velocity and location of the particle according formula (7) and (8)

Step6: When the terminal condition is coincident, it comes to end. Otherwise, return to step 2.

Step7: output PID gains K_P, K_I, K_D and then send these gains into the PID controller..

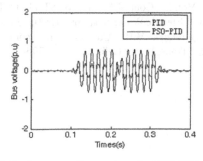

Fig. 3. Performance Comparison for RFA-PID and PID

5   Test System and Results

After the study of the DSTATCOM and PSO-PID, simulation of the DSTATCOM system has been performed using MATLAB and results are figre3, the performance of the PID tuned DSTATCOM controller and the PSO-PID based adaptive controller are compared with each other connected to it. A conclusion is founded that the PSO-PID DSTATCOM controller is performing far better than the traditional PID system.

6   Conclusion

This paper has presented the application of DSTATCOM to improve the power quality in a ship power system during and after pulsed loads. In addition, typical control strategy optimal parameters of PID controller are found by using immune genetic algorithm. The simulation results show that the voltage regulation at the point of common coupling is much better with a DSTATCOM. It is found that the PSO-PID tuned optimal controller based DSTATCOM is performing satisfactorily which establishes its effectiveness in ship power system.

Acknowledgment. This work is supported by Office of Science and Technology of Jilin Province(No. 20100307) and Office of Education of Jilin Province(No.2009241).

References

1. Samieni, S., Johnson, B.K., Hess, H.L., Law, J.D.: Modeling and Analysis of a Flywheel Energy Storage System for Voltage sag Correction. IEEE Transaction on Industry Applications 42(1), 42–52 (2006)
2. Domaschk, L.N., Ouroua, A., Hebner, R., Bowlin, O., Colson, W.B.: Coordination of large Pulsed Loads on Future Electric Ships. IEEE Transaction on Magnetics 43(1), 450–455 (2007)
3. Gupta, R., Ghosh, A.: Frequency-Domain Characterization of Sliding Mode Control of an Inverter Used in DSTATCOM Application. IEEE Transactions on Circuits and Systems—I: Regular Papers 53(3), 662–676 (2006)
4. Eldery, M.A., El-Saadany, E.F., Salama, M.M.A.: Sliding Mode Controller for Pulse Width Modulation Based DSTATCOM. In: Proceedings of Canadian Conference on Electrical and Computer Engineering, CCECE 2006, pp. 2216–2219 (2006)
5. Ou, J., Luo, A.: A Direct Voltage Control Method of STATCOM Based on Neuron Self-adaptive PID. Automation of Electric Power Systems 31(12), 77 (2007)
6. Ye, Y., Kazerani, M., Quintana, V.H.: Current-source converter based STATCOM: modeling and control. IEEE Trans. Power Delivery 20(2), 795–800 (2005)

Role-Based Access Control Technology for Digital Cultural Media Platform

Jingling Yuan, Jia Liu, Zhilong Liu, and Hongxia Xia

Computer Science and Technology School,
Wuhan University of Technology, Wuhan, 430070, China
liujiangel@yahoo.com.cn

Abstract. In order to ensure the access security of digital cultural heritage protection and dissemination and prevent illegal users from entering the system and the validated users causing the data leakage and damage, a kind of Role-Based Access Control (RBAC) technique for different applications of media platform is proposed in this paper. Firstly, the access control architect based on the unified authorization mechanism to realize a global management of user authorization is discussed. Then, RBAC model is applied to design and allocate different roles to control various users' permissions. It shows that this double-protection method realizes the access control and security management of the applications, services, information and resources in the digital cultural media platform.

Keywords: digital cultural heritage protection and dissemination, access security, Role-Based Access Control (RBAC), unified authorization.

1 Introduction

With the development of digital publishing, the construction of digital cultural heritage protection and dissemination platform is increasingly important to digital cultural heritage resource integration, creation, transformation, dissemination and appreciation. In the design and development process of digital cultural heritage protection and dissemination platform, the multifaceted security protection of the platform is the basis of ensuring it running safely and efficiently.

In 1989, the International Standardization Organization (ISO) put forward the hierarchical security architecture in the network security architecture standards (ISO 7498-2), and also defined five security service functions include: identification authentication service, access control service, data protection service, data integrity service, and non-repudiation service[1]. The access control service can not only limit access to key resources, but also prevent illegal users from entering the system and the validated users causing the data leak and damage.

Some common access control technology include Discretionary Access Control (DAC), Mandatory Access Control (MAC) and Role-Based Access Control (RBAC). DAC, also known as random access control, is a kind of access control defined by the

Trusted Computer System Evaluation Criteria as a means of restricting access to objects based on the identity of subjects and/or groups to which they belong. The controls are discretionary in the sense that a subject with certain access permission is capable of passing that permission (perhaps indirectly) on to any other subject. Illegal users are easy to avoid security protection provided by DAC to get access. DAC cannot provide adequate protection to system resources and resist Troy Trojan attack.

Compared with the DAC, MAC mechanism cannot be bypassed. In MAC, each user and document have been given a certain level of security, users can not change their security level or the access rights of any object by themselves, only the system administrators can determine the access permissions of users and groups. By comparing the user and security level of the files which he wants to access, the system determines whether the user can access the files. In addition, MAC does not allow a process to generate the shared files, thereby preventing the dissemination of information from one process to another through shared files. MAC uses sensitive labels to enforce security policies of users and resources, that is, do the mandatory access control.

RBAC is based on the users' operation in the system and the kind of operation indicates that the user has a corresponding role. When the user visits the system, his information must be checked firstly. A user can have several roles, and it's the same in reverse. The flexible strategy of RBAC is a kind of effective and safe measure. By defining the different parts of the model can implement desirable control strategy of DAC and MAC. Current research and application in this area is still in the experimental stage. George Mason University is a leader in this area, which has designed a non-root UNIX system management and a Web server management without centralized control and so on.

2   Role-Based Access Control Model

2.1   RBAC

Access control is the key technology to realize security access of the applications. RBAC is an effective and flexible security means. Resources is the object to be managed and controlled in the application system and it's the smallest unit of permission management. The abstract of the resources is determined by the actual requirements. Authorities is the key to determine whether a user can access specific resources. In RBAC, the role is a work or position of an organization or a task and it represents a kind of qualifications, rights and responsibilities, the user is the subject to access the resources of applications independently. The user often refers to a person who uses an application system. RBAC model has been widely used for its outstanding safety and operational.

As a most recognized effective method of access control, RBAC can be used to resolve access control of enterprises' and sharing platforms' resources at present. It has two obvious features as follows. One is it can reduce the complex of authorized management and management cost. The other is it can support different security strategies flexibly and adapt to various changes.

On the above advantages of RBAC, we mainly use RBAC in this paper.

2.2 RBAC Mechanism in the Digital Media Platform

Digital cultural heritage protection and dissemination platform includes unified authorized platform, access control model, content publishing platform, distribution management platform, content dissemination platform, knowledge sharing platform and content distribution platform and so on. The platform architecture is presented in Fig.1. Digital cultural heritage protection and dissemination platform provides released Web service interface based on the HTTP protocol for the client process to use[2]. Through digital cultural heritage protection and dissemination portal by means of various media forms, all kinds of users are managed by the unified authorized platform and RBAC mechanism in order to ensure the access control security.

The main function of the unified authorized platform is to authorize each user uniformly in order to realize the unified management of all kinds of users after integration of many application systems. Each level of the applications is based on the unified authorized platform is to realize the reasonable allocation of permissions, the centralized storage and unified management of user information.

The main function of access control model is to judge the legitimacy of the visit which the user issues. According to the judgment, the system allows the user to visit or sends an error message. Each level of the applications is based on RBAC management to protect the security of the website from the multi-angle and multi-level. At the same time, it provides a management tool. On one hand, this tool manages the users of the system, on the other hand, it provides an easy method for administrators to define the relationship of role, authority, operation and each element.

Content publishing platform, distribution management platform, content dissemination platform, knowledge sharing platform and content distribution platform are application platforms of digital cultural heritage protection and dissemination platform.

If a user wants to visit the digital cultural heritage protection and dissemination platform, he should go through the unified authorized platform firstly. On this platform, he can get a role allocated by administrators according to the application he wants to access. Then the user's information and the corresponding role will be checked in the RBAC model. Once the user gets the authority to access the application, he can visit it. Otherwise RBAC model returns error information.

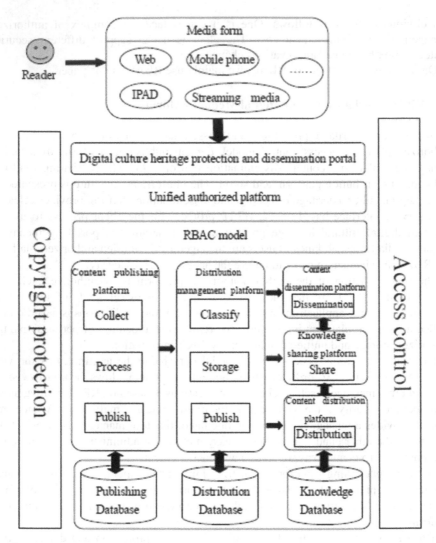

Fig. 1. The architecture of digital cultural heritage protection and dissemination platform

3   The Realization of Access Control

3.1   Unified Authorized Platform

In order to achieve the global management of user authorization, we need to achieve the global management of user information firstly. In other words, we need to construct unified authentication mechanism, which can realize centralized storage and unified management. In this way, all Web applications have a unified portal to achieve "single sign-on". At the same time, user information is consistent in each application, which helps make the user's identity fall into line in a different system and improves security.

The unified authorized platform includes five parts, database, management tool, user information self-management model, Web service interface and access control model. The database based on RBAC model is used to store access control information. Management tool and user information self-management model are both used to manage the data. Web service interface provides data access for each Web application. Access control model will be embedded in each Web application, so that users can access to pages according to the configuration of the access control. This platform can mount many applications (e.g. content publishing platform, distribution management platform and content distribution platform). It also provides RBAC management for each application and realizes cross-operating system running, cross-language information interaction and distribution, heterogeneous information reading. The process of this unified authorized platform is shown in Fig.2.

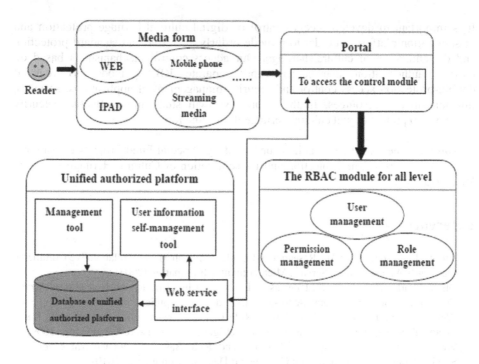

Fig. 2. Unified authorized platform

3.2   Role-Based Access Control

The roles of digital cultural heritage protection and dissemination platform are platform administrator, content administrator, disseminator, distributor, seller, reader and so on according to access and management requirements. The access control security of a system is achieved by user authentication and user authorization management. RBAC only allocates resources permissions to roles, in other words, role is a permissions set. Users can have one or more roles, but cannot allocate permissions directly to others. Authority of a role is relatively stable, while the user's is easy to change. RBAC increases the convenience and flexibility of permissions allocation and management by the separation of users and permissions.

It is necessary to build the RBAC model[3]. Users should get the corresponding role from identity authentication controller. After that, users should send the information including the current role, resource and operate permission to access control controller and ask for the decision about whether to submit access request to application server. System security administrator defines the role set, corresponding role and resources permission for each user, which is stored in role library and permission library. Role permission controller and management console provide role and its authorized management for system security administrators.

4   Conclusions

It is important to develop access control of digital cultural heritage protection and dissemination platform in order to provide orderly, effective and standard protection and dissemination of cultural heritage. Our access control mechanism is based on unified authorization management and separate RBAC model to realized double-protection access control and security management of applications, services, information and resources. In the future, we will study more effective security strategies to protect digital cultural heritage dissemination.

Acknowledgments. The research is supported by "Special Fund Projects of National Cultural Industries: The Protection and Dissemination of Cultural Heritage based on Visualization".

References

1. Fan, L.-L., Wang, S.-Y.: The Research and Implementation of The Collaborative Commerce Platform's Access Control Service. Microcomputer Information (2007)
2. Oh, S., Park, S.: Enterprise Model as A Basis of Administration on Role-Based Access Control. In: Cooperative Database Systems for Advanced Applications (2001)
3. Wen, Z., Zhou, B., Wu, D.: Three-Layers Role-Based Access Control Framework in Large Financial Web Systems. In: Computational Intelligence and Software Engineering (2009)
4. Wei, Y., Shi, C., Shao, W.: An Attribute and Role Based Access Control Model for Service-Oriented Environment. In: Control and Decision Conference (2010)

On Level-2 Condition Number
for Moore-Penrose Inverse

Shufan Wang*, Zhihui Ma, and Zizhen Li

School of Mathematics and Statistics, Lanzhou University,
Lanzhou, 730000, P.R. China
mazhh@lzu.edu.cn

Abstract. In this paper, we investigate the level-2 condition number for the Moore-Penrose inverse A^+ of the column-full rank matrix, and the condition number is in relation to F-norm of matrix. The paper is based on the results by Diao and Wei [16].

Keywords: Condition number, Level-2condition number, Linear least-squares problem, Moore-Penrose inverse, Perturbation.

1 Introduction

As for our topic, we first give the definition of Moore-Penrosegeneralized inverse A^+ [1,2] as follows.

1.1 Definition

Let $A \in C^{m \times n}$. If matrix $X \in C^{n \times m}$ satisfies

$$AXA = A, XAX = X, (AX)^H = AX, (XA)^H = XA,$$

then X is called the Moore-Penrose inverse of A, and such matrix X is unique.

In this paper, we adopt some notations as usual: A^+ denotes the Moore-Penrose inverse of matrix A; A^H denotes the conjugate transpose matrix of A. $\|\cdot\|_2$ be 2-norm; $\|\cdot\|_F$ Frobenius-norm.

Condition number measures the sensitivity of the output of a problem with respect to small perturbations of the input data [3]. The theory about condition number was given first by Rice [4] in 1966. It was applied to the condition number for computing the normal inverse A^{-1} of a nonsingular matrix A and solving the nonsingular linear system $Ax = b$ and their level-2 condition numbers by Higham [5] in 1995. The results in[5] were further extended to the condition numbers of the Moore-Penrose inverse A^+ and the linear least-squares problems and their level-2 condition numbers from some special case to the general case [7-13].

---

* The corresponding author: Zhihui Ma, Email_address: mazhh@lzu.edu.cn.

D. Jin and S. Lin (Eds.): Advances in ECWAC, Vol. 2, AISC 149, pp. 355–359.

In numerical analysis, condition number occur us a parameter in both complexity and round-off analysis and hence there is an obvious interest in their computation. In this way, a problem Φ induces a new problem, namely, the computation of its condition number $Cond_\Phi(d)$ for a given input data d. Generally, condition numbers cannot be computed exactly, so it is of interest to study the sensitivity of the problem about computing the condition number, that is, the condition number of the condition number, called the level-2 condition number. This level-2 condition number denoted by $Cond_\Phi^{[2]}$ was investigated by Demmel [14] and defined by

$$Cond_\Phi^{[2]} = \lim_{\varepsilon \to 0} \sup_{\|\Delta d\| \le \varepsilon \|d\|} \frac{\left|Cond_\Phi(d + \Delta d) - Cond_\Phi(d)\right|}{\varepsilon \left|Cond_\Phi(d)\right|} \tag{1.1}$$

Subsequently, Higham [5] studied the level-2 condition numbers for the normal inversion of nonsingular matrix and linear systems solver. Subsequent to the Higham's study, many math workers [11-13,15] have discussed the level-2 condition numbers for the Moore-Penrose inverse A^+ and the least-squares problems under some kinds of conditions satisfied by the matrix A.

In[16], Diao and Wei gave the expressions of Frobenius normwise condition number of the Moore-Penrose inverse A^+ and the generalized inverse A^+ solution of the least-squares problems of consistent linear system $Ax = b$, respectively,

$$Cond_F^+(A) = \frac{\left\|A^+\right\|_2^2 \left\|A\right\|_F}{\left\|A^+\right\|_F} \quad (1.2) \text{ and } Cond_F^{LS}(A,b) = \left\|A\right\|_F \left\|A^+\right\|_2 + \frac{\left\|A^+\right\|_2 \left\|b\right\|_2}{\left\|x\right\|_2} \tag{1.3}$$

In this paper, we investigate the condition numbers of the two condition numbers (1.2) and (1.3). Our results only apply into the Moore-Penrose inverse A^+ of the column-full rank matrix A and the least-squares problem of the full-column rank satisfying that the linear system is consistent. According to Definition (1.1), the two level-2 condition numbers that will be discussed are described as follows

$$Cond_F^{[2]}(A) = \lim_{\varepsilon \to 0} \sup_{\|\Delta A\|_F \le \varepsilon \|A\|_F} \frac{\left|Cond_F^+(A + \Delta A) - Cond_F^+(A)\right|}{\varepsilon Cond_F^+(A)} \tag{1.4}$$

$$Cond_F^{[2]}(A,b) = \lim_{\varepsilon \to 0} \sup_{\substack{\|\Delta A\|_F \le \varepsilon \|A\|_F \\ \|\Delta b\|_2 \le \varepsilon \|b\|_2}} \frac{\left|Cond_F^{LS}(A + \Delta A, b + \Delta b) - Cond_F^{LS}(A,b)\right|}{\varepsilon Cond_F^{LS}(A,b)} \tag{1.5}$$

The following lemmas will be used in this paper.

Lemma 1. [1,17]. Let $A \in C^{m \times n}$ with $rank(A) = n$. Then there existed $U \in C^{m \times m}$ and $V \in C^{n \times n}$ such that $A = U \begin{pmatrix} \Sigma \\ 0 \end{pmatrix} V$ and A^+ can be represented as $A^+ = V \begin{pmatrix} \Sigma^{-1} & 0 \end{pmatrix} U^H$ where $U = (u_1, u_2, \cdots, u_m) \in C^{m \times m}$ and $V = (v_1, v_2, \cdots, v_n) \in C^{n \times n}$ are two unities, $\Sigma = diag(\sigma_1, \sigma_2, \cdots, \sigma_n)$, $\sigma_1 \ge \sigma_2 \ge \cdots \ge \sigma_n > 0$ and σ_i are called the singular values of A.

Lemma 2. [18]. Let $A \in C^{m \times n}$ with $rank(A) = n$ and $rank(A + \Delta A) = rank(A)$, Then there exists a small enough number ε such that when $\|\Delta A\| \leq \varepsilon \|A\|$,

$$(A + \Delta A)^+ = A^+ - A^+ \Delta A A^+ + (A^H A)^{-1} \Delta A^H (I_m - AA^+) + O(\varepsilon^2). \quad (1.6)1.8$$

Lemma 3. [1,19]. Let $A \in C_r^{m \times n}$. $\{A_k\} \in C^{m \times n}$ is an matrix sequence and $A_k \to A$, then the necessary and sufficient condition of $A_k^+ \to A^+$ is $rank(A_k) = rank(A)$.

2   Level-2 Condition Number of the Moore-Penrose Inverse A^+

In this section, we study the condition number (1.4) and give its upper and lower bounds.

Theorem 1. Let $A \in C^{m \times n}$ with $rank(A) = n$ and $rank(A + \Delta A) = rank(A)$. Then the level-2 condition number (1.4) satisfies

$$\|A\|_F \|A^+\|_2 - 1 \leq Cond_F^{[2]}(A) \leq 1 + 6\|A\|_F \|A^+\|_2.$$

Proof. Suppose that $\Delta A \in C^{m \times n}$ and $\|\Delta A\|_F \leq \varepsilon \|A\|_F$. From Lemma2, we have

$$\|(A + \Delta A)^+\|_2 \leq \|A\|_2 (1 + 2\varepsilon \|A\|_F \|A^+\|_2) + O(\varepsilon^2), \quad \text{and}$$

$$\|(A + \Delta A)^+\|_F \geq \|A^+\|_F (1 - 2\varepsilon \|A\|_F \|A^+\|_2) + O(\varepsilon^2).$$

Hence we obtain $\|(A + \Delta A)^+\|_2^2 \leq \|A\|_2^2 (1 + 4\varepsilon \|A\|_F \|A^+\|_2) + O(\varepsilon^2)$ (1.7),   and

$$\frac{1}{\|(A + \Delta A)^+\|_F} \leq \frac{1}{\|A^+\|_F}(1 + 2\varepsilon \|A\|_F \|A^+\|_2) + O(\varepsilon^2) \quad (1.8).$$

Using $\|A + \Delta A\|_F \leq (1 + \varepsilon)\|A\|_F$ and the inequality (1.7), we can get

$$\|A + \Delta A\|_F \|(A + \Delta A)^+\|_2^2 \leq \|A\|_F \|A^+\|_2^2 (1 + 4\varepsilon \|A\|_F \|A^+\|_2 + \varepsilon) + O(\varepsilon^2) \quad (1.9).$$

Combining inequalities (1.8) and (1.9), we can easily obtain

$$Cond_F^+(A + \Delta A) \leq Cond_F^+(A)(1 + 6\varepsilon \|A\|_F \|A^+\|_2 + \varepsilon) + O(\varepsilon^2).$$

So $$\frac{Cond_F^+(A + \Delta A) - Cond_F^+(A)}{\varepsilon Cond_F^+(A)} \leq 1 + 6\|A\|_F \|A^+\|_2 + O(\varepsilon).$$

On the other hand, Similarly it is easy to show that

$$\frac{Cond_F^+(A + \Delta A) - Cond_F^+(A)}{\varepsilon Cond_F^+(A)} \geq -1 - 6\|A\|_F \|A^+\|_2 + O(\varepsilon).$$

According to the above inequalities and Definition (1.4) we obtain "$\leq$" in the Theorem.

Next, to obtain the lower bound, we choose a particular perturbation matrix ΔA. Let $\Delta A = -\varepsilon \|A\|_F u_n v_n^H$ where u_n and v_n are given by Lemma1, then

$\Delta A^H (I_m - AA^+) = 0$. From (1.6) and using $A^+ u_n = \dfrac{1}{\varsigma_n} v_n = \|A^+\|_2 v_n$, we know

$$\left\|(A+\Delta A)^+\right\|_2 \geq \left\|(A+\Delta A)^+ u_n\right\|_2 = \|A\|_2 (1+\varepsilon\|A\|_F \|A^+\|_2) + O(\varepsilon^2) \text{ , and}$$

$$\left\|(A+\Delta A)^+\right\|_F \leq \|A\|_F (1+\varepsilon\|A\|_F \|A^+\|_2) + O(\varepsilon^2) \text{ . Hence, combining with } \|A+\Delta A\|_F \geq$$

$(1-\varepsilon)\|A\|_F$, we obtain that $Cond_F^+ (A+\Delta A) = \dfrac{\|A+\Delta A\|_F \left\|(A+\Delta A)^+\right\|_2^2}{\left\|(A+\Delta A)^+\right\|_F} \geq Cond_F^+ (A)($

$1 + \|A\|_F \|A^+\|_2 - \varepsilon) + O(\varepsilon^2)$. This is implying that $\dfrac{Cond_F^+ (A+\Delta A) - Cond_F^+ (A)}{\varepsilon Cond_F^+ (A)} \geq$

$\|A\|_F \|A^+\|_2 - 1 + O(\varepsilon^2)$. Therefore, we have $Cond_F^{[2]} (A) \geq \|A\|_F \|A^+\|_2 - 1$. ∎

Acknowledgement. This work was supported by the National Natural Science Foundation of China (30970278, 30970291, 31100306) and the Fundamental Research Funds for the Central Universities (No. lzujbky-2011-48).

References

1. Wang, G., Wei, Y., Qiao, S.: Generalized Inverse: Theory and Computations. Science Press, Beijing (2004)
2. Stewart, G.W., Sun, J.: Matrix Perturbation Theory. Academic Press, NewYork (1990)
3. Higham, N.J.: Accuracy and stability of Numerical Algorithms, 2nd edn. SIAM (2002)
4. Rice, J.R.: A theory of condition. SIAM Journal on Numerical Analysis 3, 217–232 (1966)
5. Higham, D.J.: Condition numbers and their condition numbers. Linear Algebra Appl. 214, 193–213 (1995)
6. Golub, G.H., van Loan, C.F.: Matrix Computations, 3rd edn. Johns Hopkins University Press, Baltimore (1996)
7. Wei, Y., Wang, D.: Condition numbers and perturbation of the weighted Moore-Penrose inverse and weighted linear least squares problem. Appl. Math. Comput. 145, 45–58 (2003)
8. Xu, W., Wei, Y., Qiao, S.: Condition numbers for structured least squares. BIT 46, 203–225 (2006)
9. Gratton, S.: On the condition number of linear least squares problems in a weighted Frobenius norm. BIT 36, 523–530 (1996)
10. Wei, Y., Xu, W., Qiao, S., Diao, H.: Componentwise condition numbers for generalized matrix inversion and linear least squares. Numer. Math. J. Chinese Univ.(English Ser.) 14(3), 277–286 (2005)

11. Cucker, F., Diao, H., Wei, Y.: On the level-2 condition number for Moore-Penrose inversion (2005) (unpublished report)
12. Cucker, F., Diao, H., Wei, Y.: On mixed and componentwise condition numbers for Moore-Penrose inverse and linear least squares problems. Math. Comput. 76, 947–963 (2007)
13. Lin, L., Lu, T.-T., Wei, Y.: On level-2 condition number for the weighted Moore-Penrose inverse. Computers and Mathematics with Applications 55, 788–800 (2008)
14. Demmel, J.W.: On condition numbers and the distance to the nearest ill-posed problem. Numer. Math. 51, 251–289 (1987)
15. Cucker, F., Diao, H., Wei, Y.: Smoothed analysis of some condition numbers. Numer. Linear Algebra Appl. 13, 71–84 (2006)
16. Diao, H., Wei, Y.: On Frobenius normwise condition numbers for Moore-Penrose inverse and linear Least-Squares problems. Numer. Linear. Algebra. Appl. 14, 603–610 (2007)
17. Ben-Isral, A., Greville, T.N.E.: Generalized inverses:Theory and Applications, 2nd edn. Springer, New York (2002)
18. Wang, S.F., Zheng, B., Xiong, Z.P., Li, Z.Z.: The condition numbers for weighted Moore–Penrose inverse and weighted linear least squares problem. Appl. Math. Comput. 215, 197–205 (2009)
19. Wang, G.: Perturbation theory for weighted Moore-Penrose inverse. Comm. Appl. Math. Comput. Math. 1, 48–60 (1987)
20. Wei, Y., Ding, J.: Representations for Moore-Penrose inverse in Hilbert spaces. Appl. Math. Lett. 14, 599–604 (2001)
21. Wei, Y.: The representation and approximation for the weighted Moore-Penrose inverse in Hilbert space. Appl. Math. Comput. 136, 475–486 (2003)

Automatic Restoration Method Based on a Single Foggy Image

Huiying Dong[1], Dan Li[1], and Xinwei Wang[2]

[1] Information Science& Engineering College, Shenyang Ligong University , Shenyang, China
[2] Mechanical Engineering College, Shenyang Ligong University, Shenyang, China
huiyingdong@163.com

Abstract. The article introduces an automatic restoration method for images shot in foggy day. It is based on two-color atmospheric model. The method needs an image which is affected by the weather. The method also automatically calculates the brightness values of sky and the color direction of atmospheric light. Because the extraction from the sky area is more accurate and larger than the regional of the manual input, the brightness values of sky and the color direction of atmospheric light are got. They are closer to the actual values. This paper introduces this method of histogram adjustment for the restored image, and also introduces a method of adjusting the brightness based on the depths. The results of simulation show that the restored effect of this method is better for images which are degraded by weather under the environment of MATLAB.

Keywords: two-color atmospheric model, image restoration, automatic method, histogram adjustment, depth and brightness adjustment.

1 Introduction

The interference from weather exists generally in visual systems. Yet the current visual systems properly work only on sunny day, also the study of restoration for degraded images is not enough [1]. Due to lack of parameters, it is difficult to restore a single image in adverse weather to the contrast and color of the corresponding image in sunny day [2]. When the depth of a scene and the accurate information of atmospheric conditions are known, removing the weather's effects of an image has been shown to be feasible [3, 4]. By processing at least two images shot in adverse weather, it can automatically calculate the structure of the scene and also restore the color and contrast of the scene [5]. The restored methods based on a single foggy image don't have so many, such as interactive method [6]. When we calculate the brightness of sky and the color direction of atmospheric light, it need little amount of calculation during interactive computing, but it requires the operator to select the area of sky and need human's intervention. This can't get accurate values. In this paper, the method of automatic restoration is used to process single fog-degraded images. Because the area extracted from the sky is more accurate and larger than manually input, in the case of automatically getting the information, the value of the sky's brightness and the direction of atmospheric light are closer to actual value.

D. Jin and S. Lin (Eds.): Advances in ECWAC, Vol. 2, AISC 149, pp. 361–366.

2   Weather Interference Image Degradation Model

Degradation of images is formed by the degradation of imaging systems and some interference. If specially considering to the degradation of images caused by weather's interference, assuming that g(x, y) indicates a degraded image, f(x, y) indicates image without interfered by weather, n(x, y) indicates the disturbance of weather. The corresponding model of degradation can be expressed as:

$$g(x, y) = f(x, y) + n(x, y) \qquad (1)$$

In this paper, the described method is to remove n(x, y) from the function g(x, y) of degradation image, and estimates the ideal original image f(x, y). The method is based on physical models. It extends the degraded model of the weather interference in type (1). A method of correction for weather is a new method. Using this new method is to restore the degraded image.

3   Two Color Atmospheric Scattering Model

In the three-dimensional coordinate system, three alphabets R, G, B (representing the red, green and blue) represents the X, Y, Z axes. Two-color atmospheric scattering model is used to describe the color appearance of scene in low visibility conditions. Two-color atmospheric scattering model [7] shows that color $\vec{E}$ is the linear combination of color $\vec{A}$ and color $\vec{D}$. Color $\vec{E}$ is the color of point's pixel of scene point which is got by a color camera in fog or haze weather. Color $\vec{A}$ is the color of atmosphere light. Color $\vec{D}$ is the color of the scene point which is observed under a sunny day. It is shown in figure 1 below. In other words, $\vec{D}$, $\vec{A}$ and $\vec{E}$ are in the same two-color plane in color space. Form the mathematical point of view:

$$\vec{E} = \vec{D} + \vec{A} \qquad (2)$$

$\vec{D}$ and $\vec{A}$ can respectively be represented by using the unit vector $\hat{D}$ and $\hat{A}$. $\hat{D}$ is the direction of the color of scene point. $\hat{A}$ is the color's direction of atmosphere light.

$$D = s\hat{D}, \quad A = t\hat{A}, \quad p = \mathrm{Re}^{-\beta d}, \quad q = E_\infty(1 - e^{-\beta d}) \qquad (3)$$

In the type, s, t are scalars, s is the amplitude of direct transmission, t is the amplitude of atmosphere light. E_∞ is the brightness of sky; R is the radiation value of spots in a sunny day; $\beta \in (0,1)$ is the coefficient of atmosphere scattering; d is the depth of scene point.

From the following model, we can see the color of the scene decreased exponentially with the depth of the scene [8].

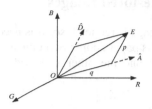

Fig. 1. This is two-color atmospheric scattering model

4   Automatic Method for Weather Degradation Image Restoration

By calculating the values, automatic method can obtain the required brightness value of sky and the direction of atmospheric light color.

4.1   Searching Algorithm of the Best Approximate Normal Distribution

First, according to the R, G, B color plane of color image, gray histograms are made respectively. Then following the steps of automatic method finds the optimum approximate normal distribution σ_R、 σ_G、 σ_B and obtains the average of μ_R、 μ_G、 μ_B .

4.2   Determination of Segmentation Threshold of Sky and Calculation of the Brightness of Sky

According to the nature of the normal distribution, when $\mu - 2\sigma \leq h \leq \mu + 2\sigma$, because its probability distribution accounts for 95% of the total distribution, so taking $[h_{R1}, h_{R2}]$、 $[h_{G1}, h_{G2}]$、 $[h_{B1}, h_{B2}]$ is the brightness level. Here, $h_{R1} = \mu - 2\sigma_R$, $h_{R2} = \mu + 2\sigma_R$, $h_{G1} = \mu - 2\sigma_G$, $h_{G2} = \mu + 2\sigma_G$, $h_{B1} = \mu - 2\sigma_B$, $h_{B2} = \mu + 2\sigma_B$. Taking $[h_1, h_2]$ is the gray distribution range of sky area, in which $h_1 = \max\{h_{R1}, h_{G1}, h_{B1}\}$ $h_2 = \max\{h_{R2}, h_{G2}, h_{B2}\}$. $\overline{E}_R, \overline{E}_G, \overline{E}_B$ are the mean values of the brightness of the sky's area in R, G, B image plane. Taking above equations into the following formula:

$$E_\infty = \sqrt{\overline{E}_R^2 + \overline{E}_G^2 + \overline{E}_B^2}\qquad(4)$$

We can calculate the sky brightness values E_∞ . Then we use the formula:

$$\hat{A} = \frac{\vec{E}}{E_\infty}\qquad(5)$$

We can calculate the direction of atmospheric light color $\hat{A}$.

5 Enhancement for Restored Images

The brightness of restored image is reduced, so, it is need to adjust image's brightness to enhance image quality. This paper adopts two enhancement methods of image. One is the histogram adjustment, and the other is based on the depth of enhancement image.

5.1 Histogram Adjustment

The brightness of the restored image is reduced. Adopting the brightness of histogram adjustment method makes the image closer to the actual image of scene under sunshine. Histogram is used to illustrate the images' brightness or the distribution of gray-scale graphics in image analysis. Each branch of histograms shows corresponding brightness or frequency of gray-scale (i.e., the ratio of appear brightness level pixel number and the total number of pixels). In order to improve the brightness, sub-brightness adjustment method is used to make the dynamic range of dark parts of the original image greatly increased, so that the effects of image after recovery is closer to the actual scene image under sunny.

5.2 Enhancing Image Based on Depth

According the brightness of the scene declines exponentially with the depth of the scene point, we present a depth-based image enhancement method. Establishing the adjustment function of depth and brightness:

$$E = (m_1 + \frac{m_2}{1+e^{-m_3 d}})E'$$

(6)

where d is the depth of scene points; E ' is the brightness of scene point in the original image; m1 is the parameter which determine the adjustment of whole brightness; m2 is the parameter which determine the brightness adjustment range.

Define the brightness adjustment coefficient as:

$$k = m_1 + \frac{m_2}{1+e^{-m_3 d}}$$

(7)

Where m3 is the parameter, it also determines the change speed of brightness adjustment coefficient k along with the depth d. Using the brightness adjustment function (equation (6)) to adjust the restored image, the effect of the image after the rehabilitation is closer to the actual scene image under sunny.

6 Simulation Results

In order to verify the automatic defogging method of eliminating the weather effect, the authors do the simulation experiment in the MATLAB environment. Figure 2 adopts automatic method to estimate the brightness of sky and adopts depth heuristic method to estimate the depth of the scene to recover the figure (a) (β=0.6). Then restored images were treated with histogram adjustment with different value of γ. Figure 3 adopts automatic method to estimate the sky brightness and adopts depth chart method to estimate the depth of the scene to recover the figure 2 (β=0.5).Then restored images were treated with histogram adjustment with different value of γ. Figure4 adopts automatic method to estimate the sky brightness, and adopts depth heuristic method to estimate the depth of the scene to recover the figure (a) (β=0.5). Then restored images were treated to adjust the depth and brightness of restored image. Figure 5 adopts automatic method to estimate the sky brightness and adopts depth chart method to estimate the depth of the scene to recover the figure (a) (β=0.5). Then restored images were treated to adjust the depth and brightness of restored image. Because the extraction from sky area is larger and more accurate than the manual input of the regional, getting the sky brightness values and atmospheric optical color direction are closer to the actual value. Automatic method is better in defogging.

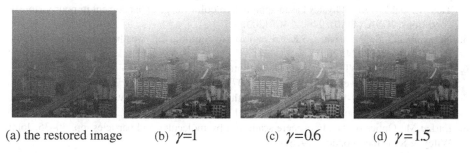

(a) the restored image   (b) $\gamma=1$ (c) $\gamma=0.6$ (d) $\gamma=1.5$

Fig. 2. Automatically calculate the sky brightness, depth of heuristics, the restored image histogram adjustment method

(a) the restored image   (b) $\gamma=1$ (c) $\gamma=0.6$ (d) $\gamma=1.5$

Fig. 3. Automatically calculate the sky brightness, depth map method, the restored image histogram adjustment method

Fig. 4. Automatically calculate the sky brightness, depth of heuristics, the restored image histogram adjustment method

Fig. 5. Automatically calculate the sky brightness, depth map method, the restored image histogram adjustment method

7 Conclusions

Atmospheric model based on two-color automatic method requires an image affected by the weather and automatically calculate the values of sky brightness and atmospheric light color direction then restore the image. The brightness of restored images is reduced and it needs to be enhanced to close the actual scene image under sunny. MATLAB simulation results show that this method is simple, effective and recovery effect is obvious. Because the extraction from sky area is larger and more accurate than the manual input of the regional, getting the brightness values of sky and atmospheric optical color direction are closer to the actual value. In future studies, effective selection input area can reduce the amount of computation and more effectively improve the image restoration results.

References

1. Narasimhan, S.G., Nayar, S.K.: Chromatic framework for vision in bad weather. In: Proceedings of the IEEE Conference on Computer Vision and Pattern Recognition, South Carolina, pp. 1598–1605 (2000)
2. Dong, H.-Y., Fang, S., Xu, X.-H., Wang, X.-W., et al.: A Method Based on Physical Model to Restore Degraded WeatherImages and Its Application. Journal of Northeastern University 26(3), 217–219 (2005)
3. Oakly, J.P., Satherly, B.L.: Improving images quality in poor visibility conditions using a physical model for degradation. IEEE Trans. on Image Processing 7(2), 167–179 (1998)
4. Tan, K., Oakley, J.P.: Physics based approach to color image enhancement in poor visibility conditions. JOSAA 8(10), 2460–2467 (2001)
5. Narasimhan, S.G., Nayar, S.K.: Contrast restoration of weather degraded images. PAMI 25(6), 713–724 (2003)
6. Ouyang, J.-Q., Li, J.-T., Zhang, Y.-D.: An Interactive Key Video Object Selection Model. J. Infrared Millim. Waves 18(3), 547–554 (2007)
7. Zhang, L.-J., Fu, Z.-L.: An image restoration method for unknown degradation model. Computer Applications 21(1), 144–148 (2001)
8. Earl, J.M.: Optics of the atmosphere scattering by molecules and particles, pp. 30–35. John Wiley & Sons Inc., London (1976)

Experiment Design of Coupled Water Tanks Based on Network Control System-4000

Lifeng Wei[1], Yuzhang Li[1,2], Xiaomei Liu[1,3], and Mingzhe Yuan[2]

[1] College of Information Engineering, Shenyang University of Chemical Technology
Shenyang, Liaoning, China
[2] Shen Yang Institute of Automation Chinese Academy of Sciences
Shenyang, Liaoning, China
weilifeng62@sina.com, li_yuzhang@126.com

Abstract. The liquid level accurate control of multiple water tanks system has become a research focus due to the nonlinear process and multiple variables. The experimental platform of multiple water tanks system is designed based on the network control systems-4000 (NCS4000) in order to study and implement the advanced PID control for this system. The mathematical model of the system is established. Finally the Lyapunov approach is used to design the PID controller for the multiple water tanks system. Therefore the experiments proved the effectively and accuracy of the PID control method for the liquid level control of platform.

Keywords: Process control, Level control, PID algorithm, Coupled water tanks.

1   Introduction

With the wide application of information and automation technology in the industry process domain, process control system plays an increasingly important role in the process industry. In process control, Level, temperature, pressure, flow are the four most popular parameters, among which, the liquid level control technology is highly significant in the process control system. Therefore advanced level control methods and strategies are sorely needed to ensure the safe, reasonable and effective production. In [1], a tuning method was proposed to design the PID controller through the Lyapunov approach. In [2], a new design method of the robust PID controller was proposed via a new interval polynomial robust stable criterion. In [3], a sliding mode controller was designed for a water tank liquid level control system to realize the level position regular and tracking control. In [4], a novel PID controller based on improved polyclonal selection algorithm and RBF network is put forward to realize the adaptive control of liquid level of a 2-container water tank system. In [5], a novel intelligent design method for PID controller with optimal self-tuning parameters is proposed based on the modified Ant Colony System (ACS) algorithm.

In order to research the process control and the level control more effectively, a distributed control system (DCS), network control system-4000(NCS4000), is used to

D. Jin and S. Lin (Eds.): Advances in ECWAC, Vol. 2, AISC 149, pp. 367–373.

realize design and operation on the experimental platform system of coupled water tank. Control results indicate that the NCS4000 can control the main parameters of the experimental platform precisely. Consequently, this control system is practical, reliable and effective in the process control and level control and raises the precision of process control and level control .

2   NCS4000 Summarize

NCS4000 control system is a new distributed control system ,which is researched by Shen Yang Institute of Automation Chinese Academy of Sciences and Shen Yang Zhong ke bo micro automation technology Co., LTD, including the man-machine interface, control station, locale equipment, applies to medium and large industrial control environment. It provides process control, logic control and batch control function. It has be widely used in metallurgy, cement, coking, sewage treatment, petroleum, chemical industry, etc.

NCS4000 has the following characteristics: support medium and large-scale control system applied, the I/O connection capacity of the 16384 points; support FF, PROFIBUS DP, AS-I, HART domainbus standards and so on; support controller redundancy, network redundancy, I/O modules redundancy, power supply redundancy, etc redundancy method; support management, diagnosis and maintenance function of equipment; support distributed configuration, offline configuration, simulation, etc; support standard of IEC61131-3 programming; support standard of OPC; Integration of advanced process control technology, and provide industry solutions.

NCS4000 consists of control station, workstations, locale devices and control network. There are 32 control stations at most in a NCS4000.The workstations consist of several industrial control computers, quantity of which is not strictly limited. Its structure is shown in **Fig.1**.

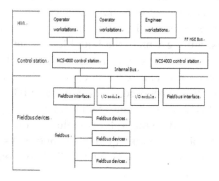

Fig. 1. NCS4000 structure

3 Experiment Design of Coupled Water Tanks

3.1 Component of Level Control System

The control system is the multi-functional experimental platform, which is integrated of measuring liquid level, flow, temperature parameters and so on. It is composed of water tank, electric control unit, detection unit, execution unit, manual valve, pump and a few parts, etc. System structure is shown in **Fig.2**.

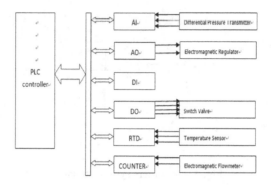

Fig. 2. Experimental platform structure of coupled water tank

(1) Water tank mainly includes three single water tanks, one sink and some water pipes connected with water tank and sink.
(2) Electric control unit is the core part of the platform, including PLC, analog input/output module (AI/AO), digital input/output module (DI/DO), bus module, counter module, power modules, relay protection devices, etc.
(3) The detection unit is consisted of the pressure sensor and the temperature sensor in the bottom of the each water tank, and the flow sensor in the pipeline. Then tested data will be transmitted to the electrical control unit through the fieldbus, as the feedback of the closed-loop system.
(4) Execution units are two electromagnetic regulators whose jawopening are controlled, their function is to control jawopening of the fluid flow between water tank and sink to the flow of fluid, and executive signal is issued by PLC with the D/A converter.
(5) Manual valve's main function is to control the flow between water tank and water tank manually, and realize the transformation between the one order system and the orders system.
(6) Pumps mainly provide power for water cycle between water tank and sink.

3.2 Establishing the Mathematical Model for the System

Control system is the level control system of coupled water tank, through changing to the manual valve , one order or two orders system can be set up. However, in order to test the experimental platform performance in an even better fashion, two orders control system is selected to establish the system model. System structure diagram is shown in **Fig.3** below

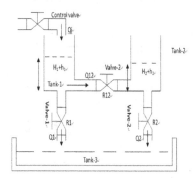

Fig. 3. Two orders system structure

It assume that Qin is the quality of the water flowing into the tank in unit of time, the Q1, Q2 and Q12 is respectively the quality of the water flowing out each tank in unit of time, the R1, R2 and R3 is respectively the liquid resistance among the first water tank, the second water tank and the pipelines, the C1, C2 andC3 is respectively the cross sectional area of each tank. So, we should keep to the following material balance principle as for each tank.

$$C\frac{dh}{dt} = Q_{in} - Q_{out}$$

According to the above relations and fluid mechanics knowledge, we can get:

$$\begin{cases} Q_{in} - Q_1 - Q_2 = C_1\frac{dh_1}{dt} \\ Q_{12} - Q_2 = C_2\frac{dh_2}{dt} \\ Q_1 = \frac{h_1}{R_1} \\ Q_2 = \frac{h_2}{R_2} \\ Q_{12} = \frac{h_1 - h_2}{R_{12}} \end{cases} \tag{1}$$

Then we can solve the transfer function G(s) of the system between the input Qin and output h1, and transform the equations (1) through the Laplace, at last we eliminate intermediate variables, we can get:

$$G(s) = \frac{H(s)}{F_{in}(s)} = \frac{b_1 s + b_2}{a_0 s^2 + a_1 s + a_2} = \frac{b_1 s/a_0 + b_2/a_0}{s^2 + a_1 s/a_0 + a_2/a_0} \tag{2}$$

Each parameter in equation (2) is:

$$b_1 = C_1 R_1 R_{12}^3,$$
$$b_2 = R_1 R_{12}^2,$$
$$a_0 = C_1^2 R_1 R_{12}^3,$$
$$a_1 = 2C_1 R_1 R_{12}^2,$$
$$a_2 = C_1 R_{12}^3 + R_{12}^2 + R_1 R_{12} - 1.$$

The mathematical model of the system can be obtained through the step response curve, and **Fig.4** shows open-loop step response curve of the liquid level h1 and h2, when jawopening of valve 1 and valve 2 are respectively in 75% and 30%.

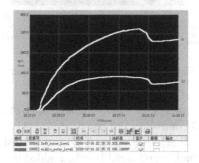

Fig. 4. The open-loop step response curve

Because the input signal u is given, and the output signal Y can be measured, then the mathematical model of the system can be calculate through system identification function ARX() of the MATLAB, equation(3) is get.

$$G_p(s) = \frac{B(s)}{A(s)} = \frac{0.0627s + 0.06024}{s^2 + 1.248s + 0.02923} \tag{3}$$

3.3   Controller Design

For control system, a stable controller is very important, so we choose PID controller, which is widely used in various and large control domain. According to the performance indexes of the system and requirements of constraint conditions(Given value $h_l \leq 350mm$, $tr \leq 500s$, $\sigma\% \leq 5\%$), and combined with Lyapunov method to determine the parameters, therefore, we can get:

$$K_p = 4.6999; K_i = 0.1387; K_d = 2.0005$$

4   The Experimental Results

Using the above method in the experimental platform of coupled tanks, we may obtain the control curve following Fig 5.

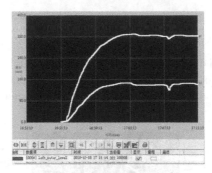

Fig. 5. Curves of the controller output

Fig.5 is shown that the controlled variable can achieve the given value quickly and steadily through using PID controller, and rise time is $tr=410s$, and system overshoot is $\sigma\% \leq 5\%$, they don't only all achieve index of system control, but also the system can recover the steady state quickly and steadily when added the disturbance of the valve changed.

5   Conclusion

The NCS-4000 not only operates simple, but also easy switch system orders. We establish the mathematical model of the liquid level control system for coupled water tanks, and Lyapunov approach are used to design PID controller for this liquid level system. It lays the good foundation for engineer to develop and propose complex control strategy, and meanwhile it provides a good teaching and research platform for students and teachers in university.

Acknowledgements. The authors would like to thank Dr. Li Xianhong for his assistances and recommendations.

References

[1] Li, X., Yu, H., Yuan, M.: Design of an Optimal PID Controller Based on Lyapunov Approach. In: Proceedings of International Conference on Information Engineering and Computer Sciences, December 19-20, pp. 1–5 (2009)
[2] Li, X., Yu, H., Yuan, M., Wang, J.: Design of robust optimal proportional–integral–derivative controller based on new interval polynomial stability criterion and Lyapunov theorem in the multiple parameters' perturbations circumstance. IET Control Theory and Application 4(11), 2427–2440 (2010)

[3] Chen, H.-M., Chen, Z.-Y., Su, J.-P.: Design of a Sliding Mode Controller for a Water Tank Liquid Level Control System. In: Proceedings of the Second International Conference on Innovative Computing, Information and Control, September 5-7, p. 335 (2007)

[4] Zuo, X.: Liquid level control of water tank system based on improved polyclonal selection algorithm and RBF network. In: International Conference on Computer Engineering and Technology (ICCET), April 16-18, vol. 2, pp. V2-528–V2-532 (2010)

[5] Zeng, Q., Tan, G.: Optimal Design of PID Controller Using Modified Ant Colony System Algorithm. In: Proceedings of 3th International Conference on Natural Computation, August 24-27, vol. 5, pp. 436–440 (2007)

[6] Tong, M.D.: Linear system theory and design. University of Science and Technology of China press (2004) (in Chinese)

Robotic Fish Wireless Communication and Positioning System Design Based on ZigBee

Qingsong Hu[1], Shuxin Zhou[1], and Lihong Xu[2]

[1] Engineering College, Shanghai Ocean University
999# Huchenghuan Rd., Shanghai 201306, China
[2] Control Science and Engineering College, Tongji University
1239# Siping Rd. Shanghai 200092, China
qshu@shou.edu.cn, 46590594@qq.com, xulhk@163.com

Abstract. Communication and positioning system are two basic functions for the robotic fish to be applied in the aquatic plant water environment supervision. A system based on ZigBee technology is constructed which can achieve the communication and positioning function simultaneously. The System hardware design and program debugging process are introduced. The practical experiment in the aquatic plant shows the system is valid and meets the design goal. The result indicates that ZigBee system is a appropriate technology to be utilized to the robotic fish in the aquiculture area.

Keywords: ZigBee, communication, positioning, robotic fish.

1 Introduction

Significant achievements have been witnessed in recent years in the development of aquatic robots, seeing [1-4]. Intrigued by the remarkable feats in biological swimming and driven by the desire to mimic such capabilities, extensive theoretical, experimental, and computational research has been conducted to understand hydrodynamic propulsion and maneuvering. Robotic fish has various applications ranging from military operations to ocean sampling as well as pipe inspection [5]. The robotic fish designed in Shanghai Ocean University (SHOU) is as Fig. 1. One of the key factors to realize these goals is the communication and positioning of the robotic fish in water.

There are several possible techniques to be selected to achieve this goal such as GPS, ZigBee/IEEE802.15.4, RFID, WIFI (IEEE 802.11), Ultra Sound, Infrared light etc., which vary greatly both in the function and the cost [6]. Olivetti Research Institute (AT&T Cambridge Research Institute at present) has designed the active tag based on infrared ray which can be applied to position. Its drawback is the tag and reader has to be aligned. Positioning system can be constructed based on the WIFI. Several base stations are built in certain area. According to the time and strength that the base stations receive from the objective and considering the topology structure of the base stations, the moving objective can be located. However, it is not suitable for robotic fish. Ultra Sound technology can position precisely, while it is too expensive.

D. Jin and S. Lin (Eds.): Advances in ECWAC, Vol. 2, AISC 149, pp. 375–380.

RFID is not a mature technology for now. ZigBee technology proposed by TI (Texas Instrument company) can realize the communication and positioning simultaneously with a 0.25 meter of resolution [7-8]. According to the robotic fish communication distance and low power energy requirement, Zigbee is appropriate selection for the moving robotic fish in water.

Fig. 1. Robotic fish designed in SHOU

2   General Structure of the ZigBee Communication and Positioning System

TI declared the first SoC system CC2431 with the positioning hardware on board to meet the low power cost wireless sensor network requirement. The product from Chipcon can meet multiple requirements including objective following, patient supervision, long-distance control and safe supervision etc.

According to the signal receiving strength and the reference nodes, CC2431 can compute the moving node location based on RSSI (Received Signal Strength Indicator). RSSI value will decrease with the distance increasing. There are two type of nodes in the CC2431/ZigBee wireless positioning system. That is Blind node (moving node) and Reference node. Reference node is static and its coordinating value (X, Y) is unvaried. A positioning area generally composes of 8 reference nodes. Reference node can be carried out by CC2430 or CC2431. Blind node is a type of moving node and can move freely in the positioning area. Blind node will compute the coordinate by the positioning algorithm after received the RSSI value of all the reference nodes. The blind node has to be carried out by the CC2431 microchip. The smallest CC2431/ZigBee wireless network positioning system includes 1 blind node and 3 reference nodes.

The ZigBee positioning system is composed of 4 parts including supervision section, gateway, reference node and blind node.

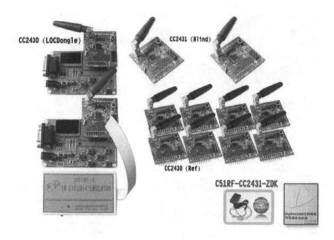

Fig. 2. Hardware utilized in the construction and debugging of the ZigBee wireless positioning system

The hardware applied in the system construction and debugging is as Fig.2 and includes USB online simulator, developing board, ZigBee wireless high frequency module (CC2430 and CC2431), 2.4G antenna, power source, battery board, RS232 connecting line. The developing board includes LCD display, small keyboard, sensor interface, CC2430 interface, variable resistance, LED, JTAG simulator interface, power interface, RS232. The evaluation software to debug the board is also provided. Battery board includes ZigBee module interface, JTAG simulator interface, resetting button, battery box. The board mainly provides power or the interface to download program to the ZigBee module. The CPU includes 128K Flash and 8K RAM, strengthening 8051 core, 8 input ADC and DMA processor, AEC-128 safe co-processor.

3 Program Design and Debugging

IAR Embedded Workbench (EW) is a compiling and debugging system supporting C/C++. Setting up the network is basic for the whole system. Take serial communication as example to introduce the networking construction process, seeing Fig.3. Start a network coordinator. If the network is built successfully, the LCD will display the coordinator and network ID No. Then open a reference node power and this node will join to the network automatically. After join successfully, the node will display the network address of itself and its father node, then the binding can begin. If the node needs to bind with the coordinator, just press UP button and press the module button to bind. If the G of the two-side lights which means the binding is successful, then the serial communication can be conducted. Connect the module to the computer and configure the baud rate as 38400 without parity check and one stop bit. Now send data from one module and same data will be received from the other module.

Fig. 3. IAR Embedded Workbench serial communication program debugging interface

Node address assign is critical for the system construction. First recover the all the node address as default value. Every node in the system can only connect with one gateway (as the coordinator). After retreat from the debugging state, configure the 64 bit IEEE address of the gateway. Open *SmartRF04 Flash Programmer* is as Fig.4. When the system detects the gateway node, click *Read IEEE* to get the default IEEE address, that is *0xFFFFFFFFFFFFFFFF*. Change the value as a different address with other nodes and click *Write IEEE*.

The computer side control interface is as Fig.5. Map drawing is an import part to reflect the blind node position. The coordinate of reference node should be decided in advance in the map which is important for the position precision. Fig.5 is a configuration example when the system is debugged in the laboratory.

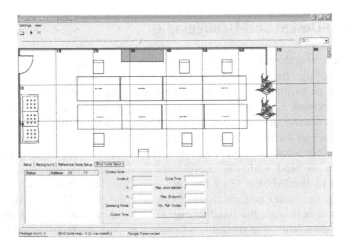

Fig. 5. Map drawing in the Z Location Engine CC2431

4   Practical Experiment

The experiment was conducted in the aquatic plants of Qingpu district, Shanghai, seeing Fig.6(a). The aquatic plant is 40 meters wide and 60 meters long. 7 reference nodes are set as Fig.6(b). The rods to lift the nodes are as Fig.6(c). Reference node should set up according to the practical distance. More precise the distance between the nodes, the positioning will be more precise. The reference node has to adapt to the Z Location Engine and guarantee the position in the software and in practical. In Fig.6(a), the distance between nodes is 20 meters.

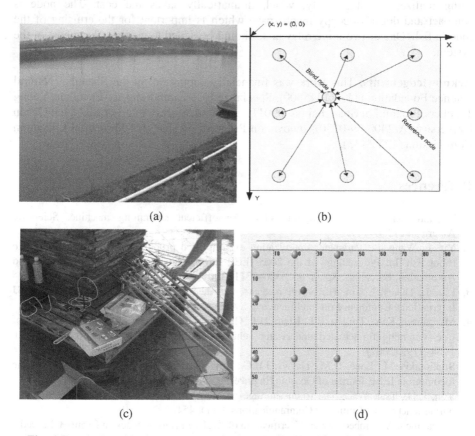

(a)　　　　　　　　　　　　　　　(b)

(c)　　　　　　　　　　　　　　　(d)

Fig. 6. Practical positioning system test in the aquatic plant of Qingpu district, Shanghai

The result is shown in Fig.6(d). The green point stands for the blind node on the robotic fish. The precision reaches less than 1 meter. The temperature sensor data is transmitted to the computer which means the signal transmission is successful.

Some problems also found in the test process. The max distance between the reference gate way node and the control computer varied with the weather. It will be short when rain. The blind node in the robotic fish requires long antenna to match the affection of the water. Relay node is required to guarantee the transmission between the system and control center.

5 Conclusion

A communication and positioning system based on ZigBee is successfully constructed. The application to the robotic fish in the aquatic plant shows it is valid in appropriate situation. Compared with the GPS, WIFI etc., it can be applied in the indoor or other special environment with the communication and positioning function being realized simultaneously, which dramatically saves the cost. The node is compact and doesn't occupy much space which is important for the cruising of the robotic fish. One or two more relay node should be added to increase the robust of the system.

Acknowledgements. This work was financially supported by the Shanghai Natural Science Foundation (11ZR1415600), Specialized Research Fund for Excellent Young Teachers Program of Shanghai (ssc09011), and Phd. Startup Fund of Shanghai Ocean University (A-2400-09-0150), Innovation Program of Shanghai Municipal Education Commission (12YZ133).

References

1. Triantafyllou, M.S., Triantafyllou, G.S.: An efficient swimming machine. Scientific American 272, 64 (1995)
2. Yu, J., Wang, L.: Parameter optimization of simplified propulsive model for biomimetic robot fish. In: Proceedings of the 2005 IEEE International Conference on Robotics and Automation, Barcelona, Spain, pp. 3306–3311 (2005)
3. Liu, J., Dukes, I., Hu, H.: Novel mechatronics design for a robotic fish. In: IEEE/RSJ International Conference on Intelligent and Systems, pp. 2077–2082 (2005)
4. Morgansen, K.A., Triplett, B.I., Klein, D.J.: Geometric methods for modeling and control of free-swimming fin-actuated underwater vehicles. IEEE Transactions on Robotics 23(6), 1184–1199 (2007)
5. Sfakiotakis, M., Lane, D.M., Davies, J.B.C.: Review of fish swimming modes for aquatic locomotion. IEEE Journal of Oceanic Engineering 24(2), 237–252 (1999)
6. Yen, L.H., Tsai, W.T.: The room shortage problem of tree-based ZigBee/IEEE 802.15.4 wireless networks. Computer Communications 33(1), 454–462 (2010)
7. Gawanmeh, A.: Embedding and Verification of ZigBee Protocol Stack in Event-B. Procedia Computer Science 5, 736–741 (2011)
8. Huircán, J.I., Muñoz, C., et al.: ZigBee-based wireless sensor network localization for cattle monitoring in grazing fields. Computers and Electronics in Agriculture 74(2), 258–264 (2010)

Some of the Problems of Wireless Transmission and the Solutions

Zhenjie Chen, Chao Zhang, and Wei Chen

Electrical Engineering & Renewable Engrgy School
China Three Gorges University
443002, Yichang, China
{Zhenjie Chen, Wei chen,Chao Zhang}heilang005@163.com

Abstract. According to the characteristics of wireless transmission technology we choose a pair of wireless transmitter and receiver module, and determine the wireless transmission scheme, and test the stationary state for the selection module. Through the experiments we observe the impact of pulse waveform frequency and duty cycle on wireless transmission, and compare the effect of wireless transmission and error of law, we have analyzed the sources of error, and finally made a number of measures to reduce the error.

Keywords: wireless transmission technology, pulse waveform, frequency, duty cycle.

1 Introduction

Brushless excitation adjustment method has been applied to many of the synchronous motors, and wireless transmission technology is also of great significance for the application of brushless excitation regulator. Wireless data transmission technology has good characteristics with low-power, low-cost, no direct connection, etc. In some places where cannot be directly connected, such as high temperature, highly corrosive environment, mobile devices, etc., the traditional wired data transmission cannot meet the need and cannot be used. But the wireless data transmission technology is not subject to these limitations by space or the external environment, showing its unique advantages and convenience. RF technology is a kind of the wireless data transmission, and it usually selects a frequency in a few specific frequencies to transfer data, and the selection of frequency can be controlled by programming .With the rapid development of integrated circuits, a simple wireless communications module formed by the RF transceiver chip, micro-controllers and peripheral devices can achieve and meet the general short-range wireless communication systems. In recent years, along with communications technology, integrated electronics and computer technology, radio frequency technology at home and abroad also tends to mature and has developed the RF transmission chip used in different areas. With wireless technology, wireless RF transmission chip size has been doing more and more small, and function is also very rich. Transmission distance is farer and the performance is more excellent and signal transmission is more stable and transfer rate is faster with appropriate use of auxiliary

D. Jin and S. Lin (Eds.): Advances in ECWAC, Vol. 2, AISC 149, pp. 381–387.
springerlink.com

components, and it is particularly suitable for all kinds of industrial control applications.

1.1 Wireless Transmission Technology Experiment

The main subject of study is the transmission between the stationary and rotating parts for the chopper drive signal by the wireless transmitter and receiver modules. We select NT-T02A and NT-R03A as the wireless transmitter and receiver module, and supply two 6V-batteries to the two modules (two modules are stationary, placed some distance apart. We programmed to MCU and let it output a frequency of 800Hz, pulse width 50% duty cycle waveform, and then make it access to the wireless transmitter module data input. At the same time, we observed waveform from wireless receiver module output with an oscilloscope and compared with the microcontroller input.   In the experiment, we see that when the MCU sends 800Hz, 50% duty cycle pulse waveform, the frequency of the waveform has not changed in the transmission process and very accurately transmits to the receiver module. The duty cycle will cause smaller changes during the transmission and there is a certain error. There are some certain but small errors in the transmission. This is consistent with the program options previously. As the pulse waveform duty cycle is equal to the conduction time and cycle ratio of the switching element of the chopper, so in fact the duty cycle is directly related to the regulation of excitation current and it needs to keep process as accurate as possible in the transmission. The following we do the wireless transmission experiment about pulse waveform of the different parameters and observe its effect ion of transmission.

1.2 The Affection of the Pulse Waveform's Duty Cycle on Wireless Transmission

Keep the frequency of 800Hz and pregame the MCU to let it output the different duty cycle pulse waveform for wireless transmission, and observe the output of the wireless receiver module with an oscilloscope. Measurement results recorded in Table1.below.

Table 1. The wireless frequency and duty cycle of different duty cycle pulse、

The waveform duty cycle by MCU (%)	5	10	20	30	40	50	60	70	80	90
Output waveform duty cycle of receiver module(%)	17.2	18.8	24.4	31.2	38.4	45.8	52.8	59.0	65.6	72.0
The output waveform frequency of wireless receiver module (Hz)	800	800	800	800	800	800	800	800	800	800

From Table 1.data, frequency keeps 800Hz unchanged after the wireless transmission for the pulse waveform which the frequency is fixed at 800Hz. It indicates that the wireless transmission has no effect on the pulse waveform frequency .However, the duty cycle of the waveform has taken place great changes, and the duty cycle errors caused by the wireless transmission will change when the waveform duty cycle sanded by the MCU changes. When the input waveform duty cycle is 90%, this error is the highest; close to about 20%.Figure 1 is based on data in Table 1 in the curve. It shows the waveform duty cycle changes after the wireless transmission when the duty cycle waveform changes sanded by MCU. In the figure, horizontal axis is for the waveform duty cycle by the MCU and vertical axis is duty cycle of the waveform after the wireless transmission. Dotted line is the actual duty cycle by pulse waveforms of different duty cycles after the wireless transmission at 800Hz frequency. The solid black line is the ideal duty cycle when the transmission should be, and it is a straight line with a slope of 1.

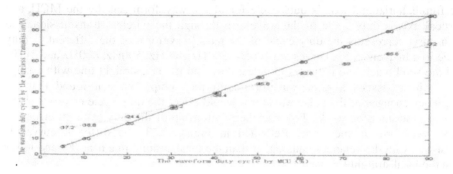

Fig. 1. The radio duty cycle curve of different duty cycles of the waveform after transmission at800Hz

From the Figure 1 we can clearly see that the curve has the deviation from the ideal straight line and the magnitude of deviation is different. The curves on both sides have a great difference from the straight line; while in the middle part is closer with the straight line. This also shows that, for different duty cycle pulse waveform, the duty cycle error generated by the transmission is different. When the waveform duty cycle is small, the duty cycle waveform from wireless transmission is large than the theoretical; When the waveform duty cycle is large, the duty cycle waveform from wireless transmission is small than the theoretical. The closer the original waveform's duty cycle to the limit (0% or 100%), the greater the transmission error is; the closer duty cycle to the middle range (40%), the more close the transmission to the theoretical situation, the more errors small.

1.3   The Affection of the Pulse Waveform's Frequency on Wireless Transmission

We select different pulse waveform frequency, such as 100Hz, 200Hz, 400Hz, 500Hz, 800Hz and 1kHz.We repeat the above experiment in each frequency, and observe the received signal and record the duty cycle.

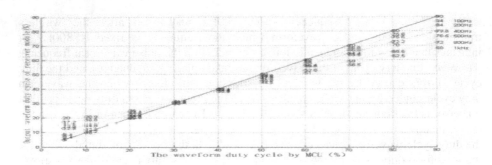

Fig. 2. The duty change of pulse waveforms of different frequencies cycle after the wireless transmission

Make the corresponding curve shown in Figure 2 based on the data in Table 1. In the figure, horizontal axis is duty cycle for of the waveform sent by the MCU, and vertical axis is duty cycle of the waveform through the wireless transmission, and each curve represents the duty cycle of the pulse waveform of the different   duty cycle in a frequency. Six curves are 100Hz, 200Hz, 400Hz, 500Hz, 800Hz and 1kHz, and the solid black line indicates the ideal duty and it   is a straight line with a slope of 1 which   issues a same waveform with the   chip. We can found that the frequency changes of the pulse waveform would affect the duty cycle error when the wireless transmission works, however the relationship of the duty cycle of error and pulse waveform is the same. Reflected in Figure 2.23, the different curves is consistent with the general trend, higher than the ideal straight line near 0% and lower than the ideal straight line near 100%.

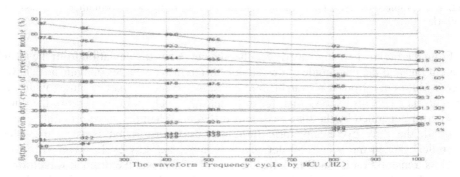

Fig. 3. The duty change of pulse waveforms of different frequencies cycle after the wireless transmission

From the effects of different frequencies on the duty cycle error, we can use the data in Table 1 and make a map, as shown in Figure 3. The horizontal axis is the frequency of the waveform given by MCU, and the vertical axis is the output waveform's duty cycle given by receiver module. Each curve represents the relationship between duty cycle and waveform frequency after the pulse waveform transmission at a certain duty cycle. Ten curves represent the duty cycle of 5%, 10%, 20% ... 90%.Through the map, we can see that the lower the frequency of pulse waveform, the smaller the duty cycle errors caused by wireless transmission ;the higher the frequency, the greater the duty cycle error. When the frequency is 100Hz, the duty cycle errors of each transmission are very small, if they are all less than 5% we can consider the signal transmission is very accurate. In Figure 2 the curve 100Hz is very close to the ideal straight line .When the frequency is 1 kHz, the near 0% or 100% duty cycle error becomes very large, even up to about 20%, and then signal transmission is only of 40% duty cycle in the vicinity, otherwise, the duty cycle of transmission waveform will be severely distorted.

1.4 The Analysis and Interpretation on the Duty Cycle Errors Resulted by Wireless Transmission

Do the experiment on the effect ion of the delay time of the waveform on duty cycle, and observe the rise delay time and fall delay time under different duty waveform and get the delay curve of the waveform on duty cycle time.

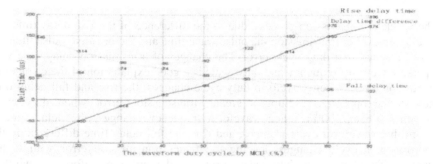

Fig. 4. The delay curve of Waveform duty cycle to delay time (800Hz bellow)

Fix duty cycle of pulse waveform 50%, observe delay time under different frequency wireless transmission and get the delay curve of the waveform frequency on delay time

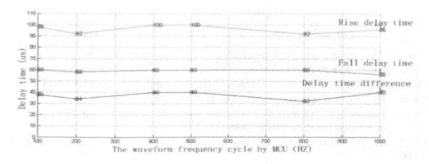

Fig. 5. The delay curve of Waveform freguecy to delay time (50% duty cycle)

It can be seen from the above two figures that the reason resulting in duty cycle error in the wireless transmission is the delay time. As at the wireless transmission the rise delay time and fall delay time is difference, thus it result that positive and negative pulse width time has changed after a transmission of the pulse waveform and a duty cycle error. When the frequency is fixed, the rise and fall delay time will change with the change in duty cycle waveform. With the lower duty cycle, the smaller rise delay time, the greater fall delay time, the greater pulse time, the duty cycle is too large; with the higher duty cycle, the greater rise delay time, the smaller fall delay time, the smaller pulse time, the duty cycle is too small. When duty cycle is fixed, rise and fall delay time does not change with the change of wave frequency, and delay time difference remains unchanged. However, the higher the frequency of the waveform, the shorter the period is, the greater the difference of the proportion of delay, the greater the duty cycle error is; the lower the frequency of the waveform, the greater the period is, the shorter the difference of the proportion of delay, the smaller the duty cycle error is.

2   Conclusion

In the wireless transmission process, due to the difference of the rise delay time and fall delay time, after transmission the pulse wave time and the negative pulse time has changed resulted in a duty cycle error. The changes of transmission delay time with the pulse waveform frequency and duty cycle is a good explanation of the changes of transmission error. When waveform duty cycle changes, the rise and fall delay time trends in the opposite, resulting a greater transmission error when duty cycle approaches 0% or 100%; When waveform frequency changes, the delay time is constant, but waveform cycle changes, and rise and fall delay time difference in the proportion of a cycle changes, leading to greater transmission error with higher frequency. To reduce the duty cycle errors of the pulse waveform in wireless transmission we can take the following measures: Appropriate to reduce the frequency of pulse waveform, or mount antenna for wireless transmitter and receiver module. This can effectively reduce the duty cycle errors caused by transmission resulting. In the various frequencies, the transmission accuracy of duty cycle has greatly been improved to meet the needs of the wireless transmission parameters.

References

1. Sun, D., Huang, W., Hu, Y.: Inductive power transfer technology based on new rotary converter. Power Technology (2005)
2. Huang, X.: A study of synchronous motor brushless excitation based on rotary converter. Nanjing University of Aeronautics and Astronautics, Nanjing (2008)
3. Huang, S., Zhang, W.: Detection of brushless synchronous motor excitation current based on RF. Automation Instrumentation (2006)
4. Liu, Y.: Design of wireless data transmission system based on RF. Xidian University, Xi'an (2009)
5. Jia, Q., Wang, D., Zhang, Z.: Wireless data transmission system basede on nRF905. International Electronic Elements (2008)

6. Wang, J., Jin, Z., Han, Y.: Development of RF technology and application. Electronic Technology (2007)
7. Che, Y.: A study of motor data acquisition system based on wireless technology. Taiyuan University of Technology, Taiyuan (2009)
8. Li, J.: DC chopper controlled by microprocessor. Baoji College of Arts and Science (Natural Science), Baoji (2004)
9. Wang, Z., Liu, J.: Power electronics technology. Machinery Industry Press (July 1, 2009)
10. Chen, J.: Electric power of the study——Power electronic conversion and control technology. Higher Education Press (December 1, 2004)

A Color Management Method for Scanning Input Image Files

Hankun Ye

International School of Jiangxi University of Finance and Economics, Nanchang,
330013, China
liyi7611@163.com

Abstract. Color management for scanning input image files is a key and difficult technique in the color reproduction in information optics. Through analyzing color rendering principle of scanned images, the paper advances a new color management method using subsectional fitting. First, rendering principle of subtractive process of scanned images is analyzed in theory; Then when collecting datum to deduce final model, the method takes standard color target for experimental sample and substitutes color blocks in color shade district for complete color space; Third, the color management method based on subsectional fitting algorithm to correct color conversion error is deduced; Finally the experimental results indicates that the method can improve color conversion accuracy for scanning images and can be used in color conversion of scanning input images practically.

Keywords: Information optics, Color reproduction, Color management, Subsectional fitting algorithm.

1 Introduction

In the computer and multimedia technology, scanner is one of main input equipments to input the graph and image information and a conversion tool of image reconstruction digitally. Generally speaking, color scanner is to produce three images corresponding to three color primaries of red (R), green (G) and blue (B) initially and then synthesize them to complete the color image scanning. The color space that represents the color characteristics of scanner is the RGB space that follows the rendering principle of additive process while the one that represents the color characteristics of scanning object rendering is the YMC (Yellow, Magenta and Cyan) space that follows the rendering principle of subtractive process. Both RGB space and YMC space depend on their respective equipment materials and it is difficult to transform directly between them. Consequently, the quality control of image color becomes one of the most difficult pivotal technologies as well as one of the research hotspots. The major task of color management is to solve the question of transforming images among different color spaces with the view to minimize the image color distortion during the whole duplication process. The basic approach includes three steps: first, a referential color space independent of equipments is selected; second,

D. Jin and S. Lin (Eds.): Advances in ECWAC, Vol. 2, AISC 149, pp. 389–393.
springerlink.com   © Springer-Verlag Berlin Heidelberg 2012

the equipment are characterized; finally, a relationship between the color space of each equipment and the referential color space independent of any equipment is established to provide a definite approach for data files when transforming among different equipments. The main focus here is to study the realization of precise conversion between RGB space that is dependent on equipments and the XYZ space that is independent through the means of scanner color management. XYZ color space mentioned is a universal standard color space that is independent of equipments recommended by CIE[1,2].

The current ways of color management include parameter method, interpolation method and machine learning method. Optical parameter and interpolation method can not ensure the accuracy of the color conversion[3,4]. Machine learning method is mainly BP neural network (BPNN) algorithm, which is fairly good in accuracy but leaves behind the question of slow convergence speed of BPNN so as to be hard to put into effect in the actual project[5,6].

The paper uses subsectional fitting algorithm to control the accuracy of color conversion for scanning input image base on theoretical analysis for scanning image error . In so doing, not only the problem of convergence speed of BPNN has been solved, but also the simplicity of the model structure and the accuracy of the transformation are ensured.

2   Rendering Principle of Scanning Images and Data Collection

Rendering Principle of Scanning Images. The In the color mixture, the three primary colors of R, G and B can be mixed to more colors and have the largest color domain. Therefore, yellow is selected to control the blue light because it is a complementary color of blue to control (absorb) the blue light effectively; in a similar way, the green's complementary color--- Magenta is selected to control the green light; the red's complementary color--- Cyan is selected to control the red light. Changing the thickness (or density) of yellow, magenta and cyan can easily change the absorbing capacity of three primary color lights of red, green and blue and complete the quantity of stimulus value of red, green and blue controlling to enter human eyes. Therefore, yellow, magenta and cyan are called three subtractive primary colors. The rendering mechanism of measurement sample is like this: the mixture of magenta(M) and yellow(Y) produces red(R),the mixture of cyan(C) and yellow(Y) produces green(G),the mixture of magenta(M) and cyan(C) produces blue(B),the mixture of yellow(Y), magenta(M) and cyan(C) produces black(BK).

Measuring Equipment. Here uses Colortron Equipment as color measuring equipment and Epson 3170 scanner with 3200*6400dpi resolution. The standard test calibration target for scanner---IT/2 calibration target used in this experiment was made by AgFa Corp. in 2006 with the serial number of 6x7c60103xx. Its rendering material is qualified for ISO12641 standard reflective color calibration target, as is shown in Fig. 2. Among its color scale area, there are three lines (column 13~15) of the color blocks of three subtractive primary colors of yellow (Y), magenta (M) and cyan (C) extending from light to dark, three lines (column 17~19) of the color blocks of red (R), green (G) and blue (B) extending from light to dark and still one line (column 16) of neutral color with increasing ash. Column 17~19 are the mixtures of two of Y, M and C three subtractive primary colors while column 16 are mixtures of Y, M and C three subtractive primary colors.

Data Measurement. To make measurement more close to the objective reality, the measured results of both calibration target and color values on scanning images are recorded to reduce various errors resulted from impersonal measuring conditions. (1) After 15 minutes of scanner preheating and white porcelain board demarcation, the picture of the whole calibration target is taken to measure its RGB and XYZ values. (2) Use the default setting to measure the RGB values of color blocks in the color scale area of the calibration target with Colortron Equipment. And after the normalization, a conversion RGB value database (3) Use Colortron Equipment to measure the XYZ values of color blocks in the color scale area of the calibration target to build up a conversion XYZ value database for intermediate conversion. is built up for intermediate conversion to derive the color management model.

3   Algorithm Design

Assume that there are N data of $z_1, z_2, z_3, ..., z_N$, four data points can determine a cubic curves from general mathematical knowledge. The consecutiveness of adjacent curves is considered after the subsection in the process of subsectional point selection, that is, the border point is bound by this constraint condition, so five data points is fitted to a cubic curve. The fitting method is as follows: the first step is the subsection of data. The first to the fifth data is another section, the fifth overlapping data point ensures the continuity of two subsections, and others are on the analogy of it; the second is the fitting of cubic curve for each subsectional data. It is easily learned that $4n+1$ is the number of the data from fitting n sections of cubic curves.

To illustrate the detailed method, if the function of cubic fitting curve in certain section of data is $w_t = a + bt + ct^2 + dt^3$ $(t = -2,-1,0,1,2)$, the curve function can be divided into odd function $v_t = bt + dt^3$ and even function $u_t = a + ct^2$. The factor of cubic fitting curve is derived from the basic principle of the least squares method as follows.

The first and final points in each section of data are fitted two times, so the derived fitting variance in a section of curve needs being weighted. In accordance with the principle of average allocation, the weight of variance $\lambda_{-2} = \lambda_2 = 1/2, \lambda_{-1} = \lambda_0 = \lambda_1 = 1$ is derived. See the derived curve fitting variance of this section in Formula 1.

$$S^2 = \sum_{t=-2}^{2} \lambda_t (w_t - z_t)^2 \tag{1}$$

Curve is the form of odd and even functions in Formula 2.

$$w_t = u_t + v_t, \qquad u_{-t} = u_t, \qquad v_{-t} = v_t \tag{2}$$

Formula 3 can be deduced by Formula 1.

$$u_t = 1/2(w_t + w_{-t}), \qquad v_t = 1/2(w_t - w_{-t}) \tag{3}$$

If $z_t = x_t + y_t, x_{-t} = x_t, y_{-t} = y_t$ Formula 4 can be derived.

$$x_t = 1/2(z_t + z_{-t}), \qquad y_t = 1/2(z_t - z_{-t}) \tag{4}$$

Therefore, Formula 5 can show fitting variance through Formula 2 and Formula 4.

$$S^2 = \sum_{t=-2}^{2} \lambda_t (u_t - x_t)^2 + \sum_{t=-2}^{2} \lambda_t (v_t - y_t)^2 = S_{odd}^2 + S_{even}^2 \tag{5}$$

That is, the smoothing of w_t and z_t can be seen as the summation of respective smoothing from odd function and even function. The fitting variance of odd and even functions can be derived by simple substitution operation through Formula 4. See Formula 6.

$$S_{odd}^2 = 2(b + d - y_1)^2 + (2b + 8d - y_2)^2 \tag{6}$$

If $b + d - y_1 = 0$ and $2b + 8d - y_2 = 0$, $b = (y_2 - 2y_1)/6$, $d = (8y_1 - y_2)/6$ can be derived. Therefore, $S_{odd} = 0$, that is, fitting variance of fitting function is 0 to reach the best approximation. After a similar operation, Formula 7 can be derived.

$$S_{even}^2 = (e - x_2)2 + 1/17(3x_0 + e - 4x_1)^2 \tag{7}$$

If $\partial S_{even}^2 / \partial e = 0$, e, a, c can be calculated separately. See Formula 8 for the formula of the final derived cubic fitting curve.

$$w_t = (3x_0 + 4x_1 - x_2) + (y_2 - 2y_1)t/6 + (-3x_0 - 2x_1 + 5x_2)t^2/18 + (8y_1 - y_2)t^3/6 \tag{8}$$

4   Experiment Confirmation

The proposed model is realized with C language. According to the scanner's rendering principle in the detailed realization, the model adopts the three-layer structure of three inputs and outputs. 84 color blocks in the color scale area represent the color management model of the whole color space, RGB value and XYZ measurement value are the input and output values of the model separately.

Table.1 shows conversion accuracy of the model which provides a conversion accuracy statistics for all the 288 color blocks in the calibration target through the algorithm of this paper, the polynomial fitting algorithm which is widely used and has its relatively high conversion accuracy and the BPNN algorithm in this experiment[5,6]. *NBS* aberration unit is the one adopted by American National Standards Institute. According to the research results of colorimetry, visual equivalency can be acceptable when $\Delta E < 5NBS$ units. From this, horizontal and vertical contrast can show that the model of this paper offers satisfactory color conversion accuracy, and is able to manage the scanner colors with regard to different situations.

Table 1. Conversion accuracy statistics for different algorithms

Algorithm	Accuracy (Unit: *NBS*)	
The algorithm of the paper	Average error	4.69
	Maximum error	10.12
	Number of blocks with an error larger than 5 *NBS*	13
Polynomial fitting algorithm	Average error	11.99
	Maximum error	27.34
	Number of blocks with an error larger than 5 *NBS*	70
BPNN algorithm	Average error	8.34
	Maximum error	14.99
	Number of blocks with an error larger than 5 *NBS*	30

5 Conclusion

Through fully applying the rendering mechanism, a model based on subsectional fitting algorithm is put forward using combination of IT calibration target. The experimental results show that new model not only solves the problem of convergence speed of BPNN, but also improves the conversion accuracy and is applied into the actual scanner color management tasks for the project.

References

1. Wang, Y.N., Liu, M.L., Zhu, H.H.: Color Space Changing from CRT Color Space to CIE Standard Color Space. J. Pre. Prin. 2, 45–46 (1996)
2. David, L.P., Christopher, S.C.: An Evaluation of Methods for Producing Desired Colors on CRT Monitors. J. Col. Res. & Appl. 14, 172–186 (1989)
3. Noriyuki, S.: Suppression of Noise Effects in Color Correction by Spectral Sensitivies of Image Sensors. J. Opti. Rev. 9, 81–88 (2006)
4. Guo, M.Z., Wang, Y.D., Su, X.H.: Research of Color Matching Method Based on BP Network. J. Comp. 23, 819–823 (2008)
5. Qian, G.L., Chen, B., Shu, W.H.: A Color Matching Method Based on Machine Learning. J. Sof. 9, 845–850 (2005)
6. James, K., Sigfredo, N., Wil, P.: Performing Color Space Conversions with Three-Dimensional Linear Interpolation. J. Elec. Ima., 226–250 (2007)

A Composite Drag and Drop Mechanism for Visual Programming in Control Educational Software

Zheng Fang, Jie Yang, and Qichun Zhang

State Key Laboratory of Synthetical Automation for Process Industries
(Northeastern University), Shenyang, Liaoning, China
fangzheng@mail.neu.edu.cn

Abstract. Since the front-end of Human-Computer Interaction (HCI) has an important influence on the user experience, it attracts more and more attention of researchers for developing educational software in recent years. Currently, there are a number of ways to develop the front-end user interface (UI). Among these ways, how to ensure visually programming of human-machine interface efficiently and effectively is an important issue for HCI. To solve this problem, this paper describes a composite mechanism that combines drag and drop, java reflection and XML persistence technology together to realize creating UI quickly and ensuring the reusability of UI. The mechanism, which named the "Composite Drag and Drop Mechanism", is applied in the design and development of a control educational software platform - EasyControl. A practical implementation validates the effectiveness of the proposed mechanism.

Keywords: visualization, drag and drop, Java reflection, XML persistence.

1 Introduction

As an important part of control engineering education, experiment is of great significance for the validation and application of the control theory. In control experimental system, experimenters often focus more on how to write the efficient control algorithm, rather than spend too much time to program the front-end UI. However, the front-end UI concerns the operability of control educational software, so how to improve its performance and interactivity is very important. Especially, how to achieve a friendly HCI by visual programming is the focus of related researches. The key idea of visual programming is that software is designed and constructed by reusable UI components [1]. A typical advantage of visual programming is that drag and drop mechanism (DnD) is used to reduce the difficulty of programming [2]. However, in the scenario for the complex data interaction, DnD alone cannot create complex UI component and implement data persistence.

In the field of control educational software, visual programming is also widely used by many researchers to develop the front-end UI. WinCon [3] is a widely used control educational software platform for real-time control experiments. However, it only provides a few UI components and the functions of the provided UI components

D. Jin and S. Lin (Eds.): Advances in ECWAC, Vol. 2, AISC 149, pp. 395–402.
springerlink.com   © Springer-Verlag Berlin Heidelberg 2012

are very simple and limited. NetConTop proposed in [4] is a visual network-based configuration software platform, which provides a variety of standard configuration components for remote control experiments, such as charts, meters, and text input box, etc. The greatest feature is that the OOA and OOD idea are applied to design and develop the UI. ControlDesk [5] uses wxWidgets as the GUI development framework, and its main program is written in C/C++. Although the architecture of the software is flexible, there are some difficulties in its UI development. Interface Designer [6] is developed by the Speedgoat Company. As a network-based software platform for real-time control experiments, Interface Designer provides Scripts Language interfaces for customizing events and properties of each UI component. But the visual programming function is not very powerful.

The above issues make us cannot help thinking about some fundamental questions—whether the experimenter can create UI quickly in some way, whether the UI can be saved and reused in document format and whether the interaction of UI can display the data in better way and reduce the software learning curve of experimenters. To solve these problems, this paper proposes a composite drag and drop mechanism. The mechanism uses the DnD technology to achieve dragging and dropping components in design mode; uses the java reflection technology to achieve the development and creation of the complex components by DnD; uses XML serialization and deserialization technology to save and reload the corresponding UI. The proposed mechanism, which is called the 'Composite Drag and Drop Mechanism', has been successfully used in the design and development of EasyControl - an educational software platform. A practical example is finally described to show the effectiveness of the proposed mechanism.

2   Composite Drag and Drop Mechanism

2.1   Problem Description

The basic requirement of the experiment UI in control educational software is that experiment UI should be created quickly in a visual programming pattern and also can easily be saved and reloaded. According to this requirement, there are two problems to be solved in the current control educational software: 1) DnD alone cannot quickly create the UI components which have complex functionalities, 2) It's difficult to achieve the preservation and reappearance of the experiment UI just by the DnD mechanism.

2.2   Constrains

First, the existing UI components should not be altered or changed too much, because the changes of UI component's source code often trigger a series of software restructuring work and the cost is rather high.

Second, there must be a unified form of DnD operation. Although the complexity of the each components is different, but there should be no difference in the operation of DnD. Because the operational differences will cause the structure of source codes become more complex, and this is deviating from our original intention.

Third, when the function of UI components is modified or upgraded, the creation and preservation of these components must still be supported in a flexible drag and drop form.

2.3 Solution

In order to solve the problems and meet the constraints, this paper presents a composite DnD mechanism. The key idea of the solution is to combine drag and drop, java reflection and XML persistence technology together to realize creating UI quickly and ensuring the reusability of UI. By doing so, the composite DnD mechanism not only supports basic drag and drop operation, but also supports the creation and preservation of the basic and custom complex components.

DnD realizes to provide a uniform dragging and dropping operation to layout and configure various UI components, so a complex experiment interface can be created easily and quickly with DnD operation. First of all, the components in the UI library will be set as drag source, and the UI design editor panel is set as drop target. The process of dragging and dropping is that dragging the source onto the target object. And DnD provides corresponding API which can automatically create and save the components. Thus, the component library is linked to the UI panel by dragging and dropping components.

Java reflection realizes to create complex UI components at the end of dropping operation. In control educational software, the UI components not only include buttons, labels, text input boxes, panels, and other simple components, but also include some complex components such as Virtual Animation, Virtual Instrument and Scope, etc. The data structure and function of these complex components are often very complicated. Generally, these complex components are often customized by the developers and are usually not included in the system's component library. If each component needs writing the construction code, then with the increase number of components, code repetitions problem will arise. The introduction of Java reflection technology can be a good solution to solve this problem: you can get the java class name of each UI component at run-time. As a result, the UI component can be dynamically instantiated at the end of dropping operation and doesn't need to write the source code to construct this component before running the program.

XML persistence realizes to ensure each UI component can be saved and reused according to its appearance and property. First, XML persistence technology provides interface to access each component's class name and editable property. And then, the class name and property of each component can be serialized into XML stream and saved into XML files in dragging and dropping operations. Therefore, the user interface can be saved. On the other hand, when reloading the experiment UI, the UI component's information stored in the XML files can be deserialized into XML stream. By using Java reflection technology, each UI component can be generated according to its saving order and the reappearance of UI is realized.

2.4 Structure

Before describing the structure of 'Composite Drag and Drop mechanism', this paper depicts the basic process of creating a UI for a Java application. In Java application, a component is the fundamental UI object [7]. A UI in Java application is composed of many components, such as windows, buttons, menus, scrollbars, and text boxes, etc. In addition, these components must be placed in containers. Containers are often different types of panels. The role of panels is to configure the layout and location of the components. Components have interfaces to call their respective functions, so a number of different components can form a complex UI.

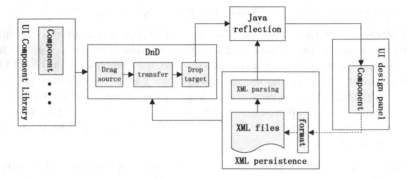

Fig. 1. The structure of 'Composite Drag and Drop Mechanism'

The structure of the 'Composite Drag and Drop mechanism' is shown in Fig. 1. The structure has three key parts: DnD, Java reflection, the serialization and deserialization of XML. The basic class diagram of the 'Composite Drag and Drop Mechanism' is shown in Fig.2. Detailed operation of creating, saving and reloading a UI can be divided into following steps.

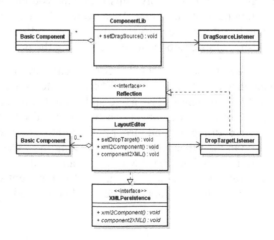

Fig. 2. The basic class diagram of the "composite drag and drop" mechanism

To create a UI, the Java class name of the UI component which will be generated is get from the dropping target firstly. Then, an object instance (using a constructor with parameter or without parameter) is generated according to the Java class name. After that, further operation of the UI component such as configuring the property can be implemented using APIs of Java reflection. Repeat the above three steps, a complex UI can be easily constructed.

To save a UI, all the UI components are scanned in the design panel firstly. Then, the Java class name of each component is written into the XML file as a node. At the same time, detailed properties and methods of component are written into corresponding node. By doing so, a UI is saved as a XML file.

To reload a UI, XML deserialization is used. First, each node in the XML file is scanned by parsing the XML file. Then, the Java class name and attributes of each node can be achieved. Finally, each component corresponding to the node is generated and configured by Java reflection.

3   Example

The proposed 'Composite Drag and Drop Mechanism' has been used in the design and development of our 'EasyControl' educational software. This section gives an example of real-time level control experiment on our Multifunctional Process Trainer [8] to show that using the proposed composite mechanism, experimenters can not only efficiently create various UI components to monitoring and interacting with the control program, but also can realize persistence of experiment UI.

Multifunctional Process Trainer is widely used experimental system for process control education, which consists of host computer, embedded real-time controller and four typical process control plant (flow, temperature, pressure, level) as shown in Fig. 3. EasyControl which is developed based on the proposed composite DnD mechanism is an educational software platform for real-time control experiment of Multifunctional Process Trainer. In order to facilitate the monitoring and interacting of the control experiment, EasyControl provides many UI components for control experiments such as Model block, Scopes, Charts, and Virtual Animations, etc.

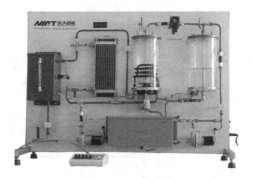

Fig. 3. Multifunctional Process Trainer

In the following, we will take a Level PID control experiment as an example to show how an experimenter can easily and efficiently create the front-end experiment UI using EasyControl. The experiment can be divided into five steps, namely creating control block diagram, compiling and downloading the control algorithm, tuning parameters of controller, monitoring control curve and saving the experiment UI.

Step1: Creating the control block diagram. Drag the No.2 pump actuator, level sensor, switch and PID controller block from the component library and drop them on the design editor panel to form a close control loop as shown in Fig. 4 (a).

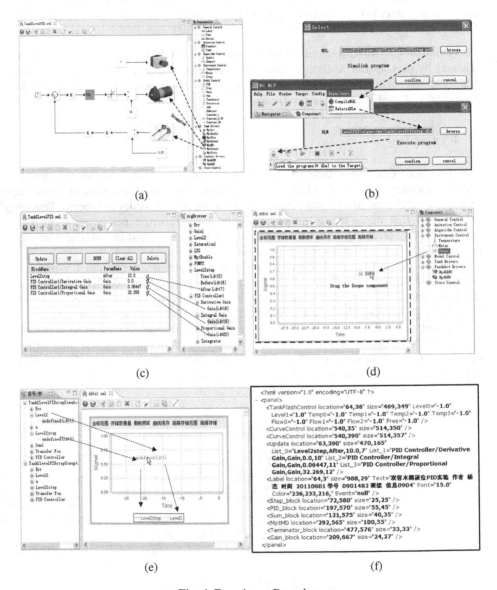

(a)   (b)

(c)   (d)

(e)   (f)

Fig. 4. Experiment Procedure

Step 2: Compiling the control block algorithm and downloading it into the embedded controller for real-time control as shown in Fig.4 (b). From the menu bar of EasyControl, choose 'CompileMdl' menu will compile the control block algorithm into binary executive code. Then, choose 'RelatedDlm' menu will download the binary executive code into the embedded hardware controller for real-time control.

Step 3: Realizing online parameter tuning function. First, Drag the Parameter Tune List from the component library and drop it onto the design editor panel. Then, open signal/parameter browser and drag the parameter variable which needs to be tuned into the Parameter Tune List as shown in Fig. 4 (c). After that, when the program is running, experimenter can realize online parameter tuning by change the parameter's value in the Parameter Tune List and click the 'Update' button.

Step 4: Monitoring the control curve using Scope component. First, find 'Scope' component from the UI component library. Then, drag and drop it onto the design editor panel and change its size by dragging the border of the component as shown in Fig. 4 (d). This paper takes observing No.2 tank level as example. Drag the setting value Level2step and actual value Level2 of No.2 tank from Signal/Parameter Browser and drop them into the Scope as shown in Fig.4 (e).

Step 5: Similarly, experimenters can drag Virtual Animation component to animate the tank system. The whole front-end UI of the Level PID control experiment is shown in Fig.5. Experimenters also can click the 'Save Experiment' menu of the design panel to save the experiment as an XML file as shown in Fig.4 (f). So, if the

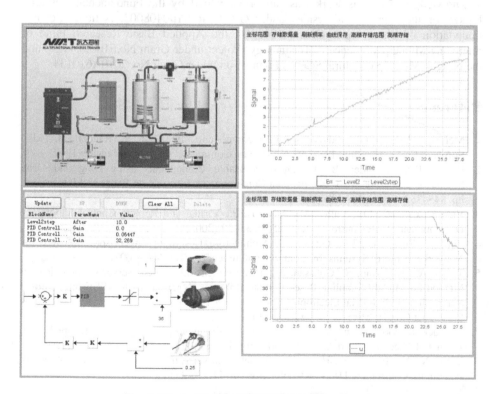

Fig. 5. Whole UI of Level PID Control Experiment

experimenter wants to re-run the experiment next time, he just needs to reload the experiment file.

From the above steps, we can see that a complex experiment UI can be easily set up using the EasyControl educational software. What the experimenter needs to do is dragging and dropping corresponding UI components to form an experiment UI. And, there doesn't need any coding work. So, the proposed composite DnD mechanism can improve the development efficiency of control experiment dramatically.

4 Conclusions

This paper presents a composite mechanism which combines the Java reflection and XML persistence into the DnD. We call it 'Composite Drag and Drop Mechanism'. DnD ensures that the UI will be created in the form of dragging and dropping operation. Java reflection is used in the dropping operation to generate any complex UI component. At the ending of dropping, the XML serialization is called to save the UI. We can also call the XML deserialization to reload the UI. The proposed composite mechanism has been applied in the design and development of EasyControl educational software platform. A practical experiment example shows the effectiveness and efficiency of the proposed mechanism.

Acknowledgments. This work was supported in part by the Fundamental Research Funds for the Central Universities under Grant No.N100408003, National Science Foundation of China under Grant 61040014, the Applied Basic Research Fund of Shenyang Municipal Science & Technology Project under Grant No.F10-205-1-50 and the Ningbo Municipal Natural Science Foundation under Grant No. 2010A610134.

References

1. Mcllroy, M.D.: Mass produced software components. In: Naur, P., Randell, B. (eds.) Software Engineering, pp. 138–155. NATO Sci. Committee, Garmisch (January 1969)
2. Reddy, P.N., Jalender, B.: Design of Reusable Components using Drag and Drop Mechanism. In: 2006 IEEE International Conference on Information Reuse and Integration, pp. 345–350 (September 2006)
3. Valera, A., Diez, J.L., Valles, M., Albertos, P.: Virtual and Remote Control Laboratory Development. IEEE Control Systems 25(1), 35–39 (2005)
4. Zhu, Y., Zheng, G., Liu, G.: System Architecture Design of Supervisory Software for Networked Control Systems. In: Control Conference, pp. 555–560 (2007)
5. Turan, A., Bogosyan, S., Gokasan, M.: Development of a Client-Server Communication Method for Matlab/ Simulink Based Remote Robotics Experiments. In: IEEE International Symposium on Industrial Electronics, vol. 4, pp. 3201–3206 (2006)
6. Speedgoat GmbH, http://www.speedgoat.ch
7. Jalender, B., Govardhan, A., Premchand, P.: A Pragmatic Approach to Software Reuse. Journal of Theoretical and Applied Information Technology (JATIT) 14(2), 87–96 (2010)
8. Yang, X., Yu, F., Xu, X.: Hierarchical variable universe fuzzy controller for double-tank system. In: Control and Decision Conference, pp. 718–722 (June 2009)

A Study on Man-Hour Calculation Model
for Multi-station and Multi-fixture Machining Center

QiaoYing Dong[1], JianSha Lu[1], and ShuLin Kan[2]

[1] College of Mechanical Engineering, Zhejiang University of Technology,
310032 Hangzhou, China
[2] School of Mechatronics Engineering and Automation, Shanghai University,
200072 Shanghai, China
dqy0910@163.com

Abstract. Aiming at a kind of key equipment in the metal machining and weld machining, namely the multi-station and multi-fixture machining center, the study on its man-hour calculation was presented to improve the accuracy in man-hour calculation in the manufacturing enterprises. The characteristics of the machining center were analyzed and then the corresponding man-hour calculation models were established. The presented models fully took the practical production situation, manual time and parallel time between man and machine into account and made the product man-hour more accurate and feasible. A primitive operation decomposition method to solve the difficulty in calculating the manual time was presented, the definitions and decomposition rules were proposed, and the familiar primitive operations of motorcar metal machining and weld machining were defined and analyzed. The presented models simplify the man-hour establishing process; realize the rapid and accurate man-hour calculation. Finally, a case study on a motorcar manufacturing company verifies the feasibility and effectiveness of the presented man-hour calculation system.

Keywords: Multi-station and multi-fixture machining center, Man-hour calculation model, parallel time, Manual time, Primitive decomposition.

1 Introduction

Man-hour quota is the important data for production planning, cost count and material requirement plan. In China, the current man-hour is also called standard time. Since Gilbreth F. B. and Gilbreth L.M. put forward motion and time study, more and more researchers paid attention to it and then increased the output, improved the quality and reduced the cost in production. Sellie[1] pointed out the development and application of predetermined time system and standard time. Freivalds and Goldberg[2] introduced the determination of various relaxation coefficients. Neter et al.[3] brought forward the linear statistics model to calculate standard time. Lee et al.[4] utilized Neural Network to estimate the assembly man-hour requirement and draw a comparison with linear regression method. Carl [5] calculated the manual task time quota based on motion

D. Jin and S. Lin (Eds.): Advances in ECWAC, Vol. 2, AISC 149, pp. 403–411.
springerlink.com © Springer-Verlag Berlin Heidelberg 2012

decomposition. Askarany and Smith [6] proposed a computer aided man-hour calculation system. Kang et al.[7] utilized the standard data method to establish the standard time of the mould production. Bin Liu and Zuhua Jiang[8] used linear regression, multivariate linear regression and Neural Network to study the man-hour estimation of semifinished product of ships. Chang et al.[9] brought forward two multivariate linear regression models to determine the standard time aimed at the multi-pattern and short life-cycle production system. Razmi et al.[10] presented the predetermined time study approach based on specific time table. Ergun[11] constructed the historical time data model based on artificial neural networks and estimated the product man-hour of the similar process.

The above studies formed the basis of man-hour(standard time)calculation. However, most of the previous man-hour estimating methods are based on the mass basic man-hour data and cannot improve the actuality that basic man-hour data scarcity. Furthermore, most of the equipments of metal and weld machining plant in the motorcar manufacturing company are the multi-station and multi-fixture machining centers, but the previous studies discussed the standard time calculation in a macroscopical view, and seldom paid attention to them. Therefore, aiming at the multi-station and multi-fixture machining center, we presented a man-hour calculation system.

2   The Man-Hour Calculation System of Multi-station and Multi-fixture Machining Center

2.1   The Man-Hour Components Analysis of Multi-station and Multi-fixture Machining Center

The characteristics of the multi-station and multi-fixture machining center are that the machining process dominates over the whole process and the working procedure man-hour consists of machine time, manual time, assistant time and relaxation allowance.

The machine time $T_{machine}$ is constant when the equipment and machining parameters are determined. It can be obtained by the control set of the machining center. If there is no control set, the machine time can be got by stopwatch measurement, appraisement and relaxation. The manual time T_{man} includes load materials, orientation, mark, test and unload materials etc. The fixtures should be replaced, and the programs and machines should be debugged when changing the product types. Therefore, these assistant time T_{assis} should be considered in the man-hour(standard time). For multi-station equipments, we should analyze man-machine operation to estimate whether the man and machine parallel time $T_{parallel}$ existed.

The assistant time Tassis is difficult to calculate, we use work sample, appraisement and relaxation to get the total time and then split it into a batch of products.

$$T_{assis} = t_{assis} / B .$$ (1)

where T_{assis} is the assistant time split into a batch of products, t_{assis} is the total assistant time, B is the production batch.

The analysis and calculation of parallel time are presented in succedent section. And the manual task time is established based on primitive decomposition. The relaxation allowance[12] is the relaxation for normal time considering the practical and human factors.

2.2 The Manual Time Calculation Based on Primitive Operation Decomposition

In order to establish the manual time rapidly and accurately, the primitive operation decomposition method is presented.

The Manual Task Decomposition Based on Primitive Operations. We found that the manual tasks include load materials, orientation, mark, test, unload and cleaning etc commonly. The manual tasks also can be regard as the combination of operations, and the operation is the combination of motion group, and the motion group is the combination of motion. The daedal and diversiform manual tasks can be decomposed to basic motion through this kind of decomposition. Therefore, the basic unit of the manual time is motion time.

However, this method is difficult and complex to perform because of the small granularity. Some familiar and functional motion combinations appear repeatedly in the manual time calculation process and the repeated calculation of these combinations will make us tired and tend to make errors. Therefore, we presented the primitive operation decomposition to solve this problem and make the manual time calculation rapid and efficient.

To avoid the problem that too big granularity makes the operation description difficult or too small granularity makes the operation description complex, the proper granularity should be defined to describe the technics accurately, simplify the manual time establishing process and eliminate function across.

Primitive Operation(PO) is the operation unit can realize certain function on the machining object, which consists of basic motion .

In order to ensure the integrality and proper granularity of primitive operation decomposition, the decomposition rule is proposed.

Function independence rule: The function independence and integrality of primitive operation should be ensured when tasks decomposition.

The Definition and Analysis of the Familiar Primitive Operations. The definitions and analysis of familiar primitive operations in motorcar parts and assemblies metal machining and weld are as follows.

(1) Load materials: The operation 'Load materials' answer for moving the workpiece to the work station. This kind of operation is divided into suspending load, common load and simple load according to whether the assistant tool is used and what tool to use.

Suspending load(SuL) is the primitive operation which needs to use the hook to load the workpieces for the large weight and volume. Its operational time is related to the distance(d) between the shelves and the work station (this distance is measured by the steps the worker takes), the suspending(dropping) time(t_1, t_2, t_3) and the workpiece weight(w). The primitive operation time can be denoted as fivevariate expression SuL (d, t_1, t_2, t_3, C_w), where C_w is the weight influence coefficient, when $W < 20N$, $C_w = 0$, when $20N \leq W < 60N$, $C_w=1$, when $60N \leq W < 100N$, $C_w=2$, and the rest may be deduced by analogy.

Common Load (*CL*) is the primitive operation carry out by one or two workers for the smaller weight and volume and relate to the distance(*d*) between the shelves and the work station(this distance is measured by the steps the worker takes) and the workpiece weight(*w*), $CL(d, C_w)$;

Simple load(*SiL*) is the primitive operation that the worker needn't to walk and only to stretch out and fetch the parts, its operational time relates to the motion range.

(2)Orientation: The operation 'Orientation' is the subsequence operation of 'Load materials'. To avoid too big granularity, we divide them into different primitive to facilitate the manual time establishment. This kind of primitive operation is divided into clamp orientation, loosening orientation and assembled orientation.

Clamp orientation(*CO*) is the primitive operation which needs to assemble, regulate and clamp, and its operational time relates to the complexity of clamping motion and the numbers of clamping location (n_{fix}), $CO(n_{fix})$;

Loosening orientation(*LO*) is the primitive operation which needs to loosen, related to the numbers of the loosening places(n_{loose}), $LO(n_{loose})$;

Assembled orientation(*AO*) is the primitive operation which needs to assemble and regulate but no clamping.

The above-mentioned primitive operations are the most familiar operations in the multi-station and multi-fixture machining center; we didn't give the details of other possible primitive operations. The users also can define new primitive operations themselves.

The Primitive Operation Time. The manual time is difficult to be measured accurately and objectively because the difference in proficiency and personal habit. We adopt MODAPTS to establish the primitive operation time. Without measuring directly, it can get the various operation time by the predetermined time standard for each motion. It is propitious to constitute the manual time, also the primitive operation time.

The principle of the presented primitive operation decomposition is shown in Fig 1. Firstly, we decompose the manual tasks according to the definition and decomposition rule of primitive operation. Secondly, we write the MODPTS expression of motion(s) and calculate the normal time of the primitive operation and get the standard time by proper relaxation. Finally, we add up all the primitive operation standard time in a working procedure(s) to get the working procedure manual standard time.

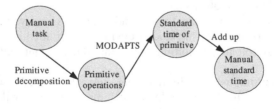

Fig. 1. Primitive operation decomposition and working procedure manual time

2.3 The Parallel Time Analysis and Man-Hour Calculation Model of Multi-station and Multi-fixture Machining Center

Supposed that the equipment of a working procedure has N work stations and M fixtures, the fixtures should be replaced when the product changed. According to the amount of work stations and fixtures and machining way, it can be divided into four types:

(1) $N=1$, $M=1$: The simplest type is that equipment has single work station and single fixture. If there is no cleaning or deburring in the processes, then the parallel time is not existed, or else. The time consumption of man-hour is shown in Fig.2.

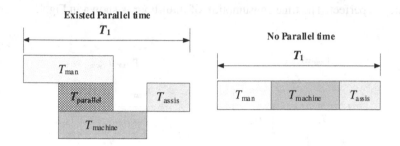

Fig. 2. The time consumption analysis of T_1

We presented the calculation models of working procedure(s) man-hour T_1, man availability ratio P_{1man} and machine availability ratio $P_{1machine}$ as follows:

$$T_1 = T_{man} \times (1+rc_1) + T_{machine} \times (1+rc_2) - T_{parallel} + T_{assis} \times (1+rc_3). \qquad (2)$$

where rc_1, rc_2, rc_3 is the manual relaxation coefficient, machine relaxation coefficient and assistant relaxation coefficient respectively.

(2) $N=1, M>1$: This type is that equipment has single work station and multi-fixture. For this type, the operations such as load materials, unload materials, test can be performed on a fixture when the products are processed on other fixtures. So there is parallel time but the production efficiency is not perfect. The time consumption of man-hour is shown in Fig.3.

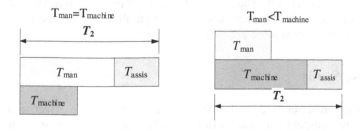

Fig. 3. The time consumption analysis of T_2

We presented the calculation models of working procedure(s) man-hour T_2, man availability ratio P_{2man} and machine availability ratio $P_{2machine}$ are as follows:

$$T_2 = \max(T_{machine} \times (1 + rc_2), T_{man} \times (1 + rc_1)) + T_{assis} \times (1 + rc_3). \tag{3}$$

(3) $N>1, M=1$: This type is that equipment has multi-station and single fixture. In this situation, the production efficiency is relatively low, it is seldom seen in reality.

(4) $N>1, M>1$: The type is typical equipment has multi-station and multi-fixture which can process the product on a fixture continuously. For this type, the operations such as load materials, unload materials, test can be performed on a fixture when the products are produced on other fixtures. So there is parallel time and the production efficiency is perfect. The time consumption of man-hour is shown in Fig.4.

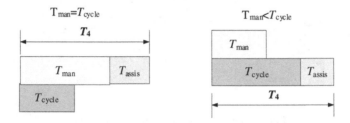

Fig. 4. The time consumption analysis of T_4

We presented the calculation models of working procedure(s) man-hour T_4, man availability ratio P_{4man} and machine availability ratio $P_{4machine}$ are as follows:

$$T_4 = \max(T_{cycle} \times (1 + rc_2), T_{man} \times (1 + rc_1)) + T_{assis} \times (1 + rc_3). \tag{4}$$

where T_{cycle} is the production cycle which can be obtained by the control set of the machine, if there is no control set, the machine time can be got by stopwatch measurement, appraisement and relaxation. This kind of equipment is working continuously with heavy load and we should pay attention to its maintenance.

The traditional man-hour establishing method seldom considers the parallel time of man and machine, and then its product man-hour is not accurate and can not measure the manual work effectively. The presented man-hour calculation model fully took the equipment characteristics, manual task time and parallel time into account and made the man-hour calculation more accurate.

3   A Case Study

The presented man-hour calculation system has been put into effect in a motorcar manufacturing corporation and performs well. In the motorcar manufacturing corporation, the weld machining plant produces the various motorcar component assemblies. Take the welded engine cradle assembly as a case to verify the feasibility and effectiveness of the presented man-hour calculation system. Its working procedure 60 is processed on a multi-station and multi-fixture machining center and the corresponding technics are shown in Table 1. According to the definition and decomposition rule of the primitive operation, and combined with the standard technics and work station arrangement to decompose the manual tasks. The primitive operation decomposition is listed in the last row in Table 1.

Table 1. The technics and primitive decomposition of working procedure 60

Name	Working steps	Primitive operation decomposition
spot welding and stud arc welding of upper and lower front girder	Fix the upper front girder and the lower front girder.	Common Load — Assembly Orientation-Common Load — Assembly Orientation — Clamp Orientation — Loosen Orientation — Common Unload — Test — Common Unload
	The work station is rotated to spot welding station and then the stud arc welding station.	
	Load the finished assembly, examine the quality and inspect whether the welding splash on the screw thread surface.	

In this working procedure, the manual task includes several primitive operations, such as Common Load, Assembly Orientation, Clamp Orientation, Loosen Orientation, Common Unload and Test. In order to ascertain the primitive operation time, we can write out their motion expression according to MODAPTS, as shown in Table 2. Motions are disassembled to left hand motion and right hand motion. Since they are synchronous, take the maximum one as the motion time. The primitive operation time is calculated in the last row in Table 2.

Table 2. A Part of Motion Analysis of Working Procedure 60

Primitive operation	Motion analysis		Primitive operation time[S]
	Right(left) hand	Remarks	
1(CL(3,4))	d*(W5)+M4G1M4+d*(W5) +Cw*L1 = 3*(W5)+M4G1M4+3*(W5) +4*L1 =43	d steps to shelves: d*(W5), fetch work piece: M4G1M4, d steps to work station: d*(W5), weight: Cw*L1	43×0.129=5.547
AO	M4P5R2 = 11	Hold the work piece and regulate:M4P5R2	11×0.129=1.419

The equipment of working procedure 60 is a two work station arc welding automaton. The amount of work stations $N=2$ and fixtures $M=3$. The different technics are processed at two work stations. Manual tasks "load" and "unload" can be executed accompany with the synchronous processing of two work stations, as shown in Fig 5.

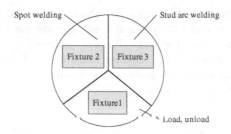

Fig. 5. The diagram of two work station arc welding automaton

In order to analyze the product machining process on the equipment clearly, the man-machine operation analysis are given in Table 3.

Table 3. Man-machine operation analysis

Man		Machine	(work station A/ work station B)		
		A		B	
Description	Time[S]	Description	Time [S]	Description	Time [S]
Load	11.868				
Unload	8.514				
Orientation	7.353	Spot welding	120	Stud arc welding	120
Test	21.93				
Idle	67.698				
Assistant time	2.637	Idle	2.637	Idle	2.637

According to the equations (4) to calculate man-hour, man availability ratio and machine availability ratio of each working procedure(group), where $rc_1=0.25, rc_2=0.1, rc_3=0.1$, $T_{cycle}=120S$, $T_{assis}=2.637S$, $T_{man}=52.302S$.

$$T_4 = T_{cycle} \times (1 + rc_2) + T_{assis} \times (1 + rc_3) = 120 \times (1 + 0.1) + 2.637 \times (1 + 0.1) = 134.9S$$

The comparison between the existing man-hour and the new man-hour are listed in Table 4. The existing man-hour is constituted by corporation through manual estimation and simple test. The new man-hour is established through presented man-hour calculation model. Man-hour change ratio is computed by equation: man-hour change ratio=(new man-hour-existing man-hour)/existing man-hour×100%.

Table 4. The comparison between the existing man-hour and the new man-hour

Number	Existing man-hour[S]	New man-hour[S]	Man-hour change ratio	Man availability ratio	Machine availability ratio
60	210	135	-35.7%	0.506	0.978

From the table, we can see that the existing man-hour is cut 35.7%, because the presented man-hour calculation system can improve the work method and man-hour level accompany the man-hour establishing process and get more accurate and reasonable data.

4 Conclusions

The man-hour calculation models on the multi-station and multi-fixture machining center and a case study about an arc welding automaton were proposed. The models simplify the man-hour establishing process and provide an excellent theory model. It has following distinct characteristics: (1) The man-hour calculation models are established according to the equipment identities and fully take the practical production situation, manual time and parallel time into account and make the product man-hour more accurate and feasible. (2) For the manual task standard time, a primitive operation decomposition method was proposed to realize its rapid and accurate calculation.

Acknowledgements. Acknowledgments to the National Natural Science Foundation of China (Grant No. 70971118) for supplied support.

References

1. Sellie, C.N.: Predetermined Motion-Time Systems and the Development and Use of Standard Data. John Wiley & Sons, New York (1992)
2. Freivalds, A., Goldberg, J.: Specification of Base for Variable Relaxation Allowance. J. Mtm 14, 2–29 (1988)
3. Neter, J., Wasserman, M., Kutner, M.H., Nachstheim, C.J.: Applied Linear Statistical Models. McGraw-Hill, New York (1996)
4. Lee, J.K., Lee, K.J., Park, H.K.: Developing Scheduling Systems for Daewoo Shipbuilding. Eur. J. Oper. Res. 2(2), 380–395 (1997)
5. Carl Lindenmeye, R.: How to Design and Conduct a Computer-Integrated Time Study with Active Element Performance Rating (2001), http://www.c-four.com
6. Askarany, D., Smith, M.: A Critical Evaluation of the Diffusion of Cost and Management Accounting Innovations. In: Proceedings of the 2000 IEEE International Conference on Management of Innovation and Technology, pp. 59–64. IEEE Press, Singapore (2000)
7. Kang, K.S., Kim, T.H., Rhee, I.K.: The Establishment of Standard Time in Die Manufacturing Process Using Standard Data. Comput. Ind. Eng. 27(1-4), 539–542 (1994)
8. Liu, B., Jiang, Z.H.: The Man-Hour Estimation Models & its Comparison of Interim Products Assembly for Shipbuilding. Int. J. Oper. Re. 2(1), 9–14 (2005)
9. Chang, S.K., Myung, S.C., Jae, J.R.: A Case Study for Determining Standard Time in a Multi-Pattern and Short Life-Cycle Production System. Comput. Ind. Eng. 53(2), 321–325 (2007)
10. Jafar, R., Shakhs-Niyaee, M.: Developing a Specific Predetermined Time Study Approach: an Empirical Study in a Car Industry. Prod. Plan. Control 19(5), 454–460 (2008)
11. Eraslan, E.: The Estimation of Product Standard Time by Artificial Neural Networks in the Molding Industry. Math. Probl. Eng. 1155–1167 (2009)
12. Wang, A.H., E, M.C, Ye, F., Zeng, M.G.: Methods, Standards, and Work Design (Eleven Edition), Tsinghua University Press, Beijing (2007) (in Chinese)

Application of Genetic Algorithm in Polarization Radar Receiving System

Xi Su[1,*], Feifei Du[2], Peng Bai[1], and Yanping Feng[1]

[1] Science Institute of Air Force Engineering University
710051 Xi'an China
[2] Military Transportation University
Tianjin China
suxi60@163.com

Abstract. Besides amplitude, phase and Doppler frequency shift, polarization state is another key characteristic of radar received waves. Setting radar receiving system to optimal polarization state is significant to anti-interference and target enhancement. If the receiving polarization state is set to the orthogonal state of interference's polarization, impact of interference can be significantly lessened. An optimal polarization setting method based on genetic algorithm is put forward in this paper. Due to its largely cut computation quantity, the presented method is much more agile and easy to handle, compared to lagrange multipliers method as well as enumeration method.

Keywords: polarization ratio, genetic algorithm, orthogonal polarization, poincare sphere.

1 Introduction

Taking advantage of polarization state, another key characteristic of received waves, polarization technology can significantly promote radar system's performance through help radar system acquire independent polarization shift ability in transmitting and receiving systems. With high speed continuous polarization shift in full polarization field, radar system can be evidently improved in these aspects such as anti-interference, anti-penetration, anti-stealth as well as anti-missile[1-2]. Different polarization states of receiving antenna have obvious impact on anti-interference performance in the background of highly polarized incoherent interference [3-5]. The multi-notch logic-product filter can be applied to polarization filtering when there are multiple active interference sources in radar's detection field. In 1981, British scholar A.J.Poelman originally proposed multi-notch logic product filter which actually is a group of polarization filters with similar polarization suppression points in parallel channels. However, to cope with multiple interference sources MLP's computation quantity could be tremendous. On the purpose of reducing computation complexity, Poelman improved MLP through taking

---

* Corresponding author.

polarization vector translation on every single interference source to shift them to one single suppression point of a polarization filter [8-9]. Accordingly, how to rapidly ascertain the average polarization state of active interference sources plays a key role in interference suppression. On the basis of optimal anti-interference target function proposed in this paper, genetic algorithm is applied to ascertaining optimal receiving polarization. Compared to enumeration or Lagrange algorithm, the computation process of this new algorithm is much simpler.

2   Received Waves' Polarization Theory

Every received wave with some original polarization state can be expressed by two orthogonal polarizations which usually are horizontal and vertical polarizations. The wave's polarization components can be expressed as follow:

$$[E_H, E_V]^T = |E| \cdot [\cos \alpha, \sin \alpha e^{j\varphi}]^T \tag{1}$$

In expression (1), E_H , E_V and E denote horizontal component, vertical component and the wave's electric field intensity respectively. And another key parameter is polarization ratio which can be expressed as follow:

$$p = \frac{E_V}{E_H} = \tan \alpha e^{j\varphi} \tag{2}$$

Polarization field is a continuous field and all polarization states can be found on Poincare sphere in Fig.1. The parameters α and φ of polarization point P are also presented on the Poincare sphere.

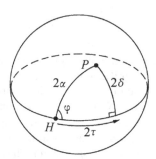

Fig. 1. Poincare Sphere

When an antenna with a height vector h receives a wave with a Jones vector E, then the received signal voltage is:

$$V = E^T h = |E||H|\cos \beta \tag{3}$$

β in expression (3) denotes the angle contained by the two vectors. The ratio of actually received signal power and the signal power under polarization match is called polarization match factor which be denoted as:

$$\rho = \frac{|V|^2}{V_m^2} = \frac{|E^T h|}{|E|^2 \cdot |h|^2} = \cos^2 \beta \le 1 \tag{4}$$

2β is the included angle between sphere radiuses of points on Poincare sphere which correspond transmitting and receiving polarizations respectively. Accordingly, when $p_T = p_R^*$ the two polarizations match while they are orthogonal as $p_T = -1/p_R$.

3 Anti-interference Performance Analysis of Polarization Technology

Application of polarization in anti-interference is to attain the receiving polarization orthogonal to interference polarization to promote radar's performance. Practically, there are many interference sources and therefore setting the receiving antenna's polarization state orthogonal to one of these interferences with various polarizations can not achieve the best performance. In this case, the key point is to find out the average polarization of interferences and set the receiving polarization orthogonal to it. For example, there are two incoherent interference sources P_1 and P_2 in same frequency band. The included angle of sphere radiuses corresponding to P_1 and P_2 is β and the energy coefficient ratio is $\omega_1 : \omega_2 = r:1$. According to expression (4), P_1' and P_2' should be chose as the receiving polarization to completely suppress interference P_1 and P_2 respectively. Consequently, to achieve optimal anti-interference, polarization state P between must be ascertained. If β_1 and β_2 denote the included angles of P and P_1', P_2' respectively, obviously in optimal case they can be expressed as:

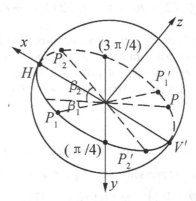

Fig. 2. Optimal Polarization under Two Random Interference Polarizations

$$\beta_1 = 2\arctan\left(\frac{-\cos\beta \cdot \omega_2 - \omega_1 + \sqrt{(\cos\beta \cdot \omega_2 + \omega_1)^2 + \sin^2 \beta \omega_2^2}}{\omega_2 \sin\beta}\right) \tag{4}$$

According to the expression upon, the value of β_2 is $\beta_2 = \beta - \beta_1$. Here is a new definition to introduce: polarization anti-interference gain which means the gain of radar system acquired through polarization technology. Its definition is expressed below:

$$G_p = 10\lg\frac{E_{Iopt}}{E_{Imat}} \qquad (5)$$

In expression (5), E_{Iopt} denote received interference energy under optimal polarization state and E_{Imat} indicates received interference energy under match polarization which doesn't use polarization filter. G_{p1} and G_{p2} denote received interference gain when receiving polarization is orthogonal to P_1 and P_2 respectively. Assuming the power of P_1 is much bigger than P_2's, therefore, G_{p1} is much less than G_{p2}. As the angle contained between sphere radiuses of P_1 and P_2 on Poincare sphere is 180°, the optimal anti-interference polarization must be in vicinity of one of the two interferences which must lead to anti-interference gain's reduction.

4   Optimal Polarization Decision Method Based on Genetic Algorithm

Supposing there are n incoherent interference sources (P_1, P_2, ..., P_n) in a same frequency band and their polarization states are acquired, the power ratio of interferences is $\omega_1 : \omega_2 : ... : \omega_n = r_1 : r_2 : ... : r_n$. Orthogonal polarizations of each interference are P_1', P_2',..., P_n'. Accordingly, the key issue is to find out optimal receiving polarization P which can minimize received interference power.

$$W_{min} = W_1 + W_2 + \cdots + W_n$$
$$= A_e\left(\omega_1 \sin^2(\beta_1/2) + \omega_2 \sin^2(\beta_2/2) + \cdots + \omega_n \sin^2(\beta_n/2)\right) \qquad (7)$$

β_i denotes included angle between optimal polarization P and polarization P_i'. It is an optimization issue to get a global optimum in a continuous field and genetic algorithm is always good at resolving this kind of problem. There are four steps of genetic algorithm's application in searching optimal polarization P (α ϕ).

4.1   Gene Expression

Through encoding variables to be optimized, variable $\alpha \in (0° \ 360°)$ and $\varphi \in (-90° \ 90°)$ can be expressed as α_1, α_2 ;$\cdots$, α_{10} $(\alpha_i = 0,1)$ and φ_1, φ_2 ;$\cdots$, φ_{10} $(\varphi_i = 0,1)$ respectively. Then an integral gene is α_1, α_2 ;$\cdots$, α_{10}, φ_1, φ_2 ;$\cdots$, φ_{10}. Phenotype α and φ are defined as:

$$\alpha = \frac{\sum_{i=1}^{10}\alpha_i \times 2^{10-i}}{2^{10}} \qquad \varphi = \frac{\sum_{i=1}^{10}\varphi_i \times 2^{10-i}}{2^{10}} \qquad (8)$$

The computation precisions of α and φ are 0.35° and 0.18° respectively.

4.2 Fitness Functions

Since the optimal value of the optimization function is the minimum and the target function's value is always below zero, the fitness function can be assumed as:

$$W' = B_{max} - W$$
$$= B_{max} - A_e \left(\omega_1 \sin^2(\beta_1/2) + \omega_2 \sin^2(\beta_2/2) + \cdots + \omega_n \sin^2(\beta_n/2) \right) \mathrm{W} < B_{max}$$
$$W' = 0 \quad W \geq B_{max} \tag{9}$$

$B_{max} = A_e(\omega_1 + \omega_2 + \cdots + \omega_n)$, to simplify computation process, value of B_{max} can be defined as 1.

4.3 Evolution Parameter Decision and Operations

To avoid prematurity and local optimization, evolution parameters are proposed below: group multitude M=90, number of generations N=110, cross probability P_C= 0.7, mutation probability P_M=1/1000. There are three main procedures: choose, roller mass loss method is adopted in this step to choose the group according to probability in proportion to fitness; cross, single point cross method is applied to generate new individuals; mutation, change values of random appointed individuals in line with P_M.

4.4 Evolution Parameter Decision and Operations

There are 7 incoherent interference sources $P_i(l_i, g_i, r_i)$ i=1,2,...,7. l_i, g_i and r_i denote P_i's polarization latitude, longitude on Poincare sphere and energy ratio coefficient

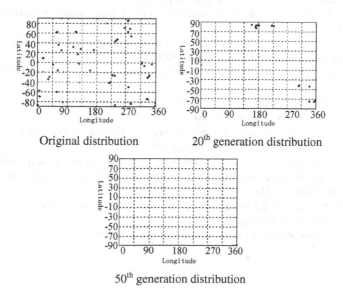

Original distribution   20th generation distribution

50th generation distribution

Fig.3. Original, 20th and 50th Generation Distribution

respectively. The 7 interference sources are $P_1(-90, 0, 0.20)$, $P_2(-70, 12.34, 0.14)$, $P_3(-60, 25.5, 0.16)$, $P_4(-45, 36.88, 0.08)$, $P_5(-35, 48.2, 0.20)$, $P_6(-10, 55, 0.12)$, $P_7(0, 67, 0.10)$. Individual polarization distributions of original generation, 20th generation and 50th generation are demonstrated in Fig3. In 20th distribution, optimal individuals gradually concentrate to two local optimums. Finally, the global optimum $P(0, -90)$ with a fitness value 0.766 is ascertained according to 50th generation distribution.

5 Conclusions

To fulfill increasingly new needs of modern electric wars, polarization technology gradually plays a key role in modern radar systems. Adopting the right polarization state in receiving system can evidently promote radar's anti-interference ability. A new optimal polarization search method based on genetic algorithm is put forward in this paper. Compared to traditional algorithms such as enumeration, this new method can not only enormously reduce computation complexity but also achieve excellent precision. It can provide sufficient precision as well as speed to radar system to cope with multiple interference sources and complicated electric environment.

References

1. Poelman, A.J.: Virtual polarization adaptation, a method of increasing the detection capabilities of a radar system through polarization vector processing. IEEE Proc. Communication, Radar and Signal Processing 128(10), 465–474 (1981)
2. Zeng, Q., Li, R., Guo, H.: Effect analysis on polarization converter in electronic countermeasures. Modern Radar 23(5), 72–76 (2001)
3. Pan, J., Chen, X., Liu, B.: A Simulation study of radar varied polarization. Modern Radar 28(4), 53–59 (2005)
4. Li, Y., Xiao, S., Wang, X.: Anti-interference technology based on radar polarization. National Defense Industry Press, Beijing (2010)
5. Jian, P., Bo, L., Erke, M.: The Method and Application of Adaptive Receiving Polarization Process. Modern Radar 26(1), 53–55, 61 (2004)
6. Zhaowen, Z., et al.: Polarization information processing and application in radar. National Defense Industry Press, Beijing (2005)
7. Zeng, Q.: Radar Polarization Technology and Application of Polarization Information. National Defense Industry Press, Beijing (2006)
8. Poelman, A.J., Guy, J.R.F.: Multi-notch logic product polarization suppression filters, A Typical Design Example and Its Performance in a Rain Clutter Environment. Proc. IEEE, pt F 131(4), 78–84 (1984)
9. Wang, B.: Polarization Theory and Application in Radar System. 14th Research Institution of Electric Industry Ministry, Nanjing (1994)
10. Liu, Z.: Genetics. People's Education Press, Beijing (1982)

Quality Assessment Model and Improvement Model for Screen Printing Process in Manufacturing of Touch Panels

Ching-Hsin Wang[1,*], Kuen-Suan Chen[2], Shun-Chieh Wu[2], and Peng-Hsiang Chang[3]

[1] Department of Logistics Management, National Defense University,
Taipei City 12258, Taiwan, R.O.C.
[2] Department of Industrial Engineering and Management,
National Chin-Yi University of Technology, Taichung City 41170, Taiwan, R.O.C.
[3] Department of Management Sciences, R.O.C. Military Academy, Kaohsiung City 83059,
Taiwan, R.O.C.
thomas_6701@yahoo.com.tw

Abstract. With the rapid development of technology, mobile consumer electronics such as cell phones, digital cameras, and laptops are receiving increasing consumer attention. Touch panels are already an essential part of many electronic devices, and functionality and quality of touch panels is a critical issue. This study investigated the screen printing process in the manufacture of touch panels and extracted essential critical to quality (CTQs). The process capability index C^{*}_{pmj} was employed to develop a process capability analysis model suitable for symmetric and asymmetric characteristics. The model enabled identification of CTQs that resulted in poor process capacity. A C^{*}_{pmj} index testing method and multiple comparison analysis method were also developed to obtain optimal settings for the CTQs resulting in poor capacity, thereby increasing process stability. The instruments developed in this study can assist touch panel manufacturers to enhance product quality as well as product competitiveness.

Keywords: touch panel, screen printing, process capability index, asymmetric tolerance.

1 Introduction

In the face of global competition, innovation has become the most crucial tactic to increase corporate competitiveness. The adoption of touch panel technology has led to the development of a range of applications found in cell phones, laptops, MP3 players, PDAs, GPS, and ultra-mobile PCs (UMPC) (Lyn and Chen, 2008)[14]. Apple, a company synonymous with innovation, announced the first iPhone in early 2007, creating tremendous business opportunities for touch panels. According to research by the Topology Research Institute, the iPhone has accelerated the

---

* Corresponding author.

D. Jin and S. Lin (Eds.): Advances in ECWAC, Vol. 2, AISC 149, pp. 419–428.
springerlink.com   © Springer-Verlag Berlin Heidelberg 2012

penetration of touch panels in global cell phone markets. In 2011, the output value of touch panels will reach USD 4.4 billion, and it is estimated to reach USD 9 billion in 2015. The technique of screen printing is widely applied in touch panel manufacturing due to the need for process enhancement and cost considerations and has become an essential critical process.

Hoshimura and Takafumi (2002)[10], Murata and Iwase (2005)[15], Chen and Chen (2008)[7], and Lyn and Chen (2008)[14] have pointed out that quality enhancement is the focus of touch panel manufacturers, as the market is now vast, and quality requirements are high. However, the manufacturing process of touch panels is complex (shown in Fig. 1), and defects are common. In particular, the quality of screen printing has major impact on subsequent processes and resulting products. In addition to the precision and stability of the screen printing machine itself, substrate thickness (off contact), lift-off speed (peel off), squeegee speed (printing speed), and squeegee angle (angle of attack) are all essential critical to quality (CTQs). Flaws in the process will lead to uneven printing thickness, smeared ink, incomplete pattern lines and poor resolution. This affects the functionality of the product, which further increases costs. This study focuses on four of these CTQs in the process of screen printing.

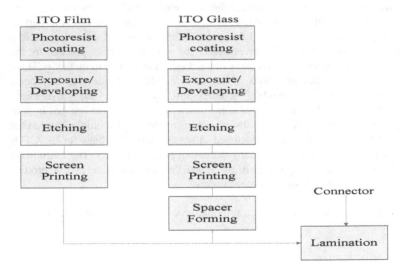

Fig. 1. Manufacturing process of touch panels.

Many instruments have been employed in assessment of processing quality. Chen and Pearn (2001)[5] and Wang et al. (2011)[18] indicated that process capability indices (PCI) are the most effective and convenient instruments in measurement of process capability. A number of researchers have conducted studies on PCIs, including Kane (1986)[12], Chan et al. (1988)[2], Boyles (1994)[1], Pearn et al. (1992)[16], Huang et al. (2002)[11], Chen et al. (2003)[6], Chen and Chen (2004)[3], Chen and Chen (2008)[7], and Wang et al. (2011)[18]. This study developed a quality assessment model for touch panel screen printing based on the four CTQs mentioned above. This model enables measurement of whether processes achieve quality standards and also provides an improvement model directed at the shortcomings of processes that don't. This model will be valuable for touch panel manufacturers wishing to enhance process quality.

2   Process Capability Assessment Model for Screen Printing Process

As mentioned above, the process conditions related to screen printing are very complex. Most products cannot be redone if there are any defects in the process; therefore, the quality of each stage of the process is essential. In terms of off contact, peel off, squeegee speed, and squeegee angle, poor quality will cause problems in the functionality of the product, thereby increasing costs. The specifications for these CTQs are listed in Table 1.

Table 1. Specifications of CTQs.

Process	CTQs	Specification attribute	Specification
Screen printing	Off contact (Substrate thickness)	NTB	6 ± 0.2 mm
	Peel off (Lift-off speed)	NTB	14 ± 2 m
	Printing speed (Squeegee speed)	NTB	250 ± 50 mm/sec
	Angle of attack (Squeegee angle)	NTB	$(15^{+2}_{-1})°$

2.1   Setting of Overall Performance Indicator for Screen Printing C_T^{SP}

It is apparent from Table 1 that the four CTQs are all NTB (nominal-the-best) type quality characteristics. The angle of attack is specified with asymmetric tolerance. Some studies have observed common use of C_{pm} in cases with asymmetric tolerance (Kushler and Hurley, 1992[13]; Franklin and Wasserman, 1992[19]). However, Boyles (1994)[1] pointed out that this approach often understates or overstates process capability. To address the issue of asymmetric specification ranges, Chen et al. (1999)[4] modified C_{pm} and proposed C^*_{pmj}, a PCI suitable for the assessment of asymmetric tolerance. This study employed this PCI to develop an assessment model capable of evaluating the process capability of each CTQ. C^*_{pmj} is defined as:

$$C^*_{pmj} = \frac{d^*_j}{3\sqrt{\sigma^2_j + A^2_j}}.$$ (1)

where $A_j = \max\{d^*_j (\mu_j - T_j)/D_{uj}, d^*_j (T_j - \mu_j)/D_{lj}\}$, $D_{uj} = USL_j - T_j$, $D_{lj} = T_j - LSL_j$, and $d^*_j = \min\{D_{uj}, D_{lj}\}$; μ_j, T_j, d^*_j, USL_j, and LSL_j are the mean, target, specification tolerance, upper specification limit, and lower specification limit of the jth CTQ, respectively. It is apparent that when $T = m$ (symmetric case), $C^*_{pmj} = C_{pm}$. Based on the concept proposed by Chen et al. (1999)[4], we can deduce the relationship between C^*_{pmj} and p_j (rejection rate involving the CTQs).

$$p_j \leq 2 - 2\Phi\left(3C^*_{pmj}\right). \tag{2}$$

A greater process capability C^*_{pmj} indicates a lower rejection rate. As mentioned above, the previous and subsequent processes of the product are highly dependant on each other. Using P to represent the rejection rate of the entire screen printing process, the relationship between P and p_j is:

$$Max\{p_1, ..., p_j\} \leq p \leq p_1 + p_2 + ... + p_j = \sum_{j=1}^{n} p_j. \tag{3}$$

where n is the number of CTQs in the process. As $p_j \leq 2 - 2\Phi\left(3C^*_{pmj}\right)$ the relationship between the rejection rate of the entire screen printing process P and the index C^*_{pmj} is shown below.

$$P \leq \sum_{j=1}^{n} p_j \leq \sum_{j=1}^{n} \left[2 - 2\Phi\left(3C^*_{pmj}\right)\right]. \tag{4}$$

Based on the relationship above, this study referred to the approach adopted by Chen et al. (2003)[6] and established an overall performance indicator to reflect the rejection rates of CTQs in the screen printing process. The indicator, C_T^{sp}, is defined as:

$$C_T^{SP} \geq \left(\frac{1}{3}\right)\Phi^{-1}\left\{\left[-\sum_{j=1}^{n}\left[2 - 2\Phi\left(3C^*_{pmj}\right)\right] + 2\right] \div 2\right\}. \tag{5}$$

By multiplying both sides of the inequality by 3 and selecting Φ, we can obtain the relationship between P and C_T^{SP} as follows:

$$P \leq 2 - 2\Phi\left(3C_T^{SP}\right). \tag{6}$$

From this inequality, it is apparent that a greater indicator C_T^{SP} results in a lower rejection rate (P). For example, when $C_T^{SP} = 1$, it ensures that the total rejection rate of the screen printing process is $P \leq 2 - 2\Phi(3 \times 1) = 0.269$ %. In other words, the total defect-free rate of the screen printing process would be greater than 99.73 %.

2.2 Setting of Process Index for C^*_{pmj}

Chen et al. (2003)[6] perceived that because manufacturing processes comprise multiple minor processes, the rejection rate of each minor process must be far lower than that of the entire manufacturing process in order to reach the standards required by customers and to guarantee product quality. Based on this concept, this study set an index C^*_{pmj} to indicate the process capacity of CTQs in screen printing. If we stipulate that $C_T^{sp} \geq v$, the overall performance indicator C_T^{SP} can be written as below.

$$C_T^{SP} \geq \left(\frac{1}{3}\right)\Phi^{-1}\left\{\left[-\sum_{j=1}^{n}\left[2-2\Phi\left(3C^*_{pmj}\right)\right]+2\right]\div 2\right\} \geq v .\tag{7}$$

Supposing that the minimum value of the index for each CTQ in the process is the same ($C^*_{pm1} = C^*_{pm2} = ... = C^*_{pmj} = w$), we can derive the following:

$$C_T^{SP} = \left(\frac{1}{3}\right)\Phi^{-1}\left(\left(\left(-n\left(2-2\Phi(3w)\right)\right)+2\right)\div 2\right) \geq v .\tag{8}$$

By solving the inequality, it can be established that when the required overall performance of the screen printing process is $C_T^{SP} \geq v$, then the minimum value of the index for each CTQ is w, which is defined as below.

$$w \geq \left(\frac{1}{3}\right)\Phi^{-1}\left(\frac{n+\Phi(3v)-1}{n}\right).\tag{9}$$

For example, the quality standard of the screen printing process is $C_T^{SP} = 1$, and there are four CTQs in the process; using Eq. (9), it is calculated that the PCI (C^*_{pmj}) of each characteristic must be at least 1.13320 to ensure product quality.

3 Analysis Graph for PCI of Multiple Quality Characteristics in Screen Printing

As the squeegee angle is specified using asymmetric tolerance (15^{+2}_{-1}), the capability analysis graph for symmetric quality characteristics proposed by Chen and Chen (2008)[7] cannot effectively reflect the capability of processes with quality characteristics that have asymmetric ranges. Therefore, this study referred to the methods adopted by Chen et al. (1999)[4] and proposed an analysis graph suitable for processes with symmetric or asymmetric ranges. Using Eq. (1), suppose $C^*_{pmj} = v$; by rewriting $A_j = max\{d_j^*(\mu_j - T_j)/D_{uj}, d_j^* (T_j - \mu_j)/D_{lj}\}$, we can obtain the deviation rate corresponding to the target value:

$$\frac{A_j}{d_j} = \max\left\{ \frac{\left(\mu_j - T_j\right)}{D_{uj}}, \frac{\left(T_j - \mu_j\right)}{D_{ij}} \right\}.$$ (10)

Eq. (1) can be rewritten as:

$$C^*_{pmj} = \frac{1}{\sqrt{\left(\frac{\sigma_{ij}}{d^*_{ij}}\right)^2 + \left(\frac{\Lambda_{ij}}{d^*_{ij}}\right)^2}} = 3v.$$ (11)

In order to present the PCIs of the different quality characteristics on the same graph, this study standardized the values, adopting the deviation rate of the process as X and corresponding tolerance variance as Y, both of which are expressed as:

$$X = \left\{ \begin{array}{ll} \frac{\left(\mu_j - T_j\right)}{D_{uj}}, & \mu_j \geq T_j \\ \frac{\left(T_j - \mu_j\right)}{D_{ij}}, & T_j > \mu_j \end{array} \right\}.$$ (12)

$$Y = \frac{\sigma_j}{d^*_j}.$$ (13)

Thus, Eq. (11) can be modified into the equation of a circle of which the radius is $1/3v$, as shown below.

$$X^2 + Y^2 = \left(1/3v\right)^2.$$ (14)

In the circumstances where $\mu_j > T_j$, $\mu_j = T_j$, or $\mu_j < T_j$, the solutions for X are $X > 0$, $X = 0$, or $X < 0$; when $\mu_j = USL$ or $\mu_j = LSL$, then $X = 1$ or $X = -1$. For instance, if the required overall performance indicator $C^{sp}_T = 1$, Eq. (9) can be employed to calculate that C^*_{pmj} equals 1.13320, at which $r = 1/(3v) = 1/3.3996 = 0.294$. The PCI curve for the screen printing process can then be plotted on the graph, as shown in Fig. 1. As can be seen, a lower r value results in a greater C^*_{pmj}, thereby indicating higher accuracy and precision in the process and better quality in the resulting products.

4 Process Improvement Model for Screen Printing

Crevelling et al. (2003)[8] stated that there are many instruments for process improvement, including experimental design, failure modes and effect analyses, and hypothesis testing. After measuring and analyzing product quality, whether the target of improvement was achieved must be confirmed, in which hypothesis testing is an appropriate method. This study referred to the Hartley's test method proposed by

Pearson and Hartley (1966)[17] for $C^{*}{}_{pmj}$. F_{max} was employed as the test statistic, the definition of which is:

$$F_{max} = \frac{\min\left\{\hat{C}^{*2}{}_{pm1}, \hat{C}^{*2}{}_{pm2}, ..., \hat{C}^{*2}{}_{pmc}\right\}}{\max\left\{\hat{C}^{*2}{}_{pm1}, \hat{C}^{*2}{}_{pm2}, ... \hat{C}^{*2}{}_{pmc}\right\}} .$$ (15)

When the null hypothesis (H_o) is true, then:

$$F_{max} = \frac{\max\left\{\left(C^{*}{}_{pm1}/\hat{C}^{*}{}_{pm1}\right)^2, \left(C^{*}{}_{pm2}/\hat{C}^{*}{}_{pm2}\right)^2, ..., \left(C^{*}{}_{pmc}/\hat{C}^{*}{}_{pmc}\right)^2\right\}}{\min\left\{\left(C^{*}{}_{pm1}/\hat{C}^{*}{}_{pm1}\right)^2, \left(C^{*}{}_{pm2}/\hat{C}^{*}{}_{pm2}\right)^2, ..., \left(C^{*}{}_{pmc}/\hat{C}^{*}{}_{pmc}\right)^2\right\}} .$$ (16)

where $\left(C^{*}{}_{pm1}/\hat{C}^{*}{}_{pm1}\right)^2$ is distributed according to a chi-square distribution with v_i degrees of freedom; $v_i = n_i \times \left[\left(1 + \lambda_i/n_i\right)^2 / \left(1 + 2\lambda_i/n_i\right)\right]$, and $\lambda_i = n_i \times \left(\mu_i - T_i\right)^2/\sigma_i^2$. In addition, F_{max} is distributed according to Hartley's distribution with $\bar{v} - 1$ and c degrees of freedom, in which $\bar{v} = \sum_{i=1}^{c} v_i / c$; c is the selected level number and v is the number of samples within a single standard. If the level of significance is set as α , and F_{max} is greater than the critical value (C_o), then $P\left(F > C_o \middle| H_o \ is \ true\right) = \alpha$, thereby rejecting H₀. On the contrary, when $F_{max} < C_o$, H₀ is accepted. Using this method, we can verify the effects of improvement on the process.

5   Application

This study applied the proposed models to the screen printing process of projected capacitive touch panels manufactured in Taiwan to demonstrate their applications. The quality standard of the company in this study case was $C_T^{sp} = 1$. With Eqs. (9) and (14), we calculated that $r = 1/(3v) = 0.294$ and drew the PCI curves as shown in Fig. 1. Within a stable process, a total of $m = 30$ sample groups were taken, each with

Table 2. Parameters of CTQs.

CTQs	Specification	Accuracy (X_i)	Precision (Y_i)	Process capability
Off contact (Substrate thickness)	6 ± 0.2mm	0.03	0.25	Meets process standards
Peel off (Lift-off speed)	14 ± 2m	0.17	0.21	Meets process standards
Printing speed (Squeegee speed)	200 ± 50mm/sec	0.048	0.36	*Insufficient accuracy*
Angle of attack (Squeegee angle)	$(15^{+2}_{-1})°$	0.12	0.27	Meets process standards

a sample size of $n = 11$. X and S were calculated, respectively, and the indices X_j and Y_j were derived based on tolerance d_j and target value T_j. The results (listed in Table 2) were plotted onto the analysis graph, as shown in Fig. 1.

Among the quality characteristics plotted in Fig. 1, off contact, peel off, and angle of attack met the quality requirements, whereas printing speed did not. The accuracy of the speed was sufficient, but precision was not (σ was too high). Therefore, this quality characteristic was selected as the focus for improvement and analyzed to enhance precision.

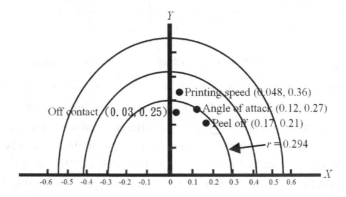

Fig. 2. PCI of multiple quality characteristics in screen printing

Discussions with process engineers and quality control personnel revealed that the speed of the servo motor (mm/sec) was the key factor in the precision of printing speed. This study investigated the speed of the servo motor, dividing it into three levels: 150 mm/sec, 200 mm/sec, and 250 mm/sec. Thirty samples were taken for each level, and the indices $\hat{C}^*_{pm1}$, $\hat{C}^*_{pm2}$, and $\hat{C}^*_{pm3}$ were derived as 1.006, 1.288, and 1.140. Next, we obtained $F_{max[3,29]} = 3.385$ from Eq. (15). From a table look up, $F_{max[3,29]} > C_o = 2.4$, indicating that differences exist between levels. The confidence intervals C^*_{pm1}/C^*_{pm2}, C^*_{pm1}/C^*_{pm3}, and C^*_{pm2}/C^*_{pm3} were calculated, the results of which are listed in Table 3. The results from multiple comparison analysis of Table 3 showed that the precision of Level 2 was higher than those of Levels 1 and 3 (a smaller C^*_{pmj} indicated higher precision). Thus setting the speed of the servo motor at 200 mm/sec induces optimal precision in printing speed.

Table 3. Confidence interval analysis of precision indices for servo motor

	Lower limit of confidence	Upper limit of confidence	Analysis results
C^*_{pm1}/C^*_{pm2}	0.1201	0.7212	$C^*_{pm1} < C^*_{pm2}$
C^*_{pm1}/C^*_{pm3}	0.3476	1.8905	$C^*_{pm1} = C^*_{pm2}$
C^*_{pm2}/C^*_{pm3}	1.1578	6.5634	$C^*_{pm1} > C^*_{pm2}$

6 Conclusion

With the substantial expansion of the touch panel market and demanding quality requirements, enhancing product quality is the priority of many touch panel manufacturers. Screen printing is a crucial procedure in an exceedingly complex process, and its quality has a profound impact on subsequent processes. Flaws in quality may lead to uneven print thickness, smeared ink, incomplete pattern lines, and poor resolution. These defects affect the functionality of the final product and thereby increase costs. This study applied the approach used by Chen et al. (2003)[6] and utilized the assessment index proposed by Chen et al. (1999)[4] to develop an assessment model and an analysis graph suitable for multiple quality characteristics of screen printing, enabling evaluation of process quality. The Hartley's test method, proposed by Pearson and Hartley (1966)[17], was then employed to construct a C^*_{pmj} index testing model and a multiple comparison method using F_{max} as the test statistic. Optimal parameters were identified for CTQs that were found to result in an insufficient process capability for the increase of process stability. The instruments developed in this study can assist touch panel manufacturers to enhance process efficiency and quality.

Acknowledgements. The authors would like to thank the National Science Council of the Republic of China for financially supporting this research.

References

1. Boyles, R.A.: Process Capability with Asymmetric Tolerances. Communications in Statistic: Computer & Simulation 23(3), 615–643 (1994)
2. Chan, L.K., Cheng, S.W., Spring, F.A.: A new Measure of Process Capability C_{pm}. Journal of Quality Technology 20(3), 162–175 (1988)
3. Chen, J.P., Chen, K.S.: Comparison the Capabilities of Two Process Using C_{pm}. Journal of Quality Technology 36(3), 329–335 (2004)
4. Chen, K.S., Pearn, W.L., Lin, P.C.: A new generalization of Cpm for processes with asymmetric tolerances. International Journal of Quality & Safety Engineering 6(4), 383–398 (1999)
5. Chen, K.S., Pearn, W.L.: Capability Indices for Processes with Asymmetric Tolerances. Journal of The Chinese Institute of Engineers 24(5), 559–568 (2001)
6. Chen, K.S., Pearn, W.L., Lin, P.C.: Capability Measures for Processes with Multiple Characteristics. Quality & Reliability Engineering International 19, 101–110 (2003)
7. Chen, K.S., Chen, W.T.: Multi-process Capability Plot and Fuzzy Inference Evaluation. International Journal of Production Economics 111(1), 70–79 (2008)
8. Crevelling, C.M., Jeffrey, L.S., David, A.J.: Design for Six Sigma in Technology and Productn Development. Prentice Hall, New Jersey (2003)
9. Franklin, L.A., Wasserman, G.: Bootstrap lower confidence limits for capability indices 24, 196–210 (1992)
10. Hoshimura, T.: Displays for business machines: operation liquid crystal displays of photocopier machines. Displays 23(1-2), 25–29 (2002)
11. Huang, M.L., Chen, K.S., Hung, Y.H.: Integrated Process Capability Analysis with an Application in Backlight. Microelectronics Reliability 42, 2009–2014 (2002)

12. Kane, V.E.: Process Capability Indices. Journal of Quality Technology 18(1), 41–52 (1986)
13. Kushler, R.H., Hurley, P.: Confidence bounds for capability indices. Journal of Quality Technology 24, 188–195 (1992)
14. Lynch, J.J., Chen, M.N.: A Six Sigma Approach to Touch Panel Quality Improvement. Journal of Quality 15(4), 271–281 (2008)
15. Murata, A., Iwase, H.: Usability of touch-panel interfaces for older adults. Human Factors 47(4), 767–776 (2005)
16. Pearn, W.L., Kotz, S., Johnson, N.L.: Distributional and Inferential Properties of Process Capability Indices. Journal of Quality Technology 24(4), 216–231 (1992)
17. Pearson, E.S., Hartley, H.O. (eds.): Biometrika Tables for Statisticians 3rd edn. By permission of the Biometrika Trustees (1966)
18. Wang, C.C., Chen, K.S., Wang, C.H., Chang, P.H.: Application of 6-sigma Design System to Developing an Improvement Model for Multi-process Multi-characteristic Product Quality. Proceedings of the Institution of Mechanical Engineers, Part B, Journal of Engineering Manufacture 225, 1205–1216 (2011)

The Application of Web Mining Ontology System in E-Commerce Based on FCA

LiuJie He[*]

Modern Education Technology Center, Huanghe Science and Technology College,
Zhengzhou, 450063, Henan, China

Abstract. Currently, FCA techniques are revealing interesting in supporting difficult activities that are becoming fundamental in the development of the Semantic Web. As the foundation of the semantic web, ontology is a formal, explicit specification of a shared conceptual model. Web mining techniques also play an important role in e-commerce and eservices, proving to be useful tools for understanding how ecommerce and e-service Web sites and services are used. Finally, web mining ontology information system in E- Commerce based on formal concept analysis is proposed based on integrating of ontology and formal concept analysis. The experimental results indicate that this method has great promise.

Keywords: formal concept analysis(FCA), ontology, web mining, E-Commerce.

1 Introduction

Ontology aims at an extension of the current Web by standards and technologies that help machines to understand the information on the Web so that they can support richer discovery, data integration, navigation, and automation of tasks[1]. Formal concept analysis (FCA) proposed by Wille, which provides a theoretical framework for the design and discovery concept hierarchies from relational information system. By using this semantic web technique, the information necessary for resource discovery could be specified as computer-interpretable.

E-commerce (electronic commerce or EC) is the buying and selling of goods and services on the Internet, especially the World Wide Web. Ontology is one of the branches of philosophy, which deals with the nature and organization of reality. Applied to the study of artificial intelligence (AI) ontology is a logical theory that gives an explicit, albeit, partial account of a conceptualized real-life system. In other words, data storage, data management, data transmission and even analysis rely on computer and network technology. This technology has enabled ecommerce to do personalized marketing, which eventually results in higher trade volumes. Web data mining is the process of extracting structured information from unstructured or

---

[*] Author Introduce: LiuJie He(1976.9-),Male, Han, Master of The PLA Information Engineering University, Research area: ontology, web mining, FCA.

D. Jin and S. Lin (Eds.): Advances in ECWAC, Vol. 2, AISC 149, pp. 429–432.
springerlink.com   © Springer-Verlag Berlin Heidelberg 2012

semi-structured web data sources. Web Extraction also referred as Web Data Mining or Web Scraping.

Formal concept analysis (FCA) is an order-theoretic method for the mathematical analysis of scientific data, pioneered by R.Wille[2]. Currently, FCA techniques are revealing interesting in supporting difficult activities that are becoming fundamental in the development of the Semantic Web. Although information and communication technology marks a new era and enables complicated e-business, it faces difficulty in the distributed and heterogeneous information sources. Ontology building is a task that pertains to ontology engineers, an emerging expert profile that requires the expertise of knowledge engineers (KEs) and domain experts (DEs).

Finally, web mining ontology information system in E- Commerce based on formal concept analysis is proposed based on integrating of ontology and formal concept analysis. With help of web data mining we can connect to a website's web pages and request ontology information or a pages, exactly as the browser would do. The web server will send back the html web page which you can then extract specific information from the web pages. The paper offers a methodology for building ontology and carries on ontology information system for knowledge sharing and reusing based on concept lattice in E-Commerce. The web mining execution ontology system supports common B2B and B2C (Business to Consumer) scenarios, acting as an information system representing the central point of a hub-and-spoke architecture based on FCA.

2   Web Mining Information System in E-Commerce Based on Ontology

Ontology is also a popular research topic in knowledge management, cooperative information systems, electronic commerce, information retrieval, intelligent information integration and medicine, among others[3]. The ontology is a computational model of some portions of the world. It is a collection of key concepts and their inter-relationships collectively providing an abstract view of an application domain.

Web-mining techniques also play an important role in e-commerce and eservices, proving to be useful tools for understanding how ecommerce and e-service Web sites and services are used. The implementation of business-to-business (B2B) eCommerce systems is fully realised in extended enterprises. It is the seamless exchange of information on top of an existing long-term relationship that distinguishes the extended enterprise from other forms of long-term collaboration such as a supply chain relationship.

Ontologies are widely used in different domains to give standard representations and semantics to concepts, predicates and actions of a particular domain. Thus, previously, authors proposed the problem-driven case representation to capture the problem solving steps of problem solvers. Our method uses this attribute to compute the degree of complementary to one product.

Domain ontology can help users locate and learn related information more effectively. A *generation* relation between a domain and its corresponding category means the *"is-kind-of"* relationship. The formula 1 which calculates the similarity of data acquisition is as follows.

$$\text{Confidence of A}\rightarrow\text{B} =\{(K_1,W_{i1}),(K_2,W_{i2}),\ldots(K_j,W_{ij})\ldots(K_n,W_{in})\} \tag{1}$$

Ontology is a conceptualization of a domain into a human understandable, machine-readable format consisting of entities, attributes, relationships, and axioms. Ontology provides common understanding of the domain knowledge and confirms common approbatory vocabulary in the domain, as well as gives specific definition of the relation between these vocabularies from formal model of different levels. E-commerce and E-services are claimed could be killer applications for web mining, and web mining now also plays an important role for E-commerce website and E-services to understand how their websites and services are used and to provide better services for their customers and users.

Enterprises are getting more knowledge intensive, and the integration of various types of knowledge becomes a challenge. The information and knowledge is in different formats, e.g., electronic documents, databases, and hardcopy documents, scattered in various systems such as Product Lifecycle Management (PLM). The class at the tail of an arrow is the domain of the relation and the class at the head of the class is the range of the relation. The ontology is a computational model of some portions of the world. It is a collection of key concepts and their inter-relationships collectively providing an abstract view of an application domain.

3   Using FCA Model to Build Ontology System

The transformation from two-dimensional incidence tables to *concept lattices* structure is a crucial paradigm shift from which FCA derives much of its power and versatility as a modelling tool. Given a domain, a concept in FCA is a pair of sets: a set of objects, which are the instances of the concept in that domain, and a set of attributes, which are the descriptors of the concept [4].

In FCA a concept is defined within a *context*. A context is a triple (O,A,R), where O and A are two sets of elements called *objects* and *attributes*, respectively, and R is a binary relation between O and A. The formula 2 which calculates the similarity of Agents is as follows.

$$\begin{aligned} A' &:= \{m \in M | gIm \text{ for all } g \in A\} \\ B' &:= \{g \in G | gIm \text{ for all } m \in B\}. \end{aligned} \tag{2}$$

The triple $K=(G,M,I)$ is called a *(formal) context*. If $A \subseteq G$, $B \subseteq M$ are arbitrary subsets, then the *Galois connection* is given by the following *derivation operators*. The aim of ontology is to obtain, describe and express the knowledge of related domain. Ontology provides common understanding of the domain knowledge and confirms common approbatory vocabulary in the domain, as well as gives specific definition of the relation between these vocabularies from formal model of different levels.

4 The Application of Web Mining Ontology System in E-Commerce Based on FCA

Finally, web mining ontology information system in E- Commerce based on formal concept analysis is proposed based on integrating of ontology and formal concept analysis. In this paper, we apply the theory of concept lattices and ontology to automatically construct the concept hierarchy of mining in e-business and to match up the binary relation matrix of documents and terms to express the independence, intersection and inheritance between different concepts to form the concept relationship of ontology. Fig.1 shows the detailed the application of Web Mining Ontology System in E-Commerce based on FCA. The experimental results indicate that this method has great promise.

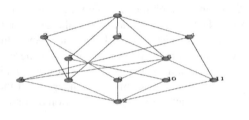

Fig. 1. Web Mining Ontology System Structure in E-Commerce based on FCA.

5 Summary

Web mining essentially has many advantages which makes this technology attractive to corporations including the government agencies. In this paper, we adopt ontology and FCA to improve the performance of data mining from documents. Web mining ontology information system in E- Commerce based on formal concept analysis is proposed based on integrating of ontology and formal concept analysis.

References

1. Berners-Lee, T., Hendler, J., Lassila, O.: The Semantic Web. Scientific American 5 (2001)
2. Maedche, A., Staab, S.: Ontology Learning for the Semantic Web. IEEE Intelligent Systems 16(2), 72–79 (2001)
3. Fensel, D.: Ontologies:Silver Bullet for Knowledge Management and Electronic Commerce. Springer, Berlin (2000)
4. Formica, B.A.: Ontology-based Concept Similarity in Formal Concept Analysis. Information Sciences 176, 2624–2641 (2006)

Study of the Three-Dimensional Map about Rural Complicated Scene

Ronghua Gao[1,2], Huarui Wu[1,2,*], and Haiying Ma

[1] National Engineering Research Center for Center for Information Technology in Agriculture,
Beijing 100097, China
[2] Key Laboratory for Information Technologies in Agriculture, Ministry of Agriculture,
Beijing 100097, China
wuhr@nercita.org.cn

Abstract. At present, construction of a rural scene must rely on professional and technical personnel and professional three-dimensional mapping software. Therefore, it needs a great deal time, energy and devotion to construct a complicated scene. This paper builds a rural scene component library with rural new properties, and gets real-time external terrain data with rural mapping assistant equipment. It can build a rural three-dimensional scene graph quickly through an interactive way and using an organization technique of resource list-style. According to experimental results, comparing to the traditional modeling software, this method in this paper is out standing for its simple functions, friendly interaction, short development cycle and favorable commonality.

Keywords: three-dimensional mapping, modularization, modeling, coordinate transformation.

1 Introduction

With the wide application of computer technology and multimedia technology, digital planning and design of rural scenes is playing an important role in new rural construction [1, 2, 3], and the map with two-dimensional symbol has no longer met people's need. The realistic three-dimensional scene in rural areas and rural appearance of the village both show their virtual forms through the Internet and the computer-aided system platform, which provides the digital support for the new rural planning and construction. However, there are no digital technology solutions of realistic three-dimensional scenes facing the planning and contraction for rural scenes. Rural area is vast, terrain is complex and diverse, residence live far and separately, bulky collection equipment is not easy to carry, expensive consumption limits the ordinary users, and the data format is not open which restricts the reuse of data collection. All of these show that external data collection of rural scenes has serious constraint.The object-oriented design promotes the development of modeling techniques [4, 5], the researchers build the base class of model object based on

---

* Corresponding author.

similarity, then a generic object model hierarchy forms [6,7]. However, if the "class" data structure of object-oriented methods were used as the basic unit of system conduction simply, the model development would be difficult to maintain. As the rural scenes have wide range and complex structure, what the object-oriented technology covers can not meet the visualization needs of the rural scene any longer. this paper offers a embedded interface for rural scene simulation which needs the three-dimensional visual performance in the future, also, it provides a strong support for the development of China's new rural simulation software, obviously,

2   Rural Scene Vector Topographic Data Generation

Transform the topographic data collected by rural mapping assistant into geometric data with point-surface structure, and get the visualization terrain vector graph. The terrain vector graphs are classified by land to define different surface structure. By capturing every facet of the corner point coordinates and connectivity, the terrain vector graph descried as follows:

```
g face 1                          g face 2
v 313.07 276.04 0.00              v 378.62 275.58 0.00
v 368.08 276.04 0.00              v 433.63 275.58 0.00
v 368.08 254.49 0.00              v 433.63 254.03 0.00
v 313.07 254.49 0.00              v 378.62 254.03 0.00
v 313.07 276.04 0.00            v 378.62 275.58 0.00
f 1 2 3 4 5                        f 6 7 8 9 10
g face 3                          g face 4
v 312.65 307.40 0.00              v 448.50 217.69 0.00
v 367.66 307.40 0.00              v 271.64 199.24 0.00
v 367.66 285.85 0.00              v 268.05 221.75 0.00
v 312.65 285.85 0.00              v 445.18 243.51 0.00
v 312.65 307.40 0.00              v 448.50 217.69 0.00
f 11 12 13 14 15                  f 16 17 18 19 20
...... .                          ......
```

Where g is the class name of surface, v is the angle point coordinates of each surface, then x and y coordinates has been converted to the local coordinates of current coordinate system by coordinate transformation matrix. Set the default value of z 0, making the terrain vector graph uniformly distributed in the grid interface of system, on which fast three-dimensional mapping of rural scenes are completed, insuring terrain data and model data get matched quickly and correctly thereby. The f is the connection order between angle and points, which is the topology information of surface. It can be achieved that all patches zoom in accordance with the proportion at one time by using three dynamic interpolation algorithms.Search all terrain geometry corner coordinates, find the values : $x_{min}, x_{max}, y_{min}, y_{max}$, set two-dimensional vector map bounding box B for:

$$(x_{min} - C, y_{min} - C), (x_{max} + C, y_{min} - C), (x_{max} + C, y_{max} + C), (x_{min} - C, y_{max} + C)$$

Rectangular area consists of four points, in which C is a constant used to set the scope of the rectangular area generated. Let ground of the scene zoom with the terrain and vector proportional while zooming, for saving running time, this paper uses zoom ratio labeling method. Add tag parameter S into the zoom function, S is the current zoom ratio. Then, it's only need to pass the parameter S to four points of the bounding box to zoom in or out S-fold rather than re-search the values of $x_{min}, x_{max}, y_{min}, y_{max}$ under current proportion. In this way, the ground of the scene can be generated faster.

3   Rural Scene Object Component Modeling

China's rural scene has the similar characters with three-dimensional model about point, line and surface, but also has its own features. According to the current goal of the new rural construction and planning, this paper concludes that the rural scene component library contains the following factors: Constructions of everyday life in the rural scene, Public buildings in the rural scene, Commercial constructions in the rural scene, Models for agricultural production in the rural scene, Infrastructures in the rural scene. All building components possess the characteristics of the current rural scene, reflecting the village appearance in the new rural development. Files in component library are organized in tree structure. The entrance of the component library is the parent node. Different types of scenes are the child nodes to connect the parent and other children, and they inherit the properties from a parent node, but also act as summaries to the next child node component. Users can input object quickly when call the component library, just based on the current characteristics of the component and through a query to the sub-tree of a tree structure. First, load the component name. And then judge whether the component is any one category among the houses owned by the villagers, government offices, agriculture, tertiary industry

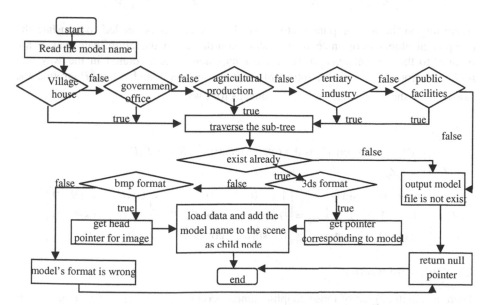

Fig. 1. Flowchart of reading component database file

and public facilities. If it does not belong to any one of a class, output that the component does not exist directly, and return null pointer. If it belongs to any one of a class, then traverse the current sub-tree under the class, and judge the existence of the component name. If the component name does not exist, then output that the component does not exist, and end the loading process. If the component name exists, then judge the current format. If the format is 3ds, then get the component pointer, and load the component data and make the component name as a child node, add it to the scene, then the whole loading process is done. If the two formats are both not exist, and then output that loading file type is error, return a null pointer, and end the component loading process, which is shown as Figure 1.

4   Coordinate Transformation and Object Interaction

4.1   Coordinate Transformation

For two Cartesian coordinates that the axes of the two are parallel correspondingly. The translation from the coordinate in one of the two to the coordinate in the other can be achieved by calculating the coordinate translation parameter (X_t, Y_t, Z_t). The coordinate conversion from the original coordinate (X_o, Y_o, Z_o) to the target point coordinate (X_p, Y_p, Z_p) is shown as follows:

$$\begin{bmatrix} X_p \\ Y_p \\ Z_p \end{bmatrix} = \begin{bmatrix} X_o + X_t \\ Y_o + Y_t \\ Z_o + Z_t \end{bmatrix}, \text{Where} \begin{bmatrix} X_t \\ Y_t \\ Z_t \end{bmatrix} = \begin{bmatrix} X_p - X_o \\ Y_p - Y_o \\ Z_p - Z_o \end{bmatrix} = \begin{bmatrix} \Delta X \\ \Delta Y \\ \Delta Z \end{bmatrix}.$$

According to the topographic vector data of the scene, it is needed to translate the component object coordinate from system coordinate to the coordinate that is not parallel to the current axis. If make the component object rotated in the way that rotate φ around the axis Y, rotate ω around the axis X and rotate θ around the axis Z, then the coordinates would become as follows:

$$\begin{bmatrix} X_p \\ Y_p \\ Z_p \end{bmatrix} = R \begin{bmatrix} X_o \\ Y_o \\ Z_o \end{bmatrix}, \text{ where } R \text{ is the rotation matrix, } R = R_1 R_2 R_3,$$

$$R_1 = \begin{bmatrix} \cos\varphi & 0 & -\sin\varphi \\ 0 & 1 & 0 \\ \sin\varphi & 0 & \cos\varphi \end{bmatrix}, R_2 = \begin{bmatrix} 1 & 0 & 0 \\ 0 & \cos\omega & -\sin\omega \\ 0 & \sin\omega & \cos\omega \end{bmatrix}, R_3 = \begin{bmatrix} \cos\theta & -\sin\theta & 0 \\ \sin\theta & \cos\theta & 0 \\ 0 & 0 & 1 \end{bmatrix}$$

4.2   Object Interaction

There is a high degree of loose coupling among component modes. The trigger events and interactive events would take work, while the component models update their status or send interactive messages to others. According to the characteristics of trigger events and message events, rural scene object components in this paper is composed by interaction class and object classes associated with the interaction class, among which the interaction class is the connection link among these object classes. This component will dynamically monitor messages sent by other components, and then accept and process events if the message is sent by the sender to the order indeed. If only part of the object classes are aware of the message need to receive, and sender object class does not know the existence of the receiver object class, then this type of the class is classified as a trigger-type component. These components send the message, however they don't concerned about the message receivers. As long as the receivers ordered the message, then it would accept and process the message. The information interaction based on this method makes the interaction between the simulation model components more flexible, and provides better interoperability.

5 Simulation and Analysis

To verify the practicality of building block fast mapping, this paper uses Visual C++ 6.0 based on component to develop rural scene fast three-dimensional mapping system on the basis of components. Figure 2 is a flow chart of simulation, which will contain rural mapping assistant with GPS-enabled devices, and this assistant will pass the real-time collection of data to the external space of Agriculture data server to generate points, surface description of the terrain data and to realize vector data visualization through the rapid three-dimensional terrain mapping system. Refer to the terrain vector, the component library built by the auxiliary modeling tools and the high-definition rural scene image library got by cameras combined with coordinate transformation and the triggering message, will show three-dimensional renderings of rural scene rapidly, and finally store the scene after three-dimensional mapping as a scene data. Users need not have much knowledge of professional mapping, but can complete three-dimensional mapping operations of complex rural scene in a short time interactively.

Figure 3 shows a rural scene rendering of three-dimensional mapping.The realistic visualization highlights the WYSIWYG three-dimensional scenes of rural, and gives a favor to the new rural planning and design. The menu system is divided into three parts, including menu modules, interface modules and drawing. The menu module achieves the import of the terrain data, adding lighting conditions, different object texture mapping and the modification of the current color of the background; Function module contains the quick adding of the component library and the interaction with scene editing, they are located in the bottom of the menu and the left side of the interface, as all components of the model are displayed with the button photo vividly, users are more aware of the object information of the current component library. Apart form adding two objects of standard component model and village component library, this paper also provides editing function of the graphics interface, in this way, it can be achieved that the show of different perspectives, the grid of the scene objects and the physical switching.

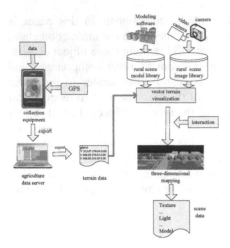

Fig. 2. Experiment simulation   **Fig. 3.** Three-dimensional rural scene

6   Conclusion

While make the digital planning and designing of the rural scenes, the existing three-dimensional mapping software shows shortages such as weak operability, poor commonality and strong professional. In order to solve these problems, this paper presents a complex rural scene building block rapid three-dimensional mapping method, aiming at the scene structure characteristics of the mixture of the current new rural life and production. Experimental results show that the proposed modular quick three-dimensional mapping method are better for that it is handled simply and friendly interactive, its development cycle is short, and it has high versatility, and the re-build of the models is not needed when the scene changes, in this way, the repetitive work is reduce.

Acknowledgments. This work was supported by the by the National Natural Science Foundation of China (60871042, 61102126), National Technology R&D Program (2010ZX01045-001-004, 2011BAD21B02), Youth Scientific Research Fund Project of Beijing Academy of Agriculture and Forestry Sciences.

References

1. Cherkasov, G.N., Nechaev, L.A., Koroteev, V.I.: Precision Agricultural System in Modern Terms and Definitions. Russian Agricultural Sciences 35(5), 334–338 (2009)
2. Zhao, C., Wang, J., Huang, W., Zhou, Q.: Spectral Indices Sensitively Discriminating Wheat Genotypes of Different Canopy Architectures. Precision Agric. 19 (November 2009) (published online)
3. Wang, F., Mei, Y., Liu, Q.: Virtual Prototyping Kinematics and Structure Analysis of Cotton picker First. Journal of Agricultural Mechanization Research 3, 58–61 (2010)

4. Lin, R., Afjeh, A.A.: An Extensible, Interchangeable and Sharable Database Model for Improving Multidisciplinary Aircraft Design. Journal of Aerospace Computing, Information, and Communication 1(3), 154–172 (2004)
5. Wu, Y., Yu, L., Piao, Y.: Aero engine numerical simulation platform based on VC++ language. Computer Engineering 34(7), 257–259 (2008)
6. Cao, Y., Jin, X., Meng, G.: Non-linear modular modeling and simulation of aeroengine. Journal of Computer-Aided Design and Computer Graphics 17(3), 505–510 (2005)
7. Reed, J.A., Afjeh, A.A.: Computational simulation of gasturbines:part II-- extensible domain framework. Journal of Engineering for Gas Turbines and Power 12(2), 377–386 (2000)

The Research of Search Engine System in e-Commerce Based on Formal Concept Analysis

XiuYing Sun*

Modern Education Technology Center, Huanghe Science and Technology College,
Zhengzhou, 450063, Henan, China
sunxiuying2011@sina.com

Abstract. Formal Concept Analysis is a branch of applied mathematics, it comes from the philosophy of understanding of the concept. According to the field with a binary relation expressed in the form of background, to extract the concept hierarchy, that concept lattice. The paper puts forward the construction and application of search engine system in e-commerce based on FCA. Therefore, the system has to narrow the search to expand content. This system can be expanded and narrow the search results reveal the contents of the scope of this class, either browse the knowledge they want to discover new knowledge can reveal deeply hidden content.

Keywords: formal concept analysis, search engine, concept lattice.

1 Introduction

Formal Concept Analysis was born in about the 1980s, is a data analysis tool or method, especially the information can be given to investigate and deal with. Each node of concept lattice is a formal concept, consists of two parts: the extension and intension. Extension is all covered by the concept of a collection of objects, meaning that these objects were owned by a collection of common features[1]. One of the most important reasons is to make decisions in a certain situation most reasonable by efficiently collecting relevant knowledge from heterogeneous domains. Navigation system to establish the main purpose of the search engine is to narrow the search, and search results in the association is established between them, so that users associated with the discovery of new knowledge of these.

The use of FCA search results for the two treatment classification method. Search result for a huge number of query words, can allow the user to narrow the search range, can more accurately find the content of their needs. Navigation of search results is the most concentrated expression of the large number of directory search engine. Navigation style search engine in the establishment of the database is based on extracts from the pages of the same or similar search terms on the organization's web pages. Construction of lattice conceptual clustering process is the

---

* Author Introduce: Xiuying Sun,Female,Han,Master of education technology of Northeast Normal University, Research area: formal concept analysis, ontology, contemporary and long-range education.

D. Jin and S. Lin (Eds.): Advances in ECWAC, Vol. 2, AISC 149, pp. 441–444.
springerlink.com   © Springer-Verlag Berlin Heidelberg 2012

process of building the concept lattice cell has a very important position. For the same data, the resulting grid is unique, that is, data or properties from the impact of the order, which is one of the advantages of concept lattice. The navigation of search results is an important way for users to gain the results they need and to narrow search scope effectively. Through the navigation of search results, users can gain the search results they need more quickly and the search scope is also narrowed effectively.

In formal concept analysis, concept lattice construction process is actually the process of conceptual clustering. And, for the same number of data, the generated lattice of the only way without data or attribute order effects, it is also one of the advantages of concept lattice. Traditional search engines are difficult to solve using technology users "to find the information difficult" issue. Essence of this difficulty is the lack of knowledge of search engine capabilities and understanding of the information is only used to retrieve the keyword matching to achieve mechanical. Through the use of concept lattice model, the user's search results are in accordance with the different attributes of the document, taking the form of their background, and then generate a concept lattice, the concept of search results in different forms presented to the user.

Finally, this paper puts forward the construction search engine navigator in e-commerce based on formal concept analysis. FCA on the search results in using the second method of dealing with the classification. For the huge number of search results for the query words, allowing users to narrow the search scope, you can more accurately find the content they need. For now, intelligent search engine is in a development stage, is developing a smart search engine applied to all kinds of artificial neural network technology, but some search engines just have individual characteristics, has not been an impact, the search engine to the user provide intelligent services there are three general strategies. System and the key in the search results made, the number of grid map to navigate, it is clear, as in the future to show the navigation tree at the root node to satisfy two conditions: (1) the relationship by looking down or must contain all most of the grid nodes; (2) when a node-level drill-down as less.

2   Formal Concept Analysis and Concept Lattices

As a branch of applied mathematics, FCA (formal concept analysis) comes of the understanding of concept in philosophical domain[2]. A software system was developed to allow automatic case-based interpretation of images and integrates the material described in the previous sections. A formal context K: = (G, M, I) consists of two sets G and M and G, M relationship between the composition $I \subseteq G \times M$, G in the form of the background elements are called objects, M elements are known as the formal context of the property, if gIm or (g, m) I, said, "the object g has attribute m".

Given a formal context assume a formal context K: = (G, M, I), where G is a collection of objects, M is the set of attributes, I is a binary relation between them, then there exists a partially ordered set with correspondence, and this produces a set of partially ordered lattice structure, which forms the background of a background (G, M, I)-induced lattice L is called a concept lattice. Lattice L of each node is an ordered pair (ie concepts) is denoted by (X, X), which is called the concept of extension XG, X M as elements of the concept. Ordered pair (X, X) on the relation R is complete, that is the nature of formula 1 and formula 2.

$$X' = \left\{ x' \in M \;\middle|\; \forall \; x \in X \; , xR \; x' \right\} \tag{1}$$

In the concept lattice nodes can establish a partial order, given C1 = (X, X), C2 = (X2, X2), then C1 <C2 X2 <X2, we can understand this partial order for the sub-concept - the concept of super-relationship. Partial order can be generated based on the Hasse diagram grid, if C1 <C2, and there is no other element C3, making C1 <C3 <C2, then there is from C1 to C2 is an edge. Both intelligent agent and semantic web service technologies are able to reach remarkable achievements and in some cases have overlapping functionalities.

$$X = \left\{ x \in G \;\middle|\; \forall \; x' \in X' \; , xR \; x' \right\} \tag{2}$$

Construction of lattice conceptual clustering process is the process of building the concept lattice cell has a very important position. For the same data, the resulting grid is unique, that is, data or properties from the impact of the order, which is one of the advantages of concept lattice. Concept lattice construction algorithm can be divided into two categories: batch processing algorithms and incremental construction algorithm.

3 Classify Search Results Based on FCA Method

Essence of this difficulty is the lack of knowledge of search engine capabilities and understanding of the information to be retrieved only by mechanical keywords match to achieve[3]. From the above analysis we can see, in the search results grid-making process, classification techniques based on FCA has the information retrieval based on keywords from the current level to knowledge-based (or concept) level, which is the solution to the problem fundamental and critical, about intelligent search engine, this is what we are talking about intelligent search engine, which is the development trend of the future search engine. Order theory and lattice theory as a practical application combined with the product, concept lattice has been considered a powerful tool for data analysis, due to the mathematical concept lattice has a good nature and suitable for batch processing, etc.

Before we know, when users search, do not want to present in front of search results is haphazard, and want to see is class specific, clear layer of search results[4]. Through the use of concept lattice model, the user's search results are in accordance with the different attributes of the document, taking the form of their background, and then generate a concept lattice, the concept of search results in different forms presented to the user.

4   FCA Search Engines in the Principles of Navigation

System module is divided into four parts, and operation of the system as follows: for the user after the search terms submitted to the data interface (Data sources are other search engines) to get back page, back page in the page properties of these concentration to obtain collection, to these pages as an object, get set for the property making the property grid, to create the grid at the interface to navigate a tree of that out, while the relationship between those pages will be mapped to the navigation tree. FCA-made frame, the construction of concept lattice database: to get the page for the object to extract the properties they are made for the property grid, the grid in the form of these pages to be organized in the database. Based on this background create a grid form as shown in Figure 1.

Fig. 1. The search engines grid of formal concept analysis.

5   Summary

In this paper, methods of formal concept analysis of the secondary processing of search results, a classification system to organize search results. In the system, a document-object to these pages for the attributes making grid search terms, these relationships through the navigation tree to map the system to page navigation. The paper discusses the construction and application of search engine navigator in e-commerce based on formal concept analysis.

References

1. Oosthuizen, G.D.: The application of concept lattice to machine learning. Technical report, University of Pretoria, South Africa (1996)
2. Ganter, B., Wille, R.: Formal Concept Analysis: Mathematical Foundations. Springer, Berlin (1999)
3. Nie, J.Y.: A general logical Approach to inferential information retrieval. Encyclopedia of Computer Science and Technology, 203–226 (2001)
4. Godin, R., Missaoui, R., April, A.: Experimental comparison of navigation in a Galois lattice with conventional information retrieval methods. International Journal of Man-Machine Studies 38, 747–767 (1993)

The Influence Mechanism and Principle by Studying Advertising on Consumer Psychology

Fan Yang

Institute of Information Technology of Guilin University of Electronic Technology
Guilin Guangxi P.R. China 541004
Thomas_yang8415@hotmail.com

Abstract. As far as we know that designing an advertisement, if we want to win the majority of consumers favor we have to think more and before than the customer does, by only according to the psychological needs of consumers. If the enterprise, moreover, wants to win in the competition, they must conquer consumers, but firstly conquering the hearts of consumers. Human psychological activities directly govern consumer buying behavior, so how to make the consumer's psychological activity to become the consuming behavior? In all other factors under the same circumstances, as well as the brand is not much difference between, what make advertising" persuade" the customer? This paper tries to illustrate the effect of advertising mechanism and principle on consumer psychology, through the customer's psychological characteristics and example analysis.

Keywords: advertising, customer psychology, the influence mechanism and principle.

1 Introduction

When the child grows up, people usually cannot notice their daily physical growth change, but after a period of time, people will suddenly find that their kids have grown up. It is quite difficult to measure the psychological impact of people by after read or watched commercial ads, which just like us, cannot notice a child had grown much in the past 24 hours. As a result of advertising on consumer psychology influence too subtle, we are unable to pay much of our attention on it, but the fact is that such little, tiny, and various advertising effects will decisively influence people to choose goods, especially in goods and all other factors under the same circumstances, as well as the brand is not much difference between. Therefore, what are your ads "persuade" your customers?

2 Feather Effect

Imagine that it is a set of scales, scales of commodity brand components are equal, then the balance in the state getting equilibrium. In this case, as long as putting on one end of the scale, with a very light feather can make the balance tilted. So likewise, for consumers, consumer choice of goods if under the brand is no difference; one thing that can make balance incline is the feather - advertising, rather than the heavy weight – brand.

Goods can be classified into two categories, high consumer involvement goods and low consumer involvement goods. Consumer involvement is refers to the consumer goods related information in the search, the time spent and consumer conscious processing commodity information and advertising effort, it decided to consumers on the categories of information selection and decision making. High involvement product is the consumer brand commodity with higher loyalty, such as automotive, education, insurance; Low involvement product is the lower consumer brand loyalty, such as daily necessities of life.

People buy low involvement goods by considering decision making is often only one second or less. Because for people, many available options of low involvement product brand are very similar, sometimes nearly the same, so there is no need to spend a lot of time on purchasing each of a commodity. For consumers, such sore of goods, are indifferent. At this time, advertising, seemingly not worth mentioning effectiveness can play a maximum effectiveness. In the influence of feather effect, even subtle changes can also cause tipped the balance. Repeat, repeat, repeat, and repeat the process through, no perception of small effect that causes the brand appeared large, obviously difference.

3 Image

The effect produced by image ads for high involvement product is more obvious. Differ from low involvement product, people always carping and carefully selecting when they choose high involvement product such as automotive, real estate, even stock. However, customers tend to drown in the sea of brand with no idea of distinguish. At this point, advertising that is committed to guide the brand image, will determine the customer's purchase. In the case of automobile, BMW car image focus on handling, comfortable the occupants enjoy, Volvo car image focus on safety performance. Image advertising effect is to cover the customer on a brand awareness gradually turned to one of its unique attributes. Just as people said BMW will think of comfort, said Volvo would think of safety. BMW comfort or Volvo safety can change the customers of the brand attributes of mental score. According to the Theory of cardinal number, if people can make the vehicle comfort scoring from 1 to 10, score of customers with who pay attention to comfort for BMW will be higher than almost any other brands of cars, perception of the brand indispensable component.

Secondly, advertisements like Coca Cola, Lux soap, Rolex watches, Lewis Vuitton, St-Laurent brands are often focusing on customer itself who used these brands. Unlike most advertising expression, this type of advertising is more focused on the display image of the brand loyal users, that is the one and only for customers to create a sense of achievement: this kind of brand merchandise is not an ordinary person can use or this kind of brand goods not in some ordinary circumstances can use. Therefore, in this kind of advertising effect, customer image of user or image of environment used will lack subjectivity, leading customers to find a completely, perfectly, reasonable, rational, fully convincing reasons when selecting of goods. To purchase decision will not be difficult as long as there is human vanity at work.

4 Persuasion

Normally, people think the function of advertising is to convince people to believe it. In fact, many of the world's mechanism investigation shows that people do not care about the content of the advertisement, even have a sore of point of view like " I won't run to buy advertising goods". Humans have a very strong ability to see things in different ways. In consumer psychology, this is called the customer sensation and perception. Quality of a brand, a product or a service is good or bad, the result will be not same after their own thinking. People associate the product attributes, greatly influence the customer perspective of goods or services. Pearl, as an example, is regarded by most people as beautiful jewelry, mostly as a gift for friends, ladies favorites, fashionable, expensive price and so on. Like any other goods, however, pearl also have features people reluctant to associate. Pearl is just cancer tumor formed in big oyster. Normally, of course, the customer will never think of pearls of these negative characteristics. When the customer is in the best of spirits selecting of pearls, can you imagine that if they has been told the fact of cancer tumor of oyster? That will definitely negative affect the customer purchases of pearls. Although this sentence is absolutely true at all, customers do not necessarily love to listen.

Therefore, advertising influences customer to evaluate a brand, a product or a service, depends entirely on how customers perceive it. The customers' attention will only focus on certain attributes of one brand or one product. Through advertising, attention shifted to the other properties, might change the product perception of the customer. In addition, the advertising text or image makes the product more prominent positive attributes, so that customers would associate the brand of those advantages characteristics when thinking of the brand. In fact, on one hand, the customer said not to buy advertising goods. On the other hand, however, they actually had been "persuaded" by the advertisement already.

5 Conformity and Crowd Psychology

Conformity refers that individuals be affected by outside the crowd behavior, and in their perception, judgment, understanding displayed conforms to public opinion or the majority of human behavior. This kind of behavior arises from some individual psychological needs. C·A·Keesler (1969) proposed conformity behavior produces four kinds of demand or desire: 1. Keep everyone in line to achieve the goal of the group; 2. in order to gain favor of the rest of the group; 3. to maintain good interpersonal relationship status; 4. Do not want to feel pressure from the group. In general, women make crowd behavior more than men; introverted inferiority person more than outgoing, confident people; low educated people more than high educated people; the younger more than the older people; social experience shallow people more than the man has rich social experience.

People hate to be thought by the others that they are different, or do not match to the trend of public, or fear of others say his new in order to be different, curry favors by claptrap. Therefore, to be on the safe side, people tend to try to guess others how to do things, at the same time also be replicated, as is to ease the group pressure. Teenagers, for instance, is the uniting groups. Adolescent young customer has the most strong herd mentality, this kind of client's lack of sense of security and confidence. On one hand, they tend to their parents outmoded values have a serious reverse psychology, at the same time they make own decisions on their small group.

The customer will receive two aspects influence when making a selection. One is what they want, and second is what other people think in their own opinion. According to the theory of Maslow's hierarchy of needs, safety is the need of customer demand need after completion of the basic physiological needs. The more lack of security, the more likely to be influenced by other people. Like Chinese Melatonin (Brain Platinum)'s advertisement, for example, said that "Just Send Melatonin as Gifts". After 10 years of repeated transmission, this advertising has formed a kind of influence character by environment of herd mentality. When a customer found his surrounding comrades sent only this Brain Platinum as gifts, all his concerns will be eliminated, because of conformity, and finally accept the ads and pick up the goods as gift as well.

6 Summary

Commodity competition, in some sense, also is the advertisement competition. In many of today's advertising war, the one who's advertising novel and clever, might seize the consumer's visual, and earn more on a return. People evaluate products, from advertising on a number of factors, to contact their visibility and to consider, bring about a total impression. Our gods, the customer, are fond of the new and tired of the old. The old advertisement in the information database of human brain is stored for a short period of time will be replaced after the new advertising appears. Only those new, interesting, artistic advertisement can long time remain in people's minds. In commercial ads operation, we must take full account to meet needs of consumers. This is an extremely important factor. Shopping motives are in many aspects. Some people out of the actual demand, some people out of psychological needs. This is about to rely on advertising function, it can induce people's desire for consumption, promote people's shopping action.

References

1. Ma, M., Wang, Y.: Theoretical Models and Empirical Research Relative to Advertising and Consumer Psychology. Brand Space, 9–27 (April 2006)
2. Xiao, Y.: Advertising Effect and Consumer Psychology. Hubei Adult Education Institute Journal 9(4) (July 2003)
3. Li, Z.: Research and Postmodern trend on Advertising and Consumer Psychology. Guangxi Social Sciences 1, 177–180 (2005)

References

1. ...
2. ...
3. ...
4. ...

The Development of Intelligent Shopping System in e-Commerce by PHP and MYSQL

Jing Li[1,*] and JianChen Wan[2]

[1] Modern Education Technology Center, Huanghe Science and Technology College,
Zhengzhou, 450063, Henan, China
[2] Department of Information Engineering, Huanghe Science and Technology College,
Zhengzhou, 450063, Henan, China
lilijingjing2011@sina.com

Abstract. The online shopping system functionality needed to be divided from the user point of view, the user functions can be divided into front and back office management functions. Develop a network of shopping system based on ASP WEB personal terminal server and MYSQL database using PHP to develop advanced, high stability and security. This paper develops online shopping system by PHP and MYSQL. System realizes the basic function of the site, including customer registered login, visit the site information, information query, shopping and fill in the order.

Keywords: intelligent system, PHP, MYSQL, ASP.

1 Introduction

Increasingly popular in the Internet today, the network is the protagonist, is the darling of the times. Large family in the network, e-commerce is a hot spot. Expression is a basic e-commerce online shopping, from a certain extent, it is the online supermarket, which is an online sales. "Network" information age, the word means that it has quick and convenient features. In fact the emergence of e-commerce, to the position of consumers, consumer attitudes have brought important changes[1]. The paper designs implementation of an online shopping site. Internet shopping carries a large amount of information, shopping process more cumbersome, and increasingly rampant Internet hackers and viruses on the network security of a higher challenge. Develop a network of shopping system based on ASP WEB personal terminal server and MYSQL database using PHP to develop advanced, high stability and security.

Internet shopping truly business sales on the Internet, making it greater market opportunities and economic benefits to a large extent to improve the core competitiveness of enterprises. The online shopping system functionality needed to be divided from the user point of view, the user functions can be divided into front and

---

* Author Introduce: Jing Li(1975-),Female, lecturer, Master, Huanghe science and technology college, Research area: computer application, e-commerce, web design.

D. Jin and S. Lin (Eds.): Advances in ECWAC, Vol. 2, AISC 149, pp. 451–454.
springerlink.com   © Springer-Verlag Berlin Heidelberg 2012

back office management functions. Function primarily to customers shopping users, including user registration, login, shopping cart, view orders, etc.; background management functions primarily for system managers to use, including user, product, order management. User registration, visit the Web site users can view online, order products, and in the forum to communicate with other users and comment and so on. Part of the design of these pages and use PHP to connect, using the database to create the relevant tables, so its view, modify or delete. Web site designs site administrators to complete maintenance and management. Permission to use the database administrator to set permissions feature, an administrator can add the information on the goods, modify, and delete, you can also order information for processing, while administrators can manage user information.

This paper expounds the development of online shopping system with design. Design USES now more popular PHP website development technology and considering website the structure characteristics of data processing and the knowledge you have learned, applied MYSQL database system as the backend database website. System realizes the basic function of the site, including customer registered login, visit the site information, information query, shopping and fill in the order and administrators for the website maintenance and update. The page design and production, the basic format for pages, use HTML language out of the large frame, then use the Zend framework Dreamweaver8 and studio inside do detailed design.

Combining the design the meaning of the Internet in the business the application and development of an overview; Then for the whole system is analyzed, including feasibility analysis, requirement analysis, system of business process analysis, data analysis, etc.; On this basis to determine the functionality of the site, will the functionality of the site target classified, and the successive subdivide the son function module; The next modules, designed and implemented in each module can complete its function under the results will be its integrated into a complete system.

2   Analysis of the Key Technologies

The system uses PHP technology development. Because PHP and similar CGI, ASP, JSP has a unique advantage compared. PHP technology, language and ASP, JSP relatively similar, but has been improved in many ways, and take lessons from the ASP and JSP some reasonable places. It's better than ASP technology platform independence, but also has good scalability[2]. With PHP technology, Web page designers can use HTML or XML tags to design and style of the Web page, use the PHP tags to generate dynamic Web pages. On the server side, PHP scripts the production content of the request, acting on the server side. This will be compared to some of the core code open source, speeds development and cross-platform, while ensuring the height of any Web browser compatibility.

MYSQL is a comprehensive and complete database and analysis product. With the browser from a database query function to achieve a rich Extensible Markup Language (XML) features can be strong evidence to support, MYSQL support for the comprehensive Web-enabled database solutions. At the same time, MYSQL scalability and reliability are still maintains a number of benchmark records, but both features are also the enterprise database system in the fierce market competition, the key to success in Kedi. MYSQL has also introduced a complex new security features: a powerful and flexible role-based server, database and application security configuration; integrated security audit tool to keep track of 18 different security events and their sub-events; precision file and network encryption support, including SSL (Secure Sockets Layer). MYSQL has passed the U.S. Government C2-level security certification, with the industry's highest level of security.

Contrast MYSQL and other large-scale database management system, MYSQL reliable security, faster memory speed, high compatibility, MYSQL as back-end database application development system provides a strong support, and later software operation provides a solid foundation.

$$confidence(X \Rightarrow Y) = P(Y \mid X) = support(X \cup Y)/support(X) \qquad (1)$$

PHP scripting language, is an object-based and event-driven and have the security of the scripting language, its purpose is to use the HTML hypertext markup language, scripting language implemented in a web page with links to more than one object, and customer interaction network, which can develop client applications. It is embedded in standard HTML or transferred to the implementation of language. CSS cascading style sheets (Cascading Style Sheet) is a set of formatting rules that control the appearance of web content. CSS can be very flexible to use and better control of the specific page appearance, from precise positioning of layout to specific fonts and styles. And can greatly facilitate the work of creating web pages. The design of the study and use of Cascading Style Sheets CSS, HTML text in the document will be a series of external storage to a CSS rule. CSS file. Use it to unify the design of the website page background and external frame styles. In a statement provided the background of the page, scroll bar color style, etc.

3   The Analysis and Design of e-Commerce Intelligent System

Overall design of the system's main function is to be completed conceptual and logical design, including conceptual design is to demand the analysis of the user needs the concept of an abstract model of the process; it is the key to the entire database design. Logical structure of the design task is to design a good conceptual design phase of the basic ER diagram into a product with the choice of DBMS data model supported by the logical structure of the line. Needs analysis will be summarized above. Come to the site two business actor, that users and administrators. And then develop their activities. Corresponding to the site's functionality can be determined, the corresponding Web site user activity front business processes; activities corresponding to the Web site administrator background business processes. In the analysis of business processes, try to consider the overall business activities of the protagonist, the function of the system as complete as possible.

Front, customers can log in after the home page link to view the merchandise and other information, you can also do the ordering of goods and other activities. To facilitate the search for interested customer product information, product search will be embedded into the main page[3]. The database will be important as part of the background site, the administrator of the main operating almost all around the database, last updated data will be displayed to the front.

4 The Implementing of Intelligent System in e-Commerce by PHP and MYSQL

The system design process in the design of the website contains page (HTML) design and server-side validation mechanism design. PHP scripting language, it is a very popular dynamic web technology. It is not only able to achieve dynamic results page, but also to a certain extent, an error is detected on the page, which to some extent to achieve security of the site. The system dynamic PHP web application server technology, and technology in support of reusable components, the combination of client-side dynamic HTML web technology and seamless integration with the design, so as to facilitate the implementation of the system. The whole system is divided into a front and back office modules function modules. The data flow diagram of the overall site as shown in Figure 1.

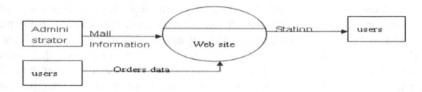

Fig. 1. Data flow diagram of the overall site.

5 Summary

This paper develops online shopping system by PHP and MYSQL. System realizes the basic function of the site, including customer registered login, visit the site information, information query, shopping and fill in the order and administrators for the website maintenance and update.

References

1. Maras, J., Stula, M., Petricic, A.: Reverse engineering legacy Web applications with phpModeler. JCIS 2(1), 82–93 (2011)
2. Dong, F., Tian, Y., Yuan, D.: Study on Cross Tabulation Query in Data Mining and Statistical Data Analysis. JDCTA 5(7), 315–319 (2011)
3. Nie, J.Y.: A general logical Approach to inferential information retrieval. Encyclopedia of Computer Science and Technology, 203–226 (2001)

Review on Policy Risk in Developing Property in the Chinese Mainland

YuHong Pan, DongMei Zhang, and JiaMei Pang

Faculty of Management, Chongqing Jiao Tong University, Chongqing, China
panyuhong3@hotmail.com

Abstract. This paper reviews the policy risks existing in the process of developing commodity houses in Mainland China. Policy risks have been identified, and their effects on the implementation of a property project across various stages have been analyzed. The study provides useful references for professionals, in particular those overseas property investors, to identify proper risk management strategies for mitigating the effects of policy risks to their property business in the Chinese market.

Keywords: risk analysis, property development policy, China.

1 Introduction

Proper policy environment is an important vehicle for developing economies to any country. Property development in those advanced countries or areas relies to large extent on the provision of proper policies, which are made under legal system. Nevertheless, legal system in China mainland used to be very poor. In the process of implementing economic reforms since early 1980s, the Chinese Government has been reforming and introducing various legal systems. such as the Constitution of the PRC in 1982, Civil Law of the PRC in 1986, Land Administration Law of the PRC in 1986, Urban Planning Law of the PRC in 1989. In fact, legal system includes laws and regulations. In China, laws are made by the National People's Congress or its Standing Committee, and regulations are usually made by the State Council for national effect and by Local Government for local effect. There are other official rules and regulations. The study by Walker et al [1] presents a law-making hierarchy in the PRC, as shown in Figure 1

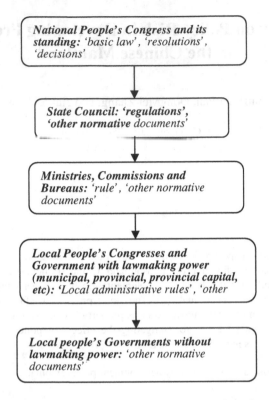

Fig. 1. Lawmaking Hierarchy in China Mainland

2   Review the Policies of Property Industry

Under the legal framework, various regulations and policies guiding the operation of property industry have been introduced, for example, the Regulation for the Implementation of Land Administration Law of the PRC issued by the State Council in 1991 [2], the Interim Regulations of the PRC on Granting and Transferring the Right to the Use of State-owned Land in Cities and Towns issued by the State Council in 1991 [3].   Local administrations can also enact local regulations for the purpose of implementing the laws of the Central Government, supplementing broad national laws, and dealing with local issues which are not covered by national legislation, for example, the Regulation for Property Management issued by Tianjin People's Government in 2002 [4]. This study focuses on discussing the policy and regulation for property development, issued by the State Council and Local administrations in China.

With its huge population, China has been considering housing as a complicated and comprehensive social problem. Housing shortage was one of the biggest social problems in China before the country adopted "open door" policy in 1980s. The Chinese property development industry was introduced accordingly when economic reform programs were introduced. The huge demand for housing and property has been the driving force for developing the property industry and the industry has become one of the major economic sectors in the Chinese economy. The report by the National Bureau of Statistics of China shows that the gross floor Space of Buildings completed in 2009 was over 726 million m2 , from null in early 1980s [5]. In line with the development of the industry, the Chinese Government has introduced various laws and regulations for guiding the development of industry. There was a shortage of regulations for supporting the industry under the old planned economy system. Under the planned economy system, construction related activities were treated as public services. Housing was centrally planned and allocated to citizens for free charge with symbolic rent. There was no competition for developing property projects among enterprises as they worked not for profit maximization but simply for the completion of the tasks allocated officially. The allocation of land did not accord to efficiency principles, thus land users did not need to improve their economic efficiency [6]. As land use was free of charge, low-rent housing could be provided to almost all the workforce. The welfare housing system was adopted over three decades in China from early 1950s and there was no property market for property transaction or lease. However, the old system caused the serious problem of housing shortage. It distorted the balance between supply and demand as land was not allocated according to the economic efficiency but largely by political criteria.

The "Open Door" policy and reform programs have brought great development to the Chinese housing industry. The per capita floor space of residential building in urban areas increased from 6.7 m2 in 1978 to 29.1 m2 by the end of 2009 [5] . In the process of this development, various legal measures have been introduced. In 1988, the land-use-rights for property development was granted to the user for the first time in Shenzhen. In 1998, the Chinese Government declared the abolishment of the old welfare house system [7], and development of commercial properties had been encouraged since then. These measures have provided business opportunities in using land for developing property. Many commodity buildings, including residential, commercial and industrial buildings have been erected in recent years. Significant proportion of these properties has been developed by overseas wholly-owned or joint venture development companies. According to China Statistic Yearbook 2010 [5], the number of overseas wholly-owned or joint venture raised from few in carly 1990s to 5, 733 in 2009.

As policy system is considered as a vehicle for the Government to ensure the implementation of its reform programs, the Chinese Government has been devoting to establish a legal system that can ensure the proper development of the property industry. Laws and policies along with the development of the property industry mainly concern a wide range of aspects including land title and supply, housing reform, urban planning, environmental protection, pre-sale and sale of the commodity house, land finance, property management, title transfer and registration. They are issued by the authorities either at central or local levels. Local Governments need to issue the policies for administrating the local property market, such as the Notice for Administration of the Commodity House Overseas or Local Sales issued by Beijing State Land Resource and Real Estate Administration Bureau in October 2000 [8], Notice for Commodity House Pre-sale relating to the Adjustment of the Construction Project Condition issued by Guangdong Provincial Construction Bureau in January 2001 [9]. However, the establishing process involves adjustment and new introduction of regulations and policies in responding to various changes, new developments and uncertainties in the industry. For example, the Regulation for Pre-sale and Sale of the Commodity House by the Ministry of Construction in 1999 was revised in 2001 [10] for reducing the illegal behavior in practice and ensuring the legal rights of purchasers. The Regulation of Supervising to the Construction Project Tending in Beijing has been revised three times since it was in effect in 1987, being revised in 1994, 1995, and 2003 respectively [11]. By adjusting policies, the Government has been improving the policy environment for ensuring that the property development in China is towards proper direction. Due to lack of experience and knowledge in operating marketing economy, the Chinese Government has to issue laws and policies on a trial basis in many cases. In this practice, revision and amendments of policies are often used as tools to adjust the mistake or error committed. It is considered that further reform on legal system in China will continue particularly in order to meet the requirements of China's entry to the WTO, thus changes and revisions on property development policies are still expected in the coming future.

Nevertheless the changes and reversions on policy present risks and uncertainties to property developers' investment plan. For example, previously relevant Government departments would be responsible for the calculation of salable areas. But the revised regulation "Administration on Pre-sale of Commodity House [10] stipulates that the developers have to compensate the purchasers if the gross floor area in the contract is different with the gross floor area surveyed. The new policy not only induces more restrictions on the sale of the commodity house, but also shifts the risk of inadequate calculation of salable areas to developers.

It is considered that the effects of policy risks are more significant to property industry in comparison to other business sectors. Property development industry is generally perceived as a higher return business. However, a high profitable investment such as property development usually carries higher risks, and this is echoed by previous study, for example, Thompson and Perry (1992) [12]. For reducing the possible loss induced by policy risks in property industry, a proper risk management strategy is important. Traditional studies have identified a comprehensive list of risk factors in the operation of property business and techniques for analyzing risk significance have been developed for managing risks [13-15]. However, these identifications were largely based on examining overall aspects including economic, political, physical, and technical aspects in developing property projects. Furthermore, these studies do not consider the difference of risk effects in different project stages. In fact, the impacts of risks on a specific project will change at different stages during the development process. In order to make an adequate investment decision on a property development project in China, investors will find valuable to establish a proper understanding about the policy risk environment where the economy system is still under transition.

3 Conclusion

The property industry is still at the infant stage in Mainland China, and uncertainties and problems are expected and forthcoming. Issuing new policies and revising old policies are unavoidable for ensuring healthy development of this industry. Various changes and revisions often happen to the existing policy system. These changes or revisions are necessary in order to regulate and supervise the development of the market particularly during economic transition process in line with the commitments made by the Chinese Government to WTO entry. However these changes bring the risks to property development across all stages.

Acknowledgements. This study is funded by a project of humanities and Social Science of Ministry of Education of the People's Republic of China (Project No. 11YJA790110), Ministry of Human Resources and Social Security of the People's Republic of China, Chongqing Social Science Association (Project No. 2010YBJJ12), Ministry of Housing and Urban-Rural Development of the People's Republic of China (Project No. 2010-R5-6).

References

1. Walker, A., Levett, D., Flanagan, R.: China Building for Joint Ventures, 2nd edn., p. 94. Hong Kong University Press (1998)
2. SC 1991a, The Regulation for the Implementation of Land Administration Law of the PRC issued by the State Council (April 1991)
3. SC 1991b, The Interim Regulations of the PRC on Granting and Transferring the Right to the Use of State-owned Land in Cities and Towns, issued by the State Council (May 1991)
4. TJPG 2002, The Regulation for Property Management, issued by Tianjin People's Government (2002)
5. NBSC 2010, China Statistical Yearbook 2010. China Statistics Press (2010)
6. Li, L.H.: Urban Land Reform in China, Antony Rowe Ltd, 24–27, 55, 58 (1999)
7. SC 1998, The Notice relating to Further Reform of Urban and Town Housing System and Speeding up Residential Construction, issued by the State Council (July 1998)
8. BJSR 2000, The Notice for Administration of the Commodity House Overseas or Local Sales, issued by Beijing State Land Resource and Real Estate Administration Bureau (October 2000)
9. GDCB 2001, The Notice relating to Process Condition of the Project Imagine Adjustment for the Pre-sale of Commodity House in the Province, issued by Guangdong Construction Bureau (January 2001)
10. MOC 2001, The Regulation of Commodity House Sales, issued by the Ministry of Construction (2001)
11. BJPG 2003, The Regulation of Supervising to the Construction Project Tending in Beijing, issued by Beijing People's Government (March 2003)
12. Thompson, P.A., Perry, J.G.: Engineering Construction Risks, pp. 1–4. Eastern Press Ltd (1992)

13. Li, Q.M., Li, X.D., Shen, J., Lu, H.M.: Risk, Uncertainty and Decision-making in Property Investment, 1st edn., pp. 36–42, 95–126. Southeast University Publisher (1998)
14. Shen, L.Y.: Application of Risk Management to the Chinese Construction Industry, Department of Construction Management University of Reading, 12–31 (1990)
15. Shen, L.Y., George, W.C., Wu, Catherine, S.K. Ng: Risk Assessment for Construction Joint Ventures in China. Journal of Construction Engineering and Management 127(1), 76–81 (2001)

On Macro-control Mechanism of the Environmental Capacity Production Factor Market

Yanli Li[1], Lijun Li[2], and Huijing Wang[2]

[1] School of Economics and Management, Shijiazhuang Tiedao University,
17 Beierhuan East Road, Shijiazhuang, China
qhdlyl@sohu.com
[2] Shijiazhuang Tiedao University, 17 Beierhuan East Road, Shijiazhuang, China
lilj@stdu.edu.cn, llj56857@sohu.com

Abstract. Believing in that the traditional economics underestimated the importance of environment as a production factor, the paper demanded to view environmental capacity as a basic production factor. Based on such a belief, the macro-control function of environmental capacity production factor market was discussed as well as the control policy goals, methods and operation mode of this macro-control function. The macro-control policy of the environmental capacity production factor, being a normal policy, can act on the society with finances, money and other macro-policies to buck for sustainable development.

Keywords: macro-control, environmental capacity, harmonious, sustainable development.

1 Introduction

The contradiction between environmental system and economic system grows sharper and the government has the macro-control tightened both in force and in extent. Therefore it is important to explore a new macro-control mechanism which can not only regulate the relationships between the two systems but also economic growth simply Based on the environmental capacity production factor theories(Hereinafter referred to as the ECPF theories), the paper discussed the macro-control function of the ECPF market and the control policy goals, methods and operation mode.

2 The Formation of the ECPF Theories and Basic Concepts

As what economics told us Economics reveals that, production factors are the resources for business activities as well as the basic source of the value. With William Petty (1662) putting the land is the mother of the wealth while labor is the very father and dynamic factor in Theory of Taxation, the production factor theory started. Adam Smith (1776) introduced us the noted three-factor theory in An Inquiry into the Nature and Causes of the Wealth of Nations. Besides the land and the labor, he thought the factor of capital should be added. He also noted that no matter what kind of society you are in, the product price can eventually be divided into these three factors or be represented by one of them. Alfred Marshall (1890) pointed out in Principles of Economics that it seems more reasonable to see entrepreneurship as an independent factor of production and put forward four-factor theory. The contemporary research in general takes sees technology and information as the fifth production factor. Among the production factors, labor, capital, entrepreneur, technology and information are all the factors relating to people while the land is the only one to represent environment. With the environment crisis goes deeper, an increasing number of researchers ,based on the law of substance and energy flawing between environment system and economic system, thought the traditional economics overestimated the influential power coming from people and at the same time underestimated the importance of environment as a production factor. The structural capacity matters of the environment such as environment capacity, self-purification capacity of environment etc are also participated in the production of wealth and value creation, but they can hardly be found in production theory. Putting nature in the first place, Marshall himself admitted, "In a sense, there are only two factors of production, and that is nature and human beings."

Now, many scholars in China advised that environment should be seen as a factor of product factors.The environment production factor theory thought that natural environment inclued is the tangible material entities (such as nature resources) the intangible structure and capacity of the environment (such as environmental capacity and self-purification capacity),which should be regarded as the basic production resources, and be mentioned in the process of economic analysis and law exploring for their contribution in production, which also means to attach importance to their power in creation of production value and to make it join in the macroeconomic accounting system. Human beings got a clearer idea to the tangible nature resources in early times. Economics generally takes the natural resource as the production factor of land. But it is quite late to know something about environment capacity. Although the environment capacity, acting as one of the basic sources of social wealth, took part in the activities as soon as people involved in business activities, its power as the production factor was underestimated. People started looking into it to a further extent since the environment resources were almost used up and the environmental emergencies became prevalent. In traditional point of view, production emissions were received by environment capacity in a passive way. At the base of the environment viewpoint with material and non-material integrated, founded on the sustainable development's need to environmental system and economic system, at the aim of correcting the restricted knowledge of the production factor theory to environment.As an independent factor ,the environmental capacity , that effected the economic activities could be view as participating in production activities initiatively and providing self-purification capacity of environment for production consuming. Being two members of the natural factor, environmental capacity and land were together named as environmental factor.

ECPF studies mainly concept and content of factors of production of environmental capacity, necessity and feasibility, the effects of function of the production manufacturer and changes of vendor behavior, market supply and demand discipline and mechanism, environmental capacity factors of production management policies and measures, environmental capacity factor markets building and macro-control effects, green GDP and green accounting system based on environmental factors, environmental compensation on market mechanism etc.

Market of production factor of environment capacity may act on the market allocation of limited production factor of environment capacity, meet energy saving and emission reduction's request and make sure the environmental capacity to recycle. With government environmental management's total amount control system, market of production factor of environmental capacity can also activate the function of adjust and control between the two systems to make them a harmonious one.

3 Macroeconomic Objectives and Means of the Market of ECPF

The macro-control of the national economy's market mechanism in an indirect way is the basic social responsibility of the government. The normal macro-control policies such as fiscal policies and monetary policies and so on which belong to economic system itself, can hardly act on the environment system and adjust the relations between the two systems. Macro-control policies are not closed. Besides the policies mentioned above, many other sub-systems related to the whole economy and society may become the means and focus of the macro-control policy to influence the whole system. The market of production factor of environmental capacity, whose influential power is not only acting on the economic system itself but also the relationship between them, is a sub-system. The latter, aiming at economic growth and environment quality stability, by means of supply and price of the ECPF includes self-stability policy, environment-oriented policy and economic-oriented policy and so on. We will give another paper for it. This paper is mainly about the former.

3.1 Macro-control Target of the Market of ECPF

Generally speaking, macro-control target of the market of production factor of environmental capacity include the fiscal policies and monetary policies such as economic growth, price stabilization, full employment and balance of international payments. Apart from these, it can also bring us a good environment quality for economic growth with high quality when achieving the economic goals. That is to say, the targets relationship include qualified economic growth at a high speed and natural environment being along well with the production growth. Macro-control target include two kind of sub-item: steady growth of the total size of economic production and stable economic development of the internal structure. The environmental factor market can harmonize economic overheated and economic contraction and implement national structural policies to achieve a smooth and coordinated development of national economy as a whole.

3.2 Macro-control Means of the Market of ECPF

By two basic means of supply and price, ECPF market can change the environmental capacity of enterprise consumption for a better effect to environment quality, harmonizing the relationship between economy and environment. In the area of macro-control, production factor market of environment provides means as the following two main types.

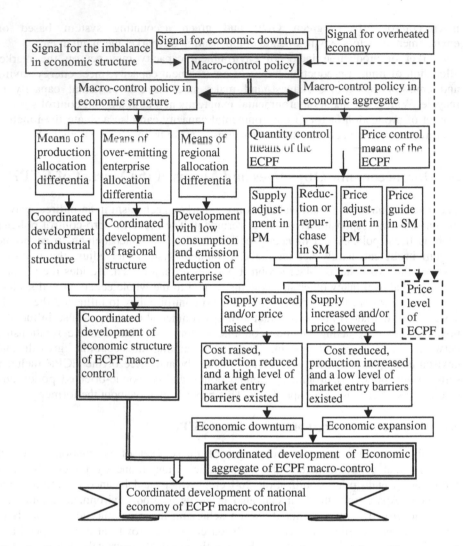

Fig. 1. Government indirect control of internal economic system from Macro-control Mechanism of the ECPF market

Means of economic structuratml adjusent policies.In the implementation of economic structural adjustment policies, three methods can be used by market of production factors of environmental capacity. That are means of production allocation differentiation, means of regional allocation differentiation and means of over-emitting enterprise allocation differentiation. These methods generally work through environmental product factor supply in primary market(PM) and will not appear differences of the supply price in the PM. Secondary market(SM) will also have no special influence in economic restructuring.

Means of economic aggregate adjustment policies.Means of economic aggregate adjustment policies include quantity control means and price control means of the ECPF. According to the market in which they are playing an important role it can also be divided into PM means and SM means of the ECPF. Specifically speaking, it is divided into four types: means of supply adjustment in PM, means of reduction and repurchase in SM, means of price adjustment in PM and means of price guide in SM.

4 Macro-control Policy Types and Discretion of the ECPF Market

In adjusting the relationship between economy and environment, control policy includes self-stability, environment-oriented and economic-oriented and others. In an aspect of macro-economic control, market of ECPF providing the principal means of control includes the following two types.

4.1 Macro-control Policy Types of the ECPF Market

According to the direction of the policy act, the macro-control policy of the ECPF market mechanism can be divided into expansion policy and tightening policy.

Expansion policy of the ECPF market is the means of policy which causes the expansion of the production capacity and the increase of the effect of the economic aggregate, mainly relating to means of increasing the supply and lowering the price of the ECPF. It can be done both in the PM and SM. When increasing the supply, enterprise's consumption can be enlarged and the other restricted production supply will be replaced at the same time with the expansion of the production directly. The production capacity therefore can be expanded further. When the price of the ECPF declined, with the production cost declined directly and the profit increased, both the cash flow of the enterprise and the ability of reproduction get better to expand the production capacity. These policies will have an influence to expande the economic aggregate.

Contrast to the expansionary policy, tightening policy of ECPF could cause the production capacity tightened and economic aggregate reduced, including means of the decline of supply and the increase of the price. The principles and processes of the function are opposite to the expansion policy mentioned above. It is no longer talked specifically here.

The combination of supply-reduction means and price-increase means, as a strong tightening combination of ECPF policy, has an apparent effect on compressing economic aggregate. Means to use one of supply-reduction means and price-increase means alone and at the same time keep the other one unchanged, being a weak tightening policy combination, is of soft influence on economic aggregate. The mix of supply-increase means and the price-reduction means, being a strong expansionary combination of ECPF policy, can help stimulating and promoting the expansion of the economic aggregate. With one means being stable, we use either supply-increase means or price-reduction means which can influence the economic aggregate lightly by forming a weak expansionary combination.

4.2   Macro-control Discretion of the ECPF Market

Macro-control mechanism of the ECPF market needs to make choices according to the situation by the authorities.

As shown in Figure 1, when there is a sign for the overheated economy in economic society, tightening policies of ECPF could reduce the supply and raise the price. The degree of tightening policy could take some adjustment according to the situation of the overheated economy. Either the supply reduction or price-increase of the ECPF will lead to the restriction of the consumption. With the production cost raised and the market entry barriers apparently exist to the emerging enterprise, the emission of the enterprise reduced as well as the production, the economic aggregate will also decline. When economic downturn coming, we can adopt the expansionary policy of the ECPF to increase market supply and lower the price. The intensity of the expansionary policy could also be adjusted to get along with the situation of the economy. As supply raised or price declined, the enterprise's enthusiasm to consume the ECPF for the expansion of the production is stimulated.

The price of the ECPF is a crucial means for macro-control as well as a signal for the change of the economic aggregate. A high price of the environment production factor can be seen as a signal for overheated economy, while a price that keeps at a low level is viewed as a signal for economic downturn.

5   Conclusions

Macro-control mechanism of the ECPF market includes auto-control function between environmental and economic systems and macro-control function. With ECPF market, the government can achieve the functions of total scale control of economic production and control of the economic internal structure.

Acknowledgments. Funded by the social science research foundation from Hebei province , NO: HB11GL027.

References

1. Petty, W., Xia, Q., Lei, Y.: Theory of Taxation. Hua Xia Press, Beijing (2006)
2. Smith, A., Xie, Z.-J.: An Inquiry into the Nature and Causes of the Wealth of Nations. New Word Press, Beijing (2007)
3. Marshall, A., Zhu, Z.-T.: Principles of Economics. The Commercial Press (1997)
4. Li, Y.-Y., Li, M.-Z.: Macroeconomic Analysis. Tsinghua Press, Beijing (2003)
5. Li, I.-J., Li, Y.-L.: Theory of Environmental Production Factors. Technology of Shijiazhuang Railway Institute (Social Science) 4, 36–40 (2009)

On Establishing the Eco-compensation Mechanism in China's Financial Policy

Yanli Li[1], Lijun Li[2], and Tengfei Jia[2]

[1] School of Economics and Management, Shijiazhuang Tiedao University,
17 Beierhuan East Road, Shijiazhuang, China
qhdlyl@sohu.com
[2] Shijiazhuang Tiedao University, 17 Beierhuan East Road, Shijiazhuang, China
lilj@stdu.edu.cn, llj56857@sohu.com

Abstract. With the increasingly serious environmental pollution, it is a necessity and a tendency to strengthen ecological and environmental protection, and go on the eco-compensation mechanism. And to bring a green sustained future, it is agreed that establishing and perfecting eco-compensation mechanism with the appropriate financial policy are imperative to a coordinated development of our economy and environment. In this way, we can safeguard the justice in society, enhance the public's initiative to protect the environment, and provide effectively the ecological service.

Keywords: environmental crisis, environmental economy, financial policy, eco-compensation.

1 Introduction

Eco-compensation mechanism, which intends to protect the eco-environment and promote the harmonious development between man and nature, is an integration of some adjustments to various interests concerned about series of administration, legislation, market...etc, regarding of ecological protection and construction. Establishing and perfecting eco-compensation mechanism contributes to not only the transform of the environmental protection in the means from administration to the sum of administration, economy, law,etc, but also the construction of the resource saving type and environment-friendly society.Generally there are three methods regarding of the eco-compensation. The first is financial transfer, the second is project support, the third is the levy of eco-compensation tax.

2 The Theoretical Basis of Eco-compensation Mechanism

The natural resources in the environment, including not only the gifted resources such an land, water, biology, mine ,but also the comprehensive environmental resources just like environmental capacity, landscape, climate, ecological balance mediation, is the foundation of human survival. From the perspective of economics, the natural environment resources as another factor of development of economy. To some extent, as people put into materialized labor and living labor in the process of the natural environmental resources, the supply of ecological environmental resources with human abstract labor has value. So providers(consumers) should get(pay) some profit(cost).

At the same time, these ecological products have the property of public goods. Samurson thought, public goods means something that will not reduce after consumption or influence others, Pure public goods usually mean items which are discompetitive and disexclusive on the consumption at the same time, and we call the items which are disexclusive but competitive the total resources. As we can get in the word 'Discompetitive', it means that one's consumption on these items would not limit any other one's consumption, so we have the chance to share these items. And the 'disexclusive' is the situation that we cannot eliminate the others in the consumption, we can say it's impossible or too hard to charge for this commodity. The ecoenvironment is both of 'diexclusive' and 'discompetitive'. The supplying of public goods has the nature of economic externalities. In production or consumption, one of the main economic activity produced outside influence on the other principal, but it cannot be reflected by trading of market and price mechanism. The use of natural resources and the protection of eco-environment have the obvious economic externalities .Produced by the ecological benefit is a kind of invisible utility ,and it cannot prevent others from enjoying .Because most of the public will choose 'a lift', so the ring by the natural environment in the market won't get due compensation and maintenance .According to pigou's theory of externality ,when the marginal cost of private (yield) and the marginal cost of society (yield) deviate from each other. The government should take appropriate economic policy to take tax on the department or individual when the marginal cost is less than private marginal social cost, at the same time subsidy the social benefit department or individual when the marginal private income is less than the private marginal social cost. Therefore, the government should set up the mechanism of ecological compensation, and through the relevant arrangement of system, make the economic externalities to be internalizational, in other word, it's the Handelnde who will bear the consequences of external diseconomy and share the fruits of external economy, which will incent the supply of ecological products .In certain conditions ,through the definition of property rights and use the way of market transactions voluntary consultation for ecological compensation ,then eliminate economic externalities.

3   Some Successful Experience of Establishing Eco-compensation

As the us government take the ecological environmental protection and construction into account seriously, it puts into the parts of environmental protection funds to set up the compensation mechanism for soil and water conservation, namely the government and residents in the downstream which is benefited from the residents contributing to the environment in the upstream must make up the monetary. From the 1970s ,the emissions trading policy, which was a set of emissions trading system including. Bubble, offset, banking, netting... etc, has had good environmental economic benefits. At the same time the open-pit mines were administrated by licenses, and the reclamation of the mining was implemented by deposit system. And the deposit that failed to complete the reclamation plan would be used to aid : the third part for reclamation. The mining firm would pay a certain number of land reclamation funds about abandoned old mining areas to reclamate the old mining area and restore the eco-environment when every ton of coalis mined. In the 1990s, the cooperative association built by German and The Czech Republic, imposed the charges for disposing pollutants of enterprises and the economic compensation from downstream to upstream, in order to improve the qualities of water in eble which is throughout the Czech Republic and German. Meanwhile salinou that has extincted for a long time now has been releascd.

For the past few years, as the projects of the mine ecological environment restoration, sandstorm source control, natural forest protection, returning farmland to forest...etc, especially the greater protection of upstream in the natural forest sime the catastrophic flood in the Yangtze river in 1998, are not into operation. It is showed that the control government carring out the ecological compensation across areas with the method of finance transfer payment. Zhejiang yiwu paid 200 million yuan one-time for the 50 million cubic mater use rights of Hengjin Reservoir in Dongyang. The pay, 0.1 yuan every ton, for the comprehensive ecological management begins the eco-compensation in China. Guangdong Province extracted each 0.01 yuan from electricity fees to conserve water and soil and help the mountain areas out of poverty around the Henan xinfengjiang reservoir in Guangdong Province. At the same time, Guangdong Province put into 150 million funds to protect the environment at Anyuan, Xunwu, Dingnan. In Jiangxi Province stopped the mining of rare-earth and wolframite, and closed down about 330 mines in the purpose of building the first eco-compensation mechanism in china to prevent the loss of the soil and water of Dongjiang river source areas.

In the support of the country, Heilongjiang province has done a lot comprehensive managements in daqing oil field and the surrounding areas ecological environment. Through the implementation of cultivating the fence, feeding grass and trees, developing irrigation facilities and so on, we can take comprehensive treatments for about 2 million mu demonstration zone, the key mine and the stopped mine; we can also control gas and recovery ecological in the four subsided coal city. In the mind of 'organized by government, operated by enterprise, market financing and financial subsidies', we can take comprehensive managements to treat the polluted gas and water in our cities, then we will change the decentralized management used many years ago into the central management.

4   Financial Policies of Building the Eco-compensation Mechanism

The establishment of the eco-compensation mechanism can contribute to the harmonious development between man and nature, and also affect the interests distribution of central and local as well as between the various places of local. First it should clear the basic principles of the compensation mechanism, then formulate workable policies and measures.

4.1   The Basic Principles of the Eco-compensation Mechanism

It should take the adjustment of the distribution relationship between environment and economic interests as the core. and the internal related ecological protection or vandalism's external costs as a benchmark, the economic incentive for the purpose, adhere to the "development, protection, and benefit from, compensation" as the principle of consistency, who protects that who exploitation, who compensates that who benefits. Who occupy ecological resources or who damage the environment should pay for the action. Who exploits and utilizes the environment resources should pay the cost, the beneficiaries of ecosystem services have the responsibility and obligation to compensate to the regions and people that provide an excellent ecological environment. The formulation of development objectives and evaluation criteria should accord to different areas of various resources and population, economic, environmental amount, and the ecologically fragile areas take more responsibility to protect ecology rather than economic development. Establishing the mechanism that downstream areas compensate to upstream areas, , development areas to the protected areas, the benefited region to the damaged areas, urban to rural, and the rich to the poor.

4.2   Perfect the Relevant Finance Policy of Eco-compensation Mechanism

4.2.1   Establishing and Perfecting Incentive Policy for the Eco-compensation

The socialist market economy system reform's important goal is to give full play to the basic role of market allocation of resources. Encouraging the society, the folk capital invest to the area of protect and control the ecological environment, reducing the threshold, breaking the monopolies, lifting the control over the accession to the market of infrastructure environment protecting. Fostering environment factor market , building the mechanism that the distribution of pollutant dischargement index is paid as soon as possible, gradually carrying out pollution-discharge right trade under government's control. Adjusting and optimizing economic structure, developing alternative industry and characteristic industry, strongly promoting clean production, developing circular economy, and promoting regional industry shifting and factors flowing, promoting the environmental to the basic turn for the better.

4.2.2   Establishing and Perfecting Tax Policy for the Eco-compensation

The mechanism of eco-compensation is the best method to solve the economic interest relationship between the environmental protection and reasonable benefit, destruction and compensation through the market rules. It is to perfect current tax policy of protection of the environment for providing the establishment of the eco-compensation with financial guarantee.

For example, to reform current resource tax system. The exploitation of natural resources will damage the resources, especially nonrenewable resources, no matter whether the enterprise is benefited from. It is to consider produced quantity as the basis of tax assessment for the enterprise. So it can make the enterprise consider the market demand, rationally develop and utilize resources according to its own interest. It is to be based on rationally divided resources tax to rationally adjust resources tax amount. To practice different taxation standards for different levels of mining, oil and other natural resources' exploitation. To stage by stage and partially collecting differential eco-compensation tax rate to the mine enterprise of different regions, according to the environmental damage degree. To levy higher tax rate and limit their exploitation for the resources extraction which seriously damage ecological environment situation. To make charges that control pollution could turn into enterprise production cost. solving the environmental cost exteriorization, and at the same time, it can enhance the ecological environmental production consciousness of the enterprise. To change the original single duty preferential form of reduction or exemption, take preferential form of accelerated depreciation and tax expenditure, develop circular and green economy, establish a resource-conserving and ecological protection type society.

In addition, but also play a consumption tax in environmental protection in the role. Fully understanding the consumption tax's function which adjust consumption structure. Playing the consumption tax in a regulatory role which "containing the ban on levy" on environmental. Recently our country has listed luxury goods and goods which seriously damage the environmental in the levy limits, such as golf, bowling, disposable chopsticks and so on. In the future mobile phones , cards that taking timber as raw material, and disposable plastic in packing materials should be listed into the levy limits of consumption tax. Thus constanting the industry and the enterprise's development , which excessively consumption resources and damage environment. Formulating consumption and production mode that contributing to save resources.

4.2.3   Further Improving the Charge Policy of Eco-compensation Mechanism

Administrative charges is a main measures which government promote the unit to control pollution and improve the environment through economic means, also commonly used in rich countries. At present it is in the form of sewage charges collect by the environmental protection departments in our country. But, the fees that enterprises pay for the environmental damage is inadequate to the cost which the environment restoration needs, and it is difficult to establish eco-compensation mechanism. We should further expand the scope of fees, list various sources of pollution in the scope, and appropriately increase sewage charges standard, while it will be charged in fiscal budget management, and are mainly used for eco-compensation and protection of ecological environment. Specific include: first, implement online monitoring of major enterprise which discharge pollutants, collect fees of excess emissions for discharge exceeds. Second, it is should be introduced that the downstream which is benefited from exploiting resources provide financial compensation to the upstream where is conservation areas, or the upstream which cause pollution accidents offer compensation to the downstream that is contaminated. Third, establishing eco-compensation which is fund by the government, NGOs or individuals to support the ecological conservation projects.

4.2.4   More Strongly Supporting the Enterprise That Supply Ecological Products

It should use of the capital market financing means, support environmental protection enterprise to be listed in the capital markets for financing, form Form the green environmental protection plate on the capital market. and implement preferential policies, encourage private investment, promote the supply of the product. Financial system can provide a guarantee, give priority to provide low-interest loans, extend loan repayment period, accelerate depreciation of fixed assets, and offer certain tax breaks for the enterprise and individual that invest in environmental production industry.

4.2.5 Broadening the Financing Channels of Eco-compensation Funds

Our country has used part of the secondary treasury bonds to arrange the production of the environment, thus significantly increasing environmental protection investment in GDP year by year, and the effect is very significant. But it must be realized that china's economy is still in the process of transition, the government function scope is in expanding, and financial capacity is very limited, it is impossible to finish in a comprehensive eco-compensation which is completely depended on the government's large number of inputs. We should continue to use debt in the future, at the same time, also adopt various financial and taxation policies, and encourage investment in multi-channel, for example, first of all, we can provide low-interest loans, extend the loan repayment period, give tax breaks, allow accelerated depreciation of fixed assets and other incentives, and encourage private to actively participate in the green industry investment. Second, China's household saving which is as high as 16 trillion yuan make environmental protection financing possible through the capital market. Therefore, we can consider to issue ecological compensation fund lottery or bonds, or long-term environmental bounds, or provide various kinds of preferential policies to encourage more environmental protection enterprise to be listed in the stock market, form environmental production plate in the stock market, for raising more eco-compensation and environmental protection capital. Third, we should actively attract foreign direct investment capital in ecological projects. This requires government departments to improve financial openness, transparency and comperency and strengthening the consistency and the stability of the investment system, and create favorable conditions to introduct overseas funds.

Acknowledgments. Funded by the social science research foundation from Hebei province , NO: HB11GL027.

References

1. Yang, G., Min, Q., Li, W., Zheng, L.: Scientific issues of ecological compensation research in China. Acta Ecologica Sinica 10, 4289–4301 (2007)
2. Yu, H., Ren, Y.: Theoretical Bases of Eco-compensation: An Analytical Framework. Urban Environment & Urban Ecology 2, 32–36 (2007)
3. Wang, X., Li, J., Gao, P., Zhuang, G.: Journal of Natural Resources 1, 1–7 (1996)
4. He, C.: Theory and Application of Ecological Compensation: A Multiple Temporal-spatial Scales Perspective. PH.D thesis of Fujian normal university (2007)
5. Li, L.: A Research on Management of Environment Production Factor. PH.D thesis of Fujian normal university (2009)

Present Positions and Suggestions of Bilingual Education for High Education in China

Xin Luo

Department of Accounting
Zhengzhou Institute of Aeronautical Industry Management
Zhengzhou, Henan Province, China
luo_xin33@yahoo.com.cn

Abstract. With the economic globalization and education reform, the importance of bilingual education emerges step by step. This is just the focus of this paper: What are the present conditions in China and how to improve this new teaching method. Firstly, the growth of bilingual education in domestic and in foreign was discussed. Next, this paper illustrated some significant issues in Chinese high education, including lacking of professional teachers, inadequate appropriate text books and so on. Finally, some advices were presented, such as seeking high level lecturers. The major conclusion is that China education department should increase the input to strengthen bilingual education levels so that Chinese students can be better to serve for multinational companies, which is helpful to Chinese economic development.

Keywords: High Education, Bilingual Education, Professional Teachers of Bilingual Education.

1 Introduction

The foreign empirical testing about bilingual education started from the last century, in the end of the last century the relevant study reached maturity and brought out a series of principles which are helpful to students' growth. Bilingual education in Canada and Singapore has achieved great success, and has extensive international influence. In Canada, the government adopts official languages bilingualism (English and French) and multiculturalism policy. Also, the teaching mode is the French immersion teaching. In Singapore, the whole education system (from primary schools to universities) is just bilingual education including mother tongue and English. Since then, the "bilingual education" grows vigorously over the world.

In the early 1990s, China began to take bilingual education experiment. From primary schools to universities, the whole country started to adopt this type of teaching.

However, in many these institutions, the situations are fictitious and this phenomenon is more serious in universities. In many universities, teachers just only use these two kinds of languages (Chinese and English) to teach but cannot

D. Jin and S. Lin (Eds.): Advances in ECWAC, Vol. 2, AISC 149, pp. 473–477.

communicate these two languages. On the other hand, several universities indeed can understand the spirit of bilingual education and achieve some good results. Nevertheless, there are still several important issues in the whole education system of China.

2 The Situations of Universities' Bilingual Education in China and Major Problems

China Ministry of Education asks all universities should use bilingual education (Chinese and English) in public subjects and professional subjects in the bachelor-degree education. In addition, among three years the percentages of bilingual subjects should take 5% to 10% in all subjects. Moreover, the relevant department of government puts the level of bilingual education into the rating index system of the bachelor-degree education. However, since many reasons the results and levels of bilingual education are not very good in a lot of universities. In summary, vast majority of Chinese universities have the following issues about bilingual education in the different extent.

2.1 Lack of Professional Teachers of Bilingual Education

Bilingual Education has a high requirement to the professional teachers. They must have a series of professional abilities, such as profound professional knowledge, solid native language, outstanding English pronunciation, fluent oral English ability and good ability of using English language. However, according to the current situation of our country, the number of appropriate professional teachers is too little. Some teachers have the good oral English ability but have no the adequate relevant professional knowledge; Some people are good at the subjects but know a little about the relevant English. As a result, lack of professional teachers is one urgent problem and one approach must be found to solve this issue.

2.2 Lack of Appropriate Text Books

Bilingual education is asked to use foreign original edition books and relevant reference books. However, the different books have the different contents in levels and qualities, which lead to many difficulties when universities do the choice about appropriate text books. On the other hand, the issue of books' prices increases schools' and students' economic burdens.

2.3 Absence of Complete Management Mechanism of Bilingual Education

For one thing, the funds of using bilingual education are inadequate. Moreover, the attitudes of universities to bilingual education are different. In addition, many universities have no the relevant approaches to monitor and value the levels of bilingual education. Finally, the arrangement of relevant subjects is not reasonable in many universities.

3 Several Suggestions about Bilingual Education

3.1 Getting High-Level Professional Teachers

The teachers who use bilingual education method must have the following aspects of abilities. Firstly, they must have the ability of knowledge integration. The teachers should be able to do the organic integration of two kinds of cultures and to transform these two cultures. Also, they must understand the principle system about relevant subjects and have the ability to treat the differences between the two languages and cultures. Secondly, they should have the ability to apply modern educational technology. This aspect requires the relevant teachers can make use of multimedia technology, including courseware making. Thirdly, they should have the favorable communicable ability. The optimum interaction only can be gotten through the efficient communication between teachers and students, which can increase the quality of bilingual education. Also, the ability of thinking problems and solving problems using English can be improved by this point. Fourthly, they must have the good ability to control the teaching process. In general, the following four methods of bilingual education can be used. The first is lecture method. The second is game method. For example, such games as "to see who know more words" and "crossword" can be used to get a better classroom atmosphere. The third is self-studying and discussing method. One teaching programme can be used, that is questions, reading, discussing, presenting and summarizing. This method can train several students' abilities, including self-studying ability, analyzing ability and presenting ability. The last is writing method. This method requires students to write one essay using the relevant knowledge.

Nevertheless, not all universities have the appropriate teachers. Therefore, according to different situations, each school can choose one or all the following methods to improve the levels of professional teachers. Firstly, they can invite some foreign teachers with solid professional knowledge to teach bilingual courses. In addition, the PhDs and masters who have the oversea study experiments are good choice as well. Secondly, they can select outstanding teachers from the people who have the superior professional knowledge and English level. These teachers should be trained by several measures, such as studying the lecture of foreign experts, special training in domestic well-known universities, further special training in overseas and so on. Thirdly, they can construct source channels of professional teachers. The "double degree" method can be used to foster the inter-disciplinary talent who has the super ability in both professional knowledge and relevant foreign language.

3.2 Choosing the Reasonable Teaching Materials and Teaching Resources

Firstly, universities should know the range of choice of relevant text books. In addition, the professional teachers should pay more attention to the motivation of books' publishing and check the index of published books in some publishing house. Moreover, universities should ask the relevant teachers to analyze the contents of relevant books and study the special logical structures when they choose the original native text books. It is commonly thought that the good books should be written by famous writers, published by famous publishing house and used by well-known universities. On the other hand, if it was difficult to find the appropriate original native books, the professional teachers should be able to select the relevant contents from some original native books to construct some appropriate lecture notes. Secondly, universities should set up special funds to do bilingual education constructions, which can be used to buy some original native books and relevant references for the professional teachers. The last but not least, the relevant departments of government should support the importing and publishing of foreign text books and relevant references.

3.3 Establish and Perfect the Management Mechanism

Firstly, universities should establish incentive mechanism of bilingual education. Establish special funds of bilingual education has been put on the agenda in order to support the growth of bilingual education. In addition, universities should formulate corresponding encouraging policies and incentive mechanism. Secondly, standards of bilingual education should be set up as soon as possible. Universities should suggest explicit requirements to relevant text books, teaching frameworks, assignments and examinations. This approach can avoid the arbitrary and blindness of bilingual education. Thirdly, universities should monitor and evaluate the bilingual education effects. Universities can take several methods to value the teaching quality of relevant bilingual subjects, including constructing the valuing teams, periodic attending a lecture, and students' questionnaire. At last, universities should set up reasonable curriculum system of bilingual education. The bilingual subjects cannot be opened willfully and universities should choose the subjects which have the characteristics of rapid development of discipline and the international generality, such as Finance, Medicine, Information Technology and so on. At the mean time, universities should pay more attention to the coherence and the cohesion between bilingual teaching subjects and other courses. Prior to the bilingual subjects, the College English subjects should be strengthened. In addition, some other subjects should be opened, for example, Professional English and Professional Foreign Material selections. Moreover, universities should choose the ideal opening time of bilingual subjects. In general, the best opening time of that should be the third grades. The College English should be opened in the first two grades and the Professional English should be opened in the fourth semester, which is helpful to the formal bilingual subjects teaching.

4 Conclusion

In summary, to improve the bilingual teaching quality is a gradual process and is a long-term and arduous task. As we all know, there is a considerable gap between the current situation of bilingual education and high levels. As a result, the relevant people must know the issues of bilingual education well and try their best to take measures to improve the bilingual teaching and make it to a new level.

References

1. Zhang, J.S.: The essence and characteristics of the case studying. Zhongguo Jiaoyu Xuekan (January 2004) (in Chinese)
2. Wang, J.J.: Some problems in the bilingual education. Beijing Daxue Xuebao (May 2007) (in Chinese)
3. Zeng, Y.B.: University's bilingual teaching: mode research and definition. Journal of Nangchang College (January 2005) (in Chinese)
4. Shi, Y.J.: The thinking of university's bilingual teaching. Liaoning Education Research (July 2005) (in Chinese)

The Options of Financial Goal for Chinese Enterprises

Xin Luo

Department of Accounting
Zhengzhou Institute of Aeronautical Industry Management
Zhengzhou, Henan Province, China
luo_xin33@yahoo.com.cn

Abstract. Purpose is essential for anyone to do anything, of course including financial management, which affects all the aspects of one enterprise. This paper's focus is just to discuss what is the appropriate financial goal for the Chinese enterprises base on the current situations. Firstly, this paper illustrated some significant theory about financial goal in the history. Based on these principles, the paper did some relevant analysis and valuations. After that, "maximization of stakeholders' wealth" was pointed out as one generally accepted principle. Finally, several positive effects about "maximization of stakeholders' wealth" were related. The conclusion of this paper is that the principle "maximization of stakeholders' wealth" is the most appropriate financial goal for Chinese enterprises according to the current Chinese social and economic positions.

Keywords: Financial Goal, Maximization of Shareholders' Wealth, Maximization of Firm's Value, Maximization of Stakeholders' Wealth.

1 Introduction

1.1 Theory of Marx's Capital Motion

In the view of Marx, the purpose of capital motion is to get value added. Marx did the perfect illustration about value added of capital in the famous "Industry Capital Communication Formula", that is G--W…P …W'--G'. This formula accurately shows that the financial goal of enterprises should be the G' must be bigger than G. On the other words, from the point of finance, the results of firms' operating all are presented by this G'. Moreover, it should be as possible as greater, which is just the goal of finance.

1.2 Theory of Firm's Govern Structure

Firm's govern structure is a kind of mechanism that can reconcile the relationship between shareholders and other stakeholders. In the view of the entity, there are two forms are accepted, including "shareholder preferred" and "common govern". The basic difference between those two forms is that that enterprises should service for,

D. Jin and S. Lin (Eds.): Advances in ECWAC, Vol. 2, AISC 149, pp. 479–484.

shareholders or all stakeholders. Some different financial goals can be brought out from the latter, such as "maximization of firm's value".

1.3 Theory of Agency

The agency relationship of modern enterprises can be definited as one kind of contract relationship or indenture relationship. Under this relationship, one or several clients hire some people (as agents) and offer them some rights of decision-making to do some operations which can bring benefits to themselves. Under the corporate system, one aspect of agency relationship exists between resource suppliers (shareholders and creditors) and resource users (management), which is based on collection and consuming of resources. The other aspects exist between different hierarchies, including senior managers and medium managers, medium managers and primary managers and managers and employees, which is based on the responsibility of property's operating. The essence of this theory is the relationship of economic benefits among all parties. Clients and agents have their different individual or private benefits. In the motivation of individual benefits, they enter enterprises from market to seek maximization of individual benefits. Due to the different purposes of different parties, the unavoidable conflicts are resulted in among them. As a result, when one company plans the financial goal, it can not think about shareholders' value or managers' benefits merely. On the contrary, benefits of all stakeholders must be taken account in. Enterprises can be operated well-organized and efficiently as long as all parties' benefits are thought about.

2 The Analysis of Current Opinions

2.1 Maximization of Profits

In China, many people think that the financial goal should be "maximization of profits". However, it can not settle the conflict between returns and risks.

Firstly, the time factor of income is ignored, which can result in the short-run activities of managers and unreasonable using of firm's resources. Secondly, the risk factor of income is ignored, which can make managers to bear high risks and take firms into operating distress or financial distress. Thirdly, the reliability of income is ignored, which can make managers to take accounting profits into account merely but neglect cash flows completely. Therefore, the company possibly has troubles because of cash insolvency.

2.2 Maximization of Shareholders' Wealth

When both the opinion of "maximization of profits" and the opinion of "maximization of economic benefits" all face big challenges, some companies begin to use one western opinion—"maximization of shareholders' wealth"—as the financial goal of Chinese enterprises. It is no doubt that in companies limited by shares and corporations, using "maximization of shareholders' wealth" rather than "maximization of profits" or "maximization of economic benefits" as the financial goal can overcome the shortcoming of firms' serious short-run activities, which is helpful to treat the relationship both between returns and risks and between present benefits and prospective benefits correctly. The reason for above mentioned are that "maximization of shareholders' wealth" means enterprises must be able to grow constantly. On the other words, two conditions must be satisfied then the shareholders' wealth can touch the highest point. For one thing, income of enterprises must have sustained increase. Another, the market price of common stock must be able to rise slightly and gradually. In order to make these two conditions come true, the enterprise must treat the relationship both between returns and risks and between present benefits and prospective benefits correctly. However, just like one coin has two sides, this opinion has its own important shortcomings. Firstly, its using range is two narrow, which means only companies limited by shares and corporations can take it as the financial goal but other types of companies can not use this opinion. Secondly, the basic economic law of socialism is to make all people of one country to be rich at the same time but this opinion only seeks one party's wealth—shareholders. This point is going against removing positivity of managers and vast employees sufficiently, which is harmful to construct the internal equilibrium mechanism of profits in enterprises.

As a result, it is very appropriate to take "maximization of shareholders' wealth" as the financial goal of enterprises in western developed countries but for Chinese enterprises, this opinion has many issues to use.

2.3 Maximization of Firm's Value

In order to overcome one significant shortcoming of "maximization of shareholders' wealth" that is this opinion neglect benefits of managers and employees, some scholars argues that one opinion of "maximization of firm's value" is a better financial goal. This opinion sounds reasonable because the "maximization of firms' value" maybe can meet all relevant stakeholders' benefits, but it have much more disadvantages when people do some more careful analysis. Firstly, the definition of "firm's value" is so indeterminate, which means the people can have several understandings. In the view of nature, "firm's value" can be classified as "economic value", "social value" and "humanity value". On the other hand, in the view of time, it can be classified as "past value", "present value" and "future value". As a result, it is confused that "firm's value" here is which one of all above mentioned. Secondly, the opinion of "firm's value" does not point out that what on earth the enterprise should do, which does not conform to the basic economic law of socialism. Thirdly, the opinion of "firm's value" can not show what are the explicit purposes of each relevant economic benefit party.

3 Maximization of Stakeholders' Wealth

3.1 Theory Proof

3.1.1 Theory of Stakeholders
The entities of ownership of one enterprise should not only include shareholders but also include suppliers of funds, creditors, employees, suppliers of materials and other all stakeholders. One enterprise's financial activities and financial relations should

focus on all the different relevant parties' benefits, and achieve the final financial goal that is maximization of all stakeholders' wealth.

3.1.2 The Issue of Social Responsibility

Based on the traditional economic theory, the enterprise is only one unit of operating whose purpose is to get more and more incomes. Maximization of profits is the permanent theme that one enterprise seeks in its all life. One enterprise has no responsibility and obligation to complete some jobs that should be finished by relevant departments of government in the theory. One enterprise can use any approaches or methods to seek profits as long as these ways are not illegal. However, nowadays this kind of opinion should to be changed according to some factors.

Firstly, the levels of socialization and professionalization of commodity production are improved constantly in modern society. Secondly, the extent of depending on each other among enterprises is different with any previous periods in history. Thirdly, the extent of social credits has more and more influences to economy.

All these factors can be summarized as one sentence—"In the modern society, all activities of one enterprise are not individual activity any more, which can affect the society and entire economy possibly to some extents". Under this situation, it will be too narrow and selfish to only take the profits-seeking as the final financial goal. On the contrary, any enterprise should take some social responsibilities.

The social responsibility of one enterprise is the responsibilities and obligations which one enterprise should take for the society. When one enterprise does the operations, especially does the decisions, in addition to shareholders and itself, it must take the other relevant parties' benefits and whole society's benefits into account properly. Moreover, when one enterprise is thinking about one activity of it is good or bad to itself, at the mean time it should think about whether this activity can bring out harmful affects to other people, such as public nuisance, environmental contamination and wasting of resources.

When one enterprise does some decisions, it should take the above mentioned issues into account and take some appropriate measures to solve these issues. This activity is just to take social responsibility. Otherwise, the enterprise may get a mess of pottage in the short-run, but it will leave some hidden troubles to the growth of this enterprise in the future.

3.1.3 Foreign Research Results

In China, the socialist marketing economy has been decided as the goal of restructuring the economic system. As a result, when we face some difficulties about the operating of socialist marketing economy, we can learn some experiences from western developed countries. Nevertheless, we can not learn everything indiscriminatively because the environment between China and western countries are different. In the fact, the choice of financial goal has the very close relation to the relevant environment.

Based on the whole world, there are two main financial management modes. The first one is based on the developed capital market just like United States. Another one is based on the developed banking system just like Japan. In United States, shareholders play the leading role in the process of financial decision-making, which ask the financial managers to pay more attention to shareholders' value. Therefore, "maximization of shareholders' wealth" is looked on as the financial goal. On the contrary, in Japan, all the relevant parties play the significant role in the process of financial decision-making, including shareholders, creditors, employees and government, which means the financial goal should give considerations to all relevant parties. As a result, the financial goal of one enterprise must think about all stakeholders' benefits but not only take shareholders into account.

3.2 Basic Principles of Financial Goal

Firstly, the benefit entities of one enterprise should include investors, creditors, management, government, public and so on. When this enterprise makes the best financial goal, it must take all the entities' benefits into account and maximize their benefits.

Secondly, the best financial goal should be able to be measured reliably. Based on the reliable measurement, the financial goal can become specific and have the characteristic of maneuverability. Moreover, it also can be tested and valued due to reliable measurement. Otherwise, this goal will become difficult to capture for the relevant people.

Thirdly, the best financial goal can be achieved through one enterprise's efforts only when it can be controlled efficiently. If one enterprise takes one uncontrollable goal as the enterprise's financial goal, this goal would become one unrealizable illusory thing, which would sow the seeds of trouble in the future for the financial management of this enterprise.

Fourthly, the best financial goal should be helpful to one enterprise's sustainable growth. In detail, the best financial goal should help one enterprise to solve the following issues. Firstly, it can overcome one activity that one enterprise does any operations only taking the short-term benefits into accounts. Secondly, it can consider all stakeholders' benefits efficiently. Thirdly, it can ensure that one enterprise is able to grow constantly, steadily and quickly.

3.3 The Opinion of "Maximization of Stakeholders' Wealth" Meets the Principles

3.3.1 Considerations of Each Stakeholder

This opinion is a complex goal, which concerns all stakeholders' benefits, including investors, creditors, management, government, public and so on. However, any other goal only takes one party or several parties into account.

3.3.2 Gauge Ability and Controllability of Each Stakeholders

For one thing, this opinion can satisfy the requirement of gauge ability of the best financial goal of one enterprise. Firstly, the benefits of investors can be measured through some financial ratios such as "return on equity" (ROE). Secondly, the benefits of creditors can be measured by some financial ratios such as "debt ratio", "degree of financial leverage" and so on. Thirdly, the benefits of management can be measured by some ratios, including "wages to operating income" and "wages to operating expenses". Fourthly, the benefits of government can be measured by some ratios such as "illegal taxes ratio", "completing taxes ratio" and so on. The last but not least, the benefits of social public can be measured through some factors, including "environmental protection" and "labor supervise". In particular, on some extents, the benefits of government are the benefits of social public and the benefits of social public are the benefits of government. These two kinds of benefits can not be divided.

For another, the above mentioned ratios or factors not only can be measured but also can be controlled through one enterprise's financial activities.

3.3.3 Sustainable Expansibility of One Enterprise

For one enterprise, ignorant of any stakeholder's benefits will bring about some significant troubles.

Firstly, if investors' benefits are ignored, they will possibly replace the management (even if they do not replace the management, the agency problem will be strengthened, which will increase the agency costs). Moreover, investors may withdraw or transfer their investments, which can make one enterprise to be destroyed.

Secondly, if creditors' benefits are neglected, they will possibly protect their interests through legal procedurals. This situation can put one enterprise into the quagmire of contentiousness, which will result in the serious negative affects to one enterprise's future financing activities.

Thirdly, if government's benefits are neglected, one enterprise will possibly take the administrative penalty and legal sanctions came from the relevant departments of government.

The last but not least, if social public's benefits are neglected, one enterprise also will possibly take the administrative penalty and legal sanctions came from the relevant departments of government. Moreover, the enterprise will possibly face the social public condemnation and boycott.

All of the above mentioned factors are harmful to one enterprise's sustainable growth; even they can lead to the downfall of this enterprise. As a result, only through taking the opinion of "maximization of stakeholders' wealth" as the best financial goal of one enterprise and considering or reconciling each party's benefits, all stakeholders can get their greatest benefits from the operating activities of one enterprise. In addition, this point can promote the sustainable growth of one enterprise to the hilt and the socialist market economy of China can be launched more efficient and deeper.

References

1. Wang, H.W.: Analysis of enterprise financial management goal and responsibility. Caikuai Yanjiu (May 2006) (in Chinese)
2. Liu, Q.Y.: Thoughts of financial goal. Friends of Accounting (January 2005) (in Chinese)
3. Liu, J.Q.: Rational choice of enterprise financial goals in knowledge economy time. Caikuai Yanjiu (May 2003) (in Chinese)
4. Wang, H.C.: Again concerning financial management objectives. Finance and Accounting (April 2003) (in Chinese)

Enterprises' Financial Security in Uncertain Operating Environment

Xin Luo

Department of Accounting
Zhengzhou Institute of Aeronautical Industry Management
Zhengzhou, Henan Province, China
luo_xin33@yahoo.com.cn

Abstract. In the process of one enterprise's operation, it may face variable situations, including good things and bad things. The combination of all situations is called operating environment. This is just the focus of this paper: how does the operating environment affect one enterprise's financial security. Firstly, this paper illustrated the definitions of operating environment and financial security. Secondly, the relationships between operating environment and financial security are discussed in this paper. Thirdly, some measures were studied in this paper that are used to reduce the effects of operating environment. Finally, this paper given several suggestions about how to deal with the risks resulted from the operating environment. The conclusion of this paper is that one enterprise must pay more attention to the operating environment, which is helpful to its growth in the future.

Keywords: Uncertain Operating Environment, Financial Security, Risk Control.

1 Introduction

With the acceleration of economic globalization, the competing among enterprises is aggravated step by step. In order to live and grow in the serious competing, enterprises takes all kinds of measures to steady their market position and get a rapid growth. In this process, the financial security of one enterprise is very important, which is concerned with success or default for one enterprise in the competing. In the current financial distress, one perfect financial security can help one enterprise to construct one shield to protect itself and live better than other enterprises. On the other hand, if one enterprise bears large financial risks, it possibly will go to bankruptcy in the time of several days.

The most important factor that can affect one enterprise's financial security is just the uncertain operating environment that can bring out some large financial risks. On the other hand, some other factors can help one enterprise to hedge the financial risk and reduce the losses brought out from these risks. These factors include reasonable capital structure, scientific financial management and perfect internal control. There is only one approach can be used by one enterprise to take the precautionary measure against issues of financial security, which is to see clearly all the effects of uncertain operating environment that can lead to variable financial issues of financial security.

2   The Link between Uncertain Operating Environment and Financial Security

2.1   External Environment

External uncertain operating environment can affect all kinds of enterprises, which is out of enterprises' control. External environment include economic environment, legal environment, market environment, social culture environment, resource environment and so on, which has the characteristics of complexity and variability. For example, if the price of crude oil increase, the price of gasoline will be affected to increase also, which will lead to a series of outcomes to transporting companies, such as increased operating costs, decreased profits and unrealizable expected income. In the following contents, three kinds of major external environment that can affect one enterprise's financial security will be illustrated in detail.

2.1.1   Uncertainty of Political Environment

The factors that can lead to political uncertainty are often concerned with the important changes of entity of political right in the place where one enterprise does some business. The uncertainties of political system changing and wars can bring factual or potential losses to one enterprise. The political environment that can affect one enterprise have some characteristics of directness, unforecastability and irreversibility. Directness means national political environment can affect one enterprise's operation directly. Unforecastability means it is difficult to forecast the changing trends of one country's political environment. Irreversibility means if the political environment does affect one enterprise's operation, the enterprise will change rapidly and obviously. Also, this result is out of the enterprise's control.

2.1.2   Uncertainty of Economic Environment

Economic environment consists of two aspects. The first is the social economic situations that can affect one enterprise. The second is national economic policies. Any uncertain changes of economic factors will affect one enterprise's operations and incomes unavoidable. For example, the current financial distress that comes from United States brings serious hidden troubles to most enterprises' financial security all over the world. Moreover, many enterprises have already shutted down and some of them even gone to bankruptcy. Nowadays, there are still some experts argue that this financial distress has not ended yet, which still can result in some serious outcomes.

2.1.3   Uncertainty of Industry Features

The financial risks of different industries are different. Some industries have close links with other types of industries. These industries' market demands will be affected by other industries easily so they have higher financial risks than other industries. Some other industries, like high-technology industry, have the characteristics of high-level input, high-level output and high-level income but at the mean time, they also bear the high-level risks. On the other hand, one enterprise will have the different financial risks in the different phases of its growth. The enterprise that is in the phase of growing up will have the lower financial risks and the enterprise that is in the mature phase or the phase of recession will have the higher financial risks.

2.2 Internal Environment

2.2.1 Contingent Events
The contingent events refer the uncertain events that are resulted from the past operations or events, whose consequence is according to the occurrence or non-occurrence of some future events. The contingent events include unsettled lawsuits, unsettled arbitrations, warranty responsibilities, compensations of products' quality. All of these events can lead to potential financial risks. The reasons that result in these events are variant such as wrong financing methods, investing faults, ineffective operations and so on.

2.2.2 Ineffective Operations
Under many situations, one enterprise will grow slowly or stop at some points, which will take this enterprise into financial distress finally. The reasons that lead to the above mentioned conditions are not the saturated market but the ineffective internal control of this enterprise. When this confused managing situation is cumulated to one certain level, it is no doubt that one enterprise will go to bankruptcy.

2.2.3 Deduction of Competing Power
The levels of one enterprise' competing power are embodied by two main aspects. The first is the products' market shares of one enterprise. The second is the profitability of one enterprise. If the products' market shares of one enterprise is very high and the profitability of this enterprise is very strong, which mean the competing power of this enterprise is powerful enough. On the other hand, if the products' market shares of one enterprise is very low and the profitability of this enterprise is very weak, which mean the competing power of this enterprise is weak enough. Moreover, as long as the competing power of one enterprise is transferred from powerful to weak, this enterprise will be not able to produce according to the plans or targets. After that, this enterprise will be taken into financial distress and may go to bankruptcy finally.

2.2.4 Having No Enough Knowing to the Objectivism of Financial Risks for Financial Managers
Financial risks are existing objectively. On the other words, all financial activities possibly can bring out financial risks. However, nowadays most financial managers have no enough knowing to the objectivism of financial risks. In stead, they think the financial risks will not emerge if they can manage funds and use funds very well. At the meantime, financial activities are concerned with economic benefits directly. Under this situation, the ethics risks and the activity risks will be difficult to be avoided if there is lack of appropriate incentive mechanism and restrict mechanism. This point also can lead to significant issues of financial security easily.

3 Protective Measures against Uncertain Operating Environment

3.1 Against External Environment

Firstly, in order to reduce financial risks, one enterprise should analyze and study the variable macro environment of financial management very conscientiously. Through the analysis and study, one enterprise should be able to grasp its changing trends or changing laws and based on these trends or laws to plan some measures. As a result, the abilities of one enterprise to deal with the variable macro environment of financial management can be improved. On the other hand, one enterprise should construct the efficient institution of financial management and hire high-character financial managers, which can reduce financial risks that are resulted from changes of environment. Also, in this process, it is very important to complete rules and regulations of financial management. Secondly, on the one hand, one enterprise that is financed from debt should pay more attention to the changes of national industry policies, investment policies, financial policies and tax policies. Based on these changes, the enterprise should forecast the negative effects that are resulted from the following aspects: investment projects, operating projects, financing funds, operating costs and so on. On the other hand, this enterprise needs to be concerned with the changes of market situations, market demands and relationships between supply and demand. In addition, for the aspect of basic constructions and technical reforms, the enterprise should think about the deferrals of one project due to funds inadequacy that is resulted from compression of banks' loans. This point will affect the realization of one enterprise's expected profits. All the above mentioned factors will take one enterprise into financial distress. Therefore, managers must be able to adjust operating strategies and investing directions in time and avoid the incurrence of the above mentioned situations.

3.2 Against Internal Environment

3.2.1 Completing Financial Management System

One enterprise should recommend scientific programmer of risk management and mechanism of financial restricts, which is helpful for one enterprise to construct and complete the managing mechanism of financial risks. The Management of financial risks is a system to recognize and value risks, analyze reasons of risks, reduce and control risk and deal with losses of risks. The mechanism of financial restricts for one enterprise include two aspects, including external restricts and internal restricts. External restricts consist of regulations and laws restricts, market restricts (manager market restricts, product market restricts and capital market restricts) and so on. Internal restricts refer to management of financial budget mainly, which include budget planning, budget exercising and valuations. As a result, the internal control mechanism can be realized through the above mentioned factors.

3.2.2 Constructing and Completing Early Warning System of Financial Risks

The purpose of early warning system of financial risks is to monitor and forecast the financial risks that one enterprise possibly or certainly will bear. It exists in the whole process of one enterprise's financial activities. Based on the following aspects, this system can oversee or control the financial risks in the process of financial activities and grasp the probabilities and reasons of financial distress, which is helpful for management to prepare some measures. The first aspect is to get one enterprise's financial statements, operating plans and other relevant materials. The second aspect is to use some theories such as accounting, finance, company management, marketing and so on. The third aspect is to use some professional methods such as portion analysis and mathematics model. For one thing, one enterprise should construct short-term warning system of financial risks and prepare the cash flows budget. The key factor that can affect one enterprise's reproduction is whether this enterprise has enough cash flows for all kinds of usages but not how many profits this enterprise can get. Through the analysis of active situations of one enterprise's cash flows, several issues can be revealed, including investments, operations and financial policies. As a result, accurate cash flows budget can provide signals of financial risks to one enterprise. Based on this point, managers can take actions to reduce risks as soon as possible. For another, one enterprise should construct long-term warning system of financial risks. This system can be reflected from one enterprise's profitability ratios and debt-service ratios. The latter include current ratio, quick ratio and debt ratio. If the current ratio is too much high, which means one enterprise will lose the reinvestment opportunities of current assets. In general, it is better if this ratio can be 2. Also, it is better if the debt ratio can be the range from 40% to 60%. More debt means more interests and higher financial risks. In the long-run, it is essential for one enterprise to have the good profitability, which can help it to be far away from financial distress. Moreover, the enterprise will have the stronger financing ability and debt-paying ability.

4 Detailed Methods against the Financial Risks

4.1 Hedging Financial Risks

The first thinking about the methods against financial risks is to hedge risks. When the losses that are resulted from the risks can not be offset by the benefits that are resulted from the same risks, hedging risks is the best method. Firstly, one enterprise should refuse to do the transactions with the enterprise whose credibility is poor. Secondly, one enterprise should abandon some projects that will lead to losses obviously. Thirdly, one enterprise should shut down one new product if it probably can result in some important issues. In addition, when one enterprise plans to provide warranty to some related parties, it should do something firstly based on the conservative attitude. For one thing, it should realize that entity or that project completely. Another, it should know the relevant national regulations about that project. Without doing these things, it is better for one enterprise not to provide the warranty. In general, most these enterprises that seek external warranty can not meet the due debt. Therefore, one enterprise that provides warranty must take the responsibility of paying debt when these enterprises are in default, which will bring out the unessential financial losses to the enterprise. Being cautious can help one enterprise to reduce this kind of financial risks.

4.2 Reducing Financial Risks

One enterprise should increase the ability to prevent financial risks from several aspects such as principles, cultures, decision-makings, organizations and controls. There are two meanings of reducing financial risks. The first is to control risk factors and reduce the risks. The second is to control the frequency of risks and reduce the losses of risks. Many usual methods can be used to reduce risks. The first is to do the accurate forecasting, including exchange rate forecasting, interest rate forecasting and security rating of debtors. The second is to choose one best proposal of all proposals of one project. The third is to get the new national regulations through communicating with relevant government departments. The fourth is to do the adequate market investigation after developing some new products. The fifth is to choose some projects that can bear higher risks and take some relevant measures of protection. The last but not least is to do some investments that have the following characteristics: variable fields, variable regions, variable projects and variable varieties. Thus, the financial risks can be diversified.

4.3 Transferring Financial Risks

In order to avoid the large losses, one enterprise can take some prices and some measures to transfer the financial risks to other parties. There are two main methods can be used. The first is to transfer risks by insurance, which means one enterprise can transfer the risks of property loss to insurance company through buying some property insurance. The second is to transfer risks by non-insurance, which means one enterprise can transfer risks to some special institutions or departments. For example, one enterprise can transfer some particular business to some special companies which have experienced skills, specialized persons and special equipments. The method of transferring risks can transfer all or partly financial risks to other parties, which can reduce the financial risks of one enterprise itself in a big extent.

4.4 Accepting Financial Risks

For some risks that only bring out a little losses, one enterprise can bear these risks and relevant losses by itself if this enterprise has enough funds and abilities to do that. For this method, some measures can be used. The first is to add the losses to the costs of one enterprise or minus the losses from the profits. The second is to prepare some special accounts, such as bad-debt funds and inventory-depreciation funds.

References

1. Liu, H.D.: Analysis of enterprise financial security and uncertain environment. Guangxi Kuaiji (November 1999) (in Chinese)
2. Wang, B.: Analysis of reasons and measures of enterprise financial risks. Shangqing (May 2008) (in Chinese)

Simulation Designing Model of Kanban Production System for a Engine Assembly Workshop Based on Witness

Yan Xiao[1], Yunyun Li[1,2], and Kangqu Zhou[1]

[1] Automobile College Chongqing University of Technology, Chongqing, 400054, China
[2] College of Mechanical Engineering Chongqing University, Chongqing, 400044, China

Abstract. To kanban production system of a engine assembly shop as the empirical study, the simulation model of the Kanban production system about designing and optimizing are elaborated, which is on the basis of understanding basic theory of Kanban production system. Before the implementation of the project designed, the implementation performance of this program is predicted in advance through the simulation test.

Keywords: Kanban, production systems, Witness simulation design.

1 Overview of the Workshop and the Kanban Production System

A line of engine assembly plant total has 26 stations, the station list as shown in Table 1. Kanban production is a pull system of material flow triggered.

Table 1. Processing list for A line

A1	Punched the engine number, the right box pressd bearing	M1	Star Wheel mounted	M10	install timing sprocket
A2	the left box pressd bearing	M2	Pumps and limit leverage installed	M11	tensioner assembled
A3	control shaft assembly	M3	Install and inspect the main drive gear	M12	adjust valve clearance
A4	bearings baffle, main shaft mounted	M4	clutch assembly	M13	electric starter gear mounted
A5	assembling fork and balance shaft,shift cam combined	M5	Adjust the clutch gap	M14	Sprocket side cover mounted
A6	Coating,combining boxesand co-tank screws mounted	M6	Starter motor installed	M15	Check the valve clearance and install valve cover
A7	fixing co-tank screws	M7	fasten right cover screws and check the	M16	install spark plugs and carburetor
A8	Mounted drive sprocket	M8	valve assembly	M17	leak detection and fueling
A9	Put in the oil bolts and neutral switc	M9	fastening		

D. Jin and S. Lin (Eds.): Advances in ECWAC, Vol. 2, AISC 149, pp. 491–496.
springerlink.com

2   The Fundamental Assumptions to Modeling Kanban Production System Simulation

Designing this simulation model need to make assumptions for some parameters. (1) The time interval of demand orders arriving, the lead time for batch size and order, the time interval of machine problem and the time for maintenance are random.(2)The average stocks of WIP, the capacity utilization, output and the average waiting time for orders are selected for system performance indicators.(3)There are two system constraints: one is the production capacity constraint; another is restriction on the number of equipment maintenance staff.(4)Raw materials are supplied in time and not considering the cost of the inventory.(5)Not considering the cost of transportation between the various processes.

Based on the above assumptions, we establish the simulation models.

3   Design the Simulation Model of Kanban Production System

3.1   Design Performance Indicators for Kanban Production System

Four performance indicators are taken into account. They are output, WIP inventory, equipment utilization rate and the average waiting time of orders. (1) System's output. The purpose of this indicator is to calculate the number of finished products in a particular period. (2) Work-in-process. This indicator is applied to count the magnitude of value of raw materials, semi-finished and finished product in the production and inventory systems at some specific time point. (3) Average Waiting Time (AWT). The function of this indicator is to compute the average waiting time caused by that order could be satisfied in time. The lower values of the indicator, the higher the customer satisfaction. (4) Operation rate (OR). This function can be used to evaluate machine utilization in the production system. The statistical functions of standard state, which are provided by system itself, aim at counting the percentage of machine time for idling, running, fault and maintenance.

3.2   The Design for System Variables, Functions and Module

3.2.1   Variables Design

The variable names and their meanings are shown in Table 2. These are created in variables (Var) module in the simulation model of kanban production system.

Table 2. Variable table for simulation kanban production system

element name	meaning	element name	meaning
breakinterval	the interval time of machines broken down	PK_A1_9	NO. of kanban at station A1 to A9
repairtime	the repair time for machines	PK_M1_17	NO. of kanban at station M1 to M17
dmdwaittime	the total waiting time for orders	cv	coefficient of variation
dmdunsatify	the number of unsatified order in time	safe	safe fcator
dmdsatify	the number of satified order on time	outpart	systemtotal output
valwip	the value of WIP	totalpart	finished production

3.2.2 Functions Design

The functions used in the system are created uniformly in the fun module, and its meaning and function body are instructed as follow.

(1) Statistical function for WIP inventory. There are two functions used in the program: One is NPARTS () ,and another is NPARTS2 (). The statistics function is applied to adding all of cost value of materials in the finished buffers, raw buffers and processing on the machines for all the workstations.

(2) Statistical function for the average machine utilization. STUIL() is used in the statistical function. STUIL(element_name, state) function returns the ratio of time for specified element in a certain state during the entire simulation time.

(3) Statistical function for the average waiting time of orders. The result of this statistical function is reached as the total waiting time of orders divided by the mounting orders.

(4) Statistical function for system output. Statistical function returns the system's total output.

3. 2.3 Simulation Module Design of Kanban Production System

(1) Demand processing module. The M_order module is designed to deal with the events that order arrives and leaves at Witness simulation system. As the simulation clock developed, orders arrive at system in discrete time. Whether the order is undertaken immediately, it depends on the state of order processing machines (expressed by demand_meet in the model). The order will be discharged into the waiting queue of orders if the machine is busying. Once the buffer to finished in M17 station have enough physical units, the machine will extract appropriate products from the assembly line to meet the current needs , and then push the order with parts to ship.

(2) The design for basic module of process. We need to analyze and design the basic modules of each process after the demand processing module design is completed. As shown in the flow chart of kanban production system, we can learn that there are only five essential elements in each workstation. A module including the five elements for every station is constructed at model establishing process to simulate actual state.The basic conditions are expressed in Table3.

Table 3. The basic elements table for each process module

element name	element type	meaning
mach	machine	Simulating the processing machine
b_raw	buffer	Simulating the buffer
b_finish	buffer	Simulating the buffer
mach_get	machine	Simulating the picking machine
pro_state	variable	Judging whether there is a production kanban

(3) The control process for each production unit. The block diagram of processing unit is explained in Figure 1. Along with the simulation clock runs, system program will judge the status of processing equipment at some special time. According to the machines' status returned, it is determined what event should be carried out at the present system so that simulation could can go on smoothly.

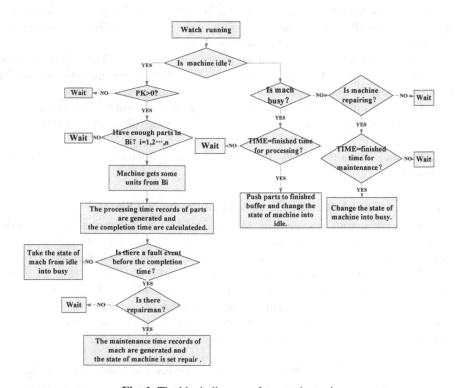

Fig. 1. The block diagram of processing unit

As above, the basic assumptions are set, the function and statistical method for system performance indicators are introduced, and the specific design process of modules are discussed in detail. All of these help us understand the whole of simulation model. Based on the research mentioned above, it becomes easier to define the concrete rule and type of machines, variables, buffers, workers etc.

4 The Establishment and Operation of Kanban Production System

According to the contents of previous section analyzed and designed, the kanban production system simulation model is established. Including warm-up time 360 minutes, the simulation model needs to run 120360 minutes. The visual interface for the model operating which is established on the basis of Witness logistics software is demonstrated in Figure 2.

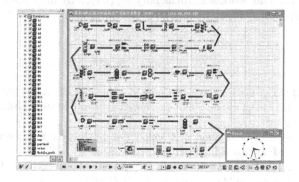

Fig. 2. The visual interface for the model operating

After researching simulation result, it shows the advantage and feasibility of the kanban production system when comparing the status before and after kanban management, as shown in Table 4. After Workshop implemented kanban production system, the number of orders cannot be satisfied in time reduced to 1, the average delay time of order reduced 315minutes, the system total output increased by 33,417, capacity utilization increased from 51% to 75%, increased of 24%. Practice shows that the kanban production system has practical effect on the workshop economic efficiency, the order could be mitten in time and customer satisfaction is improved. Besides, kanban management improves equipment utilization of workshop, the production increases at a big scale than before; the comprehensive benefit of the plant is raised remarkably.

Table 4. comparing the status before and after kanban management

	NO. of unmet order in time	Avg. delay time for orders（min.）	output	Avg. capacity utilization（%）
Prior to the implementation	109	720	98365	51%
After the implementation	1	315	131782	75%

5 Conclusions

On the basis of Witness logistics software, the process for kanban production system modeling and simulation in a engine assembly shop is described in detail. The simulation example shows the implementation performance of the kanban production system in advance, such as equipment utilization, the processing efficiency for orders and staff utilization. After both before and after contrast, we find: Kanban production system is implemented at the engine assembly shop and the operation shows good results. The capacity utilization increased by 24%, total output increased by 33,417 units, the average delay time of the order is reduced by 305 minutes.

Acknowledgment. This work was financially supported by the Foundation for Sci & Tech Research Project of Chongqing (CSTC,2009AB2051).

References

1. Qi, E., Fang, Q.: Logistics Engineering. China Machine Press, Beijing (2006)
2. Wang, Y., Ma, H.: Modeling and Simulation for Production Logistics System. China Science Press, Beijing (2006)
3. Monden, Y.: How Toyota shortened supply lot production time, waiting time, and conveyance time. J. Industrial Engineering 13, 22–29 (1981)
4. Gang, L., Peng, J., Deng, X.: Research progresses and prospection of Just-in-time. J. Industrial Engineering and Management 14, 18–20 (1997)
5. Wu, Y., Yu, H., Wu, L.: Simulating and optimizing of kanban production system based on Witness. J. Group Technology & Production Modernization 25, 9–12 (2008)
6. Moore, L.J., Clayton, E.R.: GERT: modeling and simulation: Fundamentals and applicators. Petrocelli-Charter, New York (1976)
7. Sugimori, Y., Kusunoki, K., Cho, F., Uchikawa, S.: Toyota production system and kanban system-Materialization of just-in-time respect-for-human systems. International Journal of Production Research 15, 553–564 (1977)
8. Gao, J., Bai, M., Qi, E.: Research of integration schemes for packing production line implemented Just-in-time. J. Industrial Engineering 13, 49–53 (2004)

A Compensatory Algorithm for High-Speed Visual Object Tracking Based on Markov Chain

Zhenzhong Song[1], Xiaoqing Peng[2], Huijun He[1], and Gaoang Wang[3]

[1] School of Computer Science, Fudan University, Shanghai, China
[2] Fudan University Library, Fudan University, Shanghai
[3] China Department of Electronic Engineering, Fudan University, Shanghai, China
{06300720210,xqpeng,10210240259,09300720036}@Fudan.edu.cn

Abstract. In order to effectively track high-speed objects in real time without any other instrumental methods but only based on computer vision, we propose an algorithm on camera control part by utilizing a history learning model in which a one-dimension Markov Transition Matrix synthesizes some historical information efficiently. The algorithm provides serials of conventional algorithms an effectively compensation when they perform worse in an interactive, high-speed circumstance, even lose the targets. Experimental results show that the new method is more stable and robust.

Keywords: Visual tracking, Markov Chain, machine learning, online algorithms.

1 Introduction

Real time tracking as a significant application of computer vision, especially being widely and successfully used in surveillance, vehicle tracking, pedestrian counting, etc, is more mature. But we can find that recent researches and studies achieve a good goal in a static scene, then work ineffectively or even fail for different reasons in a dynamic scene.

Generally speaking, conventional object tracking algorithms locate and track a target through a matching of some relevant and adjacent frames. But when the camera moves quickly, depending on the accurate differentiation of two or three adjacent frames, the decision velocity cannot make a judgment between an amplified noise and a subtle movement of a feature point of the object. Some algorithms such as Particle Filter and Kalman Filter[1]which based on local information extraction easily fails when the background changes quickly, because the information the Filter need is not as accurate as what extracted in a static scene or even some frames lost. Some others give a compensation by computing and matching the background of a series of different frames, as we known, then a large numbers of computing we cannot suffer cannot be avoided, such as TLD[2] and Incremental PCA[3] .

Considering all above, we propose a new algorithm based on Markov Chain[4], which provides a mechanical compensation for camera motion, and solves the problems mentioned above effectively.

D. Jin and S. Lin (Eds.): Advances in ECWAC, Vol. 2, AISC 149, pp. 497–502.

2 General PTZ Camera Control Strategy

Pan/Tilt/Zoom (PTZ) cameras are introduced into our real-time and high-speed tracking follow the increasing requirement of automatic recognition and dynamic tracing. The strategy deployed on camera can be intuitively described by an equation.

$$X_{t+1} = X_t + \vec{V}_t \tag{1}$$

X_t is the state of camera at the moment t, and $\vec{V}_t$ is the expected velocity,

For some simple cases, the above equation both looks straightforward and performs well. But in complicated circumstance, in order to lessen the effectiveness of skipped frames and noise, equation (1) should be improved by predictive velocity compensation.

3 Learning and Predicating Model Based on Markov Chain

3.1 Markov Chain

Markov Chain is a sequence of random process $\{X_i\}$, X denotes a common state at the moment i. For a one-dimension Markov Chain, the appearance probability of X_{i+1} only relates to X_i, which represents as

$$P(X_{t+1} = x \mid X_t) \tag{2}$$

If we use discrete random variables, Set S contains all the states X_i, $S \subset Z$, definition(2) will turn into a transitional matrix

$$P_{ij} = P(X_{t+1} = j \mid X_t = i) \tag{3}$$

P_{ij} is the probability of the state transfer to j from i.

3.2 Model Initialization

In order to compute the transitional matrix, we need classify all the collected data for analysis. Velocity of each moment we need as feature vector should be aggregated to a relevant cluster, the distance between each two vector can be defined as

$$d_{ij} = \sum_{t=1}^{n} \omega(k) \left| v_t^{(i)} - v_t^{(j)} \right| \tag{4}$$

$\omega(k)$ is the weight of each vector pair, simply we can set all $\omega(k) = 1$ at the first time, then justify according to.

The revised DBSCAN(Density-Based Spatial Clustering of Applications with Noise)[5] should be used to deal with the raw data records and compute the cluster according to the density-reaching which defined by the distance between two relative points. The distance threshold is given as an empirical value.

Algorithm 1. Revised DBSCAN algorithm
<hr/>

1: read a raw record
2: if the record fetched by step 1 is a kernel point(the distance between this point and at least another one is less than the threshold)
   then find all the points whose arrived distance to the kernel point is less than the threshold, forming a cluster ,the kernel point is the kernel of the cluster as well
   else the point fetched by step 1 is a marginal point, go to step 1
3: if all the records has been disposed, stop
   else go to step 1

Each cluster contains the same type aircrafts. A velocity-time curve from each cluster is at last able to be obtained. The figure below shows a representative curve of a PTZ camera tracking a high-speed aircraft.

Each curve represents a type of aircraft. We may use the extra information to obtain a Markov Transition Matrix, which will help us to overcome the judgmental dilemma between an amplified noise and a real object.

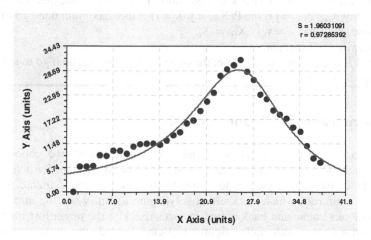

Fig. 1. X axis denotes time, and Y axis denotes PTZ angular velocities, we can simply find an equation $y = \dfrac{a + bx}{1 + cx + dx^2}$ the curve fits

3.3 Movement Predication

Based on the obtained curves and clusters above, the new unknown velocity can be predicated and classified by uni-dimension Markov Model when the normal tracking algorithm fails.

$$P_{ij} = P\left(X_{t+1} = j \mid X_t = i\right) = \frac{1}{m-1}\left(1 - \frac{\left|v_t - v_t^{(i)}\right|}{\sum_{k=1}^{m}\left|v_t \quad v_t^{(k)}\right|}\right) \tag{5}$$

$$v_t^{(k)} = f\left(t \mid a_k, b_k, c_k, d_k\right) = \frac{a_k + b_k x}{1 + c_k x + d_k x^2} \tag{6}$$

$v_t^{(m)}$ denotes the velocity of mth type of aircraft at the moment t, v_t denotes the real-time velocity, X_t denotes the most posible status of the aircraft, P_{ij} is the Transition Matrix.When the tracking algorithm fails, we will use the velocity of X_{i+1} instead.

Algorithm 2. Movement Predicated algorithm

1:compute P_{ij}
2:$X_0 = 1$, t = 0
3:obtain v_t from a normal tracking algorithm. If v_t is abnormal, then go to step 5
4: modify X_{t+1}. $X_{t+1} = j$ if the $P(X_{t+1} = j \mid X_t = i)$ is the maximum one , go to step 6
5:revise abnormal $v_t = v_t^{(Xt)}$, $X_{t+1} = X_t$
6:send v_t to PTZ camera
7:t=t+1,if t>t_{thres} reset, send a reset message to PTZ camera, then go to step 2
   else go to step 3

---

4 Experimental Result and Analysis

In order to meet the demands in complicated environment, we choose thermal imaging equipments in the experiment. In the preprocessing step, Particle Filter and Kalman Filter have been chosen to compute the features and estimate the logical velocity of a aircraft, which work effectively(figure a).But when the aircraft speeds up, object looks vague and background changes quickly, the target lost immediately, the algorithms are both failed(figure b).Being added the new compensatory algorithm , whatever the speed of the aircraft, the PTZ camera moves smoothly (figure c)and even loses the target temporally, which would relocate the target(figure d passing a pole).

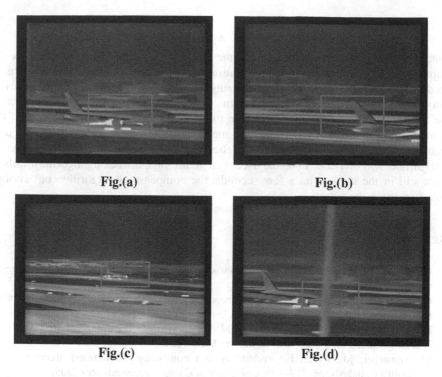

Fig.(a)   Fig.(b)

Fig.(c)   Fig.(d)

According to the figure below, the advantages of the compensatory algorithm being added to the normal algorithms is obvious. The normal algorithms often fail at the moment of 17th second (at the top of the red curve); but the algorithm mentioned above relocates the target after some trashing (at the same time of the pink curve), decreases the rate of failure less than 10%.

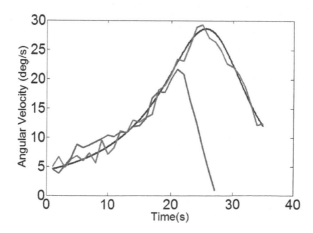

Fig. 2. X axis denotes time, and Y axis denotes PTZ angular velocities, the blue curve is the theoretical curve; the red one represents normal algorithm actual tracking and pink one represents the algorithm mentioned above.

5   Conclusion

Data modeling using the above method, the stability of the tracking algorithm greatly improved. The advantage of this algorithm is to compute a small amount of data, to predict the probability distribution utilizing historical data efficiently as empirical information, and to correct abnormal status easily. Decision-making chain, in fact, is the Markov decision chain; it is probable to return to the correct location even if the current status is error. In our specific methods described in, increasing the data records, prediction accuracy will be higher and the capacity for different aircraft recognition will increase. In actual use, even if the image tracking algorithm fails to track still in the stable after a few seconds, the compensatory algorithm has enough time to retrieve the target.

References

1. Pupilli, M., Calway, A.: Real-time camera tracking using a particle filter. In: Proceedings of the British Machine Vision (2005)
2. Kalal, Z., Matas, J.: Online learning of robust object detectors during unstable tracking. In: Computer Vision Workshops (2009)
3. Ross, D.A., Lim, J., Lin, R.-S., Yang, M.-H.: Incremental Learning for Robust Visual Tracking. International Journal of Computer Vision (2008)
4. Mitzenmacher, M., Upfal, E.: Probability and computing: randomized algorithms and probabilistic analysis, pp. 127–129. University of Cambridge, Syndicate (2005)
5. Birant, D., Kut, A.: ST-DBSCAN: An algorithm for clustering spatial–temporal dat. In: Data & Knowledge Engineering (2007)

A New Sample-Based Algorithm for Inpainting Used in Secrete Information Hiding

Yue Qin

West Anhui University, Anhui Lu'an 237000, China

Abstract. To prevent secrete information hiding in remote sensing image revealed, we propose a sample-based algorithm for inpainting used in secret information hiding in remote sensing image, the method selects the maximum region similar to secret information image, and covers the secret information block, thus attain our end of hiding the secret information. And combine with digital watermarking, realizing that users can see secret information or not according to their authorities, that makes use of remote sensing image convenient. Experimental results show that either from invisibility to the secret information or the analysis to use of remote sensing image, the effect is satisfactory.

Keywords: algorithm, inpainting, image, information hiding, digital watermarking.

1 Introduction

In recent years, increasing development of environment remote sensing technique has been used widely in every area of society.With the development of internet, digital products are spreaded extensively, which increases the possibility of having the information included in the remote sensing images stolen. For the sake of safety, it's necessary to prevent the disclosure of the information. One way is using local refinement or masking to the images, but it undoubtedly declares that there is noteworthy something in the images.In this paper, we propose a sample-based method for inpainting, and combines with digital watermarking to hide the secret information. The method has strong invisibility and doesn't attract any notice. Analysis is made to explain the influence to the remote sensing images by using of this method. Experimental results show that the method is feasible.

2 Algorithm Design and Analysis

2.1 A Sample-Based Algorithm for Inpainting

The application of digital image inpainting technology is more and more widely, and it has become a research focus of computer graphics and vision. To large-area inpainting, a commonly used algorithm is sample-based texture synthesis. That is,copy pixels from source area and fill them into the target area which contains texture and structure information, and the structure information between the areas can be got by changing the pixels' filling order. So this algorithm can not only generate texture and structural information, but also but also has the advantage of keep the efficient of the algorithm of texture synthesis[1.2.3].The algorithm advanced in this paper is just the sample-based algorithm. First, remove the secret information contained in a remote sensing image, then repair it, and combine with digital watermarking technology to realize the hiding of secret information. Paper [2.3] introduced the main process of the algorithm.

Here is the specific implementation procedure. Suppose I be an image, Ω is the target area to be filled, and its contour line is δ_Ω. p is a point on the contour line, $\Phi(\Phi = I - \Omega)$ is the source area.As show in figure 1.

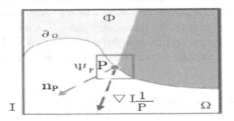

Fig. 1. Schematic diagram of inpainting area

The size of square area Ψ_p along the contour of target area, we take it 3×3, 9×9, in practical application, we should take it bigger than the maximum identifiable texel (The central point Ψ_p is on the contour line δ_Ω, template Ψ_p should include part of the synthesized pixels). In the design of the algorithm, each pixel in the template has a color and confidence value (which indicate whether the point has been filled, and 1 for filled, 0 not filled). After initialization, looply execute the following three steps until finish filling.

Step1. Computer priority

The priority level depends on two factors: one is the templates' data value, which reflects the intensity of structural information of a template, thereby ensures the preferential synthesis of the linear structure parts; another is the confidence value of the templates, give priority to those templates which contains more filled pixels, for filling such a template need rely on more known pixels. The two parts interact and form the priority together. Here the priority is calculated by the following formula:

$$P(p) = C(p)D(p) \tag{1}$$

$$C(p) = \frac{\sum_{q \in \psi_p \cap \bar{\Omega}} C(q)}{|\psi_p|}, \qquad D(p) = \frac{|\nabla I_p^{\perp} \cdot n_p|}{\alpha}$$

where

$C(p)$ is the confidence value of the template and D (p) the data value. C(q) is the confidence value of the pixels within the template, when initialization begins, each point in target area is set the value of 0, and point in source area set the value 1. That is

$$C(q) = \begin{cases} 0 & \forall \, q \in \Omega \\ 1 & \forall \, q \in L - \Omega \end{cases} \tag{2}$$

$|\psi_p|$ is the area of template ψ_p, α is a standardized parameter, for a generic gray image, $\alpha = 255$. n_p is a unit normal vector of contour line at point P, $\nabla I_p^{\perp}$ is the strength and direction of light phot line at point P. For each template, the synthesis order is according to the calculated priority. By formula (1), on the one hand, a sample containing more filled pixels has larger confidence value, and in the target area the samples forming corner and tendrils have propriety to be filled, the samples having obvious structure and containing prominent lines have large data value, which have propriety to connect. On the other hand, the samples with larger data value make priority to the sections who have prominent lines, but the samples with larger confidence value are in contrast , in this case, both will gain a certain balance.

Step2. Look for the matching template to copy as sample
Once the priority calculation is complete,take sample from the source area, look for the sample which match most with the sample's synthesis pixel, and fill this template.

Step3. Update the confidence value
Along with the rest pixels of the template finish filling, update the confidence value of the pixels again

$$C(q) = C(\hat{p}), \; \forall \, q \in \psi_p \cap \Omega \tag{3}$$

Repeat above three steps until the ready-to-fill areas finish inpainting.

2.2 Algorithm Implementation

Generally speaking, the most deficiency of the sample-based algorithm for inpainting is the completion time. In seeking of the best matching template, we usually scan and calculate the whole image, undoubtedly the completion time of inpainting for a big remote sensing images is very long. Considering the strong correlation among the textures of the remote image and according to the actual situation of the image, we can completely determine a small area near the inpainting area to search the best matching template, in this way, the completion time can be largely shorten. One need to explain is that if the image is RGB mode, synthesis can be done after treat separately.Here we deal with the gray image.The main steps of the method are as follows.

Step1. Segmented the secret image block from the remote sensing image (The location of the secret information in the remote images can be recorded).

Step2. Limit a region artificially near the blank area where extract secret image block, it's best to take the secret area locates in the center of the limited area, but the limited area must completely contain confidential the secret area.

Step3. Calculate the template's priority. First repair the template with the maximum priority. Take the size of Ψ_p 3×3, for the discrete points, the area of the template Ψ_p (denoted by $|\Psi_p|$) take the number of pixels 9. Let the coordinate of point P be (x, y).

As point P locates on the contour line of the inpainting area, it's possible that some point around P is in the inpainting area. We use difference method to calculate $\nabla I_p^{\perp}$:

let $u(x, y)$ be the pixel value of point (x, y).

$$u_x = u(x+1, y) - u(x, y) \text{ (forward difference)}$$

$u_x = u(x, y) - u(x-1, y)$ (backward difference)

Similarly, we can calculate u_y and get $\nabla I_p = (u_x, u_y)$, $\nabla I_p^{\perp} = (-u_y, u_x)$.

We adopt the following method to computer n_p: determine a vector by two points adjacent to P on the contour line of the inpainting area, n_p is just a vector perpendicular to the vector. Then we use formula(1),(2),(3)to calculate the template's priority.

Step4. Select the best matching template only in the restricted area. The source area is divided into the parts of the same size of 3×3 and then compare them with the highest priority block. Assume the target block Ψ_p has the highest priority, note $\Phi_{p'}$ be the most similar block to Ψ_p in the source area. So

$$\Phi_{p'} = \arg\min_{\Phi_p} d(\Phi_p, \Psi_p)$$

where $d(\Phi_p, \Psi_p)$ defined as the sum of squared error of filled pixels in the two mentioned blocks.

Step5. Copy the pixels from the best matching template and then fill them in the corresponding position of Ψ_p.

Step6. Update the confidence value of the newly filled pixels in template ψ_p .

Step7. Looply execute above steps until the contour line become empty.

2.3 Algorithm Improvement

We have presented a sample-based image algorithm for inpainting used in secret information hiding in remote sensing image, the method is cleaning secrete information directly. However, for different user rights, the requirements of whether the secret information can be seen are different, for some special users, they should have the rights to see the secret information[4]. How to achieve this distinction for a same image? In this section, we further improve the algorithm by digital watermarking technology, which can realize the distinction well.

Digital Watermarking technology is to make some identification information (namely, digital watermarking) embedded into digital media directly (including multimedia, documents, software, etc.), but it doesn't affect the use value of the original carrier, and it's not easy to be awarded or noticed by the perception system too. By the information hidden in the carrier, we can achieve the purpose of confirming the content creators, buyers, sending secret information or judging whether the carrier has been tampered and so on. As an effective technology to protect the copyright of digital products, digital watermarking technology has made a high achievement both in theoretical research and practical applications. The watermark embedding process and the watermark extraction (testing) process included in digital watermarking system are the key to distinguish user rights[5].

2.3.1 Realization of the Algorithm

First, extract confidential information from the remote image, and then fill the blank gray area values to get free confidential information, pseudo-sensing image. Finally, take the split information as a watermark and embedded it in the pseudo-remote image, users can extract confidential information and restore the original remote sensing images with keys.

The process of extracting confidential information should save the split image's location information in the original image, and then embedded it in the repaired remote image with DFT digital watermarking technology. After the extraction of watermarking namely confidential information, use known location information to recover remote sensing images which contain confidential information. One of the main steps of the method is image restoration; the second is the choice of watermarking algorithm.

Steps of embedded watermark and hidden the confidential information are as follows:

(1) Preprocess the remote image which contains confidential information, we can split and extract manually to extract the confidential information and saved as a watermark image. Note as

$W = \{w(i, j), 0 \le i < P, 0 \le j < Q\}$, and record its location and size in the original image.

(2) Using the sample-based image inpainting technology mentioned in section two to repair the damaged remote image, denoted by

$$I = \{g(i, j), 0 \le i < M, 0 \le j < N\}$$

(3) Embedded W as a watermark in good repaired remote sensing image.

Steps of restoring images containing confidential information are as follows:

(1) The process of extracting watermark is a inverse of embedding, and get a confidential information, denoted by

$$W' = \{w'(i, j), 0 \le i < P, 0 \le j < Q\}$$

(2) Combined the confidential information image W' with location information and then put it back to the original position, in this way a remote sensing image contain confidential information will be recovered.

From the above, we can see the algorithm can according to the user's authority to provide key to determine whether the users have the access to see the confidential information of remote sensing image.

3 Simulation Experiment and Analysis of the Results

We compared the experimental results of this process as shown in figure 2 and 3. Figure 2a is the original remote sensing images which contains confidential information, and its shade part is the confidential information. 2b is the image have been divided and extracted confidential information, 2c is the extracted confidential

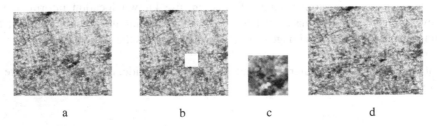

a   b   c   d

Fig. 2. Process of restoring a remote sensing image

information, 2d is the restored image which doesn't contain the confidential information. Figure 3a is a remote sensing image embedded watermark and it doesn't contain confidential information, 3b is extracted confidential information watermark, 3c for the restored remote sensing image which contains confidential information.

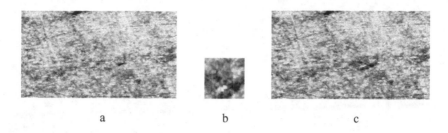

a   b   c

Fig. 3. Process of restoring a remote sensing image embedded watermark

4 Conclusions

In this paper a sample-based algorithm for inpainting used in secret information hiding in remote sensing image is proposed. The simulation results show that either from invisibility to the secret information or the analysis to use of remote sensing image, we obtain a satisfactory effect. It doesn't affect the use value of remote sensing image. All of this illustrates feasibility and effectiveness of the algorithm.

Rererences

1. Criminisi, A., Perez, P., Toyama, K.: Region filling and object removal by exemplar- based image inpainting. IEEE Transactions on Image Processing 13, 1200–1212 (2004)
2. Peng, H., Hou, W., Gong, N.: An Improved Exemplar-Based Inpainting Method for Object Removal. Journal of Computer-Aided Design and Computer Graphics 18, 1345–1349 (2006)
3. Dai, L., Wei, B.-G.: Research of image restoration algorithms. Computer Engineering Design 27, 184–187 (2006)
4. Wang, X., Wang, C., Zhou, J.: A New Technique for Information Hiding Based on Different Authorities in Remote Sensing Image. Computer Engineeering 32, 28–30 (2006)
5. Hu, Y., Chen, H., Fang, S.-B.: Application of Digital Watermarking Technology in Remote Sensing Image Copyright Protection. Computer Simulation 22, 200–202 (2005)

The Factors Affecting the Attitude of University Students towards Online Shopping
A Case Study of Students in Honghe University

Mengli Ma[1] and Rui Ma[2]

[1] Business College of Honghw University, Mengzi, Yunnan, PRC, 661100
mengli.ma@gmail.com
[2] Kun Steel Holding, Kunming, Yunnan, PRC, 650302
543118529@qq.com

Abstract. In recent years, online shopping as a new consumption method accepted by more and more people. University students are becoming the major group of online shopping. So we need have a better understanding of the university students' attitude towards online shopping, so that companies which are doing or want to do e-commerce can take advantages from it. According to the characteristics of university students, the researchers assumed that the factors of gender, grade and major, perceived safety, website construction, price and quality of products have influence on the attitude of university students towards online shopping. The data collected through the self-helped questionnaires, and Variation analysis, Pearson correlation coefficient analysis, independent sample t-test are used to analyze the hypotheses. Finally the result of analysis showed that the factors of major, perceived safety, website construction, prices and quality of products have significant influence on university students' attitude toward online shopping.

Keywords: Online Shopping; Attitude; University Students; Factors.

1 Introduction

As a new consumption form, online shopping more and more get consumers' attention. It is a part of e-commerce, and it is the process of browsing and searching products information in a virtual online environment in order to provide sufficient information for purchase decision, and then implement the purchase action. According to the data indicate from Internet Information Center of China, the number of online shopping consumers increased 14 million in 2009, and reached 87.88 million in total. Through the development in several years, many websites for online shopping are outstanding, such as Taobao, Dangdang, Alibaba, Paipai and so on. They encouraged the development of e-commerce in China.

University students have more chance to use internet and are good at using computer, it is the technical foundation for them to engage in online shopping. Moreover, university students are more curious and acceptable with new things. So that university students have become one of major group of online shopping.

The purpose of this research is to study the factors which affect the attitude of university students toward online shopping in order to provide some practical

D. Jin and S. Lin (Eds.): Advances in ECWAC, Vol. 2, AISC 149, pp. 511–515.
springerlink.com   © Springer-Verlag Berlin Heidelberg 2012

suggestions for internet or e-commerce companies to help them get advantages in the market decisions.

2 Related Theories and Research Hypothesis

2.1 Related Theories

Attitude is a learned predisposition to behave in a consistently favorable or unfavorable way with respect to a given object. It is the evaluation and behavioral tendency of people for given things that developed from their moral notion and value. Commonly it is the orientation of individual behavior, and the important index to predict individual behavior. Attitude represent the internal cognitive, affective, and conative toward external things.

Based on the previous studies, researchers found that the factors of perceived safety, website construction, price and quality of products have influence on consumers' attitude towards online shopping. Safety indicates that the hardware, software and data in internet system are protected, and away from occasional or baleful destroy, modify and leak. In another words, internet safety means safety of information in online environment. On the other side, in many people's perception they can always get cheaper goods in online shop, because of the companies which sell product online can save a lot of money compare with traditional seller from such rent fee for the store, agency fee, salaries and so on. Thirdly, online shopping is a virtual way of purchase, it is difficult for consumer to touch the real products when they make purchase decision, which makes part of consumer doubt about products quality, and then this unbelief reduced their purchase action. The last factor is website construction, the style, sort, compositor, color etc. will directly influence consumers' perception and stay time on the website, consequently influence their purchase indention and decision.

2.2 Conceptual Framework and Hypotheses

The researchers developed the conceptual framework based on the literature review and previous studies as showed below:

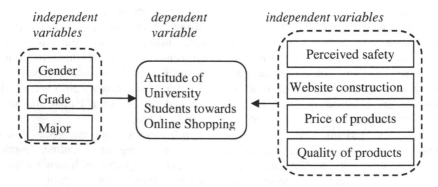

Fig. 1. Conceptual Framework

Based on the conceptual framework, 7 hypotheses are constructed in order to check the relationship between dependent variable and independent variables. The hypotheses and the statistical methods used for each hypothesis are showed in table 1.

Descriptive research design is applied in this research. Researchers conducted the survey method by using questionnaire in order to getting a feedback from respondents with efficient way, the non – probability procedure is used to select respondents. Questionnaire to gather primary data from 320 respondents recruited on a voluntary manner among the students who study in Honghe University. Pearson Product Moment Correlation Coefficient, One-Way ANOVA and Independent sample T-test was used to test the hypotheses.

Table 1. Statistical Methods Used for each Hypothesis

Hypothesis	Analysis Method
H_{1o}: There is no difference in attitude towards online shopping when segmented by gender.	Independent t-Test
H_{2o}: There is no difference in attitude towards online shopping when segmented by grade.	Analysis of Variance (ANOVA)
H_{3o}: There is no difference in attitude towards online shopping when segmented by major.	Analysis of Variance (ANOVA)
H_{4o}: There is no relationship between attitude towards online shopping and perceived safety.	Pearson correlation
H_{5o}: There is no relationship between attitude towards online shopping and website construction.	Pearson correlation
H_{6o}: There is no relationship between attitude towards online shopping and price of products.	Pearson correlation
H_{7o}: There is no relationship between attitude towards online shopping and quality of products.	Pearson correlation

3   Finding: Summary of Hypothesis Testing

Table 2. Summary of Results from the Hypotheses Testing

Hypothesis	Analysis Method	Significance	Correlation coefficient	Results
H_{1o}	Independent t-Test	0.157		Failed to Reject
H_{2o}	Analysis of Variance (ANOVA)	0.604		Failed to Reject
H_{3o}	Analysis of Variance (ANOVA)	0.002		Rejected
H_{4o}	Pearson correlation	0.000	0.479^{**}	Rejected
H_{5o}	Pearson correlation	0.000	0.538^{**}	Rejected
H_{6o}	Pearson correlation	0.000	0.329^{**}	Rejected
H_{7o}	Pearson correlation	0.000	0.528^{**}	Rejected

As showed in table 2, there are total 5 hypotheses out of 7 are rejected, means there is a difference in attitude towards online shopping when segmented by major. And the factors of perceived safety, website construction, price and quality of products have significant influence on university students' attitude toward online shopping.

4 Recommendations

It is important to know the attitude of university students in order to increase the sale volume via online medium. Therefore, several recommendations can be made for companies and venders who are willing to sell their products online through websites to university students.

First of all, according to the result of this research there is a difference in attitude towards online shopping when segmented by major. Means the students with different study or research field may pay attention on different website which related to their interested area. Then, the company should make the website more professional, while provide specialized products and service to university students. For example, try to use professional way to describe the products.

Secondly, according to the research, perceived safety has a significant influence on university students' attitude toward online shopping. The companies should improve their reputation in transaction safety. Commonly consumers have to provide the personal information during the transaction online, it increased the risk of online shopping, and many consumers are worried about the case of information reveal. Therefore, the companies should improve their technology of information protection, such as message encryption, to ensure that the website provided safe service to consumers. Moreover, the company also need communicate the method of information prospection they used with consumers in order to bring consumer reassurance in transaction.

The third one, the price of products online also has a positive relationship with attitude of university students towards online shopping. The students are very sensitive to the price fluctuate because of their consumption are more rely on the supports of their family. Then the companies should provide reasonable price to consumers by the way of cutting the cost, increasing the speed of turnover of capital, eliminating the inventory and so on.

The next one, the consumers' perception of products quality is significantly influence the attitude of university students towards online shopping. Hence, to increase these consumers' positive attitude towards online shopping, it is very important that companies prove that the products online have undergone safety tests and obtained quality certification. If consumers know that the products quality are good, and have passed all required tests, they are more likely to buy online. The companies should also introduce money back offers (returning the money the customer has spent if the customer is not satisfied with the product) on products, which shows customers that the companies care for them and this will increase the consumer's trust towards the companies.

The last one, the construction of website also influences the attitude of university students towards online shopping. Therefore, the companies should provide better designed website which can attract the consumers. The companies can redesign the website in a more reasonable, friend and interesting manner, such as increase the consumers interacting experience during the online shopping by adding more functions on the website.

Acknowledgement. At the end of this paper, I'd like to express my appreciation to all those who have lent me hands in my writing and the supports of the program funds of "The Study on E-commerce and Online shopping behavior（10BSS213）" from Honghe University.

References

1. Fishbein, M., Ajzen, I.: Belief, Attitude, Intention, and Behavior. Addison-Wesley, Reading (1975)
2. Lai, Z.H.: The Report of A Statistic Analysis on the Industry of Online Shopping in 2009, http://wenku.baidu.com/view/20758738376baf1ffc4fadbf.html (March 11, 2010/January 21, 2011)
3. Zang, L.Y., Cui, L.H.: Website Construction in E-Commerce. Shanghai University of Finance and Economics Press, Shanghai (2007)
4. Schiffman, L.G., Kanuk, L.L.: Consumer Behavior, 8th edn. Prentice Hall, Upper Saddle River (2006)
5. Zikmund, W.G.: Business Research Methods, 7th edn., Mason, OH: South–Western, USA (2003)

Author in Brief

Ma Mengli, Master of Business Administration, currently employed in Business College of Honghe University, mainly engaged in Management and Organization Behavior. Can be reached at mengli.ma@gmail.com

A Research on Cache Management for the Distributed Cache in P2P-CDN

Binjie Zhu[1] and Zhiwei Shen[2,3]

[1] School of Computer Science and Technology, Beijing University of Posts
and Telecommunications, Beijing 100876, China
[2] School of Economics and Management, Beijing University of Posts and Telecommunications,
Beijing 100876, China
[3] China Unicom Research Institute, Beijing, 100032, China
zhubinji1983@gmail.com

Abstract. Distributed cache is a key technology in content distribution network, how to management the content in the distributed cache is an important issue. This paper presents a pull method for the distributed cache group in P2P-CDN. This method improves the utilization and cache hit rate for the cache group form optimize the content deployment in the group. The analysis shows that cache space can be use effectively to improve the group hit rate and reduce the average access delay.

Keywords: Cache Management, Distributed Cache, P2P, CDN.

1 Background and Related Research

Caching technology was widely application in many areas, for example, in the three-level storage system in PC and the Internet proxy servers. Caching technology has development from single cache to the distributed cache.

In the study of the traditional single cache, the most studied is the cache replacement algorithm. There are some classic cache algorithms LFU [5], LRU [6], LRU-K [7], FIFO, SIZE, etc. FIFO is a simple cache replacement algorithm. The principle of this method is first in first out. However, this method does not take into account any factors for the value of the content object in cache. LRU and LFU are based on the principle of locality of access, leading to the request hit rate decline and response latency increases. SIZE algorithm replaces the contents base on the size of the content, but it could not represent the real importance of the objects. The advantage the classical algorithms are simple, easy to implement, but it has low hits rate, and waste the system space.

There are some improved algorithms was proposed based on the classic algorithms. However, these methods are generally application in the non-cooperative caching environment. In order to improve cache efficiency, there are many distributed cache architecture has been proposed. Some paper also refers to the content management for distributed cache [1, 2, 3].

D. Jin and S. Lin (Eds.): Advances in ECWAC, Vol. 2, AISC 149, pp. 517–521.
springerlink.com   © Springer-Verlag Berlin Heidelberg 2012

These methods are proposed for the specific areas, and have certain limitations. For example, the research in cache memory between the CPU and disk is only a single cache, rather than a cache network. This principle is to use the speed difference of disk and the cache to find a balance between the performance and price. Paper [1],[2],[3]only suitable for tree hierarchy of cooperative caching. Paper 4 proposed a distributed cooperative caching content management method called EABS. But EABS is just for general distributed cache. CDN is a technology used to increase the speed in the Internet. In this paper a content management method for distributed cache in CDN was proposed. Analysis shows that this method takes into account the CDN features and has better performance.

2   Distributed Cache Content Management Method

In this section, we will firstly introduction the cache content management issues inP2P-CDN in 2.1, and then introduction the proposed method in 2.2.

2.1   The Cache Content Management Issues in P2P-CDN

For a distributed cache group in P2P-CDN, it can store part of the contents. If the cache group has not the content requested by the user, then it can be pulled from the global scope. Content object stored position is determined HASH value of the contents. This approach ensures that the content pulled from the outside to the distributed balance in a distributed caching group server. The content first stored in the group was called original source server of this content in the cache server group. the Content object spread from the source server group to other servers within the group based on user requests in a certain way . Other servers storing the content object have become the object source server for this content. The rules for the server pull the content from the original source server is called the cache group pull method.

Most of the current study are about the issues of the cache replacement algorithm or distributed cache architecture. The research about the pull and deployment method is less.

EA is a classic content pull and deployment method. This method controls the content deployment in the cache group by the value of EA. Here the concept of SEA extends on the basis of the EA, and a new approach was proposed based on the concept of SEA, which not only use the SEA. It is more adapt to P2P-CDN optimization goals.

2.2   The Proposed Method

We propose a few concepts in new method: SEA, accelerated efficacy, the potential to accelerate efficacy between two points, network acceleration efficacy and then introduce the algorithm works.

Firstly, the concept and calculation method of SEA will be introduced. SEA is a short for service failure time. Different grade in a server is calculated separately, SEA of class N is calculated as follows:

$$SEA\ (n) = (out\ time - in\ time)\ /time \tag{1}$$

Need to define accelerated efficacy AU. Acceleration efficacy is an increase of access speed after the content cache. If a content object cache from the content source node S to destination node D, the delay between point S and point is. The arrival rate of node B is, then the acceleration efficacy is shown as below.

$$AU_{SD} = Delay_{SD} \times l_d \tag{2}$$

We will introduce the definition of potential accelerate performance EAU based on accelerate performance. There are two node C and A. The client required content stored in the node C. If the client A send request to server A, and the request content is.The node A has not the client's request content, so the request will be redirect to the node C. The delay between node A and node C is .The redirected rate is node A and node C is . The redirected rate is redirected number in time unit. Potential accelerate performance is defined as:

$$EAU_{ca} = Delay_{ca} \times Rd_{ac} \tag{3}$$

Potential acceleration performance is similar to the acceleration performance, and it is used to evaluate the cache value.

There is cache server node S, and there are N content in node S. For the content Cn,there are some request redirected form other M node. For node Cn, the NEAU is as below:

$$NEAU_{CN} = \sum_1^m Delay_{ss_m} \cdot Rd_{ms} \tag{4}$$

If a content first pulled from outside of the group, other server will pull the content from this server. The request name delay and times of the every request to content were recorded. The times of record was control by the a limit value. If the times larger than the limit value, the old record will be delete. The value of EAU and NEAU can get through the record.

If the NEAU of content object larger than the gate value, then the content will be pulled to other nodes. The nodes where to pull the content is determined by SEA. If the SEA of node larger than the source content, the node will be includes in the chosen set. The EAU of the nodes in the set and local node were caculate, than take the largest one as the pull node.Algorithm flow is as figure 1.

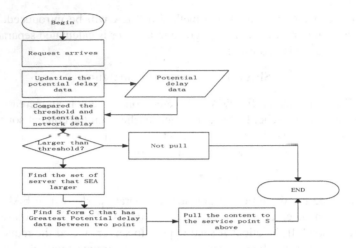

Fig. 1. Algorithm flows for the proposed method.

3 Comparative Analysis

Contrast with the EA and AD-HOC algorithm, the proposed algorithm has a higher pull requirement, thereby increasing the efficiency of collaboration in the distributed cache group. It has different groups hit rate with other method.

Assuming the cache group has n caches, each cache size is m, the size of each content object, a total of a content object.

In extreme cases in ADHOC algorithm the same content stored in each cache, there are N copies for each content object. the amount of content stored in the cache group is. Assuming an average access of each content object, hit rate is. For the EA algorithm assuming that the restrictions by EA, an average of every object has K copies. , then the group hit rate is. The proposed algorithm has L copies, store, so group hit rate is . Because L <K <Q, so the hit rate of the proposed method is the highest one.

4 Conclusions and Future Work

The proposed method is aim to the requirements of a group of cache, and it is not a single point. So it forms a collaborative relationship in the space of cache group, thereby enhancing the group's hit rate, and also reduces the overall delay.

In the future the method will be improved further, especially replacement algorithm in the cache group and pull algorithm in the global routing layer.

References

1. Che, H., Wang, Z., Tung, Y.: Analysis and design of hierarchical web caching systems. In: INFOCOM 2001, vol. 3, pp. 1416–1424 (2001)
2. Tang, X., Chanson, S.T.: Coordinated management of cascaded caches for efficient content distribution. In: Proceedings 19th International Conference on Data Engineering, pp. xviii+879 (2003)
3. Borst, S., Gupta, V., Walid, A.: Distributed caching algorithms for content distribution networks. In: INFOCOM 2010, pp. 1–9 (2010)
4. Korupolu, M.R., Dahlin, M.: Coordinated placement and replacement for large-scale distributed cache. In: Proceedings 1999 IEEE Workshop on Internet Applications, pp. 62–71 (1999)
5. Robinson, J.T., Devarakonda, M.V.: Date cache management using frequency-based replacement. In: Proceedings of SIGMETRIC on Measuring and Modeling of Computer Systems, Boulder, Colorado, USA, pp. 134–142 (1990)
6. Alghazo, J., Akaaboue, A.: SF-LRU Cache Repalcement Algorithm. In: Records of The 2004 International Workshop on 9-10, pp. 19–24 (2004)
7. Denung, W.S., Ki, Y.K., Jong, S. L.J.: LRU based small latency first replacement (SLFR) algorithm for the proxy cache. In: Proceedings of IEEE International Conference on Web Intelligence (2003), ISBN: 978-1-84334-499-5
8. Zhu, B., Xu, K., Pi, R.: A resource discovery method in P2P-CDN. Journal of Computational Information Systems 7(10), 3390–3397 (2011)
9. Shi, P., Wang, H., Gang, Y., Yuan, X.: 2011 IEEE 3rd International Conference on Communication Software and Networks, ICCSN 2011, pp. 182–187 (2011)
10. Kim, T.N., Seoul, D.-G.: A CDN-P2P hybid architecture with content/location awareness for live streaming service networks. In: Proceedings of the International Symposium on Consumer Electronics, ISCE, pp. 438–441 (2011)

The Usage of Σ-ΔA/D in Langmuir Probe

Yifeng Zhu, Xu Yang, Lei Shi[*], and Yanfu Li

Changchun University of Science and Technology, No. 7089 Weixing Road,
Changchun 130022, P.R. China
shilei@cust.edu.cn

Abstract. Langmuir probe is an important method for diagnostic of the charged particle parameters.Now the usual way is the volt-ampere characteritic curve. Probe power supply and detecting circuits of probe current and probe voltage are studied in this article, which is based on langmuir probe diagnostic principle. This paper discusses a kind of Langmuir probe collect circuit that depends on A/D conversion and give a specific implementation plan based on MCU, experiments show that the plan is simple and reliable, the test results is satisfactory.

Keywords: Langmuir probe, plasma diagnostics, electron density, Σ-ΔA/D.

1 Introduction

The research of plasma-materials interaction has become an import and area of current international research. To understand an study a thing, first it can be diagnose and characterize correctly. Plasma is also no exception, various parameters measured through plasma diagnostic method is very important for basic research and applied research in plasma, especially for production process research of plasma. Of all the ways to measure a plasma, the Langmuir is the fundermental and the most important method. Its essence is to analyze volt-ampere characteristic curve of the probe, and then some parameters of plasma can be obtained.

Process volt-ampere characteristic curve of the probe with A/D conversion, and then process it in computers, the general advantage of this digital processing mean is obvious, here the highlight of this way is that the improvement of the accuracy is benefit for researching fine structure in plasma; the short relaxation time of computer data collection make it possible for researching dynamics process of plasma.

As the charged plasma is generally stimulated by RF power in the experiment, so the curve of the probe is generally interfered by RF power and occurs distortion when plasma is in an electromagnetic environments of changing RF. Although tunable filter can curb RF interference effectively, there are also some noise caused by statistical fluctuation, so the second derivative obtained by experimental curve directly can't give an objective and inherent regularity, even it will bring big error to measurement results.

D. Jin and S. Lin (Eds.): Advances in ECWAC, Vol. 2, AISC 149, pp. 523–528.

With the emergence of A/D conversion circuit taken an oversampling Σ-Δ conversion technology, and it make the transformation of the value of the plasma probe possible by A/D conversion. Because oversampling Σ-Δ conversion technology can curb RF interference effectively, the improvement of SNR by digital filter bring assurance for the correct calibration of plasma parameter. The AD7715 is an A/D conversion chip, the chip takes advantage of oversampling Σ-Δ technology, it contains self-calibration control circuit, its programmable gain amplifier could adapt to occasions where the current of the probe changes in big range.

2 The Theory of the Langmuir Probe

The theory of the electrostatic probe is provided by Langmuir in 1924, now it's still an important means in scientific research because of its simplicity and credibility.

According to the cosine law in the kinetic theory of gases, the collision between the number of the electrons and the probe per unit area in a unit of time is:

$$N = \frac{1}{4} n_e \cdot \bar{v}_e \tag{1}$$

n_e is electron concentration closed to the probe, v_e is average speed of electrons. We assume electrons around the probe show a Marxwell distribution, let the potential applied to the probe be v_p when we conducted measurements(relative to a reference electrode), so the current density collected by probe show as:

$$I = -I_0 \cdot e^{\frac{e(V_p - V_s)}{kT_e}} + I_i \tag{2}$$

Where I_0 is the current through probe when probe at space potential, I_i is ion current, v_s is space potential of reference electrode, the above formula indicates the exponential growth of electron current should continue until $v_p = v_s$,when the electron current "saturates".

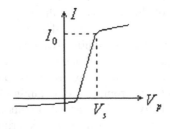

Fig. 1. The I/V curve of the probe

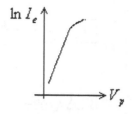

Fig. 2. Semi logarithmic volt-ampere characteristic curve

We can continue change v_p to get a set of values of I. A typical I-V curve form Langmuir probe is show in Fig.1. The traditional method of calculating the plasma electron temperature is drawing a $\ln I_e$-V_p curve(Fig.2) after we make an Ion current correction with the I-V curve in Fig.1, we could get the electron temperature by taking advantage of the slope of the part corresponding to the straight line in the semi-logarithmic plot:

$$kT_e = -\frac{1}{\dfrac{d \ln I_e}{dV_p}} \approx -\frac{1}{\dfrac{\Delta \ln I_e}{\Delta V_p}} \tag{3}$$

And we can calculate electron concentration with the saturation electron current I_0 and the measured value of T_e:

$$n_e = \frac{4I_0}{e\overline{V}_e \cdot S} \tag{4}$$

Where S is effective collection area of the probe, this method measured I_0 and T_e have been applied widely.

To curb RF interference effectively from probe power and reduce noise, we select a chip named AD7715 in the A/D conversion period, the chip employs oversampling Σ-Δ conversion technology.

3   The Probe Circuits Using Σ-Δ A/D Converter

The general block diagram of system shown in Fig.3, Σ-Δ A/D converter is the priority of the design.

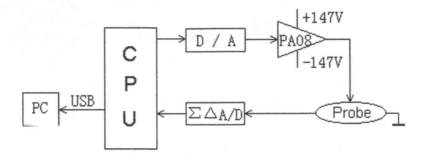

Fig. 3. System block diagram

The biggest difference between Σ-Δ A/D converter and Nyquist A/D is quantization process. In Nyquist A/D converter, the converter resolution depends on the quantification of a sampling interval, while Σ-Δ A/D converter takes advantage of a series of rough quantitative data, their sampling frequency is Fs(>>fs), then the digital decimation filter calculates high-resolution digital signal with low sampling frequency fs corresponding to analog signal. The realization of these processes are mainly based on oversampling Σ-Δ modulation and digital filtering.

The transition zone of frequency response of analog filter used by oversampling is big, the change is slow, thus the requirement of competing with aliasing filter reduced. The increase of frequency and the enlargement of scattering area of the quantization noise power reduce the baseband quantization noise.

Σ-Δ modulator is one kind circuit which has noise shaping function and works in a state of oversampling, it forms the delayed signal and the quantization noise presents high-pass form, so the noise distribution is changed, namely, the shaping function achieved. The output of the Σ-Δ modulator carries the input analog signal amplitude information, its frequency spectrum characteristic is that the frequency spectrum of the signal is in the baseband, the noise distribution focuses on the baseband, therefore we could acquire digital signal possesses high signal to noise by a digital low-pass filter.

4   Experiment and Analysis

Through the above design, implementation of the indicators are as follows:

   a. Scanning voltage: -125V to +100 V
   b. Scanning: 100 non-sampling and other steps
   c. The total sampling time: <0.15S
   d. 0 to 10mA range when zero drift error <3μA, the nonlinear error <1%

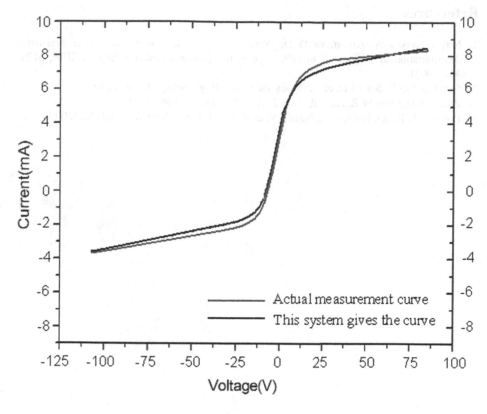

Fig. 4. Actual measurement of the I-V curve and the system gives the I-V curve

5 Conclusion

We can see from the above, Σ-Δ converter don't need hold circuit, and the anti-aliasing filter and quantizer have a low requircment, while the requirement of the digital filter is high. In the course of requirement, the analog signal turns into band-limited analog signal when it through the anti-aliasing filter, then the band-limited analog signal will turn into high-speed bit stream signal which signal frequency spectrum and noise have separated after Σ-Δ modulator, finally high resolution digital signal with Nyquist sampling frequency can be reconstruct after a digital filter. This basically fits the requirements of the design.

References

1. Nov odvorsky A, Khramova O D, Wenzel C, et al. Er osion plume char acteristics determination in ablation of metallic copper, niobium and tantalum targets. SPIE, 5121: 337(2003)
2. Sanchez Ake C, Sobral H, et al. Optics and Lasers Engineering, 39: 581(2003)
3. Zhang X D, Chen W Z, Jiang P, et al . J . Ap p l. Phys . , 93: 8842(2003)
4. Lorusso A, Krusa, Rohlena K, Nassisi V, et al . App l .Phys . L ett . , 86: 081501(2005)

Linear Cryptanalysis of Light-Weight Block Cipher ICEBERG

Yue Sun

Key Laboratory of Cryptologic Technology and Information Security,
Ministry of Education, Shandong University,
Jinan, 250100, China
yuesun@mail.sdu.edu.cn

Abstract. ICEBERG is proposed by Standaert in FSE 2004 for reconfigurable hardware implementations. ICEBERG is a fast involutional SPN block cipher and all its components allow very efficient combinations of encryption or decryption. ICEBERG uses 64-bit block size and 128-bit key and the round number is 16. In this paper, we identify the 6-round linear approximation of ICEBERG. Then we present an attack on 7 round of ICEBERG.

Keywords: Cryptography, Linear Cryptanalysis, Light-Weight, Block Cipher, ICEBERG.

1 Introduction

A block cipher based on SP-network structure has the different encryption and decryption process like AES, which will increase the hardware costs. Although the block cipher based on the Feistel structure does not have such disadvantage, its slow avalanche effect requires the large round number to guarantee the security. In this way, how to design an involutional block cipher based on SP-network structure has become an important object in the field of light-weight block cipher.

In recent years, there are many light-weight block ciphers which have been proposed, such as mCRYPTON [1], HIGHT [2] and PRESENT [3] etc. At FSE 2004, Standaert proposed a fast involutional block cipher with SP-network structure optimized for reconfigurable hardware implementations, named as ICEBERG [4]. ICEBERG uses 64-bit text blocks and 128-bit keys and the round number is 16. Specially, all components are involutional and allow very efficient combinations of encryption/decryption.

Linear cryptanalysis (LC) is introduced by Matsui in 1993 [5]. Since for a random permutation, the probability of any linear relation between the plaintext and corresponding ciphertext bits should be balanced at 1/2 while it can not always be the same in the case of block ciphers. By assuming the block cipher is Markov Cipher, linear approximations generating from each round can be concatenated to a linear approximation among plaintext bits, ciphertext bits and subkey bits, whose probability is different from 1/2. So it can be used as a distinguisher from random permutation, by which a key-recovery attack can be exploited.

D. Jin and S. Lin (Eds.): Advances in ECWAC, Vol. 2, AISC 149, pp. 529–532.
springerlink.com   © Springer-Verlag Berlin Heidelberg 2012

In this paper, we will give the concrete linear cryptanalysis for reduced-round of ICEBERG. Firstly, we analyze the property of linear layer of ICEBERG, then we give the best linear approximation for 6-round ICEBERG. Furthermore, we present the linear cryptanalysis on 7-round ICEBERG.

The paper is organized as follows. Section 2 introduces the ICEBERG algorithm. In Section 3, we identify the best 6-round linear approximation we found and present an attack on 7 round ICEBERG. Section 4 concludes this paper.

2 Description of ICEBERG

ICEBERG is a block cipher with SP-network structure. It operates on 64-bit block and uses a 128-bit key. It's composed of 16 rounds, in each round there are non-linear layer γ and linear layer ε_k. The round function can be described as $\rho_k \equiv \varepsilon_k \gamma$, while $\gamma \equiv S0 \circ P8 \circ S1 \circ P8 \circ S0$, $\varepsilon_k \equiv P64 \circ P4 \circ \sigma_k \circ M \circ P64$. S0 and S1 in non-linear layer are two 4×4 S-boxes. Pi is a permutation on i-bit. The layer γ can be viewed as one layer consisting of the application of eight identical 8×8 S-boxes. The matrix multiplication M is based on the parallel application of a simple involutional matrix multiplication, which makes each output bit equal to the exclusive-or among the three input bits, results in only 2 rounds being required to reach full diffusion. The details of the description of ICEBERG can be found in the paper [4].

3 Linear Cryptanalysis against 7-Round ICEBERG

3.1 Linear Layer P64-DP4-P64

To achieve hardware efficiency, the linear layer of light-weight block cipher is always designed to be a permutation which can be implemented by wire-crossing, sometimes to be involutional. However, permutations make diffusion slow. It is believed that the fewer rounds required to achieve full diffusion the more resistant the cipher should be. Unlike other light-weight block cipher, besides two same permutations on 64 bit named P64 is emplaced at the beginning and end of the linear layer of ICEBERG, there are 16 permutations on 4 bit named P4 and 16 diffusion boxes named D in the middle. The diffusion box D, which makes each output bit equal to the exclusive-or among the three input bits, results in only 2 rounds being required to reach full diffusion. P4 and D can be regarded as a whole, named DP4 depicted in Table 1. Now we give the following three properties of the linear layer of ICEBERG in view of linear cryptanalysis.

P1. Since P64 is involutional, the four input bits of each DP4 must come from four different S-boxes while the four output bits must go into four different S-boxes.

P2. Each two bits in each S-box must go into different bytes after P64, also in different nibbles of DP4.

P3. DP4 makes one active bit on one end to three active bits on the other end or two active bits to two active bits.

Table 1. The DP4 Linear-Layer

0	1	2	3	4	5	6	7	8	9	10	11	12	13	14	15
0	13	14	3	7	10	9	4	11	6	5	8	12	1	2	15

Because of the above three properties, the number of active S-boxes for two rounds at least is four, and there are three primary patterns ($1 \rightarrow 3$, $2 \rightarrow 2$, $3 \rightarrow 1$) and two auxiliary patterns ($2 \rightarrow 3$, $3 \rightarrow 2$) for the two-round linear approximation.

3.2 Linear Approximation for 6-Round of ICEBERG

Refer to the above five patterns, there are totally 8 possible trails for the linear approximation of 6 round of ICEBERG.

$$1 \rightarrow 3 \rightarrow 1 \rightarrow 3 \rightarrow 1 \rightarrow 3, \quad 3 \rightarrow 1 \rightarrow 3 \rightarrow 1 \rightarrow 3 \rightarrow 1,$$
$$1 \rightarrow 3 \rightarrow 1 \rightarrow 3 \rightarrow 2 \rightarrow 2, \quad 2 \rightarrow 2 \rightarrow 3 \rightarrow 1 \rightarrow 3 \rightarrow 1,$$
$$2 \rightarrow 2 \rightarrow 2 \rightarrow 2 \rightarrow 3 \rightarrow 1, \quad 1 \rightarrow 3 \rightarrow 2 \rightarrow 2 \rightarrow 2 \rightarrow 2,$$
$$1 \rightarrow 3 \rightarrow 2 \rightarrow 2 \rightarrow 3 \rightarrow 1, \quad 2 \rightarrow 2 \rightarrow 2 \rightarrow 2 \rightarrow 2 \rightarrow 2.$$

By further analysis of these 8 patterns, we obtain the best linear approximation for 6-round of ICEBERG in pattern $1 \rightarrow 3 \rightarrow 1 \rightarrow 3 \rightarrow 1 \rightarrow 3$, with bias $2^{-31.06}$, which is depicted in Table 2. Since ICEBERG is involutional, there is a trail in $3 \rightarrow 1 \rightarrow 3 \rightarrow 1 \rightarrow 3 \rightarrow 1$ with the same highest bias and the opposite order.

Table 2. 6-Round Best Linear Approximation

Round		Output Mask	Bias ε
		$S_5=40_x$	
R1	S-box	$S_5=40_x$	$2^{-3.42}$
R1	LT	$S_5=40_x, S_6=1_x, S_7=40_x$	
R2	S-box	$S_5=40_x, S_6=1_x, S_7=40_x$	$2^{-8.83}$
R2	LT	$S_5=40_x$	
R3	S-box	$S_5=40_x$	$2^{-3.42}$
R3	LT	$S_5=40_x, S_6=1_x, S_7=40_x$	
R4	S-box	$S_5=40_x, S_6=1_x, S_7=40_x$	$2^{-8.83}$
R4	LT	$S_5=40_x$	
R5	S-box	$S_5=40_x$	$2^{-3.42}$
R5	LT	$S_5=40_x, S_6=1_x, S_7=40_x$	
R6	S-box	$S_5=40_x, S_6=6_x, S_7=40_x$	$2^{-8.14}$
R6	LT	$S_0=20_x, S_1=4_x, S_4=83_x, S_6=7_x$	

3.3 Linear Cryptanalysis to 7-Round ICEBERG

By using the above 6-round linear approximation, we can not attack 7-round of ICEBERG because of the low success probability. So we use

$(00004000, 00000000) \xrightarrow{6r} (00004000, 00000000)$ with bias $2^{-31.75}$ to exploit an attack against 7-round ICEBERG. Because in the last round, there is one active S-box S_5 which has active input masks 40_x, we need to guess totally 8 bits of the subkey K_7. So $N=2^{64}$ KPs and 2^8 counters are required, and we also need $2^8 \cdot 2^{63} \cdot 2^{-2} = 2^{69}$ one round encryptions. That is about $2^{69}/7 \approx 2^{66.19}$ 7-round encryptions. The success rate is computed with the method in [6] as follows, where α is the bias of the linear approximation. The remained 120 bits of the master key can be exhaustively searched.

$$Ps = \phi(2\sqrt{N} \mid p - 1/2 \mid -\phi^{-1}(1 - 2^{-\alpha - 1})) \approx 30.6\%.$$

4   Conclusion

In this paper, we give the first linear cryptanalysis against 7-round of ICEBERG, which requires the whole code book and the time complexity is 2^{120} 7-round encryptions, the success rate is about 30.6%.

Acknowledgments. This work was supported by National Natural Science Foundation of China (Grant No. 61103237, 61070244 and 60931160442), Outstanding Young Scientists Foundation Grant of Shandong Province (No. BS2009DX030).

References

1. Lim, C.H., Korkishko, T.: mCrypton – A Lightweight Block Cipher for Security of Low-Cost RFID Tags and Sensors. In: Song, J.-S., Kwon, T., Yung, M. (eds.) WISA 2005. LNCS, vol. 3786, pp. 243–258. Springer, Heidelberg (2006)
2. Hong, D., Sung, J., Hong, S.H., Lim, J.-I., Lee, S.-J., Koo, B.-S., Lee, C.-H., Chang, D., Lee, J., Jeong, K., Kim, H., Kim, J.-S., Chee, S.: HIGHT: A New Block Cipher Suitable for Low-Resource Device. In: Goubin, L., Matsui, M. (eds.) CHES 2006. LNCS, vol. 4249, pp. 46–59. Springer, Heidelberg (2006)
3. Bogdanov, A.A., Knudsen, L.R., Leander, G., Paar, C., Poschmann, A., Robshaw, M., Seurin, Y., Vikkelsoe, C.: PRESENT: An Ultra-Lightweight Block Cipher. In: Paillier, P., Verbauwhede, I. (eds.) CHES 2007. LNCS, vol. 4727, pp. 450–466. Springer, Heidelberg (2007)
4. Standaert, F.-X., Piret, G., Rouvroy, G., Quisquater, J.-J., Legat, J.-D.: ICEBERG: An Involutional Cipher Efficient for Block Encryption in Reconfigurable Hardware. In: Roy, B., Meier, W. (eds.) FSE 2004. LNCS, vol. 3017, pp. 279–299. Springer, Heidelberg (2004)
5. Matsui, M.: Linear Cryptanalysis Method for DES Cipher. In: Helleseth, T. (ed.) EUROCRYPT 1993. LNCS, vol. 765, pp. 386–397. Springer, Heidelberg (1994)
6. Selçuk, A.A., Biçak, A.: On Probability of Success in Linear and Differential Cryptanalysis. In: Cimato, S., Galdi, C., Persiano, G. (eds.) SCN 2002. LNCS, vol. 2576, pp. 174–185. Springer, Heidelberg (2003)

The Ontology Recommendation System in E-Commerce Based on Data Mining and Web Mining Technology

TingZhong Wang[*]

College of Information Technology, Luoyang Normal University, Luoyang, 471022, China
wangtingzhong2@sina.cn

Abstract. Ontology provides common understanding of the domain knowledge and confirms common approbatory vocabulary in the domain. Web mining is the integration of information gathered by traditional data mining methodologies and techniques with information gathered over the World Wide Web. Finally, the ontology recommendation system in E-Commerce is proposed based on data mining and web mining. Our experiments showed that the proposed data mining and web mining methods to develop the ontology recommendation system can improve the efficiency of recommendation in e-commerce. The experimental results indicate that this method has great effective promise.

Keywords: ontology; data mining; web mining; e-commerce.

1 Introduction

Ontology describes a shared and common understanding of a domain that can be communicated among communities. Ontologies also play an important role in biomedical informatics and in knowledge management. Concepts and relationships are basic components in an ontology[1]. Web documents are the most important source for deriving concepts and relationships. Association rule using data mining techniques shows the purchasing association that customers generate on their purchasing behavior and determines the products that are usually accompany each other in each purchase transaction. However, such approaches also leave room for improvements in several aspects such as interpretability, modularity and accuracy. Ontologies allow web resources to be semantically enriched. Well-structured and simple problems can be solved with regular rules and principles. They have knowable and comprehensible solutions where the relationship between decision choices and all problem states is known or probabilistic.

The basic operational pattern of all the general query systems is to pass the user's query to a backend process, which is responsible for producing proper query result for the user. In order to solve this problem, semantic web technique – OWL and OWL-S, which are innovative for data discovery and service discovery respectively, have been

---

[*] Author Introduce: TingZhong Wang(1973.7-), Male, Han, Master of Henan University of Science and Technology, Research area: web mining, data mining, ontology.

D. Jin and S. Lin (Eds.): Advances in ECWAC, Vol. 2, AISC 149, pp. 533–536.
springerlink.com   © Springer-Verlag Berlin Heidelberg 2012

adopted. However, it is not easy for a tourist to search the information what exactly he really wants from a large amount of information available on the Internet. Information appliances also benefit from the specialization of function in that it allows customization in terms of operation, look, shape and feel. Ontology building and its representation in a formal language is usually carried out by knowledge engineers (KE), sometimes with the assistance of domain experts. Ontologies are formal, explicit specifications of shared conceptualizations of a given domain of discourse.

Ontology learning from texts constitutes a promising means for ontology engineers to significantly speed up the ontology building process so that several approaches have been proposed for covering the different phases it involves. With the vigorously generalization of semantic web by W3C, semantics oriented web information integrating method has been the major point of the research on web information integration technology. E-learning is an alternative concept to the traditional tutoring system.

Finally, this paper puts forward the methodology for ontology recommendation system in E- Commerce based on data mining and web mining. The system has four main characteristics. The web services execution environment supports common B2B and B2C (Business to Consumer) scenarios, acting as an information system representing the central point of a hub-and-spoke architecture. The paper offers a methodology for building ontology recommendation system for knowledge sharing and reusing based on data mining and web mining technology. Web-mining techniques also play an important role in e-commerce and eservices, proving to be useful tools for understanding how ecommerce and e-service Web sites and services are used. Web documents are the most important source for deriving concepts and relationships.

2   Data Mining-Based Ontology Information Management System in E-Commerce

Ontologies are widely used in different domains to give standard representations and semantics to concepts, predicates and actions of a particular domain. Development of an ontology for a specific domain is not yet an engineering process, but it is clear that an ontology must include descriptions of explicit concepts and their relationships of a specific domain.

Ontologies provide an unambiguous terminology that can be shared by all involved in a software development process[2]. Top-level ontologies describe very broad concepts at the information technology level. These formal ontological structures are concerned with the description of concept types and relations types and generally are not concerned with physical or process objects. Ontology is one of the branches of philosophy, which deals with the nature and organization of reality.

The web services execution environment supports common B2B and B2C (Business to Consumer) scenarios, acting as an information system representing the central point of a hub-and-spoke architecture. Recommender systems are generally two methods to formulate recommendations both depending on the type of items to be recommended and the way that user models are constructed.

Most current annotation-based transcoding systems adopt OWL-based ontologies to describe annotations. Ontology development processes, including a pre-development process (an environment study and a feasibility study), a *development process* (requirements, design, implementation), and a *post-development process* (installation, operation, support, maintenance, and retirement of an ontology). The formula 1 which calculates the ontology recommendations nodes of data mining is as follows.

$$DR=\{<ci,ck>/rel(ci,ck)\}, \ O=(C,R,Ax) \tag{1}$$

Traditional knowledge representation methods of case-based reasoning represent the case by basing it on database tables, frames or scenarios. An advantage of this representation is that it allows to declare a hierarchy in an instance data transparent way. Therefore, a key feature of XBRL is that an instance document may be analyzed according to different taxonomies according to the goal of an analytic application.

3   Using Web Mining Technology to Build Ontology Recommendation System

Web mining is the integration of information gathered by traditional data mining methodologies and techniques with information gathered over the World Wide Web. Web Mining based on Semantic Web can exploit the new semantic structures to help Web Mining, on the other hand, the results of Web Mining can also help build up Semantic Web[3].

Web-mining techniques also play an important role in e-commerce and eservices, proving to be useful tools for understanding how ecommerce and e-service Web sites and services are used. Ontologies can be used as tools for specifying the semantics of terminology systems in a well defined and unambiguous manner. The equation 2 which calculates the ontology recommendation nodes of web mining is as follows.

$$tfidf(t_k,d_j) = \#(t_k,d_j) \times \log\left(\frac{|T_r|}{\#_{T_r}(t_k)}\right). \tag{2}$$

Where t_k denotes a term, d_j denotes a document, T_r is a set of documents used for training, $\#(t_k,d_j)$ denotes the *term frequency*. The implementation of business-to-business (B2B) eCommerce systems is fully realised in extended enterprises. Firstly, this paper gives an overview of the basic idea of Web Mining based on Semantic Web, secondly, Web Content Mining, Web Structure Mining and Web Usage Mining, which based on semantic is discussed[4]. Concepts and relationships are basic components in ontology. Web documents are the most important source for deriving concepts and relationships. PageRank algorithm is explored in this article, then, a method is given to improve this algorithm on its shortcomings. Similar item is very close to the previous interested items by the user. Recommending a similar product to the user is not very efficient.

4   The Ontology Recommendation System in E-Commerce Based on Data Mining and Web Mining Technology

The aim of ontology is to obtain, describe and express the knowledge of related domain in e-commerce. In this paper, we apply the technology of data mining and web mining to automatically develop ontology recommendation system in e-commerce. The context decision agent finds the matched concepts, uses the optimization to get the semantic relation, and counts the context relation between the customer requirements and each product stored in the ontology. Fig 1 shows the application of ontology recommendation in e-commerce by data mining and web mining. The experimental results indicate that this method has great promise.

Fig. 1. The application results of ontology recommendation system.

5   Summary

As the foundation of the semantic web, ontology is a formal, explicit specification of a shared conceptual model and provides a way for computers to exchange, search and identify characteristics. In this paper, we adopt data mining and web mining to develop the ontology recommendation system in order to improve the efficiency of recommendation in e-commerce.

References

1. Berners-Lee, T., Hendler, J., Lassila, O.: The Semantic Web. Scientific American 5 (2001)
2. Green, P., Rosemann, M.: Integrated process modelling: an ontological analysis. Inf. Syst. 25(2), 73–87 (2000)
3. Missikoff, M., Navigli, R., Velardi, P.: Integrated Approach for Web Ontology Learning and Engineering. IEEE Computer 35(11), 60–63 (2002)
4. Kavalec, M., Maedche, A., Svátek, V.: Discovery of Lexical Entries for Non-taxonomic Relations in Ontology Learning. In: Van Emde Boas, P., Pokorný, J., Bieliková, M., Štuller, J. (eds.) SOFSEM 2004. LNCS, vol. 2932, pp. 249–256. Springer, Heidelberg (2004)

Design and Development of E-Learning Virtual Learning System Based on VRML and Java

Shaoliang Qi[*]

Department of Art and Design, PING DING SHAN University
Pingdingshan, 467000, P.R. China
Lq721201@163.com

Abstract. This paper proposes a novel e-lcarning hypothesized learning platform based on VRML-Java. It describes the characteristics and complementary of VRML and Java language and elaborates the architecture of the virtual learning system. Furthermore it presents a method how to use VRML to create virtual learning system. Finally this paper introduces the interaction between VRML and Java in the virtual reality system.

Keywords: VRML; Virtual Reality; Java; Learning Platform.

1 Introduction

With the high-speed development of communication technology, computer technology and network technology, distance learning is welcomed by more and more teachers and students. Distance learning has plenty of media resources, is of flexible timeliness and rationality, and is of conveniently interactive mode, so it has become a new trend in education in the 21st century.

One of the major reasons for the success of the E-learning is that it has enabled students to take courses online whenever and wherever convenient. Now, most of the e-learning systems are base on HTML which is less attractive to students due to their lack of 3D immersion and real voice interaction. The technology of virtual reality can be exploited to compensate these weaknesses [1]. We propose a realistic and interactive virtual learning system by integrating vivid 3D graphics and real-time voice communication. The distance learning without animation and interaction is bald. This paper briefly design and development the virtual learning system of VRML-java on web-based distance learning. This greatly improves people's imagination and inspires the learners' interest [2].

2 VRML and Java Characteristics

2.1 VRML

VRML is Virtual Reality Modeling Language, a file format that is the open standard for virtual reality on the Internet. Virtual Reality is an immersive, computer-generated

---

[*] Corresponding Author: Lq721201@163.com

D. Jin and S. Lin (Eds.): Advances in ECWAC, Vol. 2, AISC 149, pp. 537–542.
springerlink.com   © Springer-Verlag Berlin Heidelberg 2012

environment that can include 3D graphics, sound and more. You can access VRML worlds using a VRML browser, which can be a stand-alone or a plug-in to a regular WWW browser. VRML is an international standard on which the 3D models can be distributed and users can browse the models with common web browser with plug-in for VRML. VRML files are small. Compressed VRML files of normal scenes are about 10KBytes to 50Kbytes, so the pressure of slow network speed and narrow bandwidth has been greatly lightened. VRML's real-time 3D rendering function is of particular value to the distance learning [3]. This function brings students vivid substantial models.

2.2 Java

Java: An object-oriented programming language developed by Sun Microsystems. The widespread adoption of the "write once, run anywhere" Java language makes it possible to develop truly platform-independent client applications. Java can exist in two forms, applets and applications. VRML 2.0 will introduce a new form, based on "Script Nodes". Currently, Java applets let you do things like calculate mortgages or play Pac-Man within your Web browser.

2.3 Java vs VRML

The static scenes with the route function of VRM can easily perform simple interaction, but with the application of advanced interaction, users can not manipulate the scenes exactly and get information or data from the scenes. With combination of Java Language and VRML, scenes a new colorful virtual world is reached. The function which java language provides to interact with VRML is carried out by the accessional classes [4]. VRML plug-in installation will automatically install these classes to the specified path, for example Corona VRML Client will install these classes which are VRML.external、 VRML.external.field、 VRML.external. exception. By these accessional classes Java program can manipulate the VRML scenes, send events to or receive events from the VRML scenes, gain VRML nodes from web paces and so on.

3 Virtual Learning System Architecture

This System uses Browser/Server mechanism. virtual learning systems can be divided naturally into three subsystems: www browser subsystem, server subsystem, and data source subsystem.

In browser subsystem, HTML and VRML pages are communicated with remote client by HTTP protocol. HTTP is used to implement a Web server on the TCPI/IP stack. In the frames of the pages, VRML and HTML can communicate successfully and supplement each other. On one side, VRML can be used to display 3D scenes and nimations; on the other side, HTML can be used to add proper notes, hints and navigation mechanism [5]. All of these could make distance learning vividly. As show in the table, the servers on the platform are the VRML server and connection server and chat server. The VRML server monitors and records every event that takes place in the virtual space and reports these changes to all participant clients of the platform. Thus, the system assures that the users will have the illusion of sharing a common space. The VMRL server also maintains constantly an updated copy of the world, which is sent to the clients when they enter the system. Thus, the new users share the same updated view that the existing users already have. The Connection Server maintains a database, which the system accesses in order to authenticate the user and allow him or her to enter the virtual space of the platform. It also reports every entry or departure that takes place in the platform to all other servers. The Chat Server is responsible for the text chat support]. It allows group chat (i.e. text chatting between multiple users) or whispering (i.e. one-to-one communication between two users).

4   Useing VRML to Create Virtual Learning System

Every course is held in a 3D world, which virtually consists of all the "physical" equipment that could be found in a real classroom. We use this follow ways to build the virtually space. That is, use 3DS Max to make models, to modify the models, to export the scenes, to implement some of the interactive functions, and finally use JAVA to implement all the designed functions [6].

3DS Max is a kind of powerful 3D animation design software. It has special embedded programs for VRML export, so it can directly export 3D scenes including geometry models, materials, animations and so on. In addition, 3D Max's embedded programs can also be used to make special accessories to define interaction elements in scenes. We established virtual scene model with 3dsSmax, then export the virtual scene to VRML file[wrl format).

After that we can adjust the parameters of virtual camera to ensure consistency of the virtual scene and the real foreground; add live video components over object user selected in virtual studio background [7]. User can also insert real-time Animate into virtual studio to increase reality by costuming the moving track, velocity and direction of objects in the scene.

At last, if further interaction functions need be added to the animations, we use Java Language to implement it. VRML2.0 adds some new Script nodes, so Java Script and an external Java Applet can be used in the scenes. This absolutely enlarges the dynamic behaviors in virtual world. Further function design can implement the effective interaction between users and computers, and VRML's real-time rendering characteristic can realize the dynamic effects given by the users. As Figure1 shows, we use VRML to create virtual learning platform and Java Script to link course video, then student can select the course he/she want to learn.

Fig. 1. The Virtual classroom

5 Design Interaction between VRML and Java in the Virtual Reality System

Our goal has been to enable virtual reality based learning system, and intelligent, real-time, immersive, 3D animated explanations for E-Learning. To do that, we have chosen to use the VRML to build 3D spaces online, and realize a JAVA layer to realize shared Interaction in the 3D scene [8].

Integrating VRML, worlds with JAVA can provide dynamic access, coaching, and feedback capabilities for a practice environment. There are two methods to control VRML nodes by JAVA: Script Node or EAI (External Authoring Interface).

a). Internal Script Node

That requires the script definition in VRML scene, which designates the Event In received, Event Out to be sent out, the node to be controlled, and the processing JAVA program..

b). External Authoring Interface（EAI）

The Internal Script Node provides the Interface between scene and JAVA, while EAI defines the communication Interface between scene and Applet in HTML pages. The Applet accesses the scene using the Browser Class capsulated in the VRML external package. The base Class of Browser Class is IBrowser Interface Class which capsulates the VRML scene. The IBrowser Class indicates a unique VRML scene after establishing a Browser Object in JAVA Applet. Then the class obtains the reference of nodes defined in VRML scene, and achieved motion effects by changing the values of fields under nodes [9].

The users that participate in the virtual classroom are represented by avatars. The users' avatars are able to make various types of gestures: expressing opinions (e.g. agree, disagree), expressing feelings, mimics (e.g. happy, sad), as well as showing actions (e.g. move learning content, pick learning content). The virtual classroom is supported by audio and text chat functionality, which is available to all participating users for supporting and promoting collaboration, cooperation and communication among them [10]. As Figure2 shows, we use EAI to design User interface of the communication area. The key source code of the communication area as follows:

```
Public class JSAITextChanged extends Script{
    Public void initialize(){
    Node node= (Node)((SFNode)getfield("text")).getValue();
    textNode=(MFString)(node.getExposedField("string"));
    popupWindow p=new PopupWindow();
    }
Private void setText(String s){
    String[] str=new String[1];Str[0]=0;textNode.setValue(str);
    }
//receive the word
Class PopupWindow extends Frame{
Private Label ren=new Label("please enter the world that will be
    show in
the scene") ;
Private TextField text=new TextField("Hello");
    Private Button submit=new Button("ok")
    Public PopuWindow(){
    FlowLayout f=new FlowLayout();setLayout(f);f.setHgap(5);
    Add(red);add(text);add(submit);
    Submit.addActionListener(new ActionListener(){
    Public void actionPerformed(ActionEvent e){
    setText(text.getText));
        }
    }
    }
```

Fig. 2. User interface of the communication area

We enhanced stability through better interface with EAI, a better support of avatar and avatar's gestures, as well as server-side syntax checking of 3D spaces in order to support better and faster sharing of multi-user events.

6 Conclusions

In this paper, we introduce a novel Virtual learning System, which tries to overcome the limitations of the current platforms that based on HTML. With interactive functions like rotation, transformation of the 3-D objects, VRML can create a wonderful virtual world [11]. Using interface between Java language and VRML, more beautiful and considerate Human interface will be designed on the Internet far users. All of these greatly improve people's imagination and inspire the learners'interest. The effect of distance learning is prominent.

References

1. Tam, K., Badra, F., Marceau, R.J., Marin, M.A., Malowany, A.S.: A web-based virtual environment for operator training. IEEE Trans. Power Systems 14(3), 802–808 (1999)
2. Brutzman: The virtual reality modeling language and Java. Communications of the ACM 41(6), 57–64 (1998)
3. Thomas, J.A.: 3D visualization of multimedia content on the World Wide Web. Computer Networks and ISDN Systems 30, 594–596 (1998)
4. Zhu, Z.: VRML-JAVA Based Virtual Reality In Assembly Visualization. Journal of System Simulation 13(suppl.) (2001)
5. Mauve, M.: TeCo3 Dharing interactive and dynamic 3D models. Multimed. Tools Appl. 20(3), 283–304 (2003)
6. Chen, B.-Y., Yang, T.-J., Ouhyoung, M.: JavaGL - A 3D graphics library in Java for Internet browsers. IEEE Trans. Consumer Electronics 43(3), 271–278 (1997)
7. Mathews, G.J., Leisawitz, D., Thieman, J.: The Interactive Universe: An Application of VRML and the Web to Science. NSSDC News 13(4) (1997)
8. Dillenbourg, P.: Virtual learning environments. In: EUN Conference 2000—"Learning in the New Millennium: Building New Education Strategies for Schools" (2000)
9. Mauve, M.: TeCo3 Dsharing interactive and dynamic 3D models. Multimed. Tools Appl. 20(3), 283–304 (2003)
10. Oliveira, M., Crowcroft, J., Slater, M.: Component framework infrastructure for virtual environments. In: Proceedings of the Third International Conference on Collaborative Virtual Environments 2000 (CVE 2000), San Francisco, CA, pp. 139–146 (2000)
11. Schmid, C.: A Remote Laboratory Using Virtual Reality on the Web. Simulation (3) (1999)

Application of Pattern Recognizing Technique to Automatic Isolating Garbage Can

Zhongyan Hu[*], Zaihui Cao, and Jinfa Shi

Department of Art and Design, Zhengzhou Instiute of Aeronautical Industry Management
450015 Zhengzhou, China
huzohngyan@zzia.edu.cn, czhhn@126.com

Abstract. With the development of the social economic and the improvement of our life , the garbage has been becoming more and more difficult to deal with. It is well known that the garbage usually consists of metal, paper, plastic, glass and so on. The majority of them can be reused. But to isolate each of them is very difficult when they are mixed together. The author expatiates particularly on using the pattern recognizing technique to isolate them automatically. It is one of the best way to deal with the garbage and it is cried for by the society.

Keywords: Garbage, Pattern, Technique, Recognizing.

1 Introduction

With the globalization of economic development, improved living standards, lifestyle diversity, a variety of solid waste also will increase, and the speed is amazing. According to statistics, from 1995 to 2003, the number of EU countries increased by 19% garbage, more than economic growth. 25 member states up to 13 annual output of garbage million tons. Waste a significant increase in integrated environmental management into a very difficult problem. How to handle the increasing volume of waste, mining waste resources, turning waste into treasure, has become a focus of national attention. In advocating saving society, social and environmental recycling-oriented society of today, the waste classification, comprehensive improvement of the environment is an effective way. Garbage is a sign of human civilization, but also social progress. Increasingly scarce in today's natural resources, recycling, waste of resources is significant. However, in a wide variety of waste, how to effectively use some sort can be recycled to become the key to solve the garbage problem. This article is from the problem of waste disposal garbage sorting bottleneck that start with, the use of the relatively mature automatic identification technology, automatic classification of trash design, not only to solve the artificial problem of sorting garbage when the economy, but also to solve the recyclable waste re-use problems, and fully tap the potential waste of resources inherent in great significance.

---

[*] Corresponding author: huzhongyan@zzia.edu.cn

D. Jin and S. Lin (Eds.): Advances in ECWAC, Vol. 2, AISC 149, pp. 543–548.
springerlink.com   © Springer-Verlag Berlin Heidelberg 2012

2   Garbage Classification and Recycling

In the next century modernization, globalization, rapid development, social growth garbage faster than population growth, which in the most developed and developing countries are no exception. Municipal solid waste has become the world's major urban development in the face of a big problem. China's population, rapid urban development, urban population density, large municipal solid waste production, treatment and disposal capacity and the level is still low. China has about 2 / 3 the siege of the city into a garbage dilemma [1]. China only "municipal waste" the nearest 150 million tons of annual production, the Beijing Municipal waste is Nissan 13,000 tons, annual production of 4.95 million tons and an annual increase rate will be 8%. In each year nearly 1.5 million tons of municipal waste, discarded "renewable resources" worth up to 250 billion! According to waste experts Wang Ping of the survey, waste composition, waste recyclable resources accounted for 42.9% of the total, direct recovery efficiency should not be less than 33%. All kinds of solid waste mixed together is trash, sorting is open resources. If can fully mining can recycle trash contains resource potential, Beijing alone and you get 1.1 billion yuan a year of economic benefit [2].

In Western countries, 45% of steel production is scrap steel as raw materials, copper production of 35%, 22% of lead production, zinc production, 30%, 35% of paper production is the use of waste raw materials. According to statistics, recycling one ton of each renewable resources, representing a decrease of 4 tons the amount of waste; recycling one ton each scrap steel, mining of various minerals can save 20 tons of steel saves 1.2 tons of standard coal; paper to create good paper, saving wood; waste plastics to produce plastic products, saving the oil [3]."Garbage is a misplaced resource", has been recognized more and more countries. Seen, aspects of waste generated if the consumer timely classification, recycling is the best way to solve the garbage problem. So, garbage mountains can become a mountain of gold.

3   At Present Problems Encountered in the Garbage

Legal system is imperfect, and propaganda is not enough.Beijing, Shanghai and other cities have been garbage, but garbage collection, processing efficiency is very low. First, the legal system is imperfect, for a number of e-waste and hazardous waste, there is no specific implementation details, not bound by the corresponding laws and regulations; Second, propaganda is not enough, in our lack of a systematic study of basic education cycle of the contents of garbage, a lot of people do not understand the recycling of recyclable and non-specific categories, resulting in a lot of rubbish is required after the second classification, increasing waste disposal costs and time, did not play a proper waste separation effect.

Concentration of the high cost of waste classification.As ordinary citizens lack the necessary knowledge of garbage, so the waste classification of heavy work focuses on environmental protection departments. Not only deal with large and high cost. According to an estimate, tons of garbage each year in landfill sites, 10 people need employment; burning will need 20 to 40; sorting garbage, recycling is required 250. Sorting using the traditional manual focus can be seen the way the face of a huge number of garbage, has been powerless. But the premise is garbage recycling classification, the face of waste recycling bottlenecks, whether by means of modern scientific achievements, the use of pattern recognition technology designed to improve the trash, help people to trash garbage automatic classification, to solve the garbage the urgent needs of the line with the dummy, is worthy of our consideration and ponder a question.

4 Automatic Classification of Trash Design Technical Support [4] [5] [6]

4.1 The Basic Concept of Pattern Recognition

Mode Broadly speaking, is a basic human intelligence, exists in time and space can be observed in things, if you can distinguish whether they are identical or similar, you can call mode. Narrowly speaking, the specific pattern is observed for the individual to get things with time and space distribution of information. The model belongs to the same class category or the overall pattern is called pattern classes (or simply classes).

Pattern recognition is in some measure or observation must be based on the knowledge model is divided into the respective model class to go. With the 1940s advent of computers and the rise of the 1950s, artificial intelligence, computer people want to replace or extend part of human mental. Computer pattern recognition in the early 1960s, the rapid development and become a new discipline, refers to the use of computers and other devices on the objects, images, graphics, voice, shape and other information for automatic identification.

4.2 Pattern Recognition Method

The basic pattern recognition methods are statistical pattern recognition methods, methods of decision theory methods and syntax, and the corresponding pattern recognition system consists of two processes, namely the design and implementation. Design refers to a certain number of samples (called the training set or learning set) for classifier design. Implementation refers to the design of classifiers to classify the samples treated identify the decision-making. Shown in Figure 1.

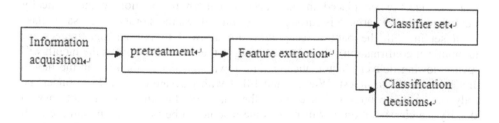

Fig. 1. Basic Composed of Pattern Recognition System

5   Automatic Classification Ash-Bin Concept Design [7] [8]

5.1   Product Positioning

Since the product to have a certain scientific and technological content, taking into account the product early in the design of the technical difficulties and cost issues, select a single garbage relatively offices office. First, relative to the interior design of the outdoor trash can, the environment is good, can reduce the adverse weather conditions on the open-air design of the special requirements of the product; the second is generally higher educational level with the crowd. Their environmental awareness, stronger than the concept of consciously, artificially reduce the risk of damage; third office office space garbage is relatively simple, mainly recyclable paper, plastic (all kinds of fast food bags, drink bottles, pens, empty pen core, etc.), glass, metal and other commonly used office supplies based, using a relatively simple pattern recognition methods can identify and reduce development and manufacturing costs.

5.2   Recognition

Put in the mouth in the pattern recognition device is installed, each identifier in accordance with the classification of the different information stored images of some categories of products (the reason for product positioning in the office the office, is stored in order to minimize the amount of information, reduce costs), when the garbage near the inlets, the recognition process automatically identify it. Such as brightness and color information of garbage, shape information, texture information, size information. This information is then extracted to classify the image to determine the category name, the image recognition.

5.3   Design concept

The trash inside the box is divided into several independent units, respectively, recycled glass, metal, plastic, paper. Effective for the better use of interior space, the trash can or cylinder designed for two kinds of rectangular shape.

Rectangular shape.The program draws on the design bar auto-sensing classification of trash design. The classification was an independent space shape arranged side by side, combined for a cuboid. Put each individual unit has a mouth, when the garbage near the inlets when placed in the inlets of pattern recognition to automatically identify and put in garbage is garbage in this unit allows the storage of the same class (each set in both the pattern recognition are allowed to put garbage images for automatic cf confirmation). Such as belonging to the same class, put in the mouth will open, allowing delivery, automatically shut down; and if not, then put in the mouth shut and refused to invest. When a unit filled with garbage can out from under the body of the dump bucket. Lined up in the shape of this encounter is not easy to identify the classification, it will have some trouble, to be put in the mouth are each try. But compared to the relatively simple office waste, there are generally garbage common sense will not encounter too much trouble.

Cylinder Shape. Break the box-type shape, overall as a cylinder, the internal fan is divided into five small space, respectively, to recycle glass, metal, plastic, paper, batteries. Structure since the beginning of this trash, outside of two parts, the outer structure of the main door is automatic rotating barrel design, the internal structure is mainly sensor device, turn the fan tray and five independent space. This trash is only one delivery port, infrared sensors and automatic pattern recognition are located on the upper spherical cap at the trash. Put in garbage, the infrared sensor automatically open inlets, automatic identification devices will be put in the open mouth at the same time, receiving spam, the control unit of the appropriate classification of garbage to put in rotation just below the mouth (which is inside the bottom of the rotating disk through the realization of a). Dumping, the available buttons to control the revolving door open, take out the trash and clear the various units within the collection.

Fig. 2. Inner structure.   **Fig. 3.** Product external form

6 Conclusion

With the increased awareness of environmental protection, human activities will be abandoned waste - garbage waste to treasure, for recycling, sustainable development of society has become a research topic. "Garbage is a misplaced resource", has been recognized more and more people. Currently, to facilitate the recycling of waste, many places are equipped with a classification of trash, even though there is some trash on the prompt recovery of recyclable and non-identity, but it is still difficult to make accurate judgments garbage, just like not all metal can be recycled and paper are the same. Therefore, it would be difficult to achieve the intended purpose of garbage.

Use of existing technology, automatic classification of waste, not only to solve people's waste classification is not clear the confusion, but also to improve the efficiency of automatic garbage sorting provides a method. Garbage finer essence, the more beneficial recycling. For most populous country like ours, only the consumer culture as the main force garbage can fundamentally solve the problem of garbage recycling. The need to mobilize the active participation of consumers in general waste classification, the other by means of science and technology to help people complete waste of automatic classification. Only multi-pronged approach to try to eliminate the garbage pollution, conducive to social and sustainable development.

Acknowledgments. This paper supported by the department of education of HeNan province natural science research program(No.2011A630062),the department of education of HeNan province humanities and social science research project (No/2011-ZX-233),Henan government decision-making tender (No.2011A630062), supported by the Innovation Scientists and Technicians Troop Construction Projects of Zhengzhou City (No.10LJRC183).

References

[1] Li, W.: China's urban recycling resource way and countermeasures, and resources. Science (3), 17–19 (2000)
[2] WSZKH, garbage many benefits (May 18, 2008), http://hi.baidu.com/nbwszkh/blog/item/3ae56ed675a0d42a06088b30.html (download date, July 20, 2008)
[3] Liu, Y.: Urbanization in our country under the background of urban living garbage melting process. The Progress of Science and Technology and Countermeasures 10(24), 76–78 (2007)
[4] Xiao, J.: Intelligent pattern recognition method. South China university of science and technology press (2006)
[5] Zhao, Q.: Edge, Pattern recognition. Tsinghua university press (2000)
[6] Zhao, C.: Pattern recognition briefly. The Public Science (scientific research and practice), 3–4 (May 2008)
[7] Wan, F.: Virtual digital model technology application in industrial design. Packaging Engineering 28, 127 (2007)
[8] Liu, Y.: City life garbage on the new ideas, city management. Science and Technology, 26–27 (May 2007)

Apply IT Curriculum Objectives Ontology to Design Aided Teaching System

YanFei Li*

College of Information Technology, Luoyang Normal University, Luoyang, 471022, China
liliyanyan2011@sina.com

Abstract. With the IT course of high school as background and ontology as basis, this thesis tries to construct IT curriculum objective ontology and aided teaching system based on ontology, then apply it to guide teaching and study.. This research constructs relations, which includes the relation of is-a, part-of, attribute-of and so on, among these concepts on the basis of roles concept theory and realizes these relations and show these relations in the form of graphics. This research tries to construct aided teaching system based on ontological analysis of IT courses. This system mainly includes lecture preparation model and individual study model. It could be used to aid the teacher in lecture preparation and teaching design, meanwhile it could be used to support students' individual study.

Keywords: Ontology, Curriculum Objectives, Information Management.

1   Introduction

With the development and maturation of internet technology, IT (Information Technology) brings about great influence to people's life and work style. Ontology plays a pivotal role in the development of the Semantic Web. It can provide a representation of a shared conceptualization of a particular domain that can be communicated between people and applications[1]. The Semantic Web is an evolving extension of the World-Wide Web, in which content is encoded in a formal and explicit way, and can be read and used by software agents. However, that has changed as web has users from all over the world now access the Internet, and web sites are available in virtually every different language.

As the application of IT in the education field, the research on teaching system which holds such technology as computer, internet, multi-media, AI and the like as its focus receives more attention than before. The W3C is currently examining various approaches with the purpose of reaching a standard for the SWS technology: OWL-S, WSMO, SWSF, WSDL-S, and SAWSDL. In other words, data storage, data management, data transmission and even analysis rely on computer and network technology. While the issues for exploring environmental sustainability are well

---

* Author Introduce: YanFei Li, Female, Han, Master of Henan Normal University, Research area: Ontology, Curriculum Objectives, Information Management System.

D. Jin and S. Lin (Eds.): Advances in ECWAC, Vol. 2, AISC 149, pp. 549–552.
springerlink.com   © Springer-Verlag Berlin Heidelberg 2012

rehearsed and known, the issues that should form the social dimension are less appreciated and addressed by stakeholders involved in the development process. As new services appear with high performance requirements, mechanisms to ensure quality of service and metrics to monitor this quality become necessary.

But current teaching system faces some problems, for example: the lack of communication between teaching systems, that is to say, the developed teaching system on the same subject always starts from the very beginning and rarely uses the resources (including functional parts) which have already been develop, so the maintenance of systems becomes difficult and it becomes hard to effectively compare and judge an already-existed system[2]. The teaching system which has already been developed, especially the system related with ITS (Intellectual Tutoring System), adopts AI and knowledge base system technology more or less, but the phenomenon that the resources could not be repeatedly utilized still exists.

With the IT course of high school as background and ontology as basis, this thesis tries to construct IT curriculum objective ontology and aided teaching system based on ontology, then apply it to guide teaching and study. According to the constructing method of ontology, this research analyzes the outline of IT course curriculum from the angle of teaching design concept and summarizes the core concepts in the field of IT. Then this research constructs relations, which includes the relation of is-a, part-of, attribute-of and so on, among these concepts on the basis of roles concept theory and realizes these relations and show these relations in the form of graphics. This research tries to construct aided teaching system based on ontological analysis of IT courses. This system mainly includes lecture preparation model and individual study model. It could be used to aid the teacher in lecture preparation and teaching design, meanwhile it could be used to support students' individual study.

2   Building Aided Teaching System Based on IT Curriculum Objectives Ontology

IT curriculum goals of Ontology Ontology building construction follow the basic sequence, namely the setting of standards, taking the concept, the concept of formalization, implementation, maintenance four stages.

According to this idea, we selected four concepts of information technology education goals: access to information teaching objectives, information processing and expression of the teaching objectives, teaching objectives of information resources management, information technology and social teaching objectives. A review of existing literature on the representation of manufacturing knowledge highlighted the need for application guidelines on the classification of knowledge for optimum reuse[3]. A concept we have refined the concept into two concepts. For two of the concept of extraction rules, we are teaching objectives based classification. This technology has enabled ecommerce to do personalized marketing, which eventually results in higher trade volumes. Web Mining is the extraction of interesting and potentially useful patterns and implicit information from artifacts or activity related to the World Wide Web. Every two to three refined the concept of the same concept, based on information technology, the content of the curriculum objectives and instructional design ideas to make this level of refinement of the concept.

This classification method is (1) above is-a relationship similar. For example, we can put information on the basic characteristics of the child to understand the concept of information access to knowledge of teaching the concept of sub-goals, that is between the two is-a relationship. For part-of and is-a distinction between the need for some explanation here.

$$\sigma(X)=\{t\in T|\forall d\in X:(t,d)\in I\} \tag{1}$$

The formula 1 which is based on the above definitions, a concept is defined. A concept is a pair of sets: a set of documents and a set of terms (X,Y) such that: $Y=\sigma(X)$ and $X=\tau(Y)$.For these teachers are familiar with the term, when we build on Ontology, the teaching methods, teaching tools, requires students to mastery of knowledge and skills to the concept of node attributes, respectively, in the form of Ontology using attribute-of relationship constraints, in order to facilitate the use of teachers. According to the theory, this study is stored in the Ontology can be used the first approach, that is, directly to the IT curriculum objectives Ontology OWL file stored. After storage, directly through the Ontology analysis tools, such as JENA, etc. on the Ontology parse, and then embedded in java code for direct search and operation.

3   Assisted Teaching System Design and Implementation

In this study, the IT curriculum aims to build Ontology, reflects the teaching of design ideas. Each concept is taken a teaching goal, each idea, there are a / o hierarchy, the class constraints are "teaching organizational strategy", the teacher's guidance through this class constraints, you can talk about their knowledge of the point teaching organization. This point relates to the teaching of knowledge organization strategies, teachers should first consider the teaching of this knowledge, teaching methods and other issues. These are defined in the Ontology concept node's children, for teachers to demonstrate specific content, selection and organization for teachers.

The system is divided into three levels, each level has a number of modules. Three levels are the storage layer, service layer, application layer[4]. Storage layer formed by the Ontology library and database, the database concept of primary storage Ontology database instance, that a specific object, Ontology library itself can also be stored in the database, where the distinction is logical taking into account both the role of different. The formula 2 which calculates the similarity of Ontology and FCA is as follows.

$$\mathscr{P}(I_1,I_2) = \{\{(a_1,b_1)\cdots(a_n,b_n)\}|a_k \in I_1, b_k \in I_2, \forall k-1,\ldots,n, \text{ and } a_k \neq a_k, b_k \neq b_l, \forall k,l \neq k\}. \tag{2}$$

Under subsection (1) step in the establishment of the Ontology and the correspondence between the database, using MySQL or Orical create database tables. Ontology of each class corresponds to a database table, field names correspond to class attributes in the Ontology. Between class and class is-a relationship, achieved through the external key. Ontology storage layer including libraries, databases, Ontology mapping between the database and database functions, Ontology service layer of the library and database access to read and write functions of four parts. Ontology mapping function described in the realization of Ontology and database objects stored in the correspondence between. Read and write functions as a storage layer for the service layer provides the interface to read and write data, access control and other functions. Realization of these functions using a standard component technology. The system frame structure shown in Figure 1.

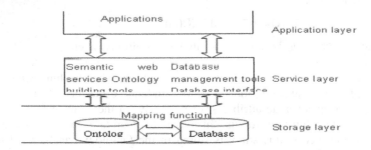

Fig. 1. System frame structure of Assisted teaching system.

4 Summary

Applied research is mainly made of Ontology-based framework for model-assisted teaching systems, information technology courses in high school goals, for example, established a goal of Ontology and information technology education teaching resource library, build a model system for teachers preparing lessons and teaching design guidance, students can also use the system for self-study, to improve the sharing and reuse of learning resources. System is the key to how to make the system to build Ontology access, access to the specific method depends on Ontology is stored, in this study to document stored, using Jena Ontology parse the file directly, the system can be accessed and manipulated on Ontology.

References

1. Chandrasekaran, B., Josephson, J.R., Benjamins, V.R.: What Are Ontologies, and Why Do We Need Them, pp. 20–25 (January/February 1999)
2. Shun, S.B., Motta, E., Domingue, J.: ScholOnto:an Ontology2based Digital Library Server for Research Documents and Discourse. Intl J. Digital Libraries 3(3), 237–248 (2000)
3. Maedche, A., Staab, S.: Ontology Learning for the Semantic Web. IEEE Intelligent Systems 16(2), 72–79 (2001)
4. Pinto, H.S., Martins, J.P.: Ontologies: How can They be Built? Knowledge and Information Systems 6, 441–464 (2004)

On the Importance of Movie Appreciation in Multimedia Network Translation Teaching

Changhong Zhai

School of Foreign Languages, Wuhan Polytechnic University, 430023 Wuhan China
zhaichanghong103@yahoo.com.cn

Abstract. The traditional translation teaching in which students are passive recipients is usually teacher-centered and not conducive to students' translation ability. Based on constructivism, with the help of multimedia network technology and by means of movie appreciation, the new translation teaching mode has many advantages: expanding the amount of teaching information; realizing the student-teacher interaction to the largest degree; stimulating students' translation interest; creating a real language environment which can help students accurately understand the original context and gradually develop their cognitive abilities and translation capabilities in the process of the interaction with the environment.

Keywords: constructivism; movie appreciation; multimedia network; translation teaching.

1 Introduction

The traditional translation teaching is usually teacher-centered and organized by some traditional ways like "evaluation", "error-correction" and the like. In the whole teaching process, teachers are the directors and judges with an authoritative voice, but students are just passive recipients so that the translation teaching has become an one-way dissemination of knowledge from teachers to students. Obviously, students can't fully play the main role in this teaching method, which is also not helpful to cultivate the students' ability to independently identify, analyze and solve problems. Meanwhile, teachers can not monitor or give guidance to students' translation and therefore can't timely find out problems or correct students' errors.

With the development of modern information technology, some multi-functional and modern teaching aids, such as projectors, audio equipment, DV, VCD, DVD, computer and so on, are increasingly used in foreign language teaching. This not only in some extent changes the traditional classroom teaching which is limited by time and space but also expands the teaching information. The application of information technology into translation teaching makes the monotonous and tiring translation teaching methods get a certain improvement. But the teaching mode is still teacher-centered. Computer is introduced to classroom in form and acts as the "blackboard" by which the teaching contents are shown to students. We can also say the teaching mode is problem-centered, that is, the teaching contents are changed into a series of

D. Jin and S. Lin (Eds.): Advances in ECWAC, Vol. 2, AISC 149, pp. 553–557.

issues through which men and machine make the simple "verbal exchanges". In essence, the above model can not solve the problem of improving students' translation ability. So it is imperative as to how to effectively use information technology achievements to reform the translation teaching mode.

2 Theoretical Foundation of the Application of Movies into Translation Teaching

The change in the teaching mode is closely related with the development of teaching theories. Both structurism and post-modernism since the 1990s provide a solid theoretical basis for the reform of the translation teaching mode.

Under the guidance of structurism, foreign language teaching is to focus on the analysis of the structure of the target language system, tending to use the grammar-translation method, namely, finishing the translation teaching by grammar analysis, pattern drills, text translation and other ways.

Influenced by this kind of mode, the translation teaching is likely to be conducted by the teachers through "providing text—analyzing text grammar—translating sentence by sentence—comparing the translated text with the reference translation—correction and comment". With teachers as the center, characteristic of infusing and imparting the systematic translation knowledge and skills, the traditional translation teaching mode attaches more weight not to the translation process but to the final results of translation, which is not conducive to cultivate students' innovative spirit and practice ability.

According to the teaching theories of post-modernism, knowledge is not a single transmission process but a dynamic and open self-regulation system which is gradually built in the process of interaction between learners and environment. Language teaching mode dominated by social constructivist was proposed and established by M • Williams (a British applied linguist) and R • Bourdon (a British psycholinguist) in the end of the last century. [1] The mode attaches importance to learners' own experience, emphasizes the students' central position and lays stress on the interactions among students, teachers, tasks and environment. The theory stresses students should be the center and teachers act as the important "mediators", namely, helpers and promoters instead of imputers of knowledge to help students take the initiative to build and promote their knowledge. That's to say, we should completely abandon the traditional teacher-centered teaching mode which simply focuses on imparting knowledge. In the new teaching mode, teaching contents and teaching methods should focus on the construction of the learning environment. That is, the new teaching mode should use a student-centered instructional design and make full use of teaching media to create some meaningful situations for students' active construction of knowledge. Multimedia computer and network-based communications technology are particularly suited to set up a constructivist learning environment because of their advantages. That is, they can serve as the ideal cognitive tools under the constructivist learning environment to effectively promote students' cognitive development. Thus, with the rapid development of modern educational technology, the constructivist learning theory is increasingly showing its strong vitality and increasing its worldwide influence. A variety of teaching practices under the guidance of the constructivist theory have gradually verified the correctness and utility of constructivism. The application of movies into translation teaching relying on multimedia network just benefits from this theory, emphasizing students' perception of their environment and their ability to interact with the virtual society and promoting the teaching mode of "perception" and "situation" in which information is found and extracted from the environment in a variety of ways.

3   The Concrete Implementation Principles of Applying Movies into Translation Teaching

In accordance with an experiment, among the information that human beings get access to, the visual information accounts for 83%, the auditory information 11%, the information gained by the olfactory receptors 3.5%, the information gained by the taste organs 1%, and the information gained by touch 1.5%. [2] In addition, people can by and large remember 20% of what they hear, 30% of what they see and 70% of what they communicate with others. [3] Therefore, participation by a variety of sensory organs will be more conducive to information acquisition and maintenance. In the teaching process, we should make full use of the characteristics of multimedia: a combination of sound, word and image; we should provide some related background knowledge with the translated text by downloading through the Internet or watching online and other forms; we should fully mobilize the students' sensory organs and create a real communication environment to stimulate students' interest, strengthen interaction and help students accurately understand the original context. Take the 2005 British film "Pride and Prejudice" (excerpts) as an example to illustrate the unique advantages of applying movies into translation teaching. The specific teaching process is as follows:

3.1   Insertion of Movie Fragment

Teachers can cut the movies into several clips and insert one of them into the courseware. Through the playback of the fragment, students see and hear the English dialogue combined with the atmosphere the film creates. Take the scene in which Darcy speaks out his love for Elizabeth for the first time as an example, students can have an immersed sense to experience the feelings of the two protagonists and have a deep understanding of the exact meaning of the original text in a specific context by means of Elizabeth and Darcy's facial expressions, movements, tone and the surrounding scene, etc.

3.2   Group Discussion and Representation

Students are required to discuss the questions and exchange their opinions among the random groups according to the situation. Teachers can monitor and participate in the discussions to help students fully understand some non-linguistic factors concerning the source, function and intention of the original text & the readers and function of the translated text. After that, each group elects a representative to give a statement of the group's understanding of the source language and the overall evaluation of the original text. Within a specified time (about 20 minutes), each student is asked to independently complete the translation of the passage. Teachers can at any time monitor students and record the difficulties students meet and the strategies they adopt in the process of translation. As to the common problems, teachers can point them out in the collective comments. After translation, group members share their translation versions, point out the advantages and disadvantages of their translation and propose some suggestions on how to promote their translation. Based on the group exchange, each team gives a comprehensive translated text which will be read by the representative from their own group.

3.3 Explanation and Comment

The explanation and comment can be done by teachers. It also means the peer assessment between the various groups. Teachers should focus intensely on the analysis and guidance of the translation process, procedures, methods and techniques as well as the common problems students go through in translation. Teachers are expected to list the different translation versions of the same novel and inspire students to reflect on the original novel from different perspectives by comparing their own translation version with the reference translation.

3.4 Performance by Dubbing Films in Chinese

Each representative should be encouraged to provide Chinese soundtracks for the characters in the film in accordance with their respective understanding. Through such imitate performance, students have a deeper understanding of the translation subject (translators), the translation object (the original text) and the nature, function and mutual relationship of the recipient (readers or society). In addition, students can also widen their horizon and firmly grasp the criteria for translation. [4]

4 The Unique Advantages of Movie Appreciation in Translation Teaching by Means of Internet and Multimedia

Through the above steps, students can deeply understand the original text and solve a series of problems in translation by virtue of the given context. For instance, when Darcy paid court to Elizabeth for the first time, Elizabeth refused Darcy by saying the sentence: "I have never desired your good opinion, and you have certainly bestowed it most unwillingly." Through the heated debate between the two leading characters in the movie segment, it can be seen that Elizabeth has the incredible self-esteem and insight although she is plain. Elizabeth intentionally or unintentionally reveals the young girl's arrogance before Darcy to combat with his pride and superiority that he shows off in public. In addition, Elizabeth hates Darcy for destroying the relationship between her sister Jane and his friend Bentley. Moreover, for the reason that Elizabeth trusts Whickham's imputation against Darcy too readily, Elizabeth is always prejudiced against Darcy. Through the appreciation of this section of the movie clips, readers are more likely to understand the story background so as to have a better understanding and general grasp of the original text and choose the correct ways of expression for the translated text. In the process of comments and appreciation of the translated text, students can more easily find out the correct translation criticism standards.

Thus, it can be demonstrated that the application of movie appreciation into translation teaching with the help of multimedia network has its unique advantages: It has both sound and shape so that students' various sense organs can be mobilized to participate in the translation activities and their translation interest is also stimulated; it can reproduce a lively verbal communication situation to help students experience and feel personally the original context and finally pick up the accurate ways of expression for the translated text; it is an important media to provide cultural knowledge reflected in movies in many aspects from historical events to customs, which is helpful to avoid the cultural mistranslation problems. [5]

5 Summary

In the translation teaching under the context of globalization and post-modernism, the organic combination of modern multimedia network technology and translation teaching not only can greatly stimulate the potential of translation teaching, but also will facilitate the all-round reform of translation teaching paradigm, which is expected to provide a new theoretical explanation for the training mode of interpreters and translators. The application of movies into constructivism-based multimedia network translation teaching can help students accurately grasp the original context, cultivate their interest in translation and give the full rein to their initiative and innovation.

References

1. Marion, W., Burden Robert, L.: Psychology for Language Teachers: a Social Constructivist Approach. Cambridge University Press, Cambridge (1997)
2. http://czdj.czedu.com.cn/yjzy/llyj/hkklw/hkklw8.htm
3. Wang, L.: The Application of Modern Multimedia Technology in Teaching Mathematics. Business Magazine (2008)
4. Yang, G.: Language Education Model of Social Constructivism and its Implications for the Teaching of Chinese-to-English Translation. Journal of Sichuan International Studies University (2004)
5. Guo, S.: The Effective Application of English Movies into Foreign Language Teaching. Movie Literature (2010)

A Parallel CRC Algorithm Based on Symbolic Polynomial

Biying Zhang

College of Computer and Information Engineering, Harbin University of Commerce,
Harbin, 150028, China

Abstract. In order to improve the computing speed and reduce the number of occupied hardware resources of CRC algorithm, a parallel CRC algorithm for arbitrary length of bit series was demonstrated based on the symbolic polynomial theory. A mathematic model for parallel computation of CRC codes was given. The proposed mathematic model and algorithm was validated with the experiments. From the comparative analysis between the proposed algorithm and the traditional LUT algorithm, it can be seen that the proposed parallel algorithm is better than the traditional one in terms of delay and area.

Keywords: CRC, Polynomial, FPGA, Design optimization.

1   Introduction

Cyclic Redundancy Check (CRC) is an error-checking technique that is widely used in data communication systems and storage systems [1]. The CRC algorithms are classified into serial ones and parallel ones. The serial algorithms implemented with a linear feedback shift register (LFSR), which deals with only one bit every time, are implemented by simpler circuit and consume very few hard resources [2]. But the speed of serial algorithms is very slow, so they can not satisfy the demands of the high speed computation. The parallel CRC algorithms compute multiple bits at a time and have higher computation speed [3].

The look-up table (LUT) was widely used in the parallel CRC algorithms because of its shorter delays [4]. In the FPGA based on the LUT structure, SRAM-based LUT is the basic programmable unit, the combination logic and ROM are mapped into LUT. For parallel CRC algorithms based on FPGA, the LUT methods are not better than the direct calculating methods in term of delays. In addition, the LUT methods consume more hardware resources. Therefore, many parallel CRC algorithms based on direct calculating methods were proposed [5-8].

2   Parallel Algorithm for Generating Check Value

2.1   Mathematic Model

Let $M_n = [m_1 m_2 \cdots m_i \cdots m_n]$ be binary series, which is composed of n bytes: m_1、 m_2、$\cdots m_i \cdots m_n$, the length of the binary series is 8×n. The series

D. Jin and S. Lin (Eds.): Advances in ECWAC, Vol. 2, AISC 149, pp. 559–565.
springerlink.com   © Springer-Verlag Berlin Heidelberg 2012

$M_{n-1} = [m_1 m_2 \cdots m_i \cdots m_{n-1}]$ is obtained by intercepting the previous n-1 bytes of M_n.

The polynomial form of M_n is

$$M_n(x) = m_1(x) \cdot x^{8(n-1)} + m_2(x) \cdot x^{8(n-2)} + \cdots$$
$$+ m_i(x) \cdot x^{8(n-i)} + \cdots + m_n(x) \qquad (1)$$

The polynomial form of M_{n-1} is

$$M_{n-1}(x) = m_1(x) \cdot x^{8(n-2)} + m_2(x) \cdot x^{8(n-3)} + \cdots$$
$$+ m_i(x) \cdot x^{8(n-i-1)} + \cdots + m_{n-1}(x) \qquad (2)$$

When computing checking codes, r 0s are appended to the back of the binary series, and the polynomial form of the whole data series is

$$M_n{}'(x) = M_n(x) \cdot x^r = m_1(x) \cdot x^{8(n-1)+r} + m_2(x) \cdot x^{8(n-2)+r} + \cdots$$
$$+ m_i(x) \cdot x^{8(n-i)+r} + \cdots + m_n(x) \cdot x^r \qquad (3)$$

The relation between $M_n{}'(x)$ and $M_{n-1}{}'(x)$ is

$$M_n{}'(x) = M_{n-1}{}'(x) \cdot x^8 + m_n(x) \cdot x^r \qquad (4)$$

For $M_{n-1}{}'(x)$

$$\frac{M_{n-1}{}'(x)}{G(x)} = Q_{n-1}{}'(x) + \frac{R_{n-1}{}'(x)}{G(x)} \qquad (5)$$

For $M_n{}'(x)$:

$$\frac{M_n{}'(x)}{G(x)} = \frac{M_{n-1}{}'(x) \cdot x^8 + m_n(x) \cdot x^r}{G(x)}$$
$$= Q_{n-1}{}'(x) \cdot x^8 + \frac{R_{n-1}{}'(x) \cdot x^8 + m_n(x) \cdot x^r}{G(x)} \qquad (6)$$

where the first term is integer which has nothing to do with the remainder. The remainder only can exist in the second term. Thus, the remainder of $R_{n-1}{}'(x) \cdot x^8 + m_n(x) \cdot x^r$ is actually the remainder of $M_n{}'(x)$, that is

$$\mathrm{Re}[\frac{M_n{}'(x)}{G(x)}] = \mathrm{Re}[\frac{R_{n-1}{}'(x) \cdot x^8 + m_n(x) \cdot x^r}{G(x)}] \qquad (7)$$

The length of the binary series denoted by $R_{n-1}'(x) \cdot x^8 + m_n(x) \cdot x^r$ is 1+p bytes, where $r = 8 \cdot p$. Set p=2, in the other words, r=16, then $R_{n-1}'(x) \cdot x^8 + m_n(x) \cdot x^{16}$ denotes the series composed of 3 bytes.

Let $R_{n-1}'(x)$ be binary series $[h_{n-1}l_{n-1}]$, $m_n(x)$ be byte m_n , then $R_{n-1}'(x) \cdot x^8 + m_n(x) \cdot x^{16}$ be a 3-bytes series as follows: left shift $[h_{n-1}l_{n-1}]$ 8 bits, supplement 0s at the back, add h_{n-1} and m_n .

2.2 Computation of Remainder

Let byte series T_{abc}, T_{a00}, T_{bc} as follows: $T_{abc} = [abc], T_{a00} = [a00], T_{bc} = [bc]$
    The relation of T_{abc}, T_{a00}, T_{bc} is denoted by polynomial as

$$T_{abc}(x) = T_{a00}(x) + T_{bc}(x) \tag{8}$$

For $T_{a00}(x)$,

$$\frac{T_{a00}(x)}{G(x)} = Q_{a00}(x) + \frac{R_{a00}(x)}{G(x)} \tag{9}$$

For $T_{abc}(x)$,

$$\frac{T_{abc}(x)}{G(x)} = \frac{T_{a00}(x) + T_{bc}(x)}{G(x)} = Q_{a00}(x) + \frac{R_{a00}(x) + T_{bc}(x)}{G(x)} \tag{10}$$

where $Q_{a00}(x)$ is integer which has nothing to do with the remainder, $R_{a00}(x)$ and $T_{bc}(x)$ are 16-bits binary series, the sum of which is still 16-bits binary series and smaller than $G(x)$, thus,

$$R_{abc}(x) = R_{a00}(x) + T_{bc}(x) \tag{11}$$

According to the above theory, the remainder of the three-bytes series $T_{abc} = [abc]$ can be computed as the following two steps:

(1)   compute the remainder R_{a00} of T_{a00};

(2)   obtain R_{abc} by exclusive OR computation of R_{a00} and T_{bc}.

The logic structure for computing the remainder of the three-bytes series is shown by Figure 1, where PCL denotes the parallel computation logic, XOR denotes the exclusive OR computation, RG denotes the register.

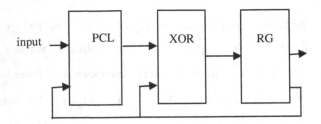

Fig. 1. Logic structure of parallel CRC

3 Parallel Computation Logic

Let $[b_1b_2b_3 \cdots b_i \cdots b_{n-1}b_n]$ be a n-bits binary series, where $b_i \in \{0,1\}$. When computing check value, r 0s are attached to the back of data series, thus, the whole data series is

$$M_n = [b_1b_2b_3 \cdots b_i \cdots b_{n-1}b_n c_1 c_2 \cdots c_j \cdots c_r] \tag{12}$$

where $b_i \in \{0,1\}$, $c_j \in \{0\}$.

The polynomial form of M_n is

$$M_n(x) = b_1 x^{n+r-1} + \cdots + b_i x^{n+r-i} + \cdots + b_n x^r \tag{13}$$

For $M_n(x)$,

$$\frac{M_n(x)}{G(x)} = \frac{b_1 x^{n+r-1}}{G(x)} + \cdots + \frac{b_i x^{n+r-i}}{G(x)} + \cdots + \frac{b_n x^r}{G(x)} \tag{14}$$

$$\frac{M_n(x)}{G(x)} = Q(x) + \frac{R(x)}{G(x)} \tag{15}$$

$$R(x) = \mathrm{Re}[\frac{M_n(x)}{G(x)}] \tag{16}$$

For $b_1 x^{n+r-1}$,

$$\frac{b_1 x^{n+r-1}}{G(x)} = Q_1(x) + \frac{R_1(x)}{G(x)} \tag{17}$$

$$R_1(x) = \mathrm{Re}[\frac{b_1 x^{n+r-1}}{G(x)}] \tag{18}$$

For $b_2 x^{n+r-2}$,

$$\frac{b_1 x^{n+r-2}}{G(x)} = Q_2(x) + \frac{R_2(x)}{G(x)} \tag{19}$$

$$R_2(x) = \mathrm{Re}[\frac{b_2 x^{n+r-2}}{G(x)}] \tag{20}$$

For $b_i x^{n+r-i}$,

$$\frac{b_i x^{n+r-i}}{G(x)} = Q_i(x) + \frac{R_i(x)}{G(x)} \qquad R_i(x) = \mathrm{Re}[\frac{b_i x^{n+r-i}}{G(x)}] \tag{21}$$

For $b_n x^r$,

$$\frac{b_n x^r}{G(x)} = Q_n(x) + \frac{R_n(x)}{G(x)} \tag{22}$$

$$R_n(x) = \mathrm{Re}[\frac{b_n x^r}{G(x)}] \tag{23}$$

For $M_n(x)$,

$$\frac{M_n(x)}{G(x)} = Q_1(x) + \cdots + Q_i(x) + \cdots + Q_n(x)$$

$$+ \frac{R_1(x) + \cdots + R_i(x) + \cdots + R_n(x)}{G(x)} \tag{24}$$

Thus,

$$R(x) = \mathrm{Re}[\frac{M_n(x)}{G(x)}] = R_1(x) + \cdots + R_i(x) + \cdots + R_n(x) \tag{25}$$

When $b_i = 0$, denote $R_i(x)$ as $R_i^{\,0}(x)$, obviously,

$$R_i^{\,0}(x) = \mathrm{Re}[\frac{0 x^{n+r-i}}{G(x)}] = 0.$$

When $b_i = 1$, denote $R_i(x)$ as $R_i^{\,1}(x)$, obviously, $R_i^{\,1}(x) = \mathrm{Re}[\frac{x^{n+r-i}}{G(x)}]$

In fact, the value of $R_i^{\,1}(x)$ is the check value $R_i^{\,1}$ of the binary series in which the ith bit(from the left to the right) is 1 and the other bits are 0s.

According the above principles, the check value of n-bits binary series $[b_1 b_2 b_3 \cdots b_i \cdots b_{n-1} b_n]$ is obtained by $R = \sum_{i=1}^{n} R_i$, where R_i is the check value of the n-bits binary series in which the ith bit is b_i and the others are 0s. If $b_i = 0$, then $R_i = 0$. If $b_i = 1$, then $R_i = R_i^1$.

4   Experimental Results

For the comparative analysis, we have implemented not only the proposed parallel CRC algorithm but also the LUT method with VHDL. In both cases, we used the Virtex2 of XILINX to carry out the algorithms. Table 1 shows the comparative results between the proposed parallel CRC algorithm and the traditional LUT method. For the proposed algorithm, the number of slices, the number of 4 input LUTs and the number of equivalent gate are 5, 8 and 48 respectively. For the traditional LUT method, the number of slices, the number of 4 input LUTs and the number of equivalent gate are 46, 57 and 423 respectively. It can be seen that the proposed algorithm occupies much fewer hard resources than LUTs. On the other hand, the maximum combinational path delay of the proposed algorithm is 7.619ns, while the maximum combinational path delay of the LUT is 9.569ns. Thus, the propose algorithm is still better than the LUT in terms of speed.

5   Conclusions

Based on the symbolic polynomial theory, a parallel CRC algorithm for arbitrary length of bits series was demonstrated. A mathematic model for parallel computation of CRC codes was given. The proposed mathematic model and algorithm was validated with the experiments. Through the comparative analysis, the proposed algorithm is better than the traditional LUTs in terms of occupied hardware resources and delays.

Table 1. Comparison between the proposed algorithm and LUT

Hardware units	The proposed algorithm	LUT
Number of Slices	5	46
Number of 4 input LUTs	8	57
equivalent gate count	48	423
Maximum combinational path delay	7.619ns	9.569ns

References

1. E. Ustunel, I. Hokelek, O. Ileri, H. Arslan: Joint optimum message length and generator polynomial selection in cyclic redundancy check (CRC) coding. In: IEEE 19th Conference on Signal Processing and Communications Applications (SIU), pp.222-225. (2011)
2. Youjie Ma, Haitao Zhang, Xuesong Zhou,et al.: The Realization of the CRC Arithmetic Which is Based on DSP. In: International Forum on Computer Science-Technology and Applications(IFCSTA '09), vol.2, pp.11-14. (2009)
3. Junho Cho, Baeksang Sung, Wonyong Sung: Block-interleaving based parallel CRC computation for multi-processor systems. In: 2010 IEEE Workshop on Signal Processing Systems (SIPS), pp.311-316. (2010)
4. Yan Sun, Min Sik Kim: A Pipelined CRC Calculation Using Lookup Tables. In: 2010 7th IEEE Consumer Communications and Networking Conference (CCNC), pp.1-2. (2010)
5. M. Ayinala, K.K. Parhi: High-Speed Parallel Architectures for Linear Feedback Shift Registers. IEEE Transactions on Signal Processing, vol.59,no.9, pp.4459-4469. (2011)
6. M. Grymel, S.B. Furber: A Novel Programmable Parallel CRC Circuit. IEEE Transactions on Very Large Scale Integration (VLSI) Systems, vol.19, no.10, pp.1898-1902. (2011)
7. C. Toal, K. McLaughlin, S. Sezer, et al.: Design and Implementation of a Field Programmable CRC Circuit Architecture. IEEE Transactions on Very Large Scale Integration (VLSI) Systems, vol.17, no.8, pp.1142-1147. (2009)
8. Jun Yang, Jun Ding, Na Li, et al.: FPGA-based multi-channel CRC generator implementation. In: 2010 International Conference on E-Health Networking, Digital Ecosystems and Technologies (EDT), pp.81-84. (2010)

Research on Intrusion Detection Based on Heuristic Genetic Neural Network

Biying Zhang

College of Computer and Information Engineering, Harbin University of Commerce
Harbin, 150028, China

Abstract. In order to model normal behaviors accurately and improve the performance of intrusion detection, a heuristic genetic neural network (HGNN) is presented. The crossover operator based on generated subnet is adopted considering the relationship between genotype and phenotype. An adaptive mutation rate is applied, and the mutation type is selected heuristically from weight adaptation, node deletion and node addition. When the population is not evolved continuously for many generations, in order to jump from the local optima and extend the search space, the mutation rate will be increased and the mutation type will be changed. Experimental results with the KDD-99 dataset show that the HGNN achieves better detection performance in terms of detection rate and false positive rate.

Keywords: intrusion detection; neural network; genetic algorithm; mutation operator.

1 Introduction

Intrusion detection is an emerging technique to protect the computer network and has become an essential tool for the network security. Recently, machine learning approaches, such as rule learning [1], hidden Markov model [2], support vector machine [3] and neural network [4], have been employed in the field of intrusion detection. Since the neural network has the capability to generalize from limited, noisy and incomplete information, the intrusion detector modeled by neural network can recognize not only previously observed attacks but also future unseen attacks. Therefore, the neural network has been considered as a promising technique for intrusion detection.

2 Heuristic Evolution of Neural Network

2.1 Subnet Crossover Operator

The generated subnet $Gen(i)$ of the node i is defined by

$$Gen(i) = \begin{cases} CO(i) & ,i \in M \\ CI(i) \cup CO(i) \cup i & ,i \in H \\ CI(i) \cup i & ,i \in P \end{cases} \tag{1}$$

D. Jin and S. Lin (Eds.): Advances in ECWAC, Vol. 2, AISC 149, pp. 567–573.
springerlink.com   © Springer-Verlag Berlin Heidelberg 2012

where M is the set of input nodes, H is the set of hidden nodes, P is the set of output nodes, $CI(i)$ is the set of all input connections of the node i, $CO(i)$ is the set of all output connections of the node i.

The generated subnet of an input node is the set of its all output connections. The generated subnet of a hidden node is the set of itself, its all input and output connections. The generated subnet of an output node is the set of itself and its all input connections. The connection information includes whether or not the connection exists and the connection weight value. The node information includes bias and activation function.

In this paper, the subnet crossover operator is employed in consideration of the relationship between genotype and phenotype. After the generated subnet of node i is crossed, two byproducts will be yielded. Firstly, connections may be generated on inoperative nodes. Because the inoperative nodes are not permitted to have connections, the new generated connections on it must be deleted. Secondly, the set of input connections or output connections on operative nodes is made empty. Here, all connections on them are deleted and they are changed to inoperative nodes. Figure 1 shows an example of the subnet crossover for a neural network that has two input nodes, three hidden nodes and one output node.

The overall procedure of crossover operator is described as follows:

Step 1. Select randomly a node $i \in M \cup H \cup P$.
Step 2. Cross the generated subnet $Gen(i)$ of node i.
Step 3. Delete the connections of inoperative nodes generated by crossing operation.
Step 4. Delete all connections of operative nodes whose set of input connections or output connections are empty and turn them to inoperative nodes.

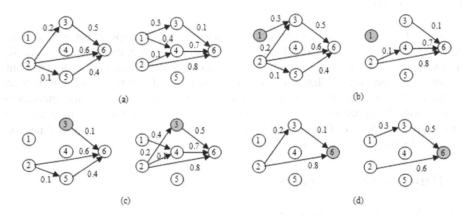

Fig. 1. Example of crossover operation. (a) Parents before crossover. (b) Children after crossover of the input node 1. (c) Children after crossover of the hidden node 3. (d) Children after crossover of the output node 6.

2.2 Heuristic Mutation Operator

2.2.1 Adaptive Mutation Rate
Adaptive mutation rate is defined by

$$P(g) = (MG+1) \cdot 0.05 \qquad (2)$$

where g denotes the current generation, MG denotes the maximum number of generations for which the population is not optimized continuously, u denotes a user-defined parameter which is used to control the increase of the mutation rate.

If there are better solutions at each generation than the previous generation, MG is equal to 0, and $P(g)$ is equal to 0.05. It can be considered that the crossover operator is working efficiently, and the mutation operator is not necessary for evolution. The more generations for which the population is not be enhanced, the larger MG is. This results in the increase of mutation rate $P(g)$. Here, as the mutation rate increases, it can be considered that the evolutionary procedure by the crossover operator may be trapped into local minima and it becomes more difficult to search the optimal solution only by the crossover operator, and the mutation operator should be used to extend the space of solutions.

2.2.2 Heuristic Selection of Mutation Operations
The mutation operator is composed of three operations: weight adaptation, node deletion and node addition.

Weight adaptation is implemented by perturbing the weights of neural network using Gaussian noise, which is described by

$$w = w + N(0, a \cdot P(g)) \qquad (3)$$

where $N(0, a \cdot P(g))$ is Gaussian random variable with mean 0 and standard deviation $P(g)$, w is a weight, $P(g)$ is the adaptive mutation rate, a is a user-defined constant used to scale $P(g)$.

Node deletion means setting all connections on the deleted node 0, and node addition means assigning connection weights between the added node and the other relative nodes. Under the encoding scheme in this paper, deleting a node is to set the corresponding columns and rows 0, and adding a node is to assign random values to the corresponding columns and rows.

The mutation operator is defined by

$$Operator = \begin{cases} WA & P(g) < m \\ ND & m \le P(g) < n \\ NA & n \le P(g) \end{cases} \qquad (4)$$

where WA denotes the adaptation of connection weights, ND denotes the node deletion, NA denotes the node addition, $P(g)$ is the adaptive mutation rate, m and n are the user-defined parameter which satisfy $0 < m < n < 1$. Obviously, which mutation operation is performed depends highly the mutation rate, and mutation rate is determined by the maximum number of generations the population do not get better. Hence, the adaptive mutation rate is the foundation of the total mutation operator.

When the subnet crossover operator is working efficiently and improving the performance of the neural network continuously, MG and $P(g)$ are very small, the weight adaptation will be performed. As the efficiency of the crossover operator becomes worse, MG and $P(g)$ becomes larger, the node deletion and addition will be carried out. The node deletion has more priority than the node addition to be apt to smaller network structure.

2.2.3   The Heuristic Mutation Procedure

The heuristic mutation procedure is shown by Figure 2. At the beginning, the adaptive mutation rate $P(g)$ is calculated. And then, the mutation operator is performed with the possibility $P(g)$ on each individual of population. For each individual, a mutation operation is selected from three candidates: weight adaptation, node deletion and node addition. For weight adaptation, a connection weight is selected uniformly and adapted with (3). For node deletion, a node, not all connection weights on which is 0, is selected uniformly, and all connections between this node and the other nodes are

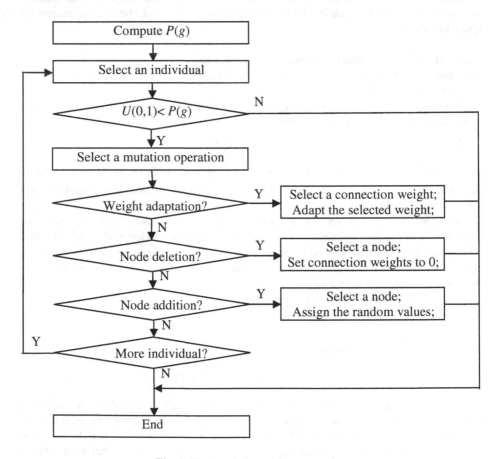

Fig. 2. The heuristic mutation procedure

set 0. For node addition, a node, all connection weights on which is 0, is selected uniformly, and the connections between this node and the other nodes are assigned uniform random values between -1 and 1.

2.3 Overall Evolutionary Framework

The overall evolutionary framework is shown by Figure 3. The initial population of the neural networks is generated with random weights and full connection. First, all initial or selected individuals are evaluated with the fitness function, and then, the best ones are selected from the parent and child individuals. Subsequently, the subnet crossover operator is performed on the selected individuals. Finally, the proposed heuristic mutation operator is carried out. The above steps are iterated until stopping conditions are satisfied.

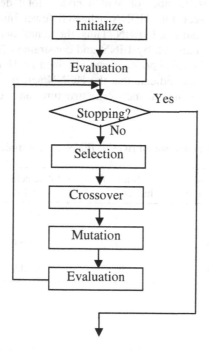

Fig. 3. Evolutionary procedure

3 Experiments

3.1 Datasets

The experiments were performed on the KDD Cup 1999 dataset to assess the effectiveness of the proposed heuristic genetic neural network(HGNN). The dataset is classified into five major categories: Normal, denial of service (DOS), remote to local (R2L), user to root (U2R) and Probe. A 10% selection from training dataset is used to train neural network. After the neural network is evolved, the best one is picked up from the population and tested with the labeled test dataset.

3.2   Results and Analysis

The proposed HGNN was compared with the JENN and the constrained JENN in [5]. The JENN evolved input features, network structure and connection weights of neural network jointly using random mutation operator. The constrained JENN, which used the same evolutionary algorithm as the JENN, only evolves network structure and connection weights simultaneously without considering feature selection by constraining the minimum number of the input nodes to be equal to 41.

Table 1 shows the comparative results of the HGNN, the JENN and the constrained JENN. We compared the performance of three neural networks from four aspects: number of input features, number of hidden nodes, total detection rates and false positive rate. It can be seen that HGNN achieved fewer input features and hidden nodes than JENN and constrained JENN. Thus, the neural network structure attained by HGNN is compacter than that by JENN and constrained JENN. Furthermore, the total detection rate and the false positive rate obtained by HGNN are better than the other two approaches. These indicate that the HGNN achieved better performance in terms of compactness of neural network structure and capability of intrusion detection.

Table 1. Comparison of HGNN, JENN and constrained JENN

Method	Num. of input features	Num. of hidden nodes	Total detection rate (%)	False positive rate (%)
constrained JENN	41	16	87.46	3.45
JENN	15	10	91.51	1.31
HGNN	14	9	93.15	1.02

4   Conclusions

In this paper, to improve the performance of intrusion detection, we proposed a heuristic genetic neural network. The subnet-based crossover operator was employed to take the relationship between genotype and phenotype into account. The heuristic mutation operator, which is composed of an adaptive mutation rate and a heuristic selecting algorithm of mutation operation, was used to jump from the local optima and extend the search space. It can be seen from the experimental results that the proposed approach achieved better performance in terms of compactness of neural network structure and capability of intrusion detection.

References

1. Li, L., Yang, D., Shen, F.: A novel rule-based Intrusion Detection System using data mining. In: 2010 IEEE International Conference on Computer Science and Information Technology (ICCSIT), vol. 6, pp. 169–172 (2010)
2. Li, L., Yang, D., Shen, F.: A novel rule-based Intrusion Detection System using data mining. In: 2010 IEEE International Conference on Computer Science and Information Technology (ICCSIT), vol. 6, pp. 169–172 (2010)
3. Zhu, G., Liao, J.: Research of Intrusion Detection Based on Support Vector Machine. In: International Conference on Advanced Computer Theory and Engineering, pp. 434–438 (2008)
4. Han, S.J., Cho, S.B.: Evolutionary neural networks for anomaly detection based on the behavior of a program. IEEE Transactions on Systems, Man, and Cybernetics, Part B 36(3), 559–570 (2006)
5. Zhang, B., Jin, X.: A joint evolutionary neural network for intrusion detection. In: 2009 International Conference on Information Engineering and Computer Science (ICIECS 2009), Wuhan, China, pp. 1–4 (2009)

Bandwidth Prediction Scheme for Streaming Media Transmission

Kexin Zhang[*], Zongze Wu, Shengli Xie, and Liangmou Feng

School of Electronic and Information Engineering, South China University of Technology,
510640 Guangzhou, China
{zhang.kexin,zzwu,adshlxie}@scut.edu.cn

Abstract. Available bandwidth which directly influences the quality of streaming media transmission is an important parameter in streaming media system, but the traditional bandwidth measurement mechanism does not solve the problem of bandwidth prediction. This paper presents a bandwidth measurement and prediction scheme for streaming media transmission. The scheme predicts available bandwidth quickly and accurately by the prediction model which is based on normal distribution, and meanwhile has no influence on the transmission of streaming media.

Keywords: Stream media transmission, available bandwidth, bandwidth measurement, bandwidth prediction.

1 Introduction

The widely use of Internet has promoted the rapid development of streaming media, which is widely used in video broadcast, online advertising, E-commerce, video on demand, remote education, medical treatment, etc., and the development of streaming media will deeply affect people's work and life in the future. Streaming media [1] means that divides the multimedia files into many packages and then transmits the packages to client continuously. In a streaming media system, users do not need to wait until the whole video file is downloaded before playing it, instead they just need to get several packages of the video file to start, and then the rest packages will be downloaded when the video is being played.

Streaming media system with good performance usually adopts scalable video coding scheme [2], the bit stream generated by this scheme can be truncated arbitrarily, and the quality of reconstructed video is proportional to the obtained bit stream, which means the more bit stream we get the better video we reconstruct. This feature enables the streaming media system to take full advantage of the available bandwidth. According to the current network condition, the server will choose suitable media stream to transmit. When the available bandwidth is not enough, the sever will select less bit stream to transmit, otherwise, more bit stream is selected. By this way, not only the network congestion is avoided but also the available bandwidth is fully used.

---

[*] Corresponding author.

D. Jin and S. Lin (Eds.): Advances in ECWAC, Vol. 2, AISC 149, pp. 575–580.

Therefore, how to obtain the available bandwidth of current network becomes the key problem of streaming media system.

The existing bandwidth measurement mechanisms such as Delphi[3] 、 Pathload [4] 、 Spruce[5], all concentrate on the measuring of current network bandwidth, which do not effectively solve the problem of predicting the available bandwidth in the next moment. What's more, the restriction in technique and authorization has also limited their application in present steaming media system. Based on these problems, this paper proposed a bandwidth measurement and prediction scheme that is suitable for streaming media transmission. The scheme predicts the current network available bandwidth quickly and accurately, and meanwhile has no influence on the transmission of streaming media.

This paper is organized as follows. Section 1 analyzed the problems of the current bandwidth measurement mechanisms. Section 2 describes the proposed bandwidth measurement and prediction scheme in detail. Section 3 gives the experimental results. Concluding remarks are given in Section 4.

2   Problems Analysis

The current bandwidth measurement mechanisms can be divided into active measurement and passive measurement according to whether the probe packages are sent or not. Via sending probe packages to the network and getting the feedback information, the active measurement techniques can predict the current network bandwidth. These techniques need to send probe packages to the network, which will increase the burden for network and meanwhile influence the network performance. Passive measurement techniques take advantage of specific software and hardware to passively monitor the data packages in network, and then predict the network bandwidth. Passive measurement techniques need not to send packages to the network, so they won't increase the burden for network, however, the needs for special equipment and authority prevent the widely use of these techniques.

Mature active bandwidth mechanisms can be divided into two kinds, which are based on the Probe Gap Model (PGM) and the Probe Rate Model (PRM) respectively.

The mechanisms based on PGM such as Spruce[5], IGI[6] and Delphi[3], need to know the capacity C of all links before the measurement, then the source end sends probe packages to the receiver in a certain speed. Suppose that the interval of sending test packages is Δin, and the interval of the packages arrive at the receiver is Δout, so the transmission speed of background flow is [(Δout-Δin)/Δin]×C, where C is the tight link capacity, then the available bandwidth A can be obtained by Formula 1.

$$A = C \times (1 - \frac{\Delta_{out} - \Delta_{in}}{\Delta_{in}}) \tag{1}$$

While the mechanisms which belong to PRM such as Pathload [4], Path chirp[7], and TOPP[8], are based on the self-congestion theory[9].In the theory, when the transmission speed of the test packages is less than the link bandwidth, the transmission delay is relatively fixed, and is decided by the network physical properties; when the transmission speed is greater than the link bandwidth, the test packages start to wait in line, and the transmission delay increases, so the speed corresponding to the turning point in transmission delay represents the maximum link available bandwidth, and is the available bandwidth of the path.

Through the analysis above, we can find that in active measurement, lot of testing packets are send, which increase the burden for network and also the measurement mechanism takes long time to get the available bandwidth; as for passive measurement, the requirement for equipment and authority is too high. Obviously, these measurement mechanisms are not suitable for the streaming media system which needs the real time transmission of large quantity media data. So this paper proposed a bandwidth measurement and prediction scheme, the scheme measures the current network available bandwidth quickly and accurately, meanwhile predicts the available bandwidth in the next moment based on the normal distribution prediction model.

3 Bandwidth Measurement and Prediction Scheme

The scheme proposed in this paper firstly measures the current network bandwidth quickly and accurately, which has no influence on the transmission of streaming media, and then predicts the available bandwidth in the next moment according to the normal distribution prediction model. The measurement and prediction process is described in detail in the following subsection.

3.1 Bandwidth Measurement

This paper presents a simple and effective method to get the available bandwidth. We get the available network bandwidth by measuring the time that a piece of media data transmitted from server to client, the measurement process is shown as follows:

1) The server chooses a set of video stream, sends the size of the video stream to client, and then begins to send video data to the client, here we assume that the size of the video stream is N bits.

2) After receiving the size of video stream, client starts receiving the video stream, and timing, when finishes the receiving, client gets the time T that used for receiving the N bits video stream.

3) Calculate the network bandwidth Bt in this time interval use Formula 2.

$$Bt=N/T \tag{2}$$

This scheme gets the available bandwidth in current time interval rapidly and precisely, meanwhile has no influence on the streaming media transmission.

3.2 Bandwidth Prediction

In this paper we collected large numbers of available bandwidth data in the process of streaming media transmission, and analyzed the data by data analysis software SPSS (Statistical Product and Service Solutions), at last we found that the data obeyed the normal distribution. Normal distribution is an important probability distribution in mathematics, physics and engineering. Its probability density function is shown as Formula 3.

$$f(x) = \frac{1}{\sqrt{2\pi}\sigma} e^{-\frac{(x-\mu)^2}{2\sigma^2}} \qquad (3)$$

From the probability density function, we can see that the distribution curve has a symmetry axis in x= μ , and inflection points in x= $\mu \pm \sigma$, the variable x distributed around μ with large probability, and away from μ with small probability; the smaller σ is, the more concentrated around μ variable x is. The probabilities of variable x distributed in each interval are obtained by evaluating integrals, which are shown in Table 1.

Table 1. The probabilities of variable x distributed in each interval

interval	probability
(μ - σ , μ + σ)	68.27%
(μ -1.96 σ , μ +1.96 σ)	95%
(μ -2.58 σ , μ +2.58 σ)	99%

Based on these, we can predict the available bandwidth in the next time interval according to the bandwidth in previous N intervals. That is to say, we calculate the parameters μ and σ of current normal distribution according to the available bandwidth in previous N intervals, and gives the probability density function, then predict the range of available bandwidth in next time interval according to reliability requirements. Here we select the range (μ -1.96 σ , μ +1.96 σ) to ensure the reliability of 95%, meanwhile consider the possibility of sudden change in available bandwidth, the value of N need to be adjusted as needed, the initial value of N is zero. The prediction process is described in detail as follows.

1) According to the bandwidth measurement process shown in Subsection A, get the available bandwidth Bt (p) in the previous time interval.
2) Compare Bt(p) with the prediction range in previous time interval. If Bt(p) is within the prediction range, go to Step 3;otherwise give an error, if the error are given three times continuously, set the value of N to zero, then go to Step 3.
3) If N<M(Here M is a positive integer, represents the number of normal distribution samples), N=N+1,then go to Step 4;otherwise, go to Step 4 directly.
4) Update the parameters μ and σ of current normal distribution according to the available bandwidth in previous N intervals.
5) Predict the range of available bandwidth in next time interval according to reliability requirements, here the range can be (μ - σ , μ + σ) to ensure the reliability of 68.27%,also it can be (μ -1.96 σ , μ +1.96 σ) to ensure the reliability of 95%.

This scheme takes full consideration of the normal distribution characteristics of the available bandwidth, meanwhile adjusts the normal distribution samples as needed, so the available bandwidth in the next time interval can be predicted precisely and quickly.

4 Experimental Results

The experiment results were given in two representative scenes: 1) the stable network environment, which means the network bandwidth is stable, has smaller fluctuation; 2) the scene that network bandwidth changes greatly. In the experiment, we used the switch TP-LINK TL-SL2210WEB which had the function of network management to realize the greater fluctuation in bandwidth, and we chose the predict range (μ -1.96σ, μ +1.96σ) to ensure the reliability of 95%,the value of M was selected as 40,the experiment results are shown in Figure 1 and Figure 2.

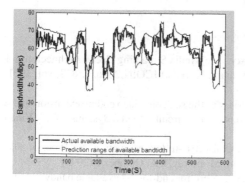 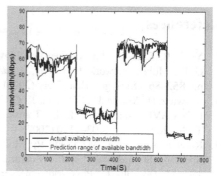

Fig. 1. Experiment results in Scene 1   **Fig. 2.** Experiment results in Scene 2

In the figures, vertical axis represents the bandwidth, with the unit of Mbps, horizontal axis represents the time, in unit of second, the thick line represents the actual available bandwidth, and the fine lines represents the predict range of available bandwidth which means (μ -1.96σ, μ +1.96σ).

From the figures we can see that the thick line which represents the actual available bandwidth is between the fine lines that represent the predict range most of the time, even in the network that bandwidth changes a lot, the proposed scheme still can adjust the predict range quickly to ensure the reliability of prediction, so the scheme can predict the available bandwidth in the next time interval accurately and quickly.

5 Conclusions

The bandwidth measurement scheme presented in this paper avoids the disadvantages of traditional measurement mechanisms which send too many test packages to the network, and has no influence on the transmission of streaming media. The proposed bandwidth prediction scheme is based on the normal distribution prediction model, can predict the available bandwidth precisely and quickly, even in the network that bandwidth changes greatly, the scheme still can adjust the predict range quickly to give the accurate prediction results.

The proposed bandwidth measurement and prediction scheme can provide good support for the media transmission in streaming media system, and can be used in the system easily and expediently.

Acknowledgements. The work is supported by the National Basic Research Program of China (973Program,No.2010CB731800), Foundation for Distinguished Young Talents in Higher Education of Guangdong, China (LYM08010) ,Key Program of National Natural Science Foundation of China (No.U0835003, 60804051), the Fundamental Research Funds for the Central Universities of SCUT (2009ZM0207, 2011ZM0032),the Doctoral Fund of Ministry of Education of China(200805611074).

References

1. Yang, B., Liao, J., Zhu, X.: Two-Level Proxy: The Media Streaming Cache Architecture for GPRS Mobile Network. In: Chong, I., Kawahara, K. (eds.) ICOIN 2006. LNCS, vol. 3961, pp. 852–861. Springer, Heidelberg (2006)
2. Schwarz, H., Marpe, D., Wiegand, T.: Overview of the scalable video coding extension of the H.264/AVC standard. IEEE Transactions on Circuits and Systems for Video Technology 17(9), 1103–1120 (2007)
3. Ribeiro, V., Coates, M., Riedi, R.: Multifractal cross-traffic estimation. In: Proc. of ITC Specialist Seminar on IP Traffic Measurement, Monterey, CA, October 1 (2000)
4. Jain, M., Dovrolis, C.: PathLoad: a measurement tool for end-to-end available bandwidth. In: Proc. of Passive and Active Measurements (PAM) Workshop, pp. 14–25 (2002)
5. Strauss, J., Katabi, D., Kaashoek, F.: A measurement study of available bandwidth estimation tool. In: Proceedings of ACM SIGCOMM Internet Measurement Conference, Karlsruhe, Germany, pp. 39–44 (2003)
6. Hu, N., Steenkiste, P.: Evaluation and characterization of available bandwidth probing techniques. IEEE Journal on Selected Areas in Communications 21(6), 879–894 (2003)
7. Ribeiro, V., Riedi, R.: PathChirp: Efficient available bandwidth estimation for network paths. In: Workshop on Passive and Active Measurement (PAM), La Jolla, California (2003)
8. Melander, B., Bjorkman, M., Gunningberg, P.: A New End - to - End Probing and Analysis Method for Estimating Bandwidth Bottlenecks, pp. 100–105. IEEE Press (2000)
9. Melander, B., Bjorkman, M., Gunningberg, P.: Regression-based available bandwidth measurements. In: Int'l Symp. on Performance Evaluation of Computer and Telecommunication Systems, San Diego, USA (2002)

Slice Partition and Optimization Compilation Algorithm for Dataflow Multi-core Processor

Biying Zhang[1,2], Zhongchuan Fu[1], Yan Wang[1], and Gang Cui[1]

[1] Department of Computer Science and Technology, Harbin Institute of Technology,
Harbin, 150001, China
[2] College of Computer and Information Engineering, Harbin University of Commerce,
Harbin, 150028, China

Abstract. To improve the performance of the multi-core processor, the slice is taken as the unit to design its architecture. The procedure is partitioned into the slices with similar sizes, such that the instructions in the slice can be executed in dataflow fashion. The instructions inside the slice are optimized according to the dataflow graph. The correctness of the proposed algorithm is validated with comparative analysis of the experimental results. The proposed algorithm of slice partition and optimization lays the foundation for establishing the whole tool chain of the multi-core compiler and programming the simulator.

Keywords: multi-core, compilation, slice, dataflow.

1 Introduction

As more cores have been integrated into a processor, the performance of processors is challenged by the communication and synchronization overheads between multiple cores. In the traditional control-flow architecture, the lower data relativity between instructions results in a mass of data communications among multiple cores, and the delay of communication between cores is much larger than that in a core. To overcome these problems, the general method is to divide a program into super instruction blocks in the compiling phase and taking the instruction block as an execution unit [1-4]. The instructions in a block are organized with the data flow, and the blocks are organized with the control flow.

For the dataflow-based execution multi-core system (DEMS), we proposed a method to divide a function into slices with similar size and optimize the slice based on the dataflow. The partitioning and optimizing algorithms were elaborated and designed. Finally, the proposed algorithm was proved to be effective by the experiment.

2 Related Work

The contradiction between the technology and the design of processor architecture has attracted heavy attention from the academe. In recent years, researches on the multi-core for high performance computation based on the tiled architecture have been performed.

D. Jin and S. Lin (Eds.): Advances in ECWAC, Vol. 2, AISC 149, pp. 581–587.
springerlink.com   © Springer-Verlag Berlin Heidelberg 2012

TRIPS (The Tera-op, Reliable, Intelligently adaptive Processing System) [1] is a revolutionary new microprocessor architecture being built in the Department of Computer Sciences at The University of Texas at Austin. The TRIPS project is developing a new class of technology-scalable, power efficient, high-performance microprocessor architectures called EDGE (Explicit Data Graph Execution) architectures. TRIPS integrate the advantage of dataflow and super-scale, and take the hyper block partitioned by the compiler as the basic unit to schedule and execute. TRIPS adopts EDGE instruction set to express the relativity of data directly, i.e., the execution is driven by dataflow in the hyper block [2].

WaveScalar [3] is a dataflow instruction set architecture (ISA) and execution model designed for scalable, low-complexity, high-performance processors. The WaveScalar ISA is designed to run on an intelligent memory system. Each instruction in a WaveScalar communicates with its dependents in dataflow fashion. WaveScalar takes the Wave partitioned by the compiler as the unit to schedule and execute. The performance is improved with simple pipeline and no advanced technique is employed [4].

MicroGrid [5] built in University of Amsterdam is a tiled multi-core architecture based on the concept of micro-threading, which has been an important part of adaptive computation in European AETHE plans [6]. MicroGrid extends the present programming model by adding parallel programming. The uTC is proposed based on the C language to make programmers be able to express the parallelism explicitly at the source code level. Essentially, the instructions in the micro-thread are executed as dataflow.

3   Slice Partition

The method-centric (Metric) multi-core processor is partitioned into procedures by the compiler, and a procedure is referred to as a coarse-grained instruction. The procedure is subdivided into fine-grained slices, in which instructions execute in dataflow fashion to solve the parallel problem.

Having extracted the control-flow graph (CFG) of the procedure, the procedure is partitioned into slices. Firstly, the algorithm searches the CFG with the deep-first traversal until meeting the start of loop, function call or the function exit. Subsequently, the basic blocks except for the header one in the result set are traveled to determine whether its all preceding blocks are in the result set. And then, the basic blocks whose preceding blocks are not in the result set are deleted. Taking the loop start, the function calling and function exit as the slice head node, a new partition is performed.

The slice partitioning algorithm is illustrated as follows:

1: Push the basic block of function entry into the stack *workList*;
2: If the stack *workList* is not empty, then pop the basic block in the top of stack, and go to step 2. Otherwise, the slice partition finishes.
3: If the partition of the basic block is completed, then go to step 2, otherwise go to step 4;
4: Mark the basic block as the head node, and push it into the stack *slice-edge-stack*;
5: If the stack *slice-edge-stack* is not empty, pop the basic block on the top of the stack, and go to step 6. Otherwise go to step 11;

6:    Determine whether the basic block is one of four types, i.e., the function call, the function end, the partitioned slice or loop start. If the basic block belongs to one of the above types, then go to step 5, otherwise go to step 11;

7:    Mark *Pblock* as a member of partitioned slices whose head node is *basicblock*, go to step 8;

8:    Travel the linked list taking the basic block *Pblock* as the exit, and push the basic blocks which are neither loop start nor loop end into the stack *slice-edge-stack*;

9:    Link *Pblock* to the end of linked list of slices whose head node is *basicblock*, go to step 5;

10:   Travel all basic blocks in the linked list of slices. If a basic block has several entry basic blocks which are not completely contained in the linked list of slices, then delete the basic block B and subsequent ones from the linked list of slices;

11:   Complete the partition of slices, save the pointer of the linked list of slices in the head node;

12:   Travel the linked list of slices. If the number of the basic block for exit is more than one, then push the basic exit blocks which are not contained in the linked list of slices into the stack *workList*, go to step 2.

The partition of slices uses two stacks: *worklist* and *slice-edge-stack*. All basic blocks of the procedure forms a bidirectional linked list, in which the entry and exit of the basic block is marked. The data structure contains the relative markers which identify whether a slice is the basic block, which slice a basic block belongs to, and so on. All instructions belonging to a basic block constitute a bidirectional linked list.

4   Dataflow Optimization

The instruction set of DEMS dataflow is obtained by extending the ALPHA, which provides the managing instructions of dataflow and memory access on the base of RISC.

4.1   Deletion of Unused Output Instruction

The data which is not accessed in the process of using a procedure will be deleted from the dataflow graph. It is assumed that *inst* is the instruction calling *function* procedure. The used information during executing *function* procedure is obtained by traveling the receiving data instructions of the exit block of *function* procedure. Then the obtained information is put in *sinkRegisters* and is compared with the output information of *inst* instruction. If the information of *inst* instruction is not in the *sinkRegisters*, it means that the information will not be used in the *function* procedure and will not propagate. Thus, this information will be deleted from the dataflow graph to avoid wasting resources due to passing unused information.

The algorithm for deleting unused output is illustrated as follows:

1:    *block= func->basicblock*, where *block* is the pointer of the basic block of the procedure;

2:    *inst=block->insts*, where *inst* is the pointer of the linked list of instructions of the basic block;

3:   If *inst->type* == *OP_TYPE_FUNCTION_CALL*, i.e., this instruction is for calling procedures, then go to step 4. Otherwise, *inst* = *inst->next*, if *inst* is not null, then go to step 3, else go to step 19;

4:   Create *sinkRegisters*, record the outputted results of the instruction, go to step 6;

5:   If *inst->functionToCall* == *NULL*, i.e., the linked list of the basic blocks of procedure called by this instruction is null, then *inst* = *inst->next*, go to step 3, otherwise, go to step 6;

6:   Set *link=inst->functionToCall->exit->insts*, i.e., *link* is the pointer of linked list of the basic output blocks of the procedure called by *inst* instruction;

7:   If *sink->type* = =*OP_sink*, i.e., this instruction is the one used to receive the type, go to step 8, otherwise, go to step 9;

8:   *input* = *instGetSrcByType(sink, INPUT_A)*;

9:   *sink=sink->next*, if *sink* is not null, go to step 7, else go to step 10;

10: *c1* = *inst->dsts*, i.e., c1 denotes the pointer of the subsequent linked list of *inst* instruction;

11: If *c1->reg* is not in the *sinkRegisters*, go to step 12, otherwise go to step 13;

12: delete *c1* instruction from the subsequent linked list of *inst* instruction;

13: *c1=c1->next*, if c1 is not null, go to step 11, otherwise go to step 14;

14: inst=inst->next, if inst is not null, go to step 3, otherwise go to step 15;

15: block=block->next, if block is not null, go to step 2, otherwise finish.

Only the information is deleted from the instructions calling procedures, and the number of instructions is not decreased. This fact increases the speed of data pass in the dataflow.

4.2 Optimization of rho Instruction

For the two branches of the *rho* instruction, if the branch 1 and the branch 2 converge to the same instruction *producer*, then the optimization can be performed based on the dataflow. The direct relationship between the preceded instructions of *rho* and *producer* is established in the dataflow, such that the information of *producer* instruction can only come from the preceded instructions of *rho*. Thus, the information can directly reach the *producer* instruction without passing the *rho* instruction, which decreases the time for transferring information and improves the executing efficiency of the procedure.

The optimization of the *rho* instruction has no adverse impact on the number of the instructions, through which the instructions at the intersected point do not necessarily wait the execution of the branch instruction.

The algorithm for optimizing *rho* instruction is illustrated as follows:

1:   *block = func->basicblock*;

2:   *inst=block->insts*;

3:   If inst is the *rho* instruction, go to step 4, else go to step 11;

4:   Obtain the two branches of the *inst* instruction: *takenArg* and *notTakenArg*;

5:   *c1=takenArg->consumers*, i.e., obtain the succeeded linked list of the *takenArg* branch;

6:   *c2= notTakenArg->consumers*, i.e., obtain the succeeded linked list of the *notTakenArg* branch;

7:    If $c1==c2$, go to step 8, else go to step 9;

8:    Delete the information from *rho* to $c1$ and $c2$, such that $c1$ can not obtain the information from *rho*;

9:    $c2=c2$->*next*, if $c2$ is not null, go to step 7, else go to step 10;

10:   $c1=c1$->*next*, if $c1$ is not null, go to step 6, else go to step 11;

11:   *inst=inst*->*next*, if next is not null, go to step 3, else go to step 12;

12:   *block=block*->*next*, if *block* is not null, go to step 2, else go to step 13;

13:   finish.

4.3 Optimization of slice_advance Instruction

A corresponding *slice_advance* instruction will be established for each data when the first basic block or local block is partitioned. If all succeeded instruction of an instruction not being *inst* in the dataflow are *slice_advanced* instructions which are not normalized, these all succeeded ones can be connected directly to the *inst* instruction, such that they can obtain directly the data information from the *inst* instruction. If all succeeded ones of the instruction not being *slice_advance* are the slice_advance instructions, the dataflow can be optimized by deleting the slice_advance instruction and replacing the *slice_advance* with the instruction not being *slice_advance*. The instructions in the basic blocks, especially the repeated ones in the dataflow, are reduced to improve the executing speed of dataflow.

The algorithm for optimizing the *slice_advance* instruction is illustrated as follows:

1:    Define the boolean variable *compressedSomething*, set its initial value be *TRUE*;

2:    If *compressedSomething* is *TRUE*, go to step 3, else go to step 11;

3:    Set *compressedSomething = FALSE, block=func*->*basicblocks*;

4:    *inst=block*->*insts*, i.e., *inst* point to the linked list of pointers in the basic block;

5:    If inst is not the instruction of *slice_advance* type, go to step 7;

6:    *dst=inst*->*dsts*, if there are instructions not being *slice_advance type*, go to step 8, else go to step 9;

7:    *inst=inst*->*next*, if inst instruction is not null, go to step 5, else go to step 10;

8:    *dst=inst*->*dsts*, create edges between *inst* instruction and the succeeded instructions of *dst* in the dataflow, delete the edges between *inst* and *dst*, go to step 9;

9:    Set *compressedSomething=TURE*, go to step 10;

10:   *block=block*->*next*, if block is not null, go to step 4, else go to step 2;

11:   Finish.

5 Experimental Results

We perform the experiments on the Red Hat Linux 7.0 as the software platform, and Intel Core 2 as the hardware platform. To validate the functional correctness of the compiler, the benchmark employs the SPEC2000 which is widely used in the world. For the comparative analysis, not only the compiler proposed in this paper but also the SimpleScalar compiler based on super-scalar are executed with SPEC2000.

The comparative results of the proposed compiler and SimpleScalar are illustrated in Table 1. These two methods are compared in terms of the number of functions and controlling instructions of dataflow. For each benchmark, the number of functions obtained by DEMS is the same as that obtained by SimpleScalar. This fact means that the functional correctness of DEMS compiler is 100%. It can be seen from Table 1 that the number of controlling instructions of DEMS compiler is more than that of SimpleScalar compiler. This is because that the slice managing instructions, the mechanism of invoking procedure and memory-accessing instructions are added in order to optimize the dataflow of processor.

6 Conclusions

(1) To improve the performance of multi-core processor, the slice is taken as the unit to design its architecture; (2) To executing the instructions inside the slice according dataflow, the procedure is partitioned into slices with similar sizes; (3) According to the dataflow, the instructions in the slice are optimized, and the corresponding optimization algorithms are proposed.

Table 1. Comparison between DEMS and SimpleScalar

Benchmark	The number of controlling instructions		The number of functions	
	DEMS	SimpleScalar	DEMS	SimpleScalar
spec2000.gzip	53.24	41.24	19	19
spec2000.vpr	43.12	37.42	21	21
spec2000.gcc	79.21	70.12	56	56
spec2000.mcf	13.14	10.23	43	43
spec2000.crafty	86.21	81.22	19	19
spec2000.parser	12.32	10.23	12	12
spec2000.eon	32.43	26.32	21	21
spec2000.perlbmk	13.41	10.31	13	13
spec2000.gap	12.11	10.22	10	10
spec2000.vortex	18.41	17.12	13	13

Acknowledgments. We thank all the staff and students of our research group for their diligence contributions. Especially, Wei Ba and Yang Gao did a lot of work for experiments. We also thank all the anonymous reviewers for their sincere support and valuable advice. This work is supported by Heilongjiang Provincial Natural Science Foundation of China under grant No.F200822, Project(HIT.NSRIF.2009066) Supported by Natural Scientific Research Innovation Foundation in HIT.

References

1. Department of Computer Sciences, University of Texas: The Tera-op, Reliable, Intelligently adaptive Processing System (2010), `http://www.cs.utexas.edu/~trips/`
2. Sankaralingam, K., Nagarajan, R., Mcdonald, R., et al.: Distributed Microarchitectural Protocols in the TRIPS Prototype Processor. In: 39th Annual IEEE/ACM International Symposium on Microarchitecture (MICRO-39), Orlando, pp. 480–491 (2006)
3. University of Washington: WaveScalar (2010), `http://wavescalar.cs.washington.edu/`
4. Swanson, S., Putnam, A., Mercaldi, M., et al.: Area-Performance Trade-offs in Tiled Dataflow Architectures. In: 33rd International Symposium on Computer Architecture (ISCA 2006), Washington, pp. 314–326 (2006)
5. Faculty of Science, University of Amsterdam: MultiProcessor System-on-Chip (MP-SoC) Design (2010), `http://csa.science.uva.nl/research`
6. Bell, I., Hasasneh, N., Jesshope, C.: Supporting microthread scheduling and synchronisation in CMPs. International Journal of Parallel Programming 34(4), 1–9 (2006)

Processing and Stationarity Analyzing of Accelerometer Parameter Based on Cubic Spline Interpolation

Yan Liu, Yuan Zhao, and Hongjie Cheng

Xi'an Research Inst. of Hi-tech, Xi'an, 710025, China

Abstract. When analyzing the performance of missile accelerometer, it's very difficult to build model for the small sample, and the accuracy and integrality of the built model are very low. Through applying the method of cubic spline interpolation, the time serial of testing data is constructed and the sample content is enlarged, thus, the difficulty of building model for small sample data is also solved. Simultaneously, difference method is adopted to carry out stationarity detection and processing, all these will offer the scientifically gist for discovering the change regularity of accelerometer's performance parameter. This study has considerable academic value as well as wide prospects.

Keywords: accelerometer, cubic spline interpolation, performance analysis, stationarity analyzing.

1 Introduction

As key element of the missile inertial navigation system, accelerometer is mainly used to measure the acceleration of the missile during the process of aviation, and its performance and technique status will influence the strike precision and effect of missile weapon directly. In order to ensure hitting the target accurately, the strict periodic test must be done in the course of readiness posture of missile and before launching. The parameters' stability of accelerometer will have an obvious change with the process of time. Nowadays, the stability period index of accelerometer is 3 months; however, it's very difficult to guarantee that test is carried out for every 3 months. Thus, the regularity of test data can't be discovered easily for the disordered data, and the difficulty of describing its time characteristic is increased for the less data[1]. The error of the model which built in small sample is a bit large, and the difficulty of describing the accelerometer's time characteristic increases greatly.

To solve this problem, cubic spline interpolation is applied to process the achieved test data, enlarge the sample size, build model for small sample and offer the gist for building time serials model of missile accelerometer.

2 Presentation of Interpolation Problem

The precise expression of the small sample serial which is constituted by all previous testing data of accelerometer is very hard to express. At present, we just use

D. Jin and S. Lin (Eds.): Advances in ECWAC, Vol. 2, AISC 149, pp. 589–595.
springerlink.com
© Springer-Verlag Berlin Heidelberg 2012

multi-position rolling test to compute a series of data which are one-to-one correspondence with different test points. This method can be described by mathematic language as a function value $y_i = f(x_i)$ in the pairwise inquality points $x_0, x_1, \cdots, x_n$ during the interval of $[a,b]$, and the achieved datasheet can be expressed as:

$$(x_0, y_0), (x_1, y_1), \cdots, (x_n, y_n)$$

Where $y_i = f(x_i)$ $(i = 0,1,\cdots,n)$, and $x_0, x_1, \cdots, x_n$ are called nodal points.

In practical application, $x_0, x_1, \cdots, x_n$ denote the x_i test of accelerometer in a given time interval, and $y_i = f(x_i)$ is used to denote the value of performance parameter of the accelerometer for the x_i test. So, the above datasheet can be used to indicate a function. But, if x is not a nodal point, the function value of this point can't be achieved by lookup the datasheet. To solve this problem, cubic spline interpolation method is introduced. According to the value of $f(x)$ in nodal point, a sufficient smoothness and quite simple function $\varphi(x)$ is constructed as a approximate expression of $f(x)$. Then, compute the function value of the random point in the interval of $[a,b]$ and adopt it as the approximate value of $f(x)$ in the same point.

When solving for the function expression of the sequence composed by all previous testing values, the achieved curve must be unsmooth for lacking enough sample data, and the calculation error will be significant[2]. To find the optimal expression of original number sequence, cubic spline interpolation method is used to enlarge the sample size. At present, there are a lot of methods in interpolating, such as Lagrange interpolation method, successive linear interpolation method, Newton interpolation method, cubic spline interpolation method, etc. In general, comparing with the other three interpolation methods, cubic spline interpolation method has the advantage of good local characteristic, high accuracy close to the end, and well-known stability, and so on. Consequently, to keep high precision of the interpolated data, cubic spline interpolation method is choused to enlarge the sample size during the data processing finally.

3   Principle of Cubic Spline Interpolation

As an indispensable tool in modern numerical analysis, spline function plays a great role in digital approximation, solving arithmetic solution of ordinary differential equation and partial differential equation for it's capable of offering smooth curve[3]. Among all the spline functions, cubic spline interpolation function is used most[4].

Within $[a,b]$, $n+1$ nodal points are given and expressed as:

$$a = x_0 < x_1 < \cdots < x_n = b \tag{1}$$

The achieved cubic spline interpolation function $\varphi(x)$ through interpolation fitting should obey the follow conditions:

(1) The power frequency of the expression $\varphi_i(x)$ within the subinterval $[x_i, x_{i+1}]$ is no more higher than 3; (2) $\varphi_i(x) = y_i$; (3) $\varphi(x)$ has continuous second derivative within the whole interval $[a,b]$.

By applying segment interpolation function constitution method, the cubic spline interpolation function has been achieved within the interval $[x_i, x_{i+1}]$ and expressed as:

$$\varphi_i(x) = \frac{y_i}{h_i^3}[2(x-x_i)+h_i](x_{i+1}-x)^2 + \frac{y_{i+1}}{h_i^3}[2(x_{i+1}-x)+h_i](x-x_i)^2$$

$$+\frac{m_i}{h_i^2}\bullet(x-x_i)(x_{i+1}-x)^2 - \frac{m_{i+1}}{h_i^2}\bullet(x_{i+1}-x)(x-x_i)^2 \qquad (2)$$

Where $h_i = x_{i+1} - x_i$

The necessary and sufficient condition which the function has continuous second derivative within the interval $[a,b]$ is defined as:

$$\varphi_{i-1}''(x_i) = \varphi_i''(x_i) \quad i = 1, 2, \cdots, n-1 \qquad (3)$$

Solving for $m_i (i = 1, 2...n-1)$ through (2) and (3), the parameters can be determined easily. But, to confirm the parameters $m_i (i = 1, 2...n+1)$, two additional conditions should be added. According to the transformation trend of the testing curve of missile accelerometer's performance parameters, the slope on ends of the curve are approximately assumed to be zero, which gives:

$$\varphi''(a) = \varphi''(b) = 0 \qquad (4)$$

Substituting (4) into (2) and solving for $m_i (i = 1, 2...n)$, the fitting parameter $\varphi(x)$ can be confirmed.

4   Preprocessing of the Testing Parameters

Following the history data, taking the coefficient of partial value (k_0) as example, 9 groups of effective data have been computed totally which are shown in table 1[5].

Table 1. Testing values before interpolation.

Date	05.08	06.02	06.06	06.09	07.01	07.08	07.12	08.04	08.07
K$_0$(mA)	-0.5351	-0.5531	-0.5098	-0.5502	-0.5388	-0.5192	-0.5086	-0.5375	-0.5006

To construct the time series model, the testing data must be preprocessed and taken 3 months as a unit for the testing time's non-uniform. Aiming at the 3 months stable period of computing result, and taking the gained performance index value as basic points, the cubic spline interpolation has been chosen to enlarge the sample size. If the testing has been carried out for no more than one time, their average value is taken as basic point. Through this way, 3 interpolation points which are illustrated in table 2 have been achieved.

Table 2. Testing values after the first interpolation.

Date	1	2	3	4	5	6	7	8	9	10	11	12
K_0(mA)	-0.535	-0.547	-0.553	-0.510	-0.550	-0.539	-0.542	-0.519	-0.509	-0.508	-0.538	-0.501

After the first interpolation, the time series of the performance parameters within 3 months stable period has been achieved while the sample size is quite small. Obviously, it's hard to gain a correct and complete model by using a small sample with few elements, and forecast the change trend of the parameters correctly. Thus, the value of the model is not very high. In order to avoid this phenomenon, the twice modified interpolation method has been adopted to enlarge the sample size. By this way, 2 interpolation points have been interpolated between the every two adjacent basic points, and a new second interpolation sequence whose sample capacity is 34 has been achieved. The interpolated data is illustrated in fig.1, where "*" stands for basic point and "+" stands for interpolation point. From fig.1 we can see, twice modified interpolation can make the achieved expression curve more smoothly, and the constructed time series model more stable.

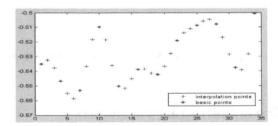

Fig. 1. The result of testing values after twice interpolation

5   Stationarity Detecting and Processing of Processed Parameters

A new time series model of accelerometer's performance parameter has been constructed by the interpolated data. In order to forecast the change trend of the performance parameter, stationarity detecting and processing must have been done to the processed testing data.

5.1   Stationarity Detecting

For the processing object of the time series algorithm is a stationarity data series, so the stationarity detecting is a necessary precondition to make corresponding handling for the dynamic data constituted by performance parameters. If the series which is constructed by all previous testing data of accelerometer is not a stationarity series, the accuracy of the calculated result may be very low.

To detect the stationarity of the performance parameters series, the run inspection method is applied[6]. In allusion to the interpolated data series x_t, supposed its mean value is $\bar{x}$, the values which are smaller than $\bar{x}$ are recorded as "-", and the others are recorded as "+". According to the above result, $\bar{x} = -0.5304$. Then, a symbol sequence corresponding to the original series is achieved and expressed as:

$$---------+++----------+++++++++++--++$$

From the above the symbol we can see that the run number of the performance parameters series is 6.

Supposed the length of the series is N, which is expressed as $N = N_1 + N_2$.

Where N_1, N_2 is used to denote the number which is larger or smaller than the mean value individually.

When N_1, N_2 are all less than 15, the run number submits to r distribution:

$$E(r) = \frac{2N_1N_2}{N} + 1 \qquad D(r) = \frac{2N_1N_2(2N_1N_2 - N)}{N^2(N-1)} + 1 \qquad (5)$$

On the contrary, if N_1, N_2 larger than 15, the sampling distribution of the statistic Z is submitted to the Gaussian distribution progressively, then the expression is given by[7]

$$Z = \frac{r - E(r)}{\sqrt{D(r)}} \propto N(0,1) \qquad (6)$$

Consequently, the discriminant rule of sample series stationarity can be described as: at the given conspicuous level (α), for the statistics r and Z, if $r_L < r < r_U$, or $|Z| < 1.96$, the sample series is considered as stationarity series, contrarily, the series is non-stationarity series.

In the missile accelerometer testing system, the sample series length of all previous testing data is 34, and $N_1 = 15$, $N_2 = 19$, substitute them into (5) and yields:

$$E(r) = 17.7647, \quad D(r) = 9.0088$$

Then

$$|Z| = \left| \frac{r - E(r)}{\sqrt{D(r)}} \right| = 3.9216 > 1.96$$

So, the series constituted by the missile accelerometer performance parameters is non-stationarity series. In other words, the stationarity processing must be done to the all previous testing data.

5.2 Stationarity Processing

Through stationarity detecting, the constructed data series is non-stationarity series, so the stationarity processing method is needed to make the series tending to stationary. In real engineering application, differential operational method is adopted to eliminate the tendency of the time series and yield stationarity sequence[8].

To data series $\{x_i\}$, supposed the achieved series is $\{\nabla x_i\}$ after the first difference, which gives

$$\nabla x_i = x_i - x_{i-1}$$

The item of the secondary difference can be expressed as:

$$\nabla^2 x_i = \nabla x_i - \nabla x_{i-1}$$

So, in common principle, high order difference can be expressed as:

$$\nabla^n x_i = \nabla^{n-1} x_i - \nabla^{n-1} x_{i-1}$$

When proceeding standardized processing to the time series which is composed by missile accelerometer performance parameters, difference equation is applied. Firstly, directed toward the linear trend of the series, first difference is used to eliminate the linear trend usually. Then, if the parameters sequence has quadratic trend, the special series can be changed into stationarity sequence through secondary difference. In other words, to a series which has multiple polynomials, it can be changed into

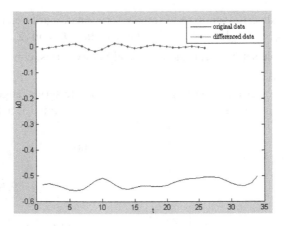

Fig. 2. The comparison curve between original and differenced data

stationarity sequence through   degree difference. But in accelerometer performance analysis system, to the non-stationarity sequence, it's hard to know the tendency item clearly. Thereafter, first differential operation is applied in the first instance before detecting its stationarity. If the processed series is still a non-stationarity series, another differential operation is needed until a relative stationarity series is achieved. From the trial result, after twice differential operation, the accelerometer static model coefficient has formed a approximately stationarity series which is detected by the run inspection method. The comparison analysis result of the original data and differenced data is shown in fig.2. Up to now, the processed series can be used to forecast the change trend of the accelerometer's performance parameters.

6 Conclusions

In this paper, for small sample data, the modeling and expansion analysis is carried out through cubic spline interpolation in the background of the missile accelerometer performance parameters forecasting. The result shows that through applying spline interpolation method, we can not only change the all previous testing data into a basic time series, but also can enlarge the sample size. Thus, the inaccuracy and uncompleted problem of building model for small sample can be solved successfully. In addition, the stationarity detection and processing have been executed to the interpolated testing data through difference method, and the reasonable gist for forecasting the change trend of accelerometer's performance parameter has been offered.

References

1. Broersen, P.M.T., Waele, S.: Frequency selective time series analysis. In: IEEE Instrumentation and Measurement Technology Conference, pp. 775–780. IM TC, Anchorage (2002)
2. Li, Q., Wang, N., Yi, D.: Numerical Analysis. Huazhong University of Science and Technology publishing company, Wu Han (1996)
3. Wang, J.: Research on Spline Interpolation Poisedness and Interpolation Approach Problems. Dalian university of technology, Dalian (2004)
4. Wang, R.: Digital Approximation. Higher Education Press, Beijing (1999)
5. Cheng, H., Zhao, Y.: Research on Performance Analyzing and Modeling of Quartzose Flexible Accelerometer. In: Proceedings of 9th International Conference on Electronic Measurement & Instruments
6. Box, G.E.P., Jenkins, G.M., Reinsel, G.C.: Time Series Analysis Forecasting and Control. China Statistic Publishing Company, Beijing (1997)
7. Yan, K.: Design and Implementation of the Algorithms on Time Series Analysis and Forecast. Beijing University of posts and telecommunications, Beijing (2008)
8. Chen, Y., Ye, Y., Sun, B.: Application of model prediction technology to bridge health monitoring. Journal of Zhejiang University (Engineering Science) 42(1) (2008)

Improving the Quality of the Personalized Collaborative Filtering Recommendation Approach Employing Folksonomy Method

Jianhua Liu

Zhejiang Business Technology Institute, Ningbo 315012, China
liujianhuaei@163.com

Abstract. Popular E-commerce systems such as Amazon and eBay require customer personal data to be stored on their servers for serving these customers with personalized recommendations. Collaborative filtering is one of the most successful technologies for building personalized recommender systems, and is extensively used in many personalized E-commerce systems. However, existing collaborative filtering algorithms have been suffering from data sparsity and scalability problems which lead to inaccuracy of recommendation. With the increase of customers and commodities, the customer rating data is extremely sparse, which leads to the low efficient collaborative filtering E-commerce recommendation system. To address these issues, many approaches of processing no-rated items in collaborative filtering recommendation algorithm have been proposed. In order to improve the quality of the personalized collaborative filtering recommendation, an item-based collaborative filtering approach employing folksonomy method is given. The approach uses the folksonomy technology to fill the no-rated items in the collaborative filtering recommendation algorithm.

Keywords: collaborative filtering, personalized recommendation, folksonomy, sparsity.

1 Introduction

Network is a huge source of information, and it contains rich and dynamic hyperlink information. Web pages are used to access and use information, and data mining provides a wealth of resources. Web mining is the use of data mining technology or services from Web documents to automatically discover and extract knowledge.

With the exponential growth of Internet knowledge, knowledge services more and more people's attention, knowledge, services, research and exploration has become a hot spot. In the service mode and service characteristics, knowledge services, greater emphasis on user-oriented goal-driven, knowledge-oriented content services, emphasizing problem solving and knowledge of the user value, so the service provided is professional and personalized service is independent and innovative services, is the dynamic integration of services. However, the traditional knowledge of services in the process of user interaction there is a "lack of service semantics" and

D. Jin and S. Lin (Eds.): Advances in ECWAC, Vol. 2, AISC 149, pp. 597–602.
springerlink.com   © Springer-Verlag Berlin Heidelberg 2012

"personalized service missing" problem, with many of the models, the lack of specific user's personalized service.

Quality of knowledge classification system and the user's personalized information services, knowledge modeling studies exist to solve the "lack of service semantics" and "lack of personalized service". Theory and practice in this regard research have yielded results. Intelligent knowledge-oriented service system is currently recommended that has been extensively studied and used in e-commerce, distance education, Web site development, information retrieval and other areas. Collaborative filtering is one of the most successful technologies for building personalized recommender systems, and is extensively used in many personalized E-commerce systems. It has also seen a number of personalized recommendation prototype system, some of the information service database products have also introduced a simple recommendation services. Although the formation of a more extensive study of the accumulation, but there are many issues to be further studied and resolved.

2   Folksonomy

The Web structure mining gives the label of the hyperlink structure of the network showing some new features. It believes that technology can be classified using Folksonomy processing in web page content and structure more closely together, combined with the Page rank algorithm. The evaluation to measure page relevance can overcome two major drawbacks of traditional structure mining, and the algorithm completely ignores the emphasis on the old website content.

Folksonomy is to define a set of labels description, and eventually can be used according to label the frequency of use of such information as the high-frequency tags class name of a method for the classification of network information. It is a bottom-up informal classification, which allows users to freely choose their own keywords used to classify the information.

Compared with the traditional classification, this classification method is characterized in that it does not use pre-established information taxonomy and vocabulary. But the user based on individual usage, the word freedom to customize the objects for digital resources annotation and classification method is closer to the user. It is easy to accept the freedom and flexibility of its outstanding advantages.

3   The Item-Based Collaborative Filtering Employing Folksonomy Method

3.1   TFxIDF Tags Matrix

Suppose the set of users U = {u1, u2, ⋯, ui, ⋯, um}, set of items I = {i1, i2, ⋯, ij, ⋯, in}, the set of tags T = {t1, t2, ⋯ tk, ⋯, tl}. Rating matrix R = (rij) m × n.

3.2 Calculating TFxIDF

The frequency Fji(k) is calculated by following formula.

$$Fji(k) = \frac{Cji(k)}{Cju(A)}$$

The inverse frequency IFji(k) is calculated by following formula.

$$IFji = \log \frac{C^I(A)}{C^I(k)}$$

Then we can get the matrix as following.

$$M = \begin{bmatrix} t(1,1) & t(1,2) & \cdots & t(1,q) \\ t(2,1) & t(2,2) & \cdots & t(2,q) \\ \vdots & \vdots & \ddots & \vdots \\ t(p,1) & t(p,2) & \cdots & t(p,q) \end{bmatrix}$$

3.3 Similarity of Tags

Using the Pearson correlation coefficient, it based on the label weight matrix to calculate the users, the similarity between items. Is different from the score matrix based on the similarity score is calculated, based on the weight label similarity matrix calculated as label similarity.

$$sim1(x, y) = \frac{\sum_{k \in Txy} (ti(k, x) - A(x))(ti(k, y) - A(y))}{\sqrt{\sum_{k \in Txy} (ti(k, x) - A(x))^2} \sqrt{\sum_{k \in Txy} (ti(k, y) - A(y))^2}}$$

3.4 Prediction

The prediction score of target user Ui on the target item Ij is calculating by the following formula.

$$Pij = \frac{\sum R iy * sim1(j, y)}{\sum sim1(j, y)}$$

3.5 Dense Matrix of User-Item Ratings

In the recommended system to the user transaction database contains a collection of m users and n items of the collection. An m*n order matrix, is as shown in Table 1.

The m rows represent m user. The n columns represent the n items. If the score is empty, then use the above formula to get the pre-score values.

Table 1. The dense user-item ratings matrix

	i_1	$\cdots$	i_k	$\cdots$	i_n
u_1	$I_{1,1}$	$\cdots$	$I_{1,k}$	$\cdots$	$I_{1,n}$
$\cdots$	$\cdots$	$\cdots$	$\cdots$	$\cdots$	$\cdots$
u_j	$I_{j,1}$	$\cdots$	$I_{j,k}$	$\cdots$	$I_{j,n}$
$\cdots$	$\cdots$	$\cdots$	$\cdots$	$\cdots$	$\cdots$
u_m	$I_{m,1}$	$\cdots$	$I_{m,k}$	$\cdots$	$I_{m,n}$

3.6 Similarity of Items

Standard cosine similarity measure: This method is by measuring the cosine of the angle between two vectors for the size, calculated as

$$sim2(i_p,i_q) = \cos(\overline{i_p},\overline{i_q}) = \frac{\sum_{k=1}^{m} I_{kp}I_{kq}}{\sqrt{\sum_{k=1}^{m}(I_{kp})^2}\sqrt{\sum_{k=1}^{m}(I_{kq})^2}}$$

Relevant similarity: Select the project ip and iq are the line score, score set to record the user U ', according to Pearson's correlation coefficient to measure the similarity between projects in order to ensure fairness, the user similarity measure can also set U 'in the appropriate.

$$sim2(i_p,i_q) = \frac{\sum_{u_i \in U'}(I_{kp}-\overline{I_p})(I_{kq}-\overline{I_q})}{\sqrt{\sum_{u_i \in U'}(I_{kp}-\overline{I_p})^2}\sqrt{\sum_{u_i \in U'}(I_{kq}-\overline{I_q})^2}}$$

3.7 Recommendation

For a target user ua, recommended system has two tasks. For users not browsing or ua score predict the user specified project it rate it; and made available to users interested in the user most likely to recommend the N items directly to the user to find a collection of items of size N.

$$I_r \subseteq I, I_r \cap I_{u_a} = \varnothing$$

Then calculate the similarity between items, the item-based collaborative filtering recommendation algorithm for the specified item it to find the most similar to the k neighbors, and their predicted weight.

$$I_{at} = \frac{\sum_{j=1}^{k}(I_{aj}sim2(i_j,i_t))}{\sum_{j=1}^{k}sim2(i_j,i_t)}$$

4 Conclusions

Popular electronic commerce systems such as Amazon and eBay require customer personal data to be stored on their servers for serving these customers with personalized recommendations. Collaborative filtering is one of the most successful technologies for building personalized recommender systems, and is extensively used in many personalized electronic commerce systems. However, existing collaborative filtering algorithms have been suffering from data sparsity and scalability problems which lead to inaccuracy of recommendation. With the increase of customers and commodities, the customer rating data is extremely sparse, which leads to the low efficient collaborative filtering electronic commerce recommendation system. To address these issues, many approaches of processing no-rated items in collaborative filtering recommendation algorithm have been proposed.

In this paper, in order to improve the quality of the personalized collaborative filtering recommendation, we give an item-based collaborative filtering approach employing folksonomy method. The approach uses the folksonomy technology to fill the no-rated items in the collaborative filtering recommendation algorithm.

References

1. Gong, S.: A Personalized Recommendation Algorithm on Integration of Item Semantic Similarity and Item Rating Similarity. Journal of Computers 6(5), 1047–1054 (2011)
2. Huang, Q.-H., Ouyang, W.-M.: Fuzzy collaborative filtering with multiple agents. Journal of Shanghai University (English Edition) 11(3), 290–295 (2007)
3. Gong, S.: Privacy-preserving Collaborative Filtering based on Randomized Perturbation Techniques and Secure Multiparty Computation. International Journal of Advancements in Computing Technology 3(4), 89–99 (2011)
4. Dotsika, F.: Uniting formal and informal descriptive power: Reconciling ontologies with folksonomies. International Journal of Information Management (29), 407–415 (2009)
5. Kome, S.H.: Hierarchical subject relationships in folksonomies. Master's thesis, University of North Carolina at Chapel Hill, Chapel Hill, NC, USA, vol. (11) (2005)
6. de Meo, P., Quattrone, G., Ursino, D.: Exploitation of semantic relationships and hierarchical data structures to support a user in his annotation and browsing activities in folksonomies. Information System 34, 511–535 (2009)

7. Chen, W.-H., Cai, Y., Leung, H.-F., Li, Q.: Generating ontologies with basic level concepts from folksonomies. In: International Conference on Computational Science, ICCS (2010)
8. Gruber, T.: Ontology of Folksonomy: A Mash-up of Apples and Oranges. AIS SIGSEMIS Bulletin (2), 3–4 (2005)
9. Degemmis, M., Lops, P., Semeraro, G.: A content-collaborative recommender that exploits WordNet-based user profiles for neighborhood formation. User Model User.-Adap. Inter. 17, 217–255 (2007)
10. Yu, L., Liu, L., Li, X.: A hybrid collaborative filtering method for multiple-interests and multiple-content recommendation in E-Commerce. Expert Systems with Applications 28, 67–77 (2005)
11. Herlocker, J.: Understanding and Improving Automated Collaborative Filtering Systems. Ph.D. Thesis, Computer Science Dept., University of Minnesota (2000)

Research on Usage Control Model in Pervasive Computing System

Haiying Wu

Dept. of Communication Engineering, Engineering University of Armed Police Force,
Xi'an ShaanXi, 710086

Abstract. With the applications development of pervasive computing which advances new demands of access control in information security. Use Usage Control model to solve problem of access control in Pervasive Computing system and provide a model of Usage Control in Pervasive Computing system (UCON_PC) and context information is added in the model, simultaneously delegate subject can delegate rights to the entrusted subject, and the model satisfies the demands of access control in pervasive computing system.

Keywords: Information Security, Pervasive Computing, Usage Control, Context.

1 Introduction

With development of mobile devices, intelligent device, wireless communication network technology and embedded system, a user needs obtain computing resources anywhere, anytime and transparently. Physical space combines with information space more and more closely[1]. Hence pervasive computing was born and is considered to be the mainstream model for future computing system, which provides user digital information services anytime, anywhere, and will bring a profound impact to the development and application of computer technology.

Characteristics of pervasive computing can be drawn [2-3]: (1) Disappearance indicates that pervasive computing system makes user forget his existence. (2) Pervasiveness indicates that large number of computing devices are arranged and embedded into the system. (3) Dynamic indicates that users are always moving in pervasive system and mobile equipments dynamically enter or exit a computing system. (4) Adaptability indicates that computing system can feel and deduce user requirement and can spontaneously provide information services which users need. (5) Receptivity indicates that pervasive computing closely relates with the context and can feel user. These characteristics determine that the pervasive computing system has a more requirement for information security. Access control is an effective method to ensure information security in pervasive computing system. Traditional access control and RBAC model is not suitable for pervasive computing system.

D. Jin and S. Lin (Eds.): Advances in ECWAC, Vol. 2, AISC 149, pp. 603–608.

2   Access Control Features in Pervasive Computing System

2.1   Requirements for Access Control in Pervasive Computing System

Pervasive system need consider continuity because there are many long-term and short-term processes of access, and authorization decision can be processed before or during access. Literature [4-5] proposed that context computing is an important feature of pervasive computing so that authorization decisions need contexts, such as time, space, usage state of resources et al. Pervasive computing system need mutability because it is open system and the attributes are changed by subject behaviors, and the changes will affect the judgment of authorization this time or next time. Through various access control models are studied, Usage Control (UCON) model is most suitable for solving the problem of access control in pervasive computing system.

2.2   Characteristics of UCON Model

UCON model is considered to be the next generation access control model. The literature [7] indicates that the UCON model includes three basic elements: subject(S), object(O), right(R) and the other three factors which are used for Usage Decision: authorization (A), obligation(B), and condition(C). The UCON model has mutability and continuity, and these two characteristics are different from other access control model[8].

3   UCON Model in Pervasive Computing System

3.1   UCON_PC Model

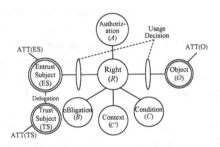

Fig. 1. The structure of UCON_PC Model. In pervasive computing system, the UCON_PC model adds context information and delegation of authority in UCON_PC which is show as Fig.1. The rights can be delegated to the trusted subject by delegate subject.

Subject and Subject Attribute (ATT(S)). Subject is an active entity which advances access operations. Subject has some attributes and some rights to operate objects. Subject attributes include status, security levels, membership, etc. These attributes are used for authorization decisions. Subject becomes entrusted subject (ES) who executes delegation. Entrusted subject attribute (ATT(ES)) is entrusted subject correlative attribute.

Trust Subject(TS). Entrusted subject judges trust subject attribute (ATT(TS)) to delegate his some rights to trust subject[9]. ATT(TS) is trusted subject correlative attribute.

Object and Object Attribute (ATT(O)). Object is operational objectives of subject and object has attributes which include security labels, ownerships, classes, and so on. These attributes can be used for authorization decision.

Right. Rights are privileges which consist of a set of usage functions that make a subject access objects. Subjects access objects according to rules of rights.

Authorization. Authorizations are functional predicates that have to be evaluated for usage decision and return whether the subject is allowed to perform the requested on the object. Authorizations evaluate subject attributes, object attributes, and requested rights together with a set of authorizations rules for usage decision.

oBligation. Obligation is that subject must finish action set before or during access. As long as there is an object which can not be satisfied by the subject then the subject have no right to access the object.

Condition. Condition evaluates current environmental of system status to check whether relevant requirements are satisfied or not before or during access.

Context(C'). Context is an information which is used to denote as entity state, such as time, location, and environment. Contexts are used to assign rights.

Usage Decision. (A,B,C,C') denotes as a set of usage decision factors. Judgment of (A,B,C,C') is satisfied which denotes that right r can be granted to subject s.

Continuity. Continuity is that the authorization can be made before, during or after the access. '*pre*','*on*' and '*post*' denote that Usage Decision is processed before, during and after the access.

Mutability. Mutability is that the attributes of subject or object can be changed. No update is denoted as '0'. Attributes update before usage which is denoted as '1'. Attributes update during usage which is denoted as '2'. Attributes update after usage which is denoted as '3'.

3.2 UCON_PC System

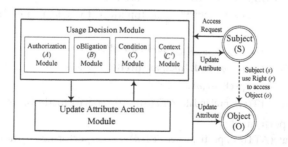

Fig. 2. Functional Module of UCON_PC System. The system includes Usage Decision Module and Update Attribute Action Module. Usage Decision Module includes Authorization Module, Obligation Module, Condition Module and Context Information Module.

Update Attribute Action Module Function
There are three update attribute actions.
preUpdate(*attribute*) : ATT(S) or ATT(O) is updated by the system before access or after denying an access.
onUpdate(*attribute*) : ATT(S) or ATT(O) is updated by the system during access phase.
postUpdate(*attribute*) : ATT(S) or ATT(O) is updated by the system after access.

Usage Decision Module Function

Authorization Module Function
$preA_0$: - allowed(s, r, o) $\Rightarrow$ preA(ATT(s),ATT(o), r) .
$preA_1$: - preUpdate(ATT(s)), preUpdate(ATT(o)) ;
    - allowed(s, r, o) $\Rightarrow$ preA(ATT(s),ATT(o), r) .
$preA_3$: - allowed(s, r, o) $\Rightarrow$ preA(ATT(s),ATT(o), r) ;
    - postUpdate(ATT(s)), postUpdate(ATT(o)) .
onA_0 : - **allowed**(s,r,o) $\Rightarrow$ *true* ;
    - stopped(s, r, o) $\Leftarrow \neg$ onA(ATT(s),ATT(o), r) .
onA_1 : preUpdate(ATT(s)), preUpdate(ATT(o)) is performed before UCON_PC$_{onA_0}$ Model is performed.
onA_2 : onUpdate(ATT(s)), onUpdate(ATT(o)) is performed during UCON_PC$_{onA_0}$ Model is performed.
onA_3 : postUpdate(ATT(s)), postUpdate(ATT(o)) is performed after UCON_PC$_{onA_0}$ Model is performed.

Obligation Module Function
$preB_0$: - preB is pre-obligation predicate set, preOBL is pre-obligation element set;
    - **preOBL $\subseteq$ OBS $\times$ OBO $\times$ OB** , OBS is obligation of subject, OBO is obligation of object, OB is obligation;
    - **preFulfilled:OBS $\times$ OBO $\times$ OB** $\rightarrow$ {*true, false*} ;
    - **getPreOBL:** $S \times O \times R \rightarrow 2^{preOBL}$, a function to select pre-obligations for a requested usage;
    - **preB**(s,r,o)= $\wedge_{(obs_i,obo_i,ob_i) \in getPreOBL(s,r,o)}$**preFulfilled**($obs_i,obo_i,ob_i$) ;
    - If **getPreOBL**(s,r,o)= ϕ is defined as **preB**(s,r,o)=*true* ;
    - **allowed**(s,r,o) $\Rightarrow$ **preB**(s,r,o) .
$preB_1$: preUpdate(ATT(s)), preUpdate(ATT(o)) is performed before UCON_PC$_{preB_0}$ Model is performed.
$preB_3$: postUpdate(ATT(s)), postUpdate(ATT(o)) is performed after UCON_PC$_{preB_0}$ Model is performed.
onB_0 : - A set of time is denoted as Time;
    - onB is ongoing-obligations predicates;
    - **onOBL $\subseteq$ OBS $\times$ OBO $\times$ OB $\times$ Time** ;

- **onFulfilled: OBS** $\times$ **OBO** $\times$ **OB** $\times$ **Time** $\rightarrow$ {*true, false*} ;
- **getOnOBL:** $S \times O \times R \rightarrow 2^{onOBL}$, a function to select ongoing-obligations for a requested usage;
- **onB**$(s,r,o)= \wedge_{(obs_i,obo_i,ob_i,time_i) \in getOnOBL(s,r,o)}$ **onFulfilled**$(obs_i,obo_i,\ ob_i,time_i)$
- If **getOnOBL**$(s,r,o)= \phi$ is defined as **onB**$(s,r,o)=true$;
- **allowed**$(s,r,o) \Rightarrow true$;
- **stopped**$(s,r,o) \Leftarrow \neg$**onB**(s,r,o) ;

onB_1 : preUpdate(ATT(s)), preUpdate(ATT(o)) is performed before UCON_PC$_{onB_0}$ Model is performed.

onB_2 : onUpdate(ATT(s)), onUpdate(ATT(o)) is performed during UCON_PC$_{onB_0}$ Model is performed.

onB_3 : postUpdate(ATT(s)), postUpdate(ATT(o)) is performed after UCON_PC$_{onB_0}$ Model is performed.

Condition Module Function

$preC_0$: - preCON is pre-conditions predicate set;
- getPreCON: $S \times O \times R \rightarrow 2^{preCON}$, a function to select pre-conditions for a requested usage;
- preConChecked:preCON $\rightarrow$ {*true, false*} ;
- preC$(s,r,o)= \wedge_{preCon_i \in getPreCON(s,r,o)}$ preConChecked$(preCON_i)$;
- allowed$(s,r,o) \Rightarrow$ preC(s,r,o) .

onC_0 : - onCON is ongoing-conditons predicate set;
- getOnCON: $S \times O \times R \rightarrow 2^{onCON}$, a function to select ongoing-conditions for a requested usage;
- onConChecked : onCON $\rightarrow$ {*true, false*} ;
- onC$(s,r,o)= \wedge_{onCon_i \in getOnCON(s,r,o)}$ onConChecked$(onCON_i)$;
- **allowed**$(s,r,o) \Rightarrow true$;
- **stopped**$(s,r,o) \Leftarrow \neg$**onC**(s,r,o) ;

Context Information Module Function

$preC'_0$: - preCONTEXT is context information set which is judged before usage;
- **getPreCONTEXT:** $S \times O \times R \rightarrow 2^{preCONTEXT}$, a function selects elements in preCONTEXT for usage decision;
- preContextChecked : preCONTEXT $\rightarrow$ {*true, false*} ;
- preC'$(s,o,r)= \wedge_{preContext_i \in getPreCONTEXT(s,o,r)}$**preContextChecked**$(preContext_i)$;
- If preC'$(s,o,r) \Rightarrow true$ then allowed$(s,o,r) \Rightarrow true$.

onC'_0 : - onCONTEXT is context information set which is judged during usage;
- getOnCONTEXT: $S \times O \times R \rightarrow 2^{onCONTEXT}$, a function selects elements in **onCONTEXT** for usage decision;
- onContextChecked : onCONTEXT $\rightarrow$ {*true, false*} ;

- $\mathrm{onC}'(s,o,r)= \bigwedge_{onContext_i \in getOnCONTEXT(s,o,r)} \mathbf{onContextChecked}(onContext_i)$;
- If $\mathrm{onC}'(s,o,r) \Rightarrow true$ then allowed$(s,o,r) \Rightarrow true$;
- $\mathbf{stopped}(s,o,r) \Leftarrow \neg\mathrm{onC}'(s,o,r)$.

4   Summarize

This paper advances UCON_PC model and system structure. To satisfy the pervasive environment, this model adds judgment of context information in UCON model besides authorization, obligation and condition. But the paper just studied the structure of UCON_PC model, system structure and not gives Formal Description of the model so that the model need further improve. User sometimes need entrust his rights to other user, so that entrusted subject can award his rights to trust subject.

Acknowledgment. This paper is supported by the Basic Research Fund Project of Engineering College of Armed Police Force (No.WJY201021).

References

1. He, J., Chen, R., Ma, K.: Self-adaptive Middleware in Ubiquitous Computing Environments. Computer Science 36(7), 103–106 (2009)
2. Xu, G., Shi, Y.: Pervasive/Ubiquitoas Computing. Chinese Journal of Computers 26(9), 1042–1050 (2003)
3. Xin, Y., Luo, C.: Context-based RBAC model in pervasive computing. Computer Engineering and Design 31(8), 1693–1697 (2010)
4. Guo, Y., Li, R.: Research on access control for pervasive computing. Joural of Central China Normal University(Nat. Sci.) 40(4), 504–506 (2006)
5. Dou, W., Wang, X., Zhang, L.: New Fuzzy Role-based Access Control Model for Ubiquitous Computing. Computer Science 37(9), 63–67 (2010)
6. Park, J., Sandhu, R.: The UCONABC usage control model. ACM Transactions on Information and Systems Security 7(1), 128–174 (2004)
7. Peng, X., Yang, P.: Survey of usage access control model. Application Research of Computers 24(9), 121–123 (2007)
8. Hu, Z., Jin, R., Yu, W.: Logical definition of usage control authorization model. Journal of Computer Application 29(9), 2360–2374 (2009)
9. Cui, Y., Hong, F.: Dynamic Context-aware Usage Control-Based Grid. Computer Science 35(2), 37–41 (2008)

A Personalized Information Filtering Method Based on Simple Bayesian Classifier

Jianhua Liu

Zhejiang Business Technology Institute, Ningbo 315012, China
liujianhuaei@163.com

Abstract. Personalized recommender systems can provide important services in a digital environment, as verified by its commercial success in book, movie, and music industries. Collaborative filtering is one of the most successful technologies in recommender systems, which predicts unknown ratings by analyzing the known ratings, and widely used in many personalized recommender areas, such as e-commerce, digital library and so on. Traditional collaborative filtering recommendation algorithm is one of the methods to solve the information overloading problem. However, with the site structure, content of the complexity and increasing number of users, there are there urgent problems in this algorithm namely data sparse, cold start and scalability. To address the sparsity problem, a personalized information filtering recommendation based on simple Bayesian classifier method is described. The algorithm used simple Bayesian classifier to smooth the rating of unrated items. It can alleviate the sparsity and improve the accurate degree of searching nearest neighbors.

Keywords: personalized, information filtering, simple Bayesian classifier, sparsity.

1 Introduction

With a wide range of products online, the customer's behavior is constantly changing. Behavior will change as the preferences constantly change. Once the recommended model is formed, the model parameters can not be any changes, thus leading to new types of data that is adding, only to re-learn that model. In the Internet and e-commerce development, e-commerce recommendation system is widely used in e-commerce site. It is to provide customers with product information and recommendations to help the user decide what goods to buy.

The basic idea of collaborative filtering algorithm with the goal of the project by reference to the neighbor set of similar high score is to predict the target project score, resulting in a final recommendation. With the changes in site structure, content, and increased complexity as more users, methods based on collaborative filtering face the following problem: how to improve the scalability of the collaborative filtering algorithm and how to improve the quality of collaborative filtering recommendation algorithms.

D. Jin and S. Lin (Eds.): Advances in ECWAC, Vol. 2, AISC 149, pp. 609–614.
springerlink.com   © Springer-Verlag Berlin Heidelberg 2012

Collaborative filtering algorithm based on user-items generally score matrix, when the matrix is very sparse, the algorithm is difficult to find similar users or items, so that the recommended quality has been greatly restricted. On the other hand, collaborative filtering does not care about products and user information; the information can effectively improve the accuracy of collaborative filtering recommendation.

In this paper, we give a personalized information filtering recommendation based on simple Bayesian classifier method. The approach used simple Bayesian classifier to smooth the rating of unrated items. It can alleviate the sparsity and improve the accurate degree of searching nearest neighbors.

2   Smoothing Based on Simple Bayesian

2.1   Simple Bayesian Classifier

Let each data sample with an n-dimensional feature vector $Y = [y1, y2, \ldots , yn]$ that were described in the n attributes A1, A2,, An on value. Assume there are m classes, respectively C1, C2, ... , Cm, said the data given a sample of unknown.

Y that is, no class label, according to Bayes' theorem:

$$P (Ci \mid Y) = P (Y \mid Ci) P (Ci) / P (Y)$$

Since P (Y) is constant for all classes, to maximize the posterior probability $P (C i \mid Y)$ can be converted to maximize the prior probability P (Y | C i) P (Ci). If the training data set has many attributes and tuples, computing P (Y | Ci) the cost may be very large. For the realization of P (Y | Ci) to estimate the effective basic Bayesian classifier usually assume that each class is independent of each other, that is to take each of the attributes are mutually independent.

This are:

$$p(Y \mid C_i) = \prod_{k=1}^{n} p(y_k \mid C_i)$$

Prior probability P (y1 | C i), P (y2 | Ci),, P (yn | Ci) can be obtained from the training data set. Under this method, a sample of an unknown type Y, Y can be calculated separately for each category Ci is the probability P (Y | Ci) P (Ci), then select one of the largest class probability as its categories, namely:

$$V_{NB}(Y) = \arg_{C_i \in c} maxP(C_i) \prod_{k=1}^{n} P(y_k \mid C_i)$$

For each attribute Ak, if Ak is a discrete attribute, then P (yk | Ci) = sik / si, which is the property of Ak sik in the class Ci with xi the number of training samples, si is the number of training samples in the Ci; If Ak is a continuous property, each row of the discretization, and then replaced with the corresponding discrete interval continuous attribute values, by calculating the class Ci yk training records into the corresponding interval to calculate the ratio of the conditional probability P (yk | Ci .).

2.2 Smoothing

Rating matrix for user missing items each rated Rij, check the score corresponding to the user attributes and project attributes and attribute values to be set

$$A = \{ Au1, Au2, ..., Aun \}.$$

The A may not be unique, there are two cases. The A only one combination, with a direct query from the classification match score value as Ri j prediction score. The ∋ A variety of combinations exist, because a property value exists d attribute the situation to take multiple values.

For example, a film is more than one category, when the class as a characteristic property, as the category attribute value is not unique, such that a is not unique, characterized by a combination of attribute values for this case there are multiple combinations, respectively, from the classification table in the query for each combination of T category and the corresponding probability score , and then choose the highest probability score prediction score as Rij.

3 Prediction Employing Collaborative Filtering

In collaborative filtering recommendation for the sparse user - information item matrix, it is usually considered from the perspective of the line that the calculation of the similarity between multiple users, while ignoring the relationship between the columns, that the similarity between the information items of. In fact, it would be difficult to avoid the sparse matrix problems. Therefore gives an algorithm, the algorithm is based on information items. The algorithm is mainly to consider the user - information item matrix information item-item the relationship between the use of these relationships for some unknown User-Item matrix values are calculated and predicted. The figure is shown as following.

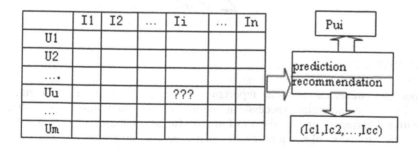

Fig. 1. Collaborative filtering method

The algorithm is described as follows:

(1) for multiple users on the evaluation of some level of information items. Mainly the establishment of the user - information item matrix, the matrix to get as much value.

(2) from the matrix to find a specific user has evaluated the information items, constitutes a set of entries Items = {i1, i2, ⋯, im}, then from the matrix to be evaluated to find those that are both predicted entry has evaluated entries focus on any an entry of the user, constitutes the user set users = {u1, u2, ⋯, un}.

(3) and being selected from the entries focus on prediction of the k most similar entries similar to entry, entry form similar to the set N = {i1, i2, ⋯, ik}, and calculate the similarity between them.

Particular: the first user to calculate the average level of concentration values for each user, and then calculate the predicted entry and similar items are similarities between {Si, 1, Si, 2, ⋯, Si, k}. As usual the cosine similarity calculation does not take into account the different number of users the size of item-level information, and so was used for similarity calculation adjusted cosine similarity Sx, y. For example:

$$S_{ij} = \frac{\sum\limits_{u \in User} (r_{ui} - \overline{r_u})(r_{uj} - \overline{r_u})}{\sqrt{\sum\limits_{u \in User} (r_{ui} - \overline{r_u})^2} \sqrt{\sum\limits_{u \in User} (r_{uj} - \overline{r_u})^2}}$$

Which, rui is the user's preferences u i evaluation of information value,. Ru is the evaluation of user u on the evaluation of information mean.

(4) predict a user's preferences for a particular item of information, the predictive value of high-information items recommended to the user. Calculate the predicted values of the methods are:

$$P_{ui} = \frac{\sum\limits_{n \in N} (S_{in} \times r_{un})}{\sum\limits_{n \in N} |S_{in}|}$$

4 Conclusions

Recommender systems can provide important services in a digital environment, as verified by its commercial success in book, movie, and music industries. Collaborative filtering is one of the most successful technologies in recommender systems, which predicts unknown ratings by analyzing the known ratings, and widely used in many personalized recommender areas, such as e-commerce, digital library and so on.

Traditional collaborative filtering recommendation algorithm is one of the methods to solve the information overloading problem. However, with the site structure, content of the complexity and increasing number of users, there are there urgent problems in this algorithm namely data sparse, cold start and scalability.

To address the sparsity problem, in this paper, a personalized information filtering recommendation based on simple Bayesian classifier method is described. The algorithm used simple Bayesian classifier to smooth the rating of unrated items. It can alleviate the sparsity and improve the accurate degree of searching nearest neighbors.

References

1. Goldberg, D., Nichols, D., Oki, B.M., Terry, D.: Using collaborative filtering to weave an information tapestry. Communications of the ACM 35(12), 61–70 (1992)
2. Yu, L., Liu, L., Li, X.: A hybrid collaborative filtering method for multiple-interests and multiple-content recommendation in E-Commerce. Expert Systems with Applications 28, 67–77 (2005)
3. Herlocker, J.: Understanding and Improving Automated Collaborative Filtering Systems. Ph.D. Thesis, Computer Science Dept., University of Minnesota (2000)
4. Sarwar, B.M., Karypis, G., Konstan, J.A., Riedl, J.: Analysis of Recommendation Algorithms for E-Commerce. In: Proceedings of the ACM EC 2000 Conference, Minneapolis, MN, pp. 158–167 (2000)
5. Sarwar, B., Karypis, G., Konstan, J., Riedl, J.: Item-Based collaborative filtering recommendation algorithms. In: Proceedings of the 10th International World Wide Web Conference, pp. 285–295 (2001)
6. Jin, R., Si, L.: A bayesian approach toward active learning for collaborativefiltering. In: Proceedings of the 20th Conference on Uncertainty in Artificial Intelligence, pp. 278–285. AUAI Press, Banff (2004)
7. Resnick, P., Iacovou, N., Suchak, M., Bergstrom, P., Riedl, J.: Grouplens: An open architecture for collaborative filtering of netnews. In: Proceedings of the ACM CSCW 1994 Conference on Computer-Supported Cooperative Work, pp. 175–186 (1994)
8. Herlocker, J.: Understanding and Improving Automated Collaborative Filtering Systems. Ph.D. Thesis, Computer Science Dept., University of Minnesota (2000)
9. Vozalis, M.G., Margaritis, K.G.: Using SVD and demographic data for the enhancement of generalized Collaborative Filtering. Information Sciences 177, 3017–3037 (2007)
10. Chien, Y.-H., George, E.I.: A bayesian model for collaborative filtering. In: Proceedings of the Seventh International Workshop on Articial Intelligence and Statistics. MorganKaufmann, San Francisco (1999)
11. Gong, S.: A Personalized Recommendation Algorithm on Integration of Item Semantic Similarity and Item Rating Similarity. Journal of Computers 6(5), 1047–1054 (2011)
12. Chickering, D., Hecherman, D.: Efficient approximations for the marginal likelihood of Bayesian networks with hidden variables. Machine Learning 29(2/3), 181–212 (1997)
13. Breese, J., Hecherman, D., Kadie, C.: Empirical analysis of predictive algorithms for collaborative filtering. In: Proceedings of the 14th Conference on Uncertainty in Artificial Intelligence (UAI 1998), pp. 43– 52 (1998)
14. Shardanand, U., Maes, P.: Social information filtering: Algorithms for automating "Word of Mouth". In: Proceedings of the ACM CHI 1995 Conference on Human Factors in Computing Systems, pp. 210–217 (1995)
15. Lekakos, G., Giaglis, G.M.: Improving the prediction accuracy of recommendation algorithms: Approaches anchored on human factors. Interacting with Computers 18, 410–431 (2006)

16. Gong, S.: Privacy-preserving Collaborative Filtering based on Randomized Perturbation Techniques and Secure Multiparty Computation. International Journal of Advancements in Computing Technology 3(4), 89–99 (2011)
17. Hill, W., Stead, L., Rosenstein, M., Furnas, G.: Recommending and evaluating choices in a virtual community of use. In: Proceedings of the CHI, pp. 194–201 (1995)
18. Karypis, G.: Evaluation of Item-Based Top-N Recommendation Algorithms. Technical Report CS-TR-00-46, Computer Science Dept., University of Minnesota (2000)
19. Dempster, A., Laird, N., Rubin, D.: Maximum likelihood from incomplete data via the EM algorithm. Journal of the Royal Statistical Society B39, 1–38 (1977)

Design and Implementation of the Digital Campus Network

Xiaomei Shang, Zheng Li, and Zhijun Chen

School of Electrical Engineering and Information Science, Hebei University of Science
and Technology, Shijiazhuang 050018, China
shangxm@hebust.edu.cn.

Abstract. With the constant expansion of higher education and the rapid
development of IT technology, the high-speed growth of information
technology can not only promote the progress of society but also promote the
development of the information technology in education. The establishment of
digital campus network is now imperative to be achieved for information
technology application and modernization of the campus. This paper mainly
analyzes the connotation and composition structure of digital campus network
and puts forward the URP (University Resources Plan) as the core solution of
the digital campus network, combined with elaborating how to use URP to
design and implement the digital campus network.

Keywords: digital campus network, design, implementation, URP.

1 Introduction

In 1990s, the professor of United States Clermont University, Keynes Green initiated
and chaired the large project "information campus plan" [1,2]. The concept of
"information campus" gradually evolved into the "digital campus" [3]. For the
problems encountered in practice, this paper mainly analyzes the connotation and
composition structure of digital campus network and puts forward the URP
(University Resources Plan) as the core solution of the digital campus network, based
on the detailed discussion of the structure of the digital campus network. Some
solutions on how to use URP to design and implement the digital campus network are
also presented.

2 Analysis on the Digital Campus

2.1 The Connotation of the Digital Campus

The so-called digital campus is based on the Web and uses the advanced information
technology tools and instruments to achieve all the digital content for the
environment, resources and activities. It can possibly build a digital space on the basis
of the traditional campus, extending the function of traditional campus and ultimately
implementing the comprehensive modernization of education process.

D. Jin and S. Lin (Eds.): Advances in ECWAC, Vol. 2, AISC 149, pp. 615–618.

2.2   The Structure of the Digital Campus

Digital Campus can be divided into the following five parts, as shown in figure 1.

The layer of network infrastructure. Network is the infrastructure of digital campus. It contains the hardware infrastructure of the campus digital information, transmission and storage. It also includes the campus network, database system, cable networks, satellite networks and server system.

The layer of network support services. The basic services of network are software-based information flow about digital campus. It provides basic services of information and supports the functional implementation of network application system, such as e-mail, file transfer, information dissemination, domain name services, unified authentication, workflow and other basic services.

The system of Data storage. This is the core resources of the digital campus and the supporting platform of various application systems. It includes information database of teachers, information database of students, electronic library, curriculum resources library, digital book library, database of research and information, databases of asset, etc.

The system of application of payment. It includes a variety of application systems. It provides the business logic and the digital information services for all types of users (such as campus management users, teachers' users, students' users, etc.). It belongs to the application layer for office automation systems, academic management systems, research management systems, e-learning systems, library resources and management systems.

The system of Information service. It is the main user interface and includes decision support systems, community services systems, information systems and e-commerce systems. It can provide office information, academic information, campus news, office telephone inquiries and other service information. In addition, it can also provide the information support for school's decision-making by integrated data and analysis.

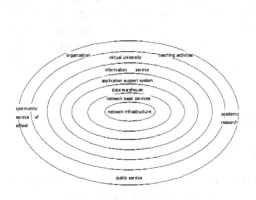

Fig. 1. The levels structure of digital campus

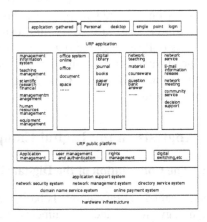

Fig. 2. URP digital campus system structure

The personalized portal. This is a total portal of digital campus. It can provide the corresponding status information and services for all types of users.

2.3   The Solutions of Digital Campus

The modern digital campus should achieve maximum sharing of resources and information. It plays a role in improving the utilization of education resource and the strength of comprehensive application. It is necessary to synthesize IT technology and modern management methods and refers it to the idea and technology of "enterprise resource is planning (ERP)" widely used by enterprises at home and abroad. We integrate the core part of university education resource management with University Resource Planning on the basis of this view. It is known by us as URP. URP can be summarized as a foundation platform, a portal and N application systems shown in figure 2.

URP portal. URP portal is called personal web portal and is the entrance to various resources of the schools that users have an access to. It is the specific performance of online information service. It includes the combination of information about all types of school. The URP portal must have functions as follows:

The user management and user access control management. It provides URP user management and the access control of user to each application system and resource.

Space management. It provides space management based on the internet and the function and service of internet storage space for URP users.

Document management. It provides function and service of document management on the basis of management in space for users of the URP.

N application systems. N application systems are related directly to all kinds of systems in URP, such as system of teaching and management. The various business systems are relatively independent. Their functions are collection, treatment and processing for the information resources of the whole school. They provide services directly for the functional departments. N application systems originated from the main business department of school, including synthesis educational system, the scientific research information management system, the human resources management system, financial management system, the assets management system of lab and equipment, document management system, office automation system, network teaching system, the decision supporting system and the other application systems. In the framework of each application system, the URP still operates independently and keeps the loose coupling between each other. Any application can call other services through standard interface, so as to realize the visit to the application system of internal information as shown in figure 3.

URP public platform. It is different in application system of the development platform, communication protocol and the data exchange in original college. Web service is a kind of architecture for service. Its outstanding advantage is to realize the real meaning of independence in the platform and language, because it uses pure text as a data carrier for which its interface has any qualification. It is a lightweight interface. Based on the web service as interface, it can achieve heterogeneous environment services integration. In order to integrate the datum of all application system, this paper introduces web service technology to construct a URP platform for the public, as shown in figure 4.

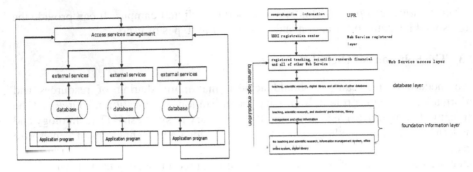

Fig. 3. Application system relationships under URP

Fig. 4. Based on the web service URP system structure schematic

3 Conclusions

The core element of the digital campus project goal is to make full use of information technology and to establish multi-layer innovative open higher level school with encouraging attempt and exploration. The latest information technology means bold attempt and exploration on this construction project. Through these work, it can meet the new requirements of talents education and efficient functioning of colleges and universities.

Acknowledgments. This work is supported by the Planned Project of Department of Science and Technology of Hebei Province and Planned Project of Department of Education of Hebei Province and the Planned Project of Department of Science and Technology of Shijiazhuang City.

References

1. Zhang, Y., Cai, R.: The digital campus URP solutions. Journal of Huzhou Normal College, 81–84 (2007)
2. Ge, G.: The key technology based for the teaching resource integration URP. Fujian Computer, 57–60 (2010)
3. Zhang, J., Wang, Y.: The construction of the digital campus URP system. Journal of Beijing University of Posts and Telecommunications (social science edition), 72–75 (2008)

Author Index